A TEXT BOOK OF

MACHINE DESIGN - II

FOR

SEMESTER – VI

THIRD YEAR (T.Y.) B. TECH COURSE IN MECHANICAL ENGINEERING

Strictly According to New Revised Credit System Syllabus of Babasaheb Ambedkar Technological University (BATU), Lonere, (Dist. Raigad) Maharashtra,

(w.e.f. June 2019-20)

Dr. R. R. GHORPADE
M. E. (Mech.) Design Engg., Ph.D, (Mech.)
Associate Professor
School of Mechanical Engineering
MIT World Peace University (MIT WPU)
Kothrud, PUNE.

L. S. UTPAT
M. E. (Mech. Engg.)
Formerly Professor & Head
Dept. of Mechanical Engg.,
D. Y. Patil College of Engineering,
Akurdi, Pune.

Dr. V. R. DEULGAONKAR
M. E. (Design Engg.) Ph. D,
Dean & Associate Professor,
Dept. of Mechanical Engg.,
MM Mandal's College of Engineering
Karvenagar, Pune.

L. V. AWADHANI
B.E. (Mech.) M. Tech. (Design Engg.),
Assistant Professor
Dept. of Mechanical Engg.,
Pimpri-Chinchwad College of Engineering,
Akurdi, Pune.

NIRALI PRAKASHAN
ADVANCEMENT OF KNOWLEDGE

AF529248

N1107

MACHINE DESIGN - II (Mechanical, BATU) **ISBN : 978-93-89686-19-7**

First Edition : **November 2019**

Published By :

NIRALI PRAKASHAN

Abhyudaya Pragati, 1312 Shivaji Nagar

Off J.M. Road, PUNE 411005

Tel : (020) 25512336/37/39

Email : niralipune@pragationline.com

➢ DISTRIBUTION CENTRES

PUNE

Nirali Prakashan (Local) : 119 Budhwar Peth, Jogeshwari Mandir Lane, Pune 411002, Maharashtra
Tel : (020) 2445 2044, Mobile : 9657703145, Email : niralilocal@pragationline.com

Nirali Prakashan (Outstation) : S. No. 28/27 Dhayari, Near Asian College, Dhayari, Pune 411041, Maharashtra
Tel : (020) 2469 0204, Fax : (020) 2469 0316, Mobile : 9657703143
Email : bookorder@pragationline.com

MUMBAI

Nirali Prakashan : 385 S.V.P. Road, Rasadhara Co-op. Hsg. Society Ltd., Girgaum, Mumbai 400004, Maharashtra
Tel : (022) 2385 6339 / 2386 9976, Fax : (022) 2386 9976, Mobile : 9320129587
Email : niralimumbai@pragationline.com

➢ DISTRIBUTION BRANCHES

JALGAON

Nirali Prakashan : 34 V. V. Golani Market, Navi Peth, Jalgaon 425001, Maharashtra
Tel : (0257) 222 0395, Mob : 94234 91860, Email : niralijalgaon@pragationline.com

KOLHAPUR

Nirali Prakashan : New Mahadvar Road, Kedar Plaza 1st Floor, Opp. IDBI Bank
Kolhapur 416012, Maharashtra. Mobile : 9850046155
Email : niralikolhapur@pragationline.com

NAGPUR

Nirali Prakashan : Above Maratha Mandir, Shop No 3, Second Floor,
Rani Jhanshi Square, Sitabuldi, Nagpur 440012, Maharashtra
Tel : (0712) 254 7129, Email : niralinagpur@pragationline.com

DELHI

Nirali Prakashan : 4593/15 Basement, Agarwal Lane, Ansari Road, Daryaganj
Near Times of India Building, New DelhiV 110002 Mobile : 8505972553
Email : niralidelhi@pragationline.com

BENGALURU

Nirali Prakashan : Maitri Ground Floor, Jaya Apartments, No. 99, 6th Cross, 6th Main,
Malleswaram, Bengaluru 560003, Karnataka
Mobile : 9449043034, Email : niralibangalore@pragationline.com

niralipune@pragationline.com | www.pragationline.com

Also find us on www.facebook.com/niralibooks

Dedicated to ....

Our Beloved Students

... Authors

PREFACE

It gives us great pleasure to present the book " **Machine Design–II** " for the students of **Semester VI Third Year (T.Y.) B. Tech. Course Mechanical Engineering of Dr. Babasaheb Ambedkar Technological University (BATU), Lonere, Dist. Raigad (Maharashtra).** This book is strictly as per the new revised syllabus 2019-20 Pattern, effective from the Academic Year July 2019-20.

In New Revised Syllabus, there will Class Assessment (CA) 20 Marks, Mid Sem. Exam. (MSE) 20 Marks and End Sem. Exam. (ESE) 60 Marks. End Sem. Exam. will be based on all Six units and each unit will carry 20 Marks.

The Theory Course will have 4 Credits.

The basic objective of this book is to bridge the gap between the vast contents of the reference books, written by the renowned International Authors and the concise requirements of Undergraduate Students. This book has been written in a comprehensive manner using Simple and Lucid language, keeping in mind students' requirements. The main emphasis has been given on exploring the basic concepts rather than merely the Information. Solved Examples and Exercises have been provided throughout the book and at the end of the Unit. Also we have given **Model Question Paper** for practice at the end of book.

Our special thanks to our family members, students and all those who directly or indirectly supported us in this project.

We also take this opportunity to express our sincere thanks to Shri. Dineshbhai Furia, Shri. Jignesh Furia, Mrs. Nirali Verma, Shri. M. P. Munde and entire team of Nirali Prakashan, namely Mrs. Deepali Lachake (Co-ordinator), and her colleagues who really have taken keen interest and untiring efforts in publishing this text.

The advice and suggestions of our esteemed readers to improve the text are most welcome and will be highly appreciated.

Pune **Authors**

SYLLABUS

Unit 1: Rolling Contact Bearings

Types, Static and Dynamic load carrying capacities, Stribeck's Equation, Equivalent load, load and life relationship, selection of bearing life, Load factor, Selection of Bearing from manufacturer's catalogue, Taper roller Bearings and their selection, Cyclic loads and speeds, Design for probability of survival other than 90% Lubrication and mountings of rolling contact bearings.

Sliding Contact Bearings : Methods of lubrication, Viscosity and its measurement, Effect of temperature, viscous flow through rectangular slot, Hydrostatic step bearing, Load capacity and energy losses, Reynolds equation, Raimondi and Boyd method, Temperature rise, Constructional details of bearing, Bearing material, Lubrication oils, Additives and Greases, Sintered metal bearings, Comparison of rolling and sliding contact bearings.

Unit 2 : Spur Gear

Gear drives, Classification of gears, Law of gearing, Terminology of spur gear, Standard system of gear tooth force Analysis, Gear tooth failures, Selection of materials Constructional, Number of teeth, Face width, Beam strength equation, Effective load on gear tooth, Estimation of module based on beam strength.

Design for maximum power capacity, Lubrication of gears.

Helical Gears: Terminology, Virtual number of teeth, Tooth proportions, Force analysis, Beam strength equation, Effective load on gear tooth, Wear strength equation.

Unit 3: Bevel Gears

Types of bevel gears, Terminology of straight bevel, force analysis, Beam and Wear strength, Effective load on gear tooth.

Worm Gears: Terminology, Proportions, Force analysis, Friction in worm gears, Vector method, Selection of materials, Strength and Wear rating, Thermal considerations

Unit 4: Belt and Chain Drives

Flat and V belts, Geometric relationship, Analysis of Belt tensions, Condition for Maximum Power, Selection of flat and V belts from manufacturer's catalogue, Adjustment of belt tensions, Roller chains, Geometric relationship, Polygonal effect, Power rating of roller chain, Sprocket wheels, and Silent chains.

Flywheel: Introduction, Types of flywheel, Stresses in disc and Armed flywheel.

Unit 5: Brakes and Clutches

Types of clutches, Torque capacity, Single and multi-plate clutches, Cone Clutch, Centrifugal Clutch, Friction materials, Types of brakes, Energy equation, Block with shoe brake, Pivoted brake with long shoe, Internal expanding shoe brake, Thermal considerations.

Unit 6: Pressure Vessel

Thin cylinders, Thick cylinders, Principal stresses, Lame's equation, Clavirino and Birnie's equation, Cylinder with external pressure, Autofrettage, Compounding of cylinders, Gasketed joint, Unfired pressure vessel, Thickness of cylindrical and Spherical pressure shells, End closure, Opening in pressure vessel, Area compensation method

CONTENTS

UNIT I - ROLLING CONTACT BEARINGS

◈◈◈

UNIT I - ROLLING CONTACT BEARINGS

CHAPTER - 1
ROLLING CONTACT BEARINGS

1.1 INTRODUCTION

- For a mechanical engineer, power transmission or conversion of motion from one form into another is of prime concern. This process of power transfer or power generation is generally a noisy operation due to the complexity of the mechanisms and involves loss of power in friction and heat. The root cause of noise or wear of component is friction between the mating components. The term rolling contact bearing (or antifriction bearing), means a mechanical component that transfers or allows transfer of power or relative motion with minimum friction. Presence of bearing in assembly ensures free rotation of shaft or axle with minimum or less friction. Bearings in addition to providing supports for shafting and axles, also take forces that are acting on them and transfer them to the foundation or frames.
- Classification of bearings is done in two ways;
  1. Direction of force that is acting on the bearing.
  2. Type of friction between the shaft and bearing surfaces.
- Depending on the direction of force acting, the bearing could be:
  (a) Radial bearing
  (b) Thrust bearing
- Depending on the type of friction between the bearing surfaces, bearings could be:
  (a) Sliding contact bearings
  (b) Rolling contact bearings
- In this chapter, we restrict our discussion to rolling contact bearings. These type of bearings are commonly used in variety of equipments such as machine tools, different machinaries, automobiles etc. They are easy to assemble, available as a standard unit and require relatively low maintenance and are suitable for applications involving frequent starts and stops.

Components and Nomenclature of Rolling Contact Bearing:

- Rolling contact bearings (Fig. 1.1) consist of four essential parts viz., the outer ring, inner ring, (inner and outer races), balls or rolling elements and the separator (retainer or cage). Balls are hardened to carry heavy loads and withstand wear within acceptable limits.
- Depending on the type of rolling element (i.e. ball, cylindrical roller or taper roller) and on the direction of load with respect to the axis of rotation (radial or axial) there are various types of rolling contact bearings. Selection of proper type of bearing is very crucial for satisfactory functioning of any equipment.

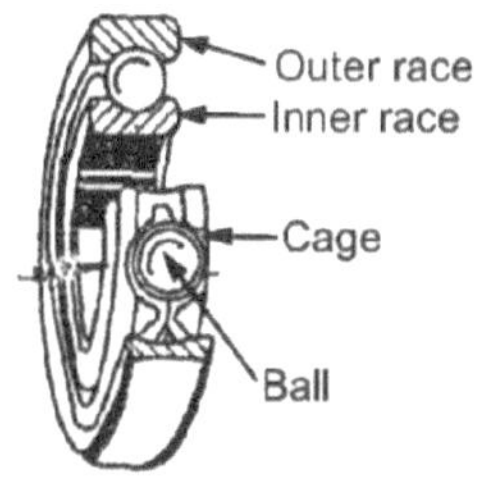

Fig. 1.1: Ball Bearing – Components

1.2 TYPES OF ROLLING CONTACT BEARINGS

- Some of the different types of rolling contact bearings are as under:

1.2.1 Deep Groove Ball Bearing

- In deep groove ball bearing (Fig. 1.2), the balls are inserted into the grooves by moving the inner ring to an eccentric position. For this insertion purpose, the size of the ball is slightly less than radius of curvature of the grooves in the races.
- It is because of this arrangement, the balls and races roll without sliding. The relative motion in a ball bearing takes place only between the ball and race. The advantages and limitations of the deep groove ball bearing are summarized as under.

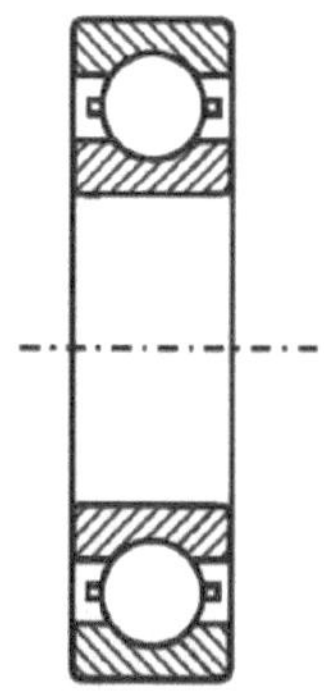

Fig. 1.2: Single Row Deep Groove Ball-bearing

Advantages of Deep Groove Ball Bearings:

- Deep groove ball bearings have high load carrying capacity. Large ball sizes used in higher series increase its load carrying capacity.
- The construction of deep groove ball bearing is such that it is able to withstand radial and axial loads.
- It can tolerate some amount of angular misalignment.
- Less frictional losses and wear due to point contact between the balls and the races and is used in high-speed applications.
- Less operational noise due to point contact.
- They are available in wide range of sizes and are used as standard parts.
- Factory grease packed bearings are available that can be used without further lubrication in service.

Disadvantages of Deep Groove Ball Bearing:

- These ball bearings demand proper alignment for bearing mounting.
- Due to point contact it is less rigid and cannot used in machine tool applications where rigidity is of prime concern.
- Load carrying capacity is limited due to point contact.

1.2.2 Cylindrical Roller Bearing

- In this type of (Fig. 1.3) bearing, there is line contact between the roller and bearing races. This enhances the load carrying capacity of the bearing. Such bearings offer following advantages and disadvantages.

Advantages of Cylindrical Roller Bearing:

- Load carrying capacity is high as compared with deep groove ball bearing due to line contact.
- This type of bearing possesses more rigidity than deep groove ball bearings.
- Line contact gives low coefficient of friction and hence low frictional losses.

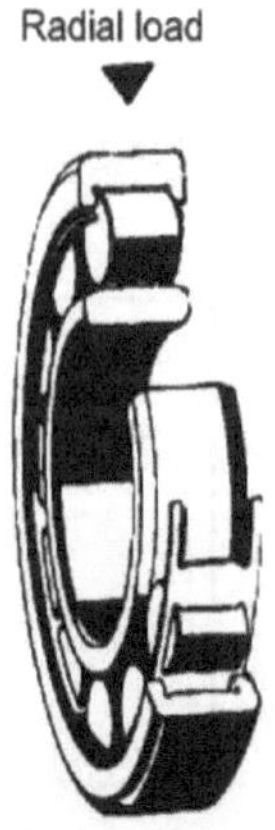

Fig. 1.3: Cylindrical Roller Bearing

Disadvantages of Cylindrical Roller Bearing:

- Unable to bear thrust load.
- Needs precise alignment between the axes of shaft and the bore as it is not self-aligning.
- Due to line contact its operation is noisy.

1.2.3 Angular Contact Bearings

- The peculiarity of angular contact bearings is that the line of reaction at contact between the balls and races makes an angle with the axis of bearing (Fig. 1.4).
- This angle of contact depends upon the shape and size of grooves in inner and outer races. As the reaction possesses two components viz radial and axial; this bearing is capable of taking both radial and thrust loads.

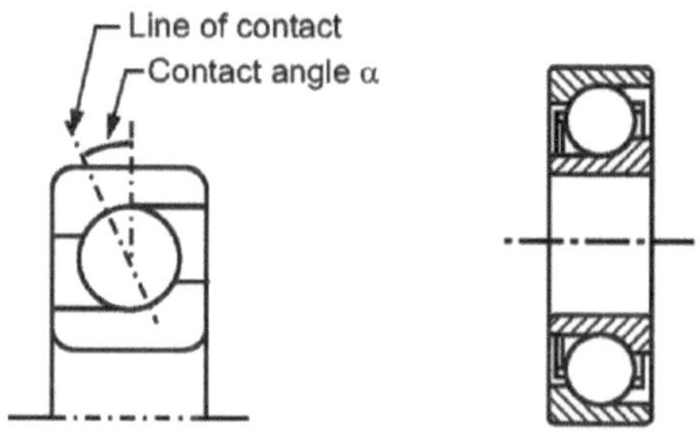

Fig. 1.4: Angular Contact Bearing

- However, these bearings can take thrust load only in one direction. Therefore, these bearings are used in pairs. The pairs are kept side by side or at the opposite ends of the shaft. These bearings are assembled with a certain magnitude of preload on them. Relative advantages and disadvantages are summarized as under;

Advantages of Angular Contact Bearing:

- Capable of withstanding both radial and axial loads.
- Capable of withstanding large axial loads than deep groove ball bearings.

Disadvantages of Angular Contact Bearing:

- In order to withstand thrust load; there is requirement of two bearings in two directions.
- This type of bearing is required to be mounted with initial amount of axial play.
- It requires pre-loading.

1.2.4 Self-Aligning Bearings

- In applications where field installation is required such as large conveyors or applications like agricultural machineries, railway axle etc., there is possibility of misalignment of power transmitting shaft with the bearing axis due to the high magnitudes of load variations involved. This action may damage the bearing oil film or surface which results into wear and breakdown. Provision is required to counter the effect of misalignment on the bearing and shaft assembly.

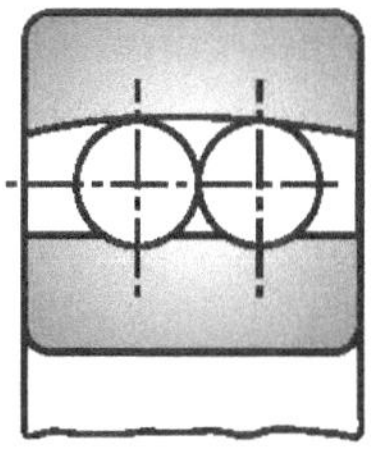

Fig. 1.5: Self Aligning Ball Bearing - Section

- This is done with the use of self aligning bearings (Fig. 1.5). In self aligning bearing the surface of outer ring is made spherical such that its centre coincides with the centre of the bearing. It is due to this arrangement that the balls along with the inner race roll freely in the seat and can carry misalignment. These self aligning bearings are of two types viz; self aligning ball bearing and spherical roller bearing. In ball bearings there are two rows of balls that are in contact with the common spherical surface while in spherical roller bearings these balls are replaced by two rows of spherical rollers that run on a common spherical surface.

1.2.5 Taper Roller Bearings

- These type of bearings include rolling elements that are in the form of frustum of a cone, arranged in such as way that the axes of individual rolling elements intersect with the common apex point on the axis of bearing. Due to this arrangement, pure rolling motion between the contacting surfaces is possible. The taper roller bearings (Fig. 1.6) can take radial and axial loads. Here also the line of resultant reaction of rolling element makes an angle with the axis of bearing as in angular contact bearings.
- Thus, even when bearing takes pure radial loads, axial load is also acting on the bearing and vice-versa. Therefore, the taper roller bearings are used in pairs to balance the thrust component. In taper roller bearings, the outer ring or cup can be separated from the rest elements. Inner ring is called cone.

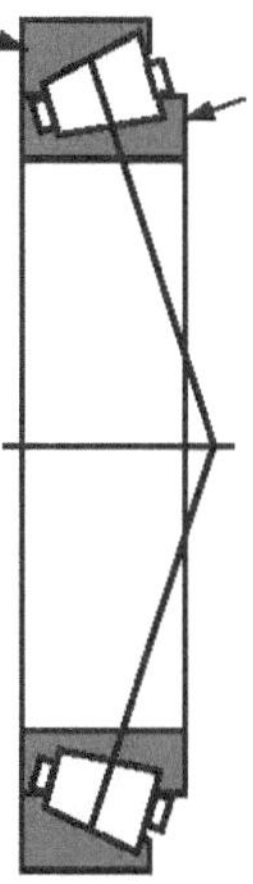

Fig. 1.6: Taper Roller Bearing

- More common applications of these taper roller bearings include automobile cars, trucks, differential gear boxes, propeller shafts, rail road axle-boxes, rolling mills, etc. Advantages and disadvantages of this bearing are summarized as under:

Advantages of Taper Roller Bearing:

- Due to cup and cone arrangement this type of bearing is able to withstand the heavy axial and radial loads.
- These bearings have more rigidity.
- As this bearing has separable construction, it allows easy assemblage and dismantling of the bearing parts.

Disadvantages of Taper Roller Bearing:

- These bearings are to be used in pairs in order to balance the axial force.
- In order to coincide the apex of cone with the apex of rolling element, it is necessary to adjust the axial location of the bearing with some pre-load.
- It cannot tolerate misalignment between shaft and housing axes.
- More costly due to versatile construction.

1.2.6 Thrust Ball Bearings

- As the name indicates this bearing is capable of withstanding thrust load in one direction. It consists of balls that run between shaft and housing rings (Fig. 1.7).
- This bearing can take thrust loads only and is unable to carry radial loads. In its construction, large numbers of balls are used in compact space and this results into large thrust load carrying capacity. These bearings are used in applications which involve large thrust loads e.g. crane hooks.

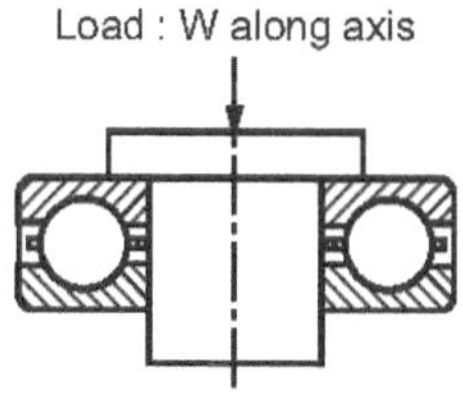

Fig. 1.7: Thrust Ball Bearing

Limitations of Thrust Ball Bearings:

- Unable to carry radial loads.
- Cannot tolerate misalignment and is not self aligning.
- Cannot operate at high speeds due to centrifugal forces and gyroscopic action generated.
- Cannot be used for horizontal shaftings.

Single Row Deep Grove (SRDG) ball bearing is the most popular type of bearing which is very commonly used in industry. It has high load carrying capacity, can take radial as well as axial load, can take misalignment and they are available in the sealed condition. Discussion here will be mainly focused on single row deep groove bearing.

Bearing Dimensions: A typical SRDG bearing size is described as 25 × 47 × 12. Here 25 is the bore diameter, 47 is the outer diameter and 12 is the width. Bore diameter is same as the shaft diameter. For a given bore diameter, bearings are available in various sizes such as (25 × 52 × 15), (25 × 62 × 17), (25 × 80 × 21). These bearings belong to different dimensional series and have higher load carrying capacities for the same bore diameter.

Bearing Designation: Bearings are designated (Table 1.1) by a 4 or 5 digit number. For example.

6308: Single row deep groove, medium series, bore 40 mm.

31306: Taper roller Bearing, medium series, bore 30 mm.

Table 1.1

Digit	CODE	Defines	Example
First OR First two	6 (2,31)	Type of bearing.	6: for SRDG 2: self aligning 31: taper roller bearing
Next	3 (0, 2, 4)	Dimension Series (Width series)	3: Medium series 0: Extra Light series 2: Light series 4: Heavy series
Last two	08 (00, 01, 02, 03…….)	Inner diameter (Bore code)	Bore = 08 × 5 = 40 mm For 00 = 10 mm, 01 = 12 mm, 02 = 15 mm, 03 = 17 mm [(From 04 onwards multiply last two digits by 05 to get diameter. For 04, Bore = 04 × 5 = 20]

1.3 STATIC AND DYNAMIC LOAD CARRYING CAPACITIES

1.3.1 Static Load Carrying Capacity: Static Load [C_o]

- This is the load which acts on the bearing when the shaft is in stationary condition. Depending upon the self weight of the shaft, its mountings and the static load acting, a permanent deformation is induced in the balls and inner/outer races of bearings. The magnitude of this deformation increases with increase in load. Permissible magnitude of permanent deformation controls the allowable static load magnitude. Experience shows that total permanent deformation of 0.0001 × d (where d is the ball or roller diameter) at most heavily stressed balls and race can be tolerated.
- For calculation of static load bearing capacity, Indian Standard I.S. 3823 is referred. These formulae are not required during bearing selection process. However manufacturer's catalogue gives static load carrying capacities of bearings based on the design, material and manufacturing by manufacturer. **Static load capacity is the maximum static radial load on the bearing that can be withstood by it without exceeding permanent deformation of 0.0001 × d on ball and race at its most heavily stressed contact.** The Stribeck's equation gives the static load carrying capacity of the bearing. This equation is discussed in later section.

1.3.2 Dynamic Load Carrying Capacity [C]

- When the ball or roller bearing is under consideration, the contact between the balls (or rollers) and the inner and outer races (or ring) generate Hertz's contact stresses between them. For a properly lubricated, clean and sealed bearing, metal fatigue is the only criterion of failure. Metal fatigue involves millions of stress alternations successfully endured. Common life measures are; number of revolutions of the inner ring until the first sign or evidence of fatigue crack and number of hours of use at a certain standard angular speed until the first sign or evidence of fatigue. Dynamic load carrying capacity of the bearing is based on the fatigue life of the bearing. Life of a single bearing is difficult to predict. For life measurement of bearing, statistical average for a group of bearings is prominent in practice.
- The rating life of a group of approximately identical bearings is defined as the number of revolutions that 90% of the bearings will complete or exceed before the evidence of first fatigue crack. Rating life is also known as minimum life, catalogue life, L_{10} life or B_{10} life. Life of individual ball bearing may or may not be different than rating life. It is proved that life which 50% of a group of bearings will exceed or complete, is five times the rating or L_{10}.
- **The dynamic load carrying capacity of bearing is defined as radial load in radial bearings or thrust**

load in thrust bearings that can be carried for a life of one million revolutions by 90% of the bearings from a group of identical bearings. This dynamic load carrying capacity is dependent on the assumption that inner race is rotating and the outer race is stationery. Most of the manufacturer's catalogue provide ready to use information or values of dynamic load capacity.

1.4 STRIBECK'S EQUATION

- This equation deals with the static load carrying capacity of the bearing. Following are the assumptions made in analysis of static load carrying capacity of the bearing.
  - ➢ The inner and outer races of the bearings are rigid. This assumption is made as circular shape of the races may deviate from the fundamental circular profile.
  - ➢ There is equal interspacing between the balls.
  - ➢ Balls of a roller bearing in the upper half do not support any load magnitude.
- The static load equation is derived based on Hertz equation for contact stresses which states that $\delta \propto (F)^{2/3}$ where F is the radial load and δ is the deflection of the ball.

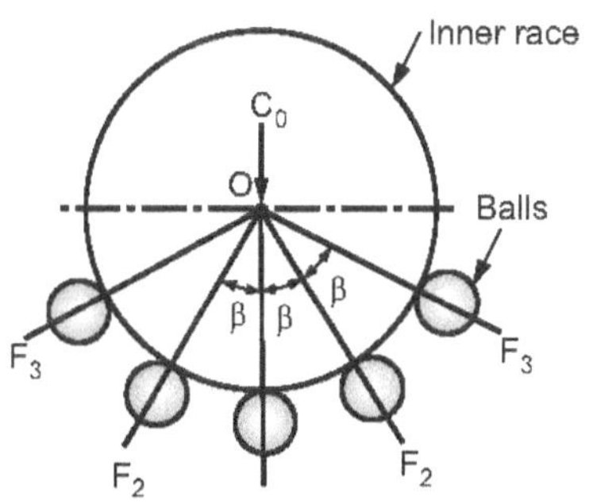

(a) Static Load on Inner Race

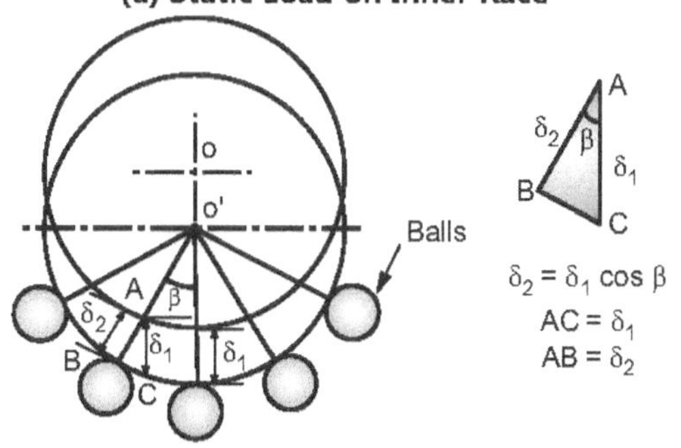

(b) Deflected Inner Race

Fig. 1.8

- Fig. 1.8 below, shows the inner race of a roller bearing and forces acting on it. These forces are acting through the rolling elements or balls, which support the static load C_0. For the analysis of static load on bearing single row of balls on races is considered. Referring the figure, the equation for equilibrium of forces in vertical direction is deduced as
- If C_0 is the total static load acting, central ball will take maximum load (= F_1) and subsequent balls will share F_2, F_3 ......
- The vertical equilibrium of forces gives

$$C_0 = F_1 + 2F_2 \cos\beta + 2F_3 \cos(2\beta) + \ldots \ldots \quad \ldots (1.1)$$

- According to first assumption made in the analysis, only the balls take deformation and there is no deformation in the inner and outer races.

Let δ_1 be the deflection that occurs in the most heavily stressed ball which lies on the axis through which C_0 is acting. It is because of this deflection, the inner race gets deflected with respect to outer race through the same amount δ_1. Due to this the centre of inner ring moves from O to O', without deformation in the fundamental circular profile of inner ring.

Now let δ_1, δ_2 and δ_3 be the radial deformations occurring in respective balls.

As shown in Fig. 1.8 (b), $\delta_2 = \delta_1 \cos\beta$. The assembly is of bearing is subjected to contact stresses. As per Hertz's theory of contact stresses $\delta \propto (F)^{2/3}$. $\delta = B(F)^{2/3}$, B = Constant of proportionality.

Hence $\delta_1 = B\,(F_1)^{2/3}$ and $\delta_2 = B\,(F_2)^{2/3}$.

$$\text{Thus,} \quad \frac{\delta_2}{\delta_1} = \left(\frac{F_2}{F_1}\right)^{2/3} ; \Rightarrow F_2 = F_1 (\cos\beta)^{3/2} \quad \ldots (1.2)$$

Similarly it can be proved that $F_3 = F_1 (\cos 2\beta)^{3/2}$ and so on...

Using these values in (1.1), we get

$$C_0 = F_1 + 2F_1 (\cos\beta)^{3/2} \cos\beta + 2F_1 (\cos 2\beta)^{3/2} \cos(2\beta) + \ldots$$

$$C_0 = F_1 [1 + 2(\cos\beta)^{5/2} + 2(\cos 2\beta)^{5/2} + 2\cos(3\beta)^{5/2} + \ldots]$$

This can be written as

$$C_0 = F_1 M; \quad \ldots (1.3)$$

$$\text{where,} \quad M = [1 + 2(\cos\beta)^{5/2} + 2(\cos 2\beta)^{5/2} + 2\cos(3\beta)^{5/2} + \ldots]$$

If Z is the number of balls, the angular relation between the ball angle and the number of balls is given as $\beta = \frac{360}{Z}$.

... (1.4)

Thus, β dpends on Z. The number of balls in a standard bearing generally is 8, 10, and 12, 15 or higher. So evaluating the β values for various values of Z, M can be calculated. It is observed that, the ratio (Z/M), is almost constant. Stribeck suggested a constant value of 5 for the ratio (Z/M).

Hence; $$M = \frac{1}{5} Z \quad \text{... (1.5)}$$

Using equation (1.5) in (1.3), we get

$$C_0 = \frac{1}{5} Z F_1 \quad \text{... (1.6)}$$

Experimental evaluation shows that the force F_1 needed to produce a given permanent deformation in the balls is given as

$$F_1 = kd^2 \quad \text{... (1.7)}$$

Where, k is a factor depending upon radii of curvature and material properties including modulii of elasticity of material; d is the ball diameter. From equations (1.6) and (1.7); we get

$$C_0 = kd^2Z/5 \quad \text{... Stribeck's equation.}$$

1.5 EQUIVALENT BEARING LOAD (F_e)

- In most of the practical situations, the force acting on the bearing possesses two components as radial and thrust. These two loads may be present in any proportions and therefore it is necessary to convert these two components into single equivalent which can be compared with the dynamic load carrying capacity. The expression for the equivalent dynamic load is given as

  $$F_e = X V F_r + Y F_a$$

  where, F_e = Equivalent dynamic load in N

  F_r = Radial load component in N

  F_a = Thrust or axial load component in N

  V = Race rotation factor

  X and Y are the radial and thrust load factors. Their values are directly given or selected from manufacturer's catalogue. The values depend upon the geometry of bearings, number of balls, ball diameter and the relative values of F_a, F_r and C_0.

  Value of race rotation factor V depends upon whether inner or outer race rotates.

  V = 1; when inner race rotates and outer race is stationery.

  V = 1.2; when outer race rotates and inner race is stationery.

- The value of V = 1.2 is actually an acknowledgement that fatigue life is reduced for these operations. In most of the applications the inner race rotates and the outer race is stationery. (For example: ceiling fan, wherein the outer race is fixed in housing and the inner race rotates with fan spindle). However the value of V is unity for self aligning bearings whether inner or outer race rotates.

- For the case when the bearing is subjected to pure radial and axial loads, the value of equivalent load becomes $F_e = F_r$ and $F_e = F_a$.

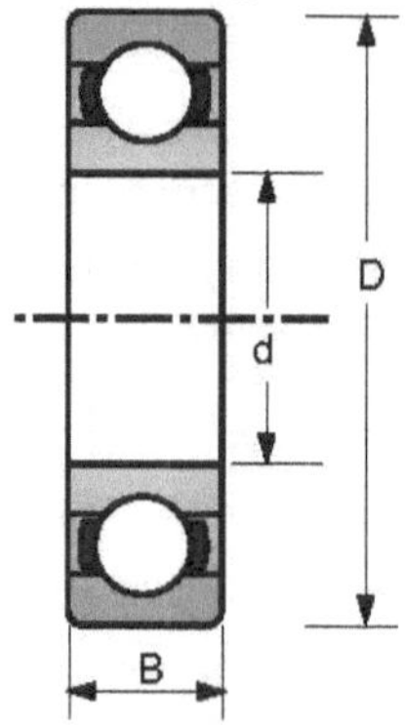

Fig. 1.9: Single Row Deep Groove Ball Bearing

Table 1.2: Details of Single Row Deep Groove Ball Bearing

d mm	D mm	B mm	C N	C_0 N	Designation
10	26	8	4620	1960	6000
	30	9	5070	2240	6200
	35	11	8060	3750	6300
12	28	8	5070	2240	6001
	32	10	6890	3100	6201
	37	12	9750	4650	6301
15	32	9	5590	2500	6002
	35	11	7800	3550	6202
	42	13	11400	5400	6302
17	35	10	6050	2800	6003
	40	12	9560	4500	6203
	47	14	13500	6550	6303
	62	17	22900	11800	6403
20	42	12	9360	400	6004
	47	14	12700	6200	6204
	52	15	15900	7800	6304
	72	19	30700	16600	6404
25	47	12	11200	5600	6005
	52	15	14000	6950	6205
	62	17	22500	11400	6305
	50	21	35800	19600	6405
30	55	13	13300	6950	6006
	62	16	19500	10000	6206
	72	19	28100	14600	6306
	90	23	4360	24000	6406

35	62	14	15900	8500	6007
	72	17	25500	13700	6207
	80	21	33200	18000	6307
	100	25	55300	31000	6407
40	68	15	16800	9300	6008
	80	18	30700	16600	6208
	90	23	41000	22400	6308
	110	27	63700	36500	6408
45	75	16	21200	12200	6009
	85	19	33200	18600	6209
	100	25	52700	30000	6309
	120	29	76100	45500	6409
50	82	16	21600	13200	6010
	90	20	35100	19600	6210
	110	27	61800	36000	6310
	130	31	87100	52000	6410
55	90	18	28100	17000	6011
	100	21	43600	25000	6211
	120	29	71500	41500	6311
	10	33	99500	63000	6411
60	95	18	29600	18300	6012
	120	22	47500	28000	6212
	130	31	81900	48000	6312
	150	35	108000	69500	6412
65	100	18	30700	19600	6013
	120	23	55900	34000	6213
	140	33	92300	56000	6313
	150	37	119000	78000	6413

1.6 LOAD-LIFE RELATIONSHIP

- The inter-relationship between the dynamic load carrying capacity, equivalent dynamic load and the bearing life is given by

$$L_{10} = (C/F_e)^a$$

L_{10} = Rated bearing life measured in million revolutions

C = Dynamic load carrying capacity in N

a = 3 for ball bearings

a = 10/3 for cylindrical and tapered roller bearings

Rearranging the above equation for rated life as;

$$C = F_e\,(L_{10})^{1/a}$$

For ball bearings $C = F_e\,(L_{10})^{1/3}$ and for cylindrical and tapered roller bearings $C = F_e\,(L_{10})^{0.3}$.

1.7 SELECTION OF BEARING LIFE

- The desired bearing life is required to be specified for the purpose of deciding the size of the bearing. Life expectancy of bearing for a given application is expressed from past experience. For vehicles and many other applications, rotational speed is not constant but varies as per load conditions and functional requirements. Hence the bearing life is expressed in million revolutions. Based on past experiences, the desired bearing life for wheeled applications are mentioned in Table 1.3 below.

Table 1.3: Values of Life in Million Revolutions for General Applications

Sr. No	Life in Million Revolutions	Applications
1.	50	Road vehicles - Medium category
2.	100	Road vehicles - Heavy vehicles, trucks
3.	500	Off road vehicles - Trolley cars, tractors

- For applicactions in which the speed of rotation is constant, the expected life is measured in terms of hours of service. The recommended bearing life for such applications is given in table 1.3 below.

Table 1.4: Guidelines for Bearing Life in Hours for Common Applications

Sr. No	Expected Service In Hours	Applications
1.	4000-8000	Intermittent machine applications: as lifting tackles, hand tools, drills, mixer - grinders, household appliances etc.
2.	12,000-20,000	Electric motors, gear drives and applications in which machines are used for stipulated hours of service as 8, 10 or 12.
3.	40,000-60,000	Pumps, compressors, and similar applications where machines are used for 24 hours of service per day and continuous operations.

- These above mentioned values provide only a general guidance. However for some specific application the designer must account for past experiences, time and cost involved in replacement, repair of bearings and sometimes even breakdown costs.

1.8 LOAD FACTOR IN ROLLING CONTACT BEARINGS

- The force analysis of bearing is done by taking into consideration the forces acting in vertical and horizontal planes. However the equations derived do not account for the dynamic loads and the other operational variations.
- To account for these and such other related factors, the elementary load values are multiplied by load application factors depending on the equipment. Guidelines for selection of load factors are given in Table 1.5 below.

Table 1.5: Load Application Factor Values for Gear, Belt and Chain Drives

Sr. No	Load Application Factor	Drives
1.		**Belt Drives**
	2	i. V- Belts
	3	ii. Single Ply leather belt
	3.5	iii. Double Ply Belt
	1.5	Chain drives
2.		**Gear Drives**
	1.2 to 1.4	Electric motors, turbo-compressors
	1.4 to 1.7	I.C. Engines and Compressors
	2.5 to 3.5	Hammer mills

1.9 BEARING SELECTION FROM THE CATALOGUE

Bearing selection process gets divided in two steps:

1. Selection of the Type of Bearing based on Design Requirement:

This depends on

- Absolute as well as relative magnitudes of radial and axial loads on the bearing.
- Provision required to take misalignment based on likely assembly and installation environment.
- Lubrication provision that can be made and ease of accesibilty.
- Speeds, working temperature and expected life.

Bearing manufacturer's catalogue gives selection guideline. For example, a single row deep groove (SRDG) ball bearing can withstand radial as well as axial load and can also take misalignment. Whereas, if the radial loads are heavy, cylindrical roller bearings could be the choice. Taper roller bearings can take radial as well as axial loads. Type of bearing is selected by considering above requirements.

2. Selection of the Size of Bearing:

Static and dynamic load carrying capablilities are published in the catalogue for every size of bearing. Required life of the bearing at specified reliability is decided by the application. The size of bearing is decided by iterative process to get optimum design. The selection process discussed here is mainly for the SRDG bearings which can be extended to cylindrical and spherical roller bearings by using prescribed tables from manufacturer's catalogue.

Selection Steps are Described Below:

1. The shaft diameter is already calculated based on Bending Moment and torque. Bearing bore diameter is thus finalized.
2. Find axial (F_a) and radial (F_r) load values for the bearing.
3. Decide the type of bearing to be used as discussed earlier.
4. Bearings are available in different series (light, medium, heavy). Select a light series of bearing from the catalogue to start with for the given bore size and find static load capacity 'Co_{cat}.' And Dynamic load capacity 'C_{cat} from catalogue for the selected bearing.
5. Referring to the table from manufacturer, find values of radial load factor (X) and axial load factor (Y). For the values of (F_a/C_o) for the selected bearing, find 'e'. Compare (F_a/F_r) value with 'e' and accordingly find X and Y using the table

Table 1.6: Single Row Deep Groove Ball Bearing

	$(F_a/F_r) < e$		$(F_a/F_r) > e$		e
F_a/C_o	X	Y	X	Y	
0.025				2.00	0.22
0.056				1.71	0.26
0.084				1.55	0.28
0.13	1.0	0.0	0.56	1.4	0.31
0.28				1.15	0.38
0.50				1.0	0.44

6. Find equivalent load $F_{eq} = V \cdot X \cdot F_r + Y \cdot F_a$

   Here V is the rotation factor =1 when inner race is rotating and = 1.2 when outer race is rotating.
7. Knowing F_{eq} and desired bearing life at 90% reliability, find required value of 'C' (C_{reqd}) using the relation. $C = F_{eq} \cdot L^a$
8. If $C_{cat} > C_{reqd}$, selection is suitable to get desired life.
9. If $C_{cat} < C_{reqd}$, try next higher series for the same diameter and repeat steps from step 4, till we reach step 8. The next higher series bearing has same bore

but larger outer diameter and ball size. Thus, it has also larger dynamic load capacity but it is costlier.

Iterative procedure is essential to arrive at optimum results.

1.10 MOUNTING OF TAPER ROLLER BEARING

- As already seen, in case of taper roller bearings, the load carrying member i.e. the taper roller has its axis at an angle w.r.t. shaft axis. Thus, even when the load is purely radial, it has components in axial direction and thus a thrust load. Moreover the construction of a taper roller bearing is such that the cone is a separable element and therefore the axial load must be carried by bearing at the other end. Thus, these bearings are used in pair. Two mounting methods are common. (Refer Fig. 1.10).
  1. Face to Face and
  2. Back to Back.
- Thus, as the name suggests either the front sides (face) of the two bearings will be facing each other or the backs will be facing each other. Back to back arrangement is preferred when distance between two bearings is short and moment loads are applied. Face to face is popular because mounting is easier when interference is essential for inner ring.

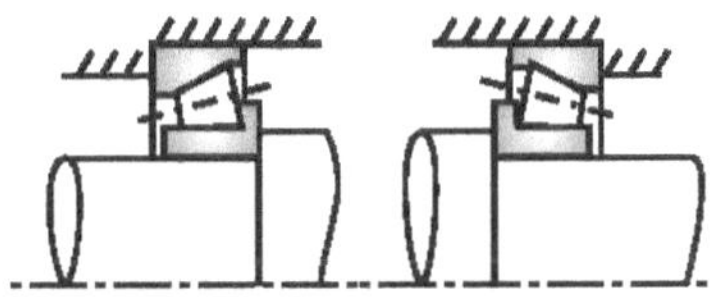

(a) Face to Face Mounting

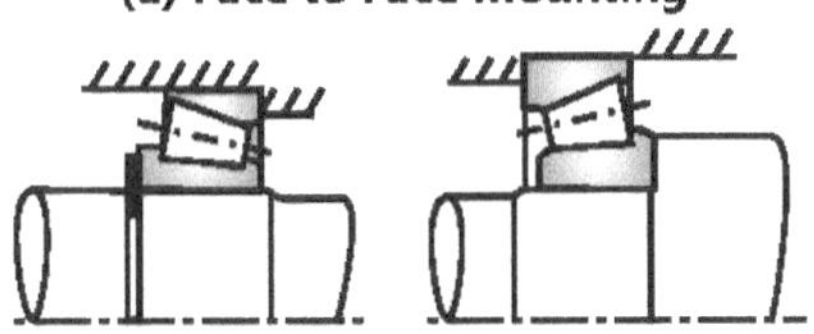

(b) Back to Back Mounting

Fig. 1.10

Table 1.7: Data on Taper Roller Bearing

d mm	D mm	B mm	C N	C_0 N	Designation	e	Y
20	42	15	22900	28700	32004X	0.37	1.6
	47	14	26000	30500	30204	0.35	1.7
	52	15	31900	36800	30304	0.3	2.0
	52	21	41300	31500	32304	0.3	2.0
25	47	15	25500	33700	32005 X	0.43	1.4
	52	15	29200	35500	30205	0.37	1.6
	62	17	41800	48800	30305	0.30	2.0
	62	17	35800	42700	31305	0.83	0.72
	62	24	56100	71800	32305	0.30	2.0
30	55	17	33600	41500	32006 X	0.43	1.4
	62	16	38000	46900	30206	0.37	1.6
	62	20	47300	61600	32206	0.37	1.6
	72	19	52800	63500	30306	0.31	1.9
	72	19	44600	53400	31306	0.83	0.72
	72	27	72100	95700	32304	0.31	1.9
35	62	18	40200	56100	32007 X	0.46	1.3
	72	17	48400	59800	30207	0.37	1.6
	72	23	61600	82800	32207	0.37	1.6
	80	21	68200	82100	30307	0.31	1.9
	80	21	57200	71800	31307	0.83	0.72
	80	31	89700	123000	32307	0.31	1.9
40	68	19	49500	73600	32908 X	0.37	1.6
	80	18	58300	71600	32208	0.37	1.6
	80	23	70400	92000	32208	0.37	1.6
	90	23	80900	103000	30308	0.35	1.7
	90	23	69300	85600	31307	0.83	0.72
	90	33	110000	153000	32307	0.35	1.7
45	75	20	55000	81000	32808 X	0.40	1.5
	85	19	62700	81000	30209	0.40	1.5
	85	23	74800	103000	32209	0.40	1.5
	100	25	101000	132000	30309	0.35	1.7
	100	25	85800	110000	31309	0.83	0.72
	100	36	132000	188000	32309	0.35	1.7
50	80	20	57200	83300	32018 X	0.43	1.4
	90	20	70400	95700	30218	0.43	1.4
	90	23	76500	105000	32210	0.43	1.4
	110	27	117000	153000	30310	0.35	1.7
	110	27	99000	128000	31310	0.83	0.72
	110	40	161000	234600	32310	0.35	7.7

For a given taper roller bearing (Table 1.7)

For $\frac{F_a}{F_r} > e \rightarrow X = 0.4$ and Y = from table

$\frac{F_a}{F_r} < e \rightarrow X = 1.0$ and $Y = 0$

Equivalent Load and Selection of Taper Roller Bearing:

Consider a back to back arrangement shown below with radial loads F_{1r}, F_{2r} on two bearings and external thrust load T.

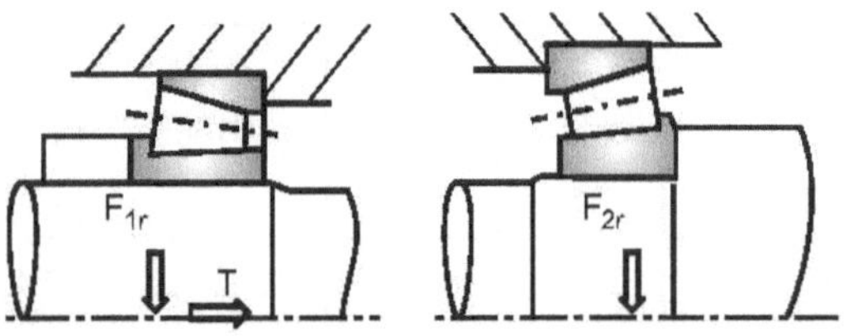

Fig. 1.11: Back to Back Mounting (Radial and Axial Load)

Radial load on bearing 1 = F_{1r}

Total axial load on bearing 1 = F_{1a}

$$F_{1a} = \text{Thrust on bearing 1 due to radial load on bearing 2 + External thrust}$$

$$F_{1a} = \frac{0.5 \times F_{2r}}{Y_2} + T$$

Similarly,

Radial load on bearing 2 = F_{2r}

Total axial load on bearing 2 =

$$F_{2a} = \text{Thrust on bearing 2 due to radial load on bearing 1 − External thrust}$$

$$F_{2a} = \frac{0.5 \times F_{1r}}{Y_1} - T$$

Having found radial loads on bearings 1 and 2, equivalent loads can be determined.

Equivalent Load on Bearing 1:

$$F_{eq1} = \left[X_1 \cdot F_{1r} + Y_1\left(\frac{0.5 \times F_{2r}}{Y_2}\right) + T\right]K_1 \text{ or } F_{1r} \times K_1$$

Whichever is larger.

Similarly,

Equivalent Load on Bearing 2:

$$F_{eq2} = \left[X_2 \cdot F_{2r} + Y_2\left(\frac{0.5 \times F_{1r}}{Y_1} + T\right)\right]K_1 \text{ or } F_{2r} \times K_1$$

Whichever is larger.

Having found equivalent loads the life of the bearing can be found as in case of SRDG bearings using the relation $L_{10} = (C/F_{eq})^3$. The method is typical and can be used for face to face arrangement by taking into consideration axial thrust directions.

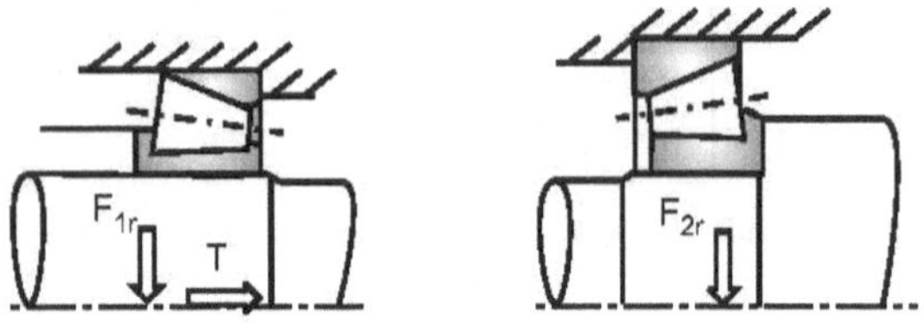

Fig. 1.12

Example: For an automobile application with shaft diameter of 35 mm and back to back taper roller bearing arrangement, the radial load on the LH side bearing is 2000 N and on the RH side bearing it is 1600 N. The external thrust is 1200 N towards RH side. Taper roller bearing 30207 is selected and the expected life is 800 million revolutions minimum. Comment on the suitability of the bearings. Take load application factor as 1.4.

F_{1r} = 2000 N, F_{2r} = 1600 N, T = 1200 N

Referring to the table 1.7 for selected bearing, 30207

C_0 = 59800 N; C = 48400 N

From the table, X = 0.4, Y = 1.6, e = 0.37

Consider bearing No. 1 on LH side

Radial load on LH side bearing 1 = F_{1r} = 2000 N

Total axial load on bearing 1 =

$$F_{1a} = \text{Thrust on bearing 1 due to radial load in bearing 2 + External thrust}$$

$$F_{1a} = \frac{0.5 \times F_{2r}}{Y_2} + T = \frac{0.5 \times 1600}{1.6} + 1200$$

$$= 1700 \text{ N}$$

$$\text{Since} \quad \frac{F_{1a}}{F_{1r}} = \frac{1700}{2000} > e$$

$$\therefore \quad X_1 = 0.4 \text{ and } Y_1 = 1.6$$

$$F_{eq1} = \left[X_1 \cdot F_{1r} + Y_1\left(\frac{0.5 * F_{2r}}{Y_2} + T\right)\right]K_1 \text{ or}$$

$$= F_{r1} \times K_1$$

$$F_{eq1} = [0.4 \times 2000 + 1.6\ (1700)]\ 1.4 = 4928 \text{ N}$$

$$\text{or} \quad F_{eq1} = 2000 \times 1.4 = 2800 \text{ N}$$

$$\text{Thus,} \quad F_{eq1} = 4928 \text{ N}$$

Consider bearing No. 2 on RH side.

Radial load on RH side bearing 2 = F_{2r} = 1600 N

Total axial load on bearing 2 =

$$F_{1a} = \text{Thrust on bearing 1 due to radial load on bearing 2 + External thrust}$$

$$F_{2a} = \frac{0.5 \times F_{1r}}{Y_2} - T = \frac{0.5 \times 2000}{1.6} - 1200$$

$$= -575 \text{ N}$$

$$F_{eq2} = \left[X_1 \cdot F_{2r} + Y_1\left(\frac{0.5 \times F_{1r}}{Y_1} + T\right)\right]K_1 \text{ or}$$

$$= F_{r2} \times K_1$$

$$Fe_{q1} = [0.4 \times 1600 + 1.6\ (-575)]\ 1.4 = -392$$

$$\text{or} \quad = 1600 \times 1.4 = 2240 \text{ N}$$

$$\text{Thus,} \quad F_{eq2} = 2240 \text{ N}$$

Taking maximum equivalent load as 4928 N

$$L = \left(\frac{C}{P}\right)^3 = \left(\frac{59800}{4928}\right)^3$$

$$= 1786.86 \text{ million revolutions.}$$

Minimum requirement is 800 million revolutions.

Hence, the bearing is adequate.

Selection of the proper type and size of bearing is the key to successful operation of machinaries.

1.11 DESIGN FOR CYCLIC LOADS AND SPEEDS

For certain applications, bearings are subjected to cyclic loads and speeds. Considering a simplified approach we may consider that, a ball bearing having dynamic load capacity = C, operates under the following conditions:

- Radial load of F_1 (N) having life = L_1 (at F_1), completes n_1 rotations
- Radial load of F_2 (N) having life = L_2 (at F_2), completes n_2 rotations
- Radial load of F_3 (N) having life = L_3 (at F_3), completes n_3 rotations

  (Where n_1, n_2, n_3 may be available as a % of total revolutions also).

Let L be the life in million revolutions under above conditions for this bearing.

Since, $L_1 = (C/F_1)^3$;

$$\frac{1}{L_1} = \frac{F_1^3}{C^3}$$

For one revolution, $1/L_1$ life is consumed

Therefore,

Fraction of life completed by F_1 is $\frac{n_1}{L_1}$ and will be

$$\frac{n_1}{L_1} = n_1\left(\frac{F_1^3}{C^3}\right)$$

Fraction of life completed by F_2 is $\frac{n_2}{L_2}$ and will be

$$\frac{n_2}{L_2} = n_2\left(\frac{F_2^3}{C^3}\right)$$

If the equivalent load = F_{eq} and total cycles covered are = n = $n_1 + n_2 + n_3$

Cumulative fraction of life completed = $\frac{n}{L} = n\left(\frac{F_{eq}^3}{C^3}\right)$

Equating effect of F_1, F_2, F_3 .... with F_{eq}

$$n_1.F_1^3/C^3 + n_2.F_2^3/C^3 + ...+ = n.F_{eq}^3/C^3 \quad \ldots (1.8)$$

Therefore,

$$F_{eq} = \sqrt[3]{\frac{n_1 \cdot F_1^3 + n_2 \cdot F_2^3 + n_3 F_3^3}{n_1 + n_2 + n_3}}$$

Once equivalent load value is known, we can find corresponding life 'n'.

When the load varies continuously with time, the above equation is rewritten as

$$F_e = \sqrt[3]{\left\{\left[\int_0^N F^3\, dN\right] \Big/ \left[\int_0^N dN\right]\right\}}$$

$$F_e = \sqrt[3]{\left\{\left[\frac{1}{N}\int_0^N F^3\, dN\right]\right\}} \quad \ldots (1.9)$$

For applications in which the bearings are subjected to combined axial and radial loads, these loads must be first converted into equivalent loads for the above calculations.

1.12 BEARING WITH A PROBABILITY OF SURVIVAL OTHER THAN 90%

For applications involving risk to human life, or where number of such bearings are used in a given system, it becomes mandatory to select a bearing that possesses reliability of more than 90% (R_{90} = 0.9). The relationship between bearing life and reliability is given by a stastical curve known as Wiebull distribution.

For Wiebull distribution,

$$R = e^{-\left(\frac{L}{a}\right)^b}$$

In the above equation R is the reliability, L is life and a and b are constants. Rearranging the above equation we get;

$$\text{Log}_e(1/R) = (L/a)^b \quad \ldots (1.10)$$

When we consider 90% reliability of the bearing i.e. R_{90}, corresponding life is L_{10}, then the above equation is rewritten as

$$\text{Log}_e(1/R_{90}) = (L_{10}/a)^b \quad \ldots (1.11)$$

Dividing equation (1.10) by (1.11), we get

$$(L/L_{10}) = \{[\log_e(1/R)]/[\log_e(1/R_{90})]\}^{1/b} \quad \ldots (1.12)$$

Values of constants a and b are obtained by imposing the condition $L_{50} = 5L_{10}$

The values are a = 6.839, b = 1.169.

1.13 LUBRICATION AND MOUNTING OF BEARINGS

1.13.1 Lubrication of Rolling Contact Bearings

Rolling motion between the balls and races generates friction and heat. Lubrication is essential in antifriction bearings in order to reduce the friction between balls and races, including the other objectives as heat dissipiation (generated due to friction), avoid corrosion and protection of bearing from other particles. Oil and grease are two major types of lubricants used for lubrication.

Advantages of Oil over Grease as Lubricant are:

- Oil, being less viscous than grease carries frictional heat more easily and effectively than grease.
- Being in liquid state, oil feeds more contact areas under bearing load.
- It flushes dirt, corrosion and foreign particles from the bearing.

However, effective oil sealing arrangement has to be provided.

Grease lubricated bearings require less maintenance cost, are not subjected to leakage and rust. Also these bearings need simple housings.

Guidelines for Selecting Lubricant are:

- Temperature at the bearings is an important criteria. For applications involving temperature less than 100°C, grease provides excellent sealing action and lubrication, whereas oils have to be used in applications involving temperature more than 100°C.
- Greases are not preferred at high linear speeds. Where the product of bore (dimensions in mm) and speed (in rpm) is less than 200,000 grease is preferred and for higher values of product, lubricating oils are recommended.
- Grease is suitable for low operational speeds and moderate load magnitudes, whereas lubricating oils are preferred for heavy-duty applications.
- For applications like gear boxes where central lubricating system is employed lubricating oil is used.

1.13.2 Mounting of Rolling Contact Bearings

Bearing mouting is a complicated process involving a wide range of considerations ranging from manufactrinng to assembley considerations. Considerable expertise is required to select optimum mountings. Bearings provide housing for shafts. Usually bearings are mounted over shafts with tighter interference fit whereas bearings are usually fitted in the housing with a push fit. Bearing catalogues recommend suitable fits for the type of application.

Mounting of Bearings Over Shaft:

(a) Cold Mounting:

Bearings up to a 100 mm outside diameter can be cold mounted using a sleeve and a hammer or a press.

(b) Temperature Mounting:

Temperature mounting is the method of obtaining an interference fit by first introducing a temperature differential between the parts to be fitted. The required temperature differential can be obtained by heating one part and/or cooling the other part.

(c) Hydraulic Mounting:

Used for cold mounting a tapered bore bearing. It is based on forcing oil between the interfering surfaces, thereby greatly reducing the required axial force.

Radial Location:

- Inner race of the bearing is fitted on shaft by means of interference fit. Similarly outer race is also fitted with interference fit. This intereference fit is responsible for prevention of wear and relative motion between inner race and the shaft. The degree of tightness of interference selected for outer race is less as compared with inner one. If the degree of tightness is insufficient than required, then the bearing is subjected to creep. For bearing creep refers to the small relative motion of the outer race relative to seat.
- The bearings therefore are required to be located axially such that neither the shaft nor the bearings can move axially. Care has to be taken at the same time to ensure that there is provision for expansion of the shaft and there by provision for one of the bearings to move axially. Thus, in case a pair of bearing is mounted on same shaft, the outer race of one of the bearings is permitted for axial shift in order to withstand axial deflection due to thrust load or temperature variation.

Methods for Axial Location for the Bearing:

There are various methods employed to locate bearings axially. (Ref. Fig. 1.13). These can be:

- A step on the shaft or in the housing.
- Use of circlip on the shaft.
- Use of locking plate on the shaft.

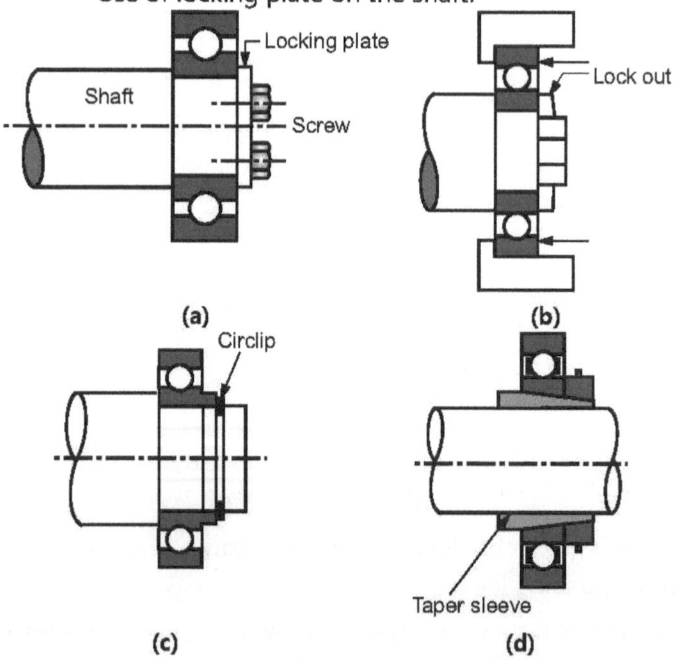

Fig. 1.13: Bearing-Axial Location Methods

- Having threads on the shaft and use of locking nut.
- Use of end cap in the housing, thus locating outer race and many more.

Preloading of Bearings:

- A bearing is fitted on a shaft and located inside a housing and has some clearance between different parts of the ball bearing. This clearance results into noise and vibrations during running. The preloading is a process wherein a thrust load is applied to remove axial or radial clearance within the bearing parts. This also helps in controlling rotational accuracy.
- Optimum preload is essential. Excessive preload results in generation of excessive heat and also reduces fatigue life. Insufficient preload causes fretting corrosion. Values of optimum preload are estimated based on contact stresses at ellipse area between ball and race.
- All applicatons do not require preloading since a positive clearance is desirable. However for applications such as machine tool spindle bearings, pinion bearings in automobils, a negative clearance is essential to increase running accuracy.

Advantages of Preloading:

- Eliminates radial and axial play.
- Eliminate unnecessary clearances, which induce rigidity to the bearings.
- By reducing the clearances, controls the rotational accuracy of the bearing.

Methods of Preloading:

- Bearings with a tapered bore are suitable for radial preloading since, by driving the bearing up on to its tapered seat reduces radial clearances and achieves preloading.

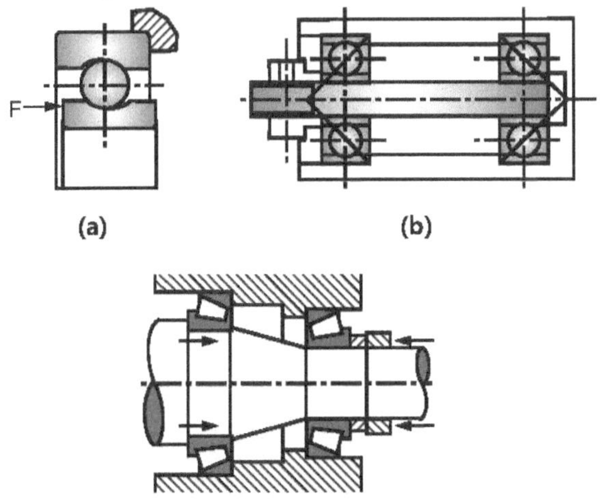

(a) (b)

(c)

Fig. 1.14: Bearing - Preloading Methods

- Figs. 1.14 (a) and 1.14 (b) shows a methodology wherein axial thrust is given on inner race while holding the outer race against a rigid surface. Reduces axial clearance and provides preloading.
- Fig. 1.14 (c) shows a typical taper roller beaing mounting whereby tightening the lock nut radial clearances will be reduced. Preloading values have to be accurately determined beforehand.

Types of Failures in Rolling Contact Beaings:

- Theoretically, point or line contact occurs in a bearing which results into large values of contact stresses. Pitting and wear failures are therefore unavoidable over a period of time. Therefore even when they are used under ideal conditions, bearings do fail by deterioration of material due to rolling fatigue. Pitting occurs due to excessive contact stresses and are basically designed to avoid this. Increasing surface hardness can extend the life.
- Scoring is another fundamental problem which can occur with lubrication failure. Excessive heat is generated resulting in stick slip phenomenon. A bearing is said to have failed with the occurance of the first failure in the inner ring, outer ring or rolling element due to repeated stresses. However in order to get a satisfactory rated life from a bearing following modes of failures and their causes should be understood.

Main Causes of Bearing Failures are:

- Wrong bearing selection (Check the type of bearing as well as its size).
- Faulty installation. (Too close or loose fittings, excessive misalignment - can lead to creep failure).
- Lubrication provided being inadequate. (lubrication method, type of lubricant, presence of corrosive additives, sealing provision – can lead to rust, corrosion wear or seizure).
- Foreign material contamination (during installation or working leading to abrasive wear).
- Extreme working conditions (speed and temperature – can lead to dry running).
- Abnormal shock loads (ball distortion – can lead to noisy operation).
- To understand bearing failure causes, experience with bearings, its lubrication and understanding of the characteristics of the equipment are required.

SOLVED EXAMPLES

Example 1.1: *A cylindrical roller bearing is subjected to a radial load of 5000 N. The desired life of the bearing with 90% reliability is 15,000 hours. The load application factor is 1.5. If the shaft rotates at 1440 rpm, calculate the required basic dynamic load rating of the bearing.*

Solution: Given data: Type of bearing: Cylindrical roller

F_r = Radial load = 5000 N;

F_a = 0 N (Axial load)

L_{10h} = 15000 hours

K_a = 1.5 (Load application factor)

n = 1440 rpm

Radial load factor,

$$X = 1 \quad \ldots (1)$$

(Since the bearing is subjected to pure radial load).

Axial load factor,

$$Y = 0$$

(Since bearing is not subjected to any axial load).

Equivalent dynamic load (F_e),

Since the shaft (inner race) rotates V = 1.0

$$F_e = (XVF_r + YF_a) K_a = (1 \times 1 \times 5000 + 0 \times 0)\, 1.5 = 7500 \text{ N}$$

Rating life of bearing,

$$L_{10} = \frac{L_{10h} \times 60 \times n}{10^6} = \frac{15000 \times 60 \times 1440}{10^6} = 1296 \text{ million revolutions}$$

Required basic dynamic load capacity, (C) is given by

$$L_{10} = \left(\frac{C}{F_e}\right)^a$$

$$1296 = \left(\frac{C}{7500}\right)^{\frac{10}{3}}$$

$$C = 64393.61 \text{ N}$$

$$C = 64.393 \text{ kN}$$

Example 1.2: *A 22 kW, 1440 rpm electric motor is directly coupled to a shaft of 25 mm diameter, which is supported by two cylindrical roller bearings. The shaft transmits power to another line shaft through the flat pulley of 300 mm diameter, which is placed mid-way between the two bearings. The coefficient of friction between the belt and pulley is 0.3, while the angle of lap is 180°. The belt is horizontal. The load factor is 1.5. If the expected life is 50,000 hours, select the bearings from manufacturer's catalogue.*

Bearing No.	NU2205	NU2305
Basic Dynamic Capacity 'C' kN	15.99	31.39

Solution: Given data:

Type of bearing = Cylindrical roller bearing

kW = Power transmitted by electric motor = 22×10^3 W

n = Speed of shaft = 1440 rpm

d = Diameter of shaft = 25 mm

D = Diameter of pulley = 300 mm

μ = Coefficient of friction = 0.3

θ = Angle of lap = 180° = π radians

K_a = Load application factor = 1.5.

L_{10h} = Rating life of bearing = 50000 hours

(i) Select Suitable Bearing for Ends of Shaft

Radial load and axial load (F_a)

$$\frac{P_1}{P_2} = e^{\mu\theta} = e^{0.3\pi} = 2.566$$

$$\therefore \quad P_1 = 2.566\, P_2$$

(ii) Belt Speed V_B:

$$V_B = \frac{\pi D n}{60 \times 1000} = \frac{\pi \times 300 \times 1440}{60 \times 1000} = 22.62 \text{ m/sec}$$

(iii) Power Transmitted by Belt Drive

$$\text{Power} = (P_1 - P_2)\, V_B$$

$$22 \times 10^3 = (2.566\, P_2 - P_2) \times 22.62$$

$$P_2 = 621.07 \text{ N}$$

$$P_1 = 2.566\, P_2 = 1593.7 \text{ N}$$

(iv) Total Horizontal Force, (F) on Pulley:

$$F = P_1 + P_2 = 1593.7 + 621.07 = 2214.8 \text{ N}$$

(v) Radial Load on each Bearing (F_r):

$$F_r = \frac{F}{2} = \frac{2214.8}{2} = 1107.4 \text{ N}$$

(vi) Axial load on each bearing F_a = 0:

$$K_a = 1.5$$

$$F_{eq} = (XVF_r + YF_a) K_a = (1 \times 1 \times 1107.4 + 0) \times 1.5 = 1661.11 \text{ N}$$

$$L_{10} = \frac{L_{10h} \times 60 \times N}{10^6} = \frac{50000 \times 60 \times 1440}{10^6} = 4320 \text{ million revolutions}$$

$$L_{10} = \left(\frac{C}{F_e}\right)^a; \quad 4320 = \left(\frac{C}{1661.1}\right)^{10/3}$$

$$\therefore \quad C = 20466 \text{ N}$$

Hence, we select bearing NU2305 having dynamic load capacity, C = 31.39 kN.

Example 1.3: *The radial load carried on ball bearing is 2500 N for the first five revolutions and reduces to 1500 N for the next ten revolutions. The load variation then repeats itself. The expected life of the bearing is 25 million revolutions. Determine dynamic load carrying capacity of the bearings:*

Solution:

Type of bearing : Single row deep groove ball bearing

F_1 = Radial load acting during first element = 2500 N

F_2 = Radial load acting during second element = 1500 N

N_1 = Number of revolution in first element = 5

N_2 = Number of revolution in second element = 10

L_{10} = Rating life of bearing = 25 million revolution

To find 'C':

Equivalent dynamic load for entire work cycle (F_e)

$$F_e = \left(\frac{N_1 (F_1)^3 + N_2(F_2)^3}{N_1 + N_2}\right)^{1/3}$$

$$= \left[\frac{5 \times (2500)^3 + 10\,(1500)^3}{5 + 10}\right]^{1/3}$$

$$= 1953.80 \text{ N}$$

Basic dynamic load capacity of bearing (C)

$$L_{10} = \left(\frac{C}{F_e}\right)^3$$

$$25 = \left(\frac{C}{1953.80}\right)^3$$

$$C = 5712.95 \text{ N}$$

Example 1.4: *A single row deep groove ball bearing is subjected to the following work cycle. Reliability is 90%, consider 4 bearings and calculate reliability of system.*

Fraction of Cycle	Radial Load F_r kN	Thrust Load F_a kN	Radial Factor 'X'	Thrust Factor 'Y'	Race Rotating	Service Factor	Speed rpm
1/10	1.5	0.25	1.0	0	Inner	1.2	400
1/4	1.0	0.75	0.56	2.0	Outer	1.8	500
1/2	5.0	1.1	0.56	2.0	Inner	1.5	600
Remaining	1.0	-	1.0	0	Outer	2.0	800

Desired life of bearing: 15000 hours.

Select bearing from the following data. Find n_{avg}.

Number	6011	6211	6311	6411
Dynamic Capacity	28.1 kN	43.6 kN	71.5 kN	99.5 kN

Solution :

Given Data:

Type of bearing = Single row deep groove ball bearing

Desired life of bearing = 15000 hours.

For 1st cycle:

$$F_1 = (X_1V_1F_{r1} + Y_1F_{a1}) \cdot K_{a1}$$

$$= (1 \times 1 \times 1500 + 0 \times 250) \times 1.2$$

$$F_1 = 1800 \text{ N}$$

Revolutions N_1

$$= \frac{1}{10} \times \left(\frac{L_{10h} \times 60}{10^6}\right) n_1$$

$$= \frac{1}{10}\left(\frac{15000 \times 60}{10^6}\right) \times 400$$

$$= \frac{1}{10} \times (0.9) \times 400$$

$$N_1 = 36 \text{ million revolutions}$$

For 2nd cycle:

$$F_2 = (X_2V_2F_{r2} + Y_2F_{a2})\, K_{a2}$$

$$= (0.56 \times 1.2 \times 1000 + 2 \times 750) \times 1.8$$

$$= 3909.6 \text{ N}$$

$$N_2 = \frac{1}{4}\frac{L_{10h} \times 60 \times n_2}{10^6} = \frac{1}{4} \times \frac{15000 \times 60 \times 500}{10^6}$$

$$= \frac{1}{4}(0.9) \times 500$$

$$N_2 = 112.5 \text{ million revolutions}$$

For 3rd cycle:

$$F_3 = (X_3V_3F_{r3} + Y_3F_{a3})\, K_{a3}$$

$$= (0.56 \times 1 \times 5000 + 2 \times 1100) \times 1.5$$

$$= 7500 \text{ N}$$

$$N_3 = \frac{1}{2} \times \frac{L_{10h} \times 60 \times n_3}{10^6}$$

$$= \frac{1}{2} \times \left(\frac{15000 \times 60}{10^6}\right) \times 600 = \frac{1}{2} \times 0.9 \times 600$$

$$N_3 = 270 \text{ million revolutions}$$

For 4th cycle:

$$F_4 = (X_4V_4F_{r4} + Y_4 + F_{a4})\, K_{a4}$$

$$= (1 \times 1.2 \times 1000 + 0) \times 2$$

$$F_4 = 2400 \text{ N}$$

$$N_4 = 0.15 \times \frac{L_{10h} \times 60 \times n_4}{10^6}$$

$$= 0.15 \times \left(\frac{15000 \times 60}{10^6}\right) \times 800$$

$$= 0.15\,(0.9)\,(800)$$

N_4 = 108 million revolutions

Equivalent dynamic load for entire working cycle,

$$F_e = \left[\frac{N_1(F_1)^a + N_2(F_2)^a + N_3(F_3)^a + N_4(F_4)^a}{N_1 + N_2 + N_3 + N_4}\right]^{1/a}$$

$$= \left[\frac{36 \times (1800)^3 + 112.5\,(3909.6)^3 + 270 \times (7500)^3 + 108 \times (2400)^3}{526.5}\right]^{1/3}$$

F_e = 6147.74 N

Rating life of bearing, (L_{10})

$L_{10} = N = N_1 + N_2 + N_3 + N_4$

= 36 + 112.5 + 270 + 108

= 526.5 million revolutions

$$L_{10} = \left(\frac{C}{F_e}\right)^3$$

$$526.5 = \left(\frac{C}{6147.74}\right)^3$$

C = 49641.56 N

∴ 6311 bearing number with dynamic capacity 71.5 kN selected for given work cycle application.

Average speed of bearing, (n_{avg})

$$n_{avg} = \frac{\text{Total life of bearing in revolution}}{\text{Total life of bearing in minutes}}$$

$$= \frac{526.5 \times 10^6}{15000 \times 60}$$

n_{avg} = 585 rpm

Reliability of total system,

$R_s = (R)^{N_s} = (0.9)^4 = 0.6561$

R_s = 65.61%

Example 1.5: *A single row deep groove ball bearing operates with the following work cycle.*

Element Time %	Radial Load (F_r) kN	Thrust Load (F_a) kN	Radial Factor X	Thrust Factor Y	Race Rotating	Service Factor	Speed r.p.m.
40	4.0	1.0	0.56	1.4	Inner	1.8	720
20	2.5	1.0	0.56	1.6	Outer	2.0	1440
40	No load	No load	–	–	Outer	–	720

If the expected life of the bearing is 15000 hours with a reliability of 95%. Calculate the basic dynamic load rating of the bearing so that it can be selected from the manufacturer's catalogue based on 90% reliability.

If there are 6 such bearings in a system, what is the probability that all bearings will survive for 15000 hours.

Solution: Type of bearing – Single row deep groove ball bearing.

L_{05h} = Rating life of bearing at 95% reliability

= 15000 hours

N_s = Number of bearings = 6

$P_1 = (X_1V_1F_{r1} + Y_1F_{a1}) \cdot K_{a1}$

$= (0.56 \times 1 \times 4000 + 1.4 \times 1000) \times 1.8$

P_1 = 6552 N

$$N_1 = 0.4 \times \left(\frac{L_{10h} \times 60}{10^6}\right) \times n_1$$

$$= 0.4 \times \left(\frac{L_{10h} \times 60}{10^6}\right) \times 720$$

$N_1 = 0.01728\ L_{10h}$ million revolutions

$P_2 = (X_2V_2F_{r2} + Y_2F_{a2}) \cdot K_{a2}$

$= (0.56 \times 1.2 \times 2500 + 1.6 \times 1000) \times 2$

P_2 = 6560 N

$$N_2 = 0.2 \times \frac{L_{10h} \times 60 \times n_2}{10^6}$$

$$= 0.2 \times \frac{L_{10h} \times 60 \times 1440}{10^6}$$

$N_2 = 0.01728\ L_{10h}$ million revolutions

$P_3 = (X_3V_3Fr_3 + Y_3F_{a3}) \cdot K_{a3} = 0$ N

$$N_3 = 0.4 \times \frac{L_{10h} \times 60 \times n_3}{10^6}$$

$$= 0.4 \times \frac{L_{10h} \times 60 \times 720}{10^6}$$

$N_3 = 0.01728\ L_{10h}$ million revolutions

$$F_e = \sqrt[3]{\frac{N_1P_1^3 + N_2P_2^3 + N_3P_3^3}{N_1 + N_2 + N_3}}$$

$$= \sqrt[3]{\frac{[0.01728 \times (6552)^3 + 0.01728\,(6.560)^3 + 0.0178\,(0)^3]\ L_{10h}}{3 \times (0.01728)\ L_{10h}}}$$

F_e = 5.7272 kN = 5727.2 N

(i) Rating Life of the Bearing for 90% Reliability (L_{10h}):

$$\frac{L_{5h}}{L_{10h}} = \left[\frac{\log_e (1/R_{95})}{\log_e (1/R_{90})}\right]^{\frac{1}{1.7}}$$

$$\frac{15000}{L_{10h}} = \left[\frac{\log_e (1/0.95)}{\log_e (1/0.90)}\right]^{1/1.17}$$

L_{10h} = 27751.44 hours

(ii) Rating Life of Bearing for 90% Reliability, (L_{10})

$L_{10} = N = N_1 + N_2 + N_3$

$= 0.01728 \times L_{10h} + 0.01728 \times L_{10h} + 0.01728\ L_{10h}$

$= 0.05184 \times L_{10h}$

$$= 0.05184 \times 27751.44$$

$$L_{10} = 1438.63 \text{ million revolutions}$$

Basic dynamic load capacity, (C)

$$L_{10} = \left(\frac{C}{F_e}\right)^a$$

$$1438.63 = \left(\frac{C}{5727.2}\right)^3$$

$$C = 64653.5 \text{ N} = 64.65 \text{ kN}$$

Reliability of the system (R_s)

$$R_s = (R)^{Ns} = (0.95)^6$$

$$R_s = 0.7351$$

$$R_s = 73.51\%$$

Example 1.6: *A single row deep groove ball bearing is used to support the lay shaft of a four speed automobile gearbox. It is subjected to the following loads in respective speed ratios.*

Gear	Axial Load (N)	Radial Load (N)	% Time Engaged
1st gear	3400	4500	2%
2nd gear	800	3400	5%
3rd gear	100	2800	23%
4th gear	Nil	Nil	70%

The lay shaft is fixed to the engine shaft and rotates at 1600 rpm. The static and dynamic load carrying capacities of the bearing are 11800 N and 17300 N respectively. The bearing is expected to be in use for 4500 hours of operation. Find out the reliability with which life could be expected.

Refer the following data.

X and Y Factors for Single Row, Deep Groove Ball Bearings

F_a/C_o	$(F_a/F_r) \leq e$		$(F_a/F_e) > e$		e
	X	Y	X	Y	
0.025	1	0	0.56	2.0	0.22
0.040	1	0	0.56	1.8	0.242
0.070	1	0	0.56	1.6	0.27
0.130	1	0	0.56	1.4	0.31
0.250	1	0	0.56	1.2	0.37
0.500	1	0	0.56	1.0	0.44

Solution: Type of bearing: Single row deep groove

n = Speed of shaft rotation = 1600 rpm

C_o = Basic static load capacity = 11800 N

C = Basic dynamic load capacity

= 17300 N

L_h = Life of bearing = 4500 hours

For 1st gear:

$$\frac{F_a}{C_o} = \frac{3400}{11800} = 0.288 \rightarrow e$$

$$= 0.37 + \frac{(0.07) \times (0.038)}{0.25}$$

e = 0.3806 (By interpolation)

$$\frac{F_a}{F_r} = \frac{3400}{4500} = 0.755 \text{ which is} > e$$

∴ From above chart, X = 0.56 and Y is obtained by interpolation.

$$Y = 1.2 - \frac{(0.288 - 0.25) \times (1.2 - 1.0)}{(0.5 - 0.25)}$$

$$= 1.1697 \simeq 1.17$$

For remaining cases interpolation is not required since $(F_a/F_r) < e$. Hence, X and Y values are read from chart. Below shown table finds X and Y for all the cases.

Radial Load (F_r)N	Axial Load (F_a)N	F_a/F_r	F_a/C_o	e	X	Y	Gear
4500	3400	0.755	0.2669	0.3747	0.56	1.17	1st gear
3400	800	0.1769	0.0487	0.2487	1	0	2nd gear
2800	100	0.0231	6.36×10^{-3}	< 0.22	1	0	3rd gear
Nil	Nil	0	0		1	0	4th gear

Equivalent dynamic (Radial) load during 1st gear (F_1)

$$F_1 = (X_1 \cdot V_1 \cdot F_{r1}) + Y_1 \cdot F_{a1}) \cdot K_a$$

$$= 0.56 \times 4500 + 1.17 \times 3400) \times 1$$

$$F_1 = 6498.0 \text{ N}$$

Revolutions of bearing during 1st gear

$$N_1 = 0.02 \times \frac{L_{10h} \times 60 \times n_1}{10^6}$$

$$= 0.02 \times \frac{L_{10h} \times 60 \times 1600}{10^6}$$

$$N_1 = 1.92 \times 10^{-3} L_{10h} \text{ million revolutions}$$

$$F_2 = (X_2 \cdot V_2 \cdot F_{r2} + Y_2 \cdot F_{a2}) \cdot K_{a2}$$

$$= (1 \times 1 \times 3400 + 0) \times 1$$

Parallely F_2 = 3400.0 N

$$N_2 = 0.05 \times \frac{L_{10h} \times 60 \times n_2}{10^6}$$

$$= 0.05 \times \frac{L_{10h} \times 60 \times 1600}{10^6}$$

$$N_2 = 4.8 \times 10^{-3}\ L_{10h} \text{ million revolutions}$$

$$F_3 = (X_3 \cdot V_3 \cdot F_{r3} + Y_3 \cdot F_{a3}) \cdot K_a$$

$$= (1 \times 1 \times 2800 + 0) \times 1$$

$$F_3 = 2800 \text{ N}$$

$$N_3 = 0.23 \times \frac{L_{10h} \times 60 \times n_3}{10^6}$$

$$= 0.28 \times \frac{L_{10h} \times 60 \times 1600}{10^6}$$

$$N_3 = 0.0221\ L_{10h} \text{ million revolutions}$$

$$F_4 = (X_4 \cdot V_4 \cdot F_{r4} + Y_4 \cdot F_{a4}) \cdot K_a$$

$$F_4 = 0$$

$$N_4 = 0.7 \times \frac{L_{10h} \times 60 \times n_4}{10^6}$$

$$= 0.7 \times \frac{L_{10h} \times 60 \times 1600}{10^6}$$

$$N_4 = 0.0672\ L_{10h} \text{ million revolutions}$$

Equivalent dynamic (radial) load during entire work cycle (F_e).

$$F_e = \sqrt[3]{\frac{N_1 F_1^3 + N_2 F_2^3 + N_3 F_3^3 + N_4 F_4^3}{N_1 + N_2 + N_3 + N_4}}$$

$$= \sqrt[3]{\frac{1.92 \times 10^{-3} \times L_{10h} \times 6498.5^3 + 4.8 \times 10^{-3}\ L_{10h} \times 3400^3 + 0.0221\ L_{10h} \times 2800^3 + 0.0672 \times L_{10h} \times 0^3}{1.92 \times 10^{-3}\ L_{10h} + 4.8 \times 10^{-3}\ L_{10h} + 0.022\ L_{10h} + 0.0672\ L_{10h}}}$$

$$F_{eq} = \sqrt[3]{12.51} = 2.32 \text{ kN} = 2322 \text{ N}$$

Rating Life of Bearing (L_{10}):

$$L_{10} = \left(\frac{C}{F_e}\right)^P = \left(\frac{17300}{2322}\right)^3$$

$$= 413.6 \text{ million revolutions}$$

$$L_{10} = 413.6 \text{ million revolutions}$$

Actual revolutions completed by bearing during entire work cycle (L).

$$L = \frac{4500 \times 60 \times 1600}{10^6} \quad \ldots \left(\frac{L_h \times 60 \times n}{10^6} = L\right)$$

$$L = 432 \text{ million revolutions}$$

Reliability of bearing (R)

$$\frac{L}{L_{10}} = \left[\frac{\log_e (1/R)}{\log_e (1/R_{90})}\right]^{1/1.17}$$

$$\frac{432}{413.6} = \left[\frac{\log_e (1/R)}{\log_e (1/0.90)}\right]^{1/1.17}$$

$$\Rightarrow \quad R = 0.895 \qquad R = 89.5\%$$

Example 1.7: *The following data is given for a belt drive.*

Diameter of pulley	=	*250 mm*
Shaft diameter	=	*20 mm*
Power transmitted	=	*5 kW*
Speed	=	*720 rpm*
Ratio of belt tensions	=	*3 : 1*
Load factor	=	*3*

Assume the pulley to be placed centrally with the belt tension acting vertically downwards. The required reliability of the bearing is 95% with life of 10,000 hours. Find the dynamic capacity of the bearing, when the bearings are selected from the manufacturer's catalogue, which lists dynamic load carrying capacity at 90% reliability.

Solution: Type of bearing – Single row deep groove

$$D = 250 \text{ mm}$$

$$d = 20 \text{ mm}$$

$$\text{Power} = 5 \text{ kW}$$

$$n = 720 \text{ rpm}$$

$$\frac{P_1}{P_2} = 3$$

$$K_a = 3$$

$$L_h = 10{,}000 \text{ hours}$$

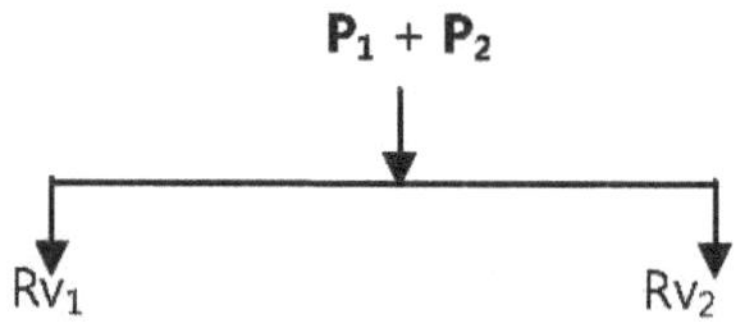

Belt tension

$$\frac{P_1}{P_2} = 3, \quad P = 3P_2$$

Belt speed (v)

$$v = \frac{\pi \cdot D \cdot n}{60000}$$

$$v = \frac{\pi \times 250 \times 720}{60000}$$

$$v = 9.4248 \text{ m/s}$$

Power transmitted by belt

$$\text{kW} = (P_1 - P_2) \cdot v$$

$$5 \times 10^3 = (3P_2 - P_2) \cdot 9.4248$$

$$P_2 = 265.26 \text{ N}$$

$$P_1 = 3P_2 = 795.77 \text{ N}$$

Radial load at each bearing

$$Rv_1 = Rv_2 = \frac{P_1 + P_2}{2}$$

$$= \frac{795.77 + 265.26}{2} = 530.52 \text{ N} = F_r$$

$$\text{Axial load} = 0 \text{ N} (= F_a)$$

$$K_a = 3$$

$$F_e = (X \cdot V \cdot F_r + Y \cdot F_a) \cdot K_a$$

... (Equivalent dynamic radial load)

$$= (1 \times 1 \times 530.52 + 0) \times 3$$

$$F_e = 1591.56 \text{ N}$$

Rating life of bearing (L_{10})

$$L = \frac{L_h \times 60 \times n}{10^6} = \frac{10000 \times 60 \times 720}{10^6}$$

$$L = 432 \text{ million revolutions}$$

Now,

$$\frac{L}{L_{10}} = \left[\frac{\log_e (1/R)}{\log_e (1/R_{90})}\right]^{1/1.17}$$

$$\frac{432}{L_{10}} = \left[\frac{\log_e (1/0.95)}{\log_e (1/0.90)}\right]^{1/1.17}$$

$$L_{10} = 799.24 \text{ million revolutions}$$

Basic dynamic load capacity (C),

$$L_{10} = \left(\frac{C}{F_e}\right)^P$$

$$799.24 = \left(\frac{C}{1591.56}\right)^3$$

$$C = 14770.06 \text{ N} = 14.8 \text{ kN}$$

Thus, bearing with basic dynamic load capacity greater than 14.8 kN should be selected for this application.

Example 1.8: *A transmission shaft is supported by two deep groove ball bearings at two ends. The C.D. between the bearings is 160 mm. A load of 300 N acts vertically downwards at 60 mm distance from the left hand bearing whereas a load of 550 N acts horizontally at 50 mm distance from right hand bearing. Shaft speed is 3000 rpm and expected life of the bearings is 7000 hours with a reliability of 95%. It is intended to use same bearing at both ends of the shaft. Calculate dynamic load rating of the bearing so that it can be selected from manufacturer's catalogue. Load application factor is 1.5.*

Solution: Type of bearing – Single row deep groove

$$n = 3000 \text{ rpm}$$

$$L_h = 7000 \text{ hours}$$

Considering vertical plane

Taking moment at point 1, i.e. $\Sigma M_1 = 0$

$$-300 \times 60 + Rv_2 \times 160 = 0$$

$$Rv_2 = 112.5 \text{ N}$$

$$\Sigma F_y = 0$$

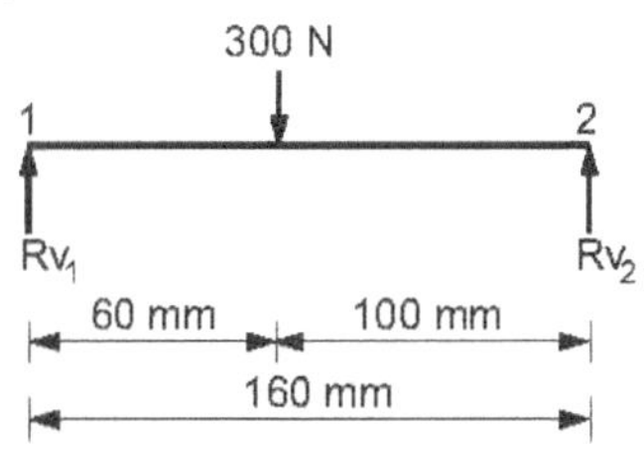

Fig. 1.15

$$\therefore \quad Rv_1 + Rv_2 - 300 = 0$$

$$Rv_1 = 300 - 112.5$$

$$Rv_1 = 187.5 \text{ N}$$

Consider horizontal plane

Taking moment at point 1

i.e. $\Sigma M_1 = 0$

$$-550 \times 110 + R_{H2} \times 160 = 0$$

$$R_{H2} = 378.13 \text{ N}$$

$$\Sigma F_x = 0$$

$$R_{H1} + R_{H2} - 550 = 0$$

$$R_{H1} = 550 - 378.13$$

$$R_{H1} = 171.88 \text{ N}$$

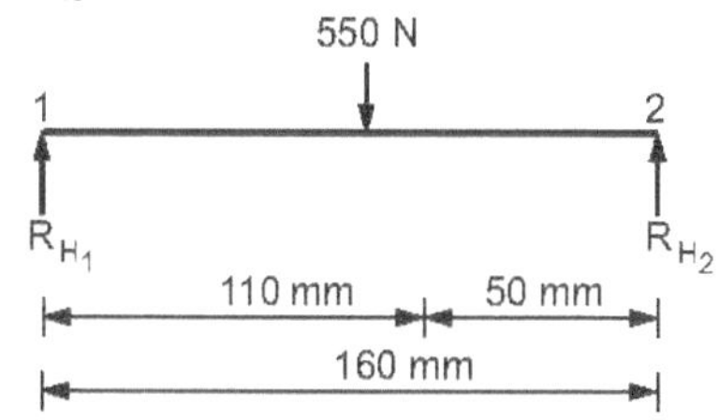

Fig. 1.16

Resultant radial load at bearing 1

$$F_{r1} = \sqrt{Rv_1^2 + R_{H1}^2} = \sqrt{187.5^2 + 171.88^2}$$

$$F_{r1} = 254.36 \text{ N}$$

Resultant radial load at bearing 2.

$$F_{r2} = \sqrt{Rv_2^2 + R_{H2}^2} = \sqrt{112.5^2 + 378.13^2}$$

$$F_{r2} = 394.51 \text{ N}$$

Axial load on bearings 1 and 2 is zero.

$$K_a = 1.5$$

Equivalent dynamic load (P_e)

At bearing 1

$$F_{e1} = (XVF_{r1} + Y \cdot F_{a1}) \cdot K_a$$

$$= (1 \times 1 \times 254.36 + 0) \times 1.5$$

$$F_{e1} = 381.54 \text{ N}$$

At bearing 2

$$F_{e2} = (X \cdot V \cdot F_{r2} + Y \cdot F_{a2}) \cdot K_a$$

$$= (1 \times 1 \times 394.51 + 0) \times 1.5$$

$$F_{e2} = 591.77 \text{ N}$$

Rating life of bearing, (L_{10})

$$L = \frac{L_h \times 60 \times n}{10^6} = \frac{7000 \times 60 \times 3000}{10^6}$$

$$L = 1260 \text{ million revolutions}$$

$$\frac{L}{L_{10}} = \left[\frac{\log_e (1/R)}{\log_e (1/R_{90})}\right]^{1/1.17}$$

$$\frac{1260}{L_{10}} = \left[\frac{\log_e (1/0.95)}{\log_e (1/0.90)}\right]^{1/1.17}$$

$$L_{10} = 2331.12 \text{ million revolutions}$$

Basic dynamic load capacity (C)

$$L_{10} = \left(\frac{C}{F_e}\right)^P$$

$$2331.12 = \left(\frac{C}{591.77}\right)^3$$

$$C = 7846.47 \text{ N}$$

Example 1.9: *A shaft is supported on two bearings A and B which are 250 mm apart. A gear is attached at a distance of 100 mm on the right from left hand side bearing A. Weight of pulley is 100 N which is attached at an overhang of 150 mm on the right of the right hand side bearing B. Horizontal belt tensions are 498 N and 166 N respectively. Horizontal tangential force component for the gear is 497 N which is directed same as belt tensions. Vertically downwards radial force component for the gear is 181 N. The load factor for application is 2.5 and expected life of bearing is 8000 hours. If the shaft speed is 720 rpm. Find dynamic load capacity for bearings A and B so that they can be selected from manufacturer's catalogue.*

Solution: Type of bearing – Single row deep groove

$$P_1 = 498 \text{ N}$$

$$P_2 = 166 \text{ N}$$

$$F_t = \text{Tangential force on gear} = 497 \text{ N}$$

$$F_r = \text{Radial force on gear} = 181 \text{ N}$$

$$K_a = 2.5$$

$$L_{10h} = 8000 \text{ hours}$$

$$n = 720 \text{ rpm}$$

$$W = 100 \text{ N} = \text{Weight of pulley}$$

Considering vertical plane

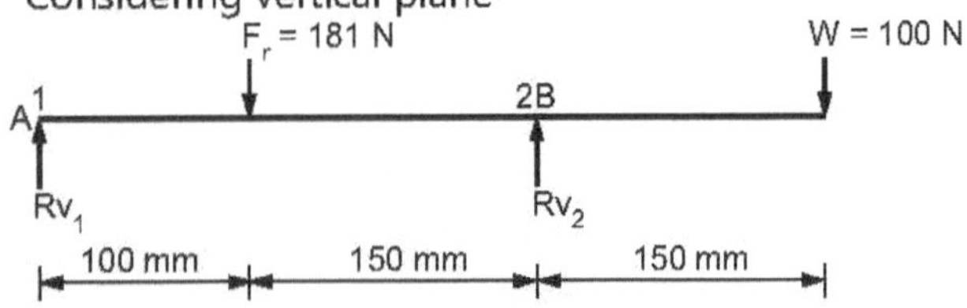

Fig. 1.17

Taking moment at point 1,

$$\Sigma M_1 = 0$$

$$-181 \times 100 + Rv_2 \times 250 - 100 \times 400 = 0$$

$$Rv_2 = 232.4 \text{ N}$$

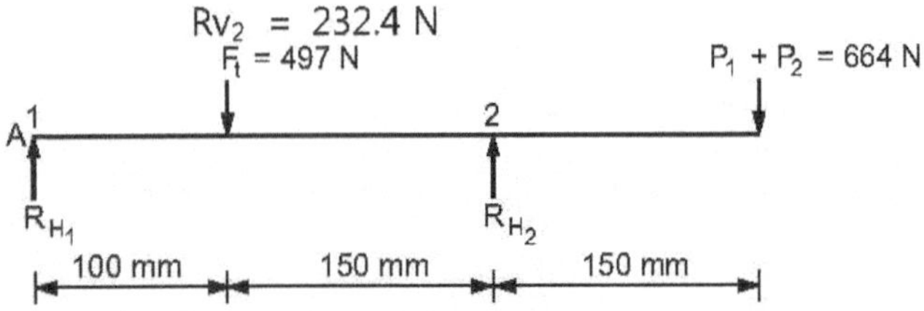

Fig. 1.18

$$\Sigma F_y = 0$$

$$Rv_1 - 181 + 232.4 - 100 = 0$$

$$Rv_1 = 48.6 \text{ N}$$

Considering horizontal plane

Taking moment at point 1,

$$\Sigma M_1 = 0$$

$$R_{H2} \times 250 - 497 \times 100 - 664 \times 400 = 0$$

$$R_{H2} = 1261.2 \text{ N}$$

$$\Sigma F_x = 0$$

$$R_{H1} + R_{H2} - 497 - 664 = 0$$

$$R_{H1} = -100.2 \text{ N}$$

$$F_{r1} = \sqrt{(Rv_1)^2 + (R_{H1})^2} = 111.36 \text{ N}$$

$$F_{r2} = \sqrt{(Rv_2)^2 + (R_{H2})^2} = 1282.43 \text{ N}$$

$$K_a = 2.5$$

$$F_{e1} = (X \cdot V \cdot F_{r1} + Y \cdot F_{a1}) \cdot K_a$$

$$= (1 \times 1 \times 111.36 + 0) \times 2.5$$

$$P_{e1} = 278.4 \text{ N}$$

$$F_{e2} = (X \cdot V \cdot Fr_2 + Y \cdot F_{a2}) \cdot K_a$$

$$= (1 \times 1 \times 1282.43 + 0) \times 2.5$$

$$P_{e2} = 3206.08 \text{ N}$$

Rating life of bearing (L_{10})

$$L_{10} = \frac{L_{10h} \times 60 \times n}{10^6} = \frac{8000 \times 60 \times 720}{10^6}$$

$$L_{10} = 345.6 \text{ million revolutions}$$

For bearing 1

$$L_{10} = \left(\frac{C_1}{F_{e1}}\right)^P \Rightarrow 345.6$$

$$= \left(\frac{C_1}{278.4}\right)^3 \Rightarrow C_1 = 1953.71 \text{ N}$$

For bearing 2

$$L_{10} = \left(\frac{C_2}{F_{e2}}\right)^P = \left(\frac{C_2}{3206.08}\right)^3$$

$$L_{10} = 345.6$$

$$C_2 = 22499.12 \text{ N}$$

Example 1.10: *Using a constant $k = 60.8 \times 10^6$ N/mm² in the Stribeck's equation, compute the static capacity of a single row deep groove ball bearing series 208 with 9 balls having diameter of 12 mm.*

Solution: As per Stribeck's equation of static capacity for ball bearing,

$$C_o = \frac{Kd^2Z}{5}$$

$$C_o = \frac{60.8 \times 10^6 \times 12^2 \times 9}{5}$$

$$C_o = 1.5759 \times 10^{10} \text{ N}$$

Example 1.11: *An equivalent radial load on a bearing varies continuously from 0 to 20 kN in a sinusoidal manner. Determine the dynamic load rating at 90% reliability, if the bearing is to have a life of 20 million revolutions at a reliability of 99%. Assume shaft speed as 1000 rpm.*

Use life reliability relationship as

$$\frac{L}{L_{10}} = \left[9.491 \log_e\left(\frac{1}{R}\right)\right]^{1/1.17}$$

Solution: Type of bearing – Single row deep groove ball bearing

$$P_{max} = 20 \times 10^3$$

$$L_{o1} = \text{Rating life of bearing}$$

$$= 20 \text{ million revolutions}$$

$$n = 1000 \text{ rpm}$$

Equation for variable load is given by,

$$P = P_{max} \cdot \sin^2 (\theta/2)$$

$$= P_{max} \cdot \left(\frac{1 - \cos \theta}{2}\right)$$

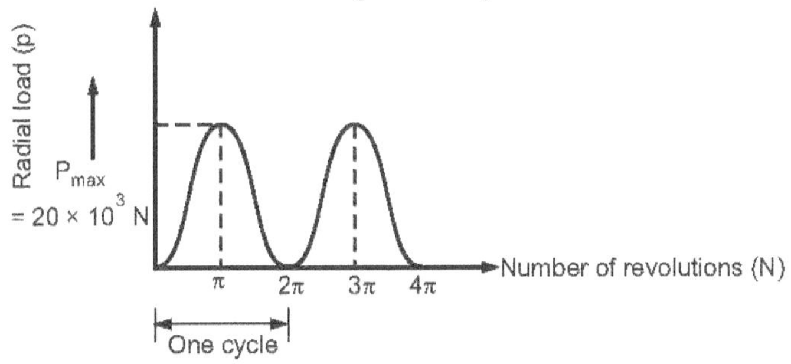

Fig. 1.19

Considering the work cycle for $\theta = 0$ to $\theta = 2\pi$

We have, $P_e = \left[\frac{1}{N} \int P^P \cdot dN\right]^{1/P}$

$$P_e = \left[\frac{1}{N} \int P^3 \, dN\right]^{1/3}$$

$$= \left[\frac{1}{2\pi} \int_0^{2\pi} (P_{max})^3 \times \frac{(1 - \cos \theta)^3}{8} d\theta\right]^{1/3}$$

$$= \frac{P_{max}}{2} \left[\frac{1}{2\pi} \int_0^{2\pi} (1 - \cos \theta)^3 \cdot d\theta\right]^{1/3}$$

$$= \frac{P_{max}}{2} \left[\frac{1}{2\pi} \int_0^{2\pi} (1 - 3 \cos \theta + 3 \cos^2 \theta - \cos^3 \theta) \cdot d\theta\right]^{1/3}$$

$$= \frac{P_{max}}{2}$$

$$\left[\frac{1}{2\pi} \int_0^{2\pi} \left\{1 - 3 \cos \theta + 3 \left[\frac{1 + \cos 2\theta}{2}\right] - \cos \theta (1 - \sin^2 \theta)\right\} \cdot d\theta\right]^{1/3}$$

$$= \frac{P_{max}}{2}$$

$$\left[\frac{1}{2\pi} \int_0^{2\pi} \left(\frac{5}{2} - 4 \cos \theta + \frac{3}{2} \cos \theta + \sin^2 \theta \cdot \cos \theta\right) d\theta\right]^{1/3}$$

$$= \frac{P_{max}}{2} \left[\frac{1}{2\pi} \left[\frac{5}{2}\theta - 4 \sin \theta + \frac{3}{2} \cdot \frac{\sin^2 \theta}{2} + \frac{\sin^3 \theta}{3}\right]_0^{2\pi}\right]^{1/3}$$

$$= \frac{P_{max}}{2} \left[\frac{1}{2\pi} \left(\frac{5}{2} \times 2\pi - 0 + 0 + 0\right)\right]^{1/3} = \frac{P_{max}}{2} \left[\frac{5}{2}\right]^{1/3}$$

$$= \frac{20 \times 10^3}{2} \left[\frac{5}{2}\right]^{1/3} \Rightarrow P_e = 13572.09 \text{ N}$$

$$\frac{L}{L_{10}} = \left[9.491 \log_e \left(\frac{1}{R}\right)\right]^{1/1.17}$$

$$\frac{20}{L_{10}} = \left[9.491 \log_e \left(\frac{1}{0.99}\right)\right]^{1/1.17}$$

$$L_{10} = 149.02 \text{ million revolutions}$$

Basic dynamic load capacity, (C)

$$L_{10} = \left(\frac{C}{P_e}\right)^P$$

$$149.02 = \left(\frac{C}{13572.09}\right)^3$$

$$C = 71955.1 \text{ N}$$

Example 1.12: *A deep groove ball bearing operating under a constant load rotates at 700 rpm for 50% of cycle time, 1000 rpm for 30% of cycle time and 1500 rpm for rest of the cycle. The outer race of the bearing rotates. Considering load factor as 1.3 and radial and axial loads to be 3.0 kN and 1.5 kN, determine expected life bearings with 95% reliability, if the dynamic load capacity of the bearings is 45 kN. Consider radial load factor X = 0.56 and axial load factor, Y = 1.71.*

Solution:

Rotation factor,

$$V = 1.2 \text{ (since outer race is rotatory)}$$

Load factor i.e. service factor = 1.3

Equivalent load

$$F_{eq} = 1.3 \,(X\,V\,F_r + Y \cdot F_a)$$

$$= 1.3\,(0.56 \times 3.0 \times 1.2 + 1.71 \times 1.5)$$

$$= 5.955 \text{ kN}$$

(This load is constant)

Average shaft speed

$$= \frac{700 \times 0.5t + 1000 \times 0.30t + 1500 \times 0.2t}{(0.5t + 0.3t + 0.2t)}$$

$$N_{av} = 950 \text{ rpm}$$

Rated life is given by

$$C = F_e L^{1/3}$$

$$\therefore \quad L = \left(\frac{C}{F_e}\right)^3 = \left(\frac{45}{5.955}\right)^3$$

$$= 431.51 \text{ million revolutions}$$

This is L_{10} life i.e. at 90% reliability.

Life at 95% reliability will be given by

$$\frac{L}{L_{10}} = \left[\frac{\log_e\left(\frac{1}{R_{95}}\right)}{\log_e\left(\frac{1}{R_{90}}\right)}\right]^{\frac{1}{1.17}} = \left[\frac{\log_e\left(\frac{1}{0.95}\right)}{\log_e\left(\frac{1}{0.9}\right)}\right]^{\frac{1}{1.17}}$$

$$= [0.4868]^{1/1.17}$$

$$\frac{L}{L_{10}} = 0.540$$

$$\therefore \quad L = 0.540 \times L_{10} = 0.54 \times 431.51$$

$$= 233 \text{ million revolutions}$$

$$\therefore \quad \text{Life in hours} = \frac{233 \times 10^6}{950 \times 60} = 4087.7 \text{ hours}$$

Example 1.13: *A shaft rotating at 1000 rpm is supported in bearings at its ends. Left side bearing has a radial load of 2500 N and axial load of 1200 N. Right side bearing has a radial load of 3000 N and axial load is zero. For commonisation it is proposed to use single row deep groove ball bearing 6307 with static load capacity as 18000 N and dynamic load capacity of 33200 N. What is expected trouble free life in hours.*

Use following table for radial and thrust factors.

$\left(\frac{F_a}{C_o}\right)$	$\left(\frac{F_a}{F_r}\right) \le e$		$\left(\frac{F_a}{F_r}\right) > e$		e
	X	Y	X	Y	
0.04	1	0	0.56	1.8	0.24
0.07	1	0	0.56	1.6	0.27
0.130	1	0	0.56	1.4	0.31

Considering L.H. bearing

$F_a = 1200$, $F_r = 2500$, $\frac{F_a}{F_r} = 0.48$, $F_a = 1200$,

$C_o = 18000$, $\frac{F_a}{C_o} = 0.066$

'e' for $\frac{F_a}{C_o} = 0.066$ is between 0.24 to 0.27

$$\therefore \quad \frac{F_a}{F_r} > e$$

$$\therefore \quad X = 0.56$$

Value of Y is found by interpolation for $\frac{F_a}{C_o} = 0.066$

$$Y = 1.8 - \frac{(1.8 - 1.6) \times (0.066 - 0.04)}{(0.07 - 0.04)}$$

$$= 1.627$$

Radial load factor,

$$X = 0.56$$

Axial load factor,

$$Y = 1.627$$

$$\therefore \quad (F_{eq})_{LH} = V \cdot X \cdot F_r + Y \cdot F_a$$

(With V = 1 for inner race rotating)

$$(F_{eq})_{LH} = 1 \times 0.56 \times 2500 + 1.627 \times 1200$$

$$= 1400 + 1952.4$$

$$= 3352.4 \text{ N}$$

Considering RH bearings:

$$(F_{eq})_{RH} = 3000 \text{ N}$$

(Since there is no axial load, X = 1)

$\therefore$ Load on LH bearing is critical

$$F_{eq} = 3352.4 \text{ N}$$

Bearing life which can be expected is given by

$$C = F_e L^{1/3}$$

$$\therefore \quad L = \left(\frac{C}{F_e}\right)^3$$

$$= \left(\frac{33200}{3352.4}\right)^3$$

$$= 971.285 \text{ million revolutions}$$

Life in hours at 1000 rpm will be

$$= \frac{971.285 \times 10^6}{60 \times 1000}$$

$$= 16188 \text{ hours}$$

EXERCISE

1. Explain types of rolling contact bearings?
2. Discuss equivalent dynamic load and load life relationship for rolling contact bearing.
3. Derive Stribeck's equation for static capacity of a rolling contact bearing.
4. What do you understand by rating and median life of the bearing.
5. What is the objective of preloading? Explain mounting and preloading of a taper roller bearing with appropriate sketch.
6. Explain the mounting and preloading of a taper roller bearing with appropriate sketch.
7. Compare ball bearing with roller bearing.
8. What are the basic principles to be followed in bearing mounting.
9. Discuss design of bearing cyclic loads and speeds.
10. Write short note on lubrication of bearing.
11. Discuss lubrication consideration for rolling contact bearing.
12. Discuss types of failures and causes of bearing failures for rolling contact bearing.

◈◈◈

CHAPTER - 2
SLIDING CONTACT BEARINGS

2.1 METHODS OF LUBRICATION [Aug. 16]

- Lubrication is the most important component of tribology as it takes care of friction and wear, in addition to performing other functions.
- In moving parts of machines, friction is undesirable. It causes loss of energy which is converted in the form of heat energy. Due to friction, an additional force is required to cause the motion of a body. To improve upon the efficiency of working machines, the frictional force is reduced to the minimum possible. The act of doing so is called as **Lubrication** and the substance used for lubrication is called a **Lubricant.**
- Laws of dry friction are not applicable to film friction.
- When the two surfaces in contact are completely separated by a film of lubricant, the friction occurs due to resistance of relative motion between the lubricant and surfaces in contact. This is known as **Film Friction, Fluid Friction** or **Viscous Friction.**
- It is thus obvious that the film friction is not due to the surfaces in contact, but it is due to **Viscosity** and **Oiliness** of the lubricant.
- Thus, lubrication is the science of reducing friction by application of suitable substance called lubricant between the rubbing surfaces of bodies having relative motion.

2.1.1 Purpose of Lubrication

Though lubrication is important in reducing friction, it also performs multiple functions as listed below :

- To reduce friction (primary function) between surfaces in contact where the relative motion exists.
- To dissipate heat, which is generated due to frictional loss of power in machines.
- To take away foreign particles.
- To reduce or prevent wear in between surfaces in contact where the relative motion exists.
- To protect the surfaces against corrosion.
- To reduce the electrical and mechanical power consumption (again by reducing friction).
- To reduce noise, vibration and shock between gear teeth and other components.

2.1.2 Basic Modes of Lubrication [Nov. 15]

For performing various functions during lubrication, the basic modes of lubrication are classified as follows :

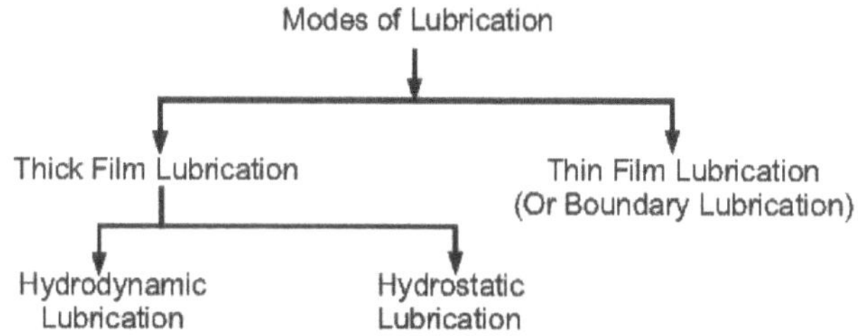

Thick Film Lubrication describes a condition of lubrication where two surfaces of the bearing in relative motion are completely separated by a film of fluid. Film fluid (oil film) always remains in between the two components. Since there is no contact between the surfaces, least wear takes place between these components. The performance of the bearing is affected by the viscosity of the lubricant. Hydrodynamic lubrication and hydrostatic lubrication are considered as Thick Film Lubrication.

Thin Film Lubrication describes a condition of lubrication where the lubricant film is relatively thin and there is partial metal to metal contact. Boundary lubrication is considered as Thin Film Lubrication.

1. Hydrodynamic Lubrication

In heavily loaded bearings such as thrust bearings and horizontal journal bearings, the fluid viscosity alone is not sufficient to maintain a film between the moving surfaces. In these bearings, higher fluid pressures are required to support the load until the fluid film is established. If this pressure is supplied by an outside source, it is called hydrostatic lubrication. If the pressure is generated internally, that is, within the bearing by dynamic action, it is referred to as hydrodynamic lubrication.

In hydrodynamic lubrication, a fluid wedge is formed by the relative surface motion of the journals or the thrust runners over their respective bearing surfaces.

The principle of hydrodynamic lubrication in journal bearing is shown in Fig. 2.1. In hydrodynamic lubrication, the load supporting fluid film is created by the shape and relative motion of the sliding surfaces.

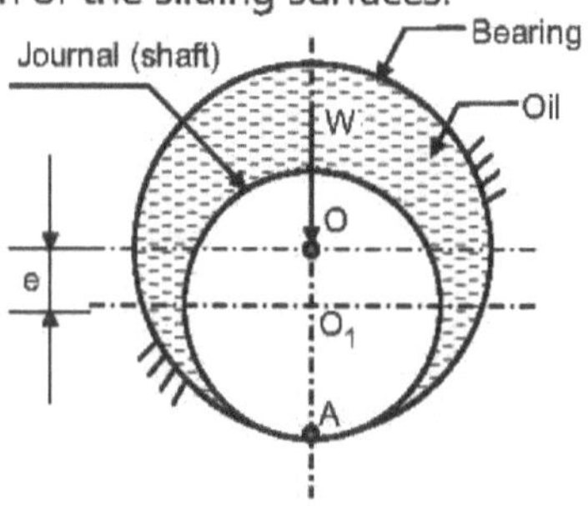

(a) Journal at rest

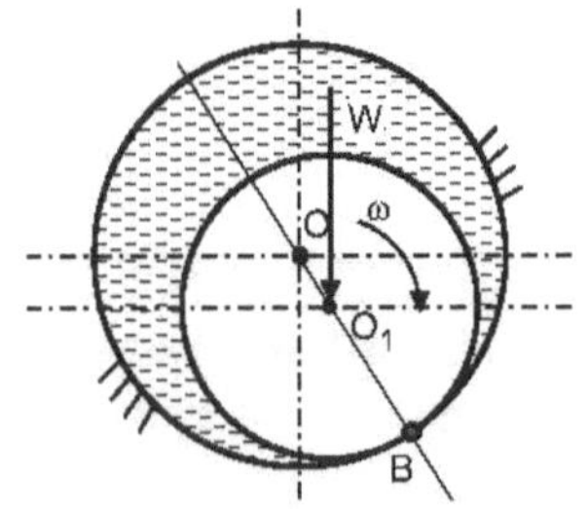

(b) Journal starts rotating

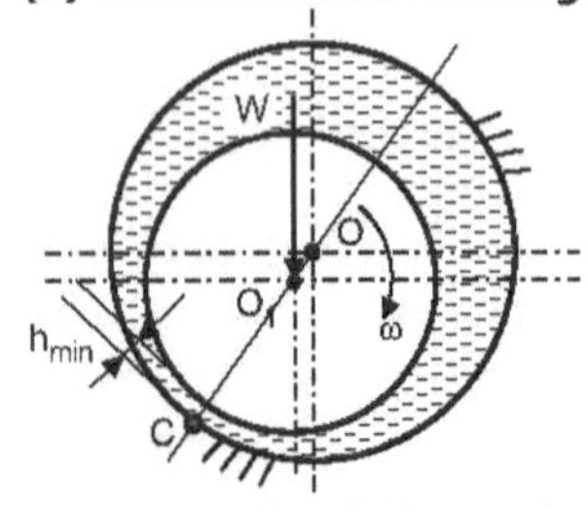

(c) Journal at full speed

e – Eccentricity
W – Radial load on journal
O – Centre of bearing
O_1 – Centre of journal
ω – Speed

Fig. 2.1 : Hydrodynamic lubrication

As shown in Fig. 2.1 (a), initially, the journal (shaft) is at rest and it rests at the bottom of the bearing A, under the action of load W. Metal to metal contact is established at A. This point of contact is called as **seat of pressure.**

The journal now starts to rotate in the clockwise direction, inside the bearing as shown in Fig. 2.1 (b). Due to friction between the two rubbing surfaces, the friction starts opposing the motion. The seat of pressure or point of contact between the two surfaces then climbs up the bearing in a direction opposite to that of rotation. The point of contact shifts from A to B. As the speed of journal is further increased, it will force the fluid into the wedge-shaped regions as shown in Fig. 2.1 (c). Since more and more fluids is forced into the wedge-shaped clearance space, sufficient pressure is generated within the system to carry the load. The pressure distribution around the periphery of the journal is shown in Fig. 2.2.

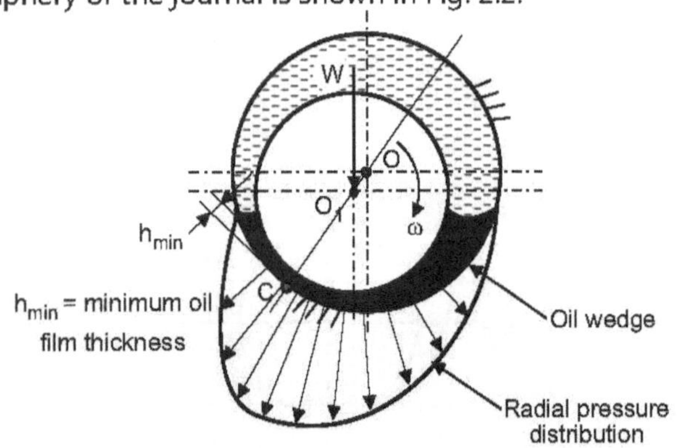

Fig. 2.2 : Radial pressure distribution in hydrodynamic lubrication

Since the pressure is created within the system due to rotation of the journal, this type of bearing is also known as **Self-Acting Bearing.** At a particular journal speed, the pressure becomes sufficient to support the load (W) and the journal is shifted to the other side [i.e. to left side as shown in Fig. 2.1 (c)] of vertical. In this situation, there is minimum clearance and oil film thickness is minimum (h_{min}) at point C. [Refer Fig. 2.1 (c)].

The amount of rotation of the line of centres of the journal and bearing from the load-line depends on :

- The magnitude of applied radial load (W).
- The journal speed (ω), and
- The viscosity of the lubricant.

Thus, in hydrodynamic bearings, it is not necessary to supply the lubricant under pressure. The only requirement is to ensure sufficient and continuous supply of the lubricant.

The minimum thickness of the fluid film increases with an increase in fluid viscosity and surface speed and decreases with an increase in load.

Examples : Main bearings, connecting rod bearings, gudgeon pin bearings, etc. mounted on engines. Also, the applications of these bearings are seen in centrifugal pumps, compressors, etc.

Hydrodynamic bearings are simple in construction, easy to maintain, initial and maintenance cost is less as compared to hydrostatic bearings.

In hydrodynamic bearings, the journal always runs eccentric in the bearing and there is a wear at the start-up and stopping of rotation of the journal.

2. Hydrostatic Lubrication

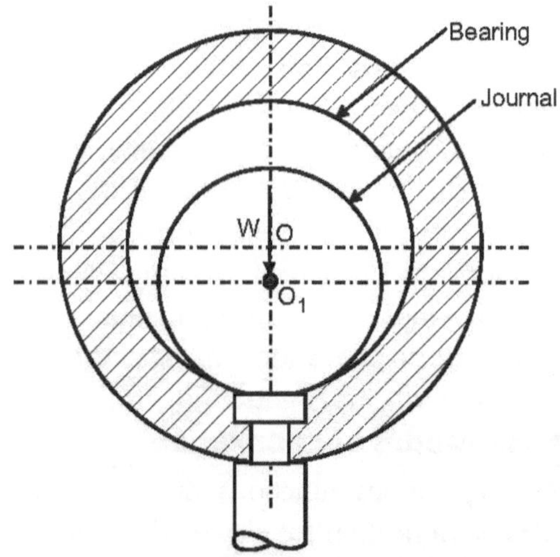

(a) Journal at rest

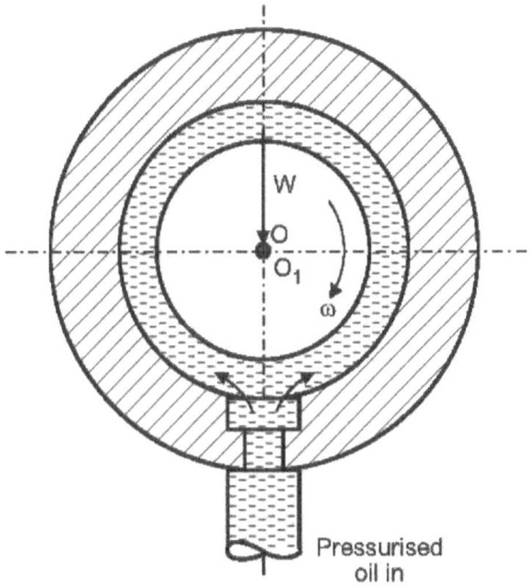

(b) Journal at full speed

Fig. 2.3 : Hydrostatic lubrication

In hydrostatic lubrication, the load supporting fluid film, separating the two surfaces is created by an external source like a pump, supplying sufficient fluid under pressure as shown in Fig. 2.3.

Since the lubricant is supplied under pressure, this type of bearing is also called as **Externally Pressurised Bearing.**

Fig. 2.3 shows the principle of hydrostatic lubrication in journal bearing. As shown in Fig. 2.3 (a), initially the journal rests on a bearing surface under the radial load W. As the pump starts, high pressure lubricant is admitted in the clearance space forcing the surfaces of the bearing and the journal to separate out as shown in Fig. 2.3 (b).

Examples :

- Vertical turbo generators, centrifuges, ball mills, etc.

  Hydrostatic bearings are costly as compared to hydrodynamic bearings. However, it offers the following advantages :
- High load carrying capacity even at low speeds.
- No starting friction.
- There is no rubbing action at any speed and load, hence wear and tear is less in hydrostatic bearings as compared to hydrodynamic bearings.

3. Boundary Lubrication [Nov. 16]

When a complete fluid film does not develop between potentially rubbing surfaces, the film thickness may be reduced to permit momentary dry contact between the wear surface high points or asperities. This condition is the characteristic of boundary lubrication. Boundary lubrication occurs whenever any of the essential factors that influence formation of a full fluid film are missing. The most common example of boundary lubrication includes bearings, which normally operate with fluid-film lubrication but experience boundary lubricating conditions during routing, starting and stopping of equipment. Other examples include door hinges, machine tool slides and guideways, etc.

Thus, in boundary lubrication, the lubricant film is relatively thin and there is partial metal to metal contact. Various causes of boundary lubrication are as follows :

- Excessive load
- Insufficient oil supply
- Low speed
- Misalignment

4. Elasto-Hydrodynamic Lubrication (EHL)

The lubrication principles applied to rolling bodies, such as ball or roller bearings, gears, cams, etc., is known as elasto-hydrodynamic lubrication (EHL).

Although lubrication of rolling objects operates on a considerably different principle than sliding objects, the principles of hydrodynamic lubrication can be applied, within limits, to explain the lubrication of rolling elements. An oil wedge, similar to that which occurs in hydrodynamic lubrication exists at the lower leading edge of bearing. Adhesion of oil to the sliding element and the supporting surface increases the pressure and creates a film between the two bodies. Because the area of contact is extremely small in a roller and ball bearing, the force per unit area, or **Load Pressure is Extremely High.** Under these pressures, it would appear that the oil would be entirely squeezed from between the wearing surfaces. However, the viscosity increase that occurs under extremely high pressure, prevents the oil from being entirely squeezed out. Consequently, a thin film of oil is maintained. The surfaces which carry this load are likely to deform. A very thin lubricant film actually supports the load. Thus, elastohydrodynamic lubrication (EHL) deals with the lubrication of elastic contacts.

All the assumptions used in the classical theory of hydrodynamic lubrication cannot be used especially because the variation of viscosity with pressure must be considered here. In addition to this, a heavy load which causes elastic deformation of solids changes the geometry of the lubricating film. Thus, *in the problems of elastohydrodynamic lubrication, it is necessary to solve simultaneously the hydrodynamic as well as elasticity equations concerned with the contact, since the shape of oil film largely controls pressure distribution.*

5. Extreme Pressure (EP) Lubrication [Nov. 16]

Extreme Pressure (EP) Lubrication is applicable to extreme boundary cases where high contact pressure and intense sliding combine to generate high localised temperature. EP additives are added to lubricants to allow chemical reactions leading to the formation of organometallic components of low shear strength. During motion, shearing takes place between the organometallic layers, thus reducing friction and wear. The shearing and regeneration of reaction products is a continuous process. As the melting points of the EP additives are high, they remain on the rubbing surfaces and work effectively at higher temperatures.

2.2 LUBRICANTS

Lubrication is the use of a substance to provide smoothness to the movement of the surfaces of machine elements, which move relative to each other. The substance used to achieve this purpose, is termed as **lubricant.**

2.2.1 Types of Lubricants

A large number of lubricants are used in industries; some are general purpose and some are tailor made, but all can be grouped into the following four categories.

1. Lubricating oils
2. Greases
3. Solids
4. Gases

The primary function of all above lubricants is to reduce friction and wear. In addition to this, they perform several other important functions such as – they act as a coolant, protect surfaces against corrosion, flush away contaminants, provide damping effects, reduce noise and act as a sealent.

1. Lubricating Oils

Any liquid having some amount of viscosity can be regarded as a lubricant. At the same time, it should be as mobile as possible to remove heat and to avoid power loss due to viscous drag. It should be stable under thermal and oxidation stresses, have low volatility, good mechanical stability, non-toxic or harmless etc. The bulk of lubricating oils today are made from crude petroleum. Lubricants are liquids varying in appearance from light yellow to black.

Lubricating oils include **natural organics** like animal fat, whale oil, shark oil, vegetable oils, mineral oils, etc.

The most commonly used lubricants are **mineral oils**. The refining of petroleum produces mineral oils. These are the crude oils or mixtures of hydrocarbons blended with chemicals to cater to various lubrication requirements for specific application called additives. The chemical compounds making up mineral oils are mainly hydrocarbons of three types :

(i) Paraffins – Open chain hydrocarbons.

(ii) Napthenes – Closed ring hydrocarbons.

(iii) Aromatics – Hydrocarbons based on benzene rings.

Oil is termed as a paraffinic oil or napthenic oil depending on the majority percentage of paraffins or napthenes in it. The aromatics, though remain as a small percentage, play a vital role in boundary lubrication. Mineral oils can be used upto a maximum temperature of 200°C whereas animal and vegetables can be used upto a maximum temperature of 120°C.

Lubricating oils also include **synthetic organics** like synthetic hydrocarbons, diesters, chlorofluorocarbon, fatty acid esters, polyglycoether, fluoroester, phosphate ester, silicate ester, polyether, etc. Synthetic hydrocarbons are prepared by the polymerisation of specific olefin monomers and these can be prepared to optimize their viscosity-temperature, stability and volatility properties. Synthetic organics have evolved as a result of the requirement of lubricants to be used at extreme temperatures, pressure, humidity, etc. These can be used upto a maximum temperature of 400°C.

Functionally, lubricating oils are basically classified into following categories :

- Automobile lubricating oils (including aviation oils).
- Industrial lubricating oils.
- Metal working oils.
- Industrial speciality oils.
- Marine lubricating oils.

Following types of lubricating oils are commonly used in automotive equipments :

- Crank case oil (engine oil).
- Gear oils.
- Transmission oils.
- Universal tractor oils.
- Two stroke engine oils.
- Preservatives and running-in oils.
- Special oils (brake oil, shock absorber oil, calibration fluid, etc.)

For standardization and variety reduction of these oils, for the ease of consumers and to maintain quality, grading of these oils are generally governed by SAE, API and BIS specifications, etc.

Industrial lubricating oils have to cater to the needs of today's modern machineries, operating at high speed or heavy loads or extreme temperatures and occasionally under tight schedules in automatic production lines.

Because of variety of applications, the industrial lubricating oil can be further classified as –

Turbine oils, circulating and hydraulic oils, industrial gear oils, spindle oils, machine tool-way oils, general purpose machinery oils, steam cylinder oils, refrigeration compressor oils, etc.

Many of above oils are graded by ISO, BIS, BS, DIN, AGMA specification etc.

Metal working oils serve the dual purpose of lubrication in cutting zones and also cooling of workpiece in machines. Cutting fluids, hot and cold rolling oils, wire drawing and deep drawing oils, etc. come under this category.

Industrial specially oils are primarily not used for lubrication but are used for other functions such as heat transfer, heat treatment and other finishing operations in order to improve the quality and serviceability of the product as well as to improve its sales appeal. Quenching oils, heat transfer fluids, electrical oils, rubber process oils and rust preventives etc. come in this category.

Marine lubricating oils are modified versions of automotive oils to make the oil sea-worthy.

2. Greases

Greases are oils that are thickened with solids to form semi-fluid products. The type of thickener determines the characteristics of the grease. Greases are preferred to liquid lubricants : (i) In cases where the application of continuous supply of lubricant is not needed or (ii) There is no sufficiently tight enclosure to retain a liquid lubricant. (iii) Greases have the great advantage is that, they remain in the bearing system, thereby the lubricating arrangements are extremely simplified. Due to this advantage, greases are common lubricants in rolling-element bearings. (iv) Here, grease acts as a seal against dirt as well as lubricant. But the use of grease is limited as they do not flow easily and cannot carry away the heat generated at high-speed operations. Grease consists of basically following components :

- Fluid (about 85-90% by volume) which can be selected from mineral oils, polyglycols or other synthetics or a combination of such fluids.
- Fatty materials (vegetable, animal or fish), about 3 to 15% of total, usually called fatty acids.
- Thickeners (in the form of base or alkali), about 1 to 4% of total, these are mainly metallic soaps of calcium, aluminium, sodium, barium and lithium components, etc.
- Additives and modifiers – Very small quantity as per desired properties, graphite, MoS_2, fluorocarbon powders, zinc oxides, etc. are used to increase tackiness, low temperature performance, water resistance, oxidation resistance, EP and anti-wear properties, etc.

One of the most important properties of a grease lubricant is its **drop point**, a temperature at which a grease melts under certain operating conditions. If a grease is used at a temperature above drop point, it loses any advantage it has over an oil.

Most lubricant greases used in industry have petroleum oils as their liquid base. The petroleum oil-based greases can be used upto 175°C depending on their thickener. The synthetic oil-based greases can be used at higher temperature. However, they are more costly.

3. Solid Lubricants

Lubrication with liquids and greases has some technological and economic limitations. e.g. for hot working of steel, no suitable additive is available for liquid lubricants and liquid evaporate and the low volatility hydrocarbons burn readily. Also, in some cases, liquid lubricants may be too expensive as they require pumps and seals, etc. and any leakage may be very costly. All these led to the development of solid lubricants. Need of solid lubricant was, it should be able to completely separate two sliding surfaces, even under very high load and adverse condition, while allowing the surface to slide.

Graphite and molybdenum disulphide (MoS_2) are the two best known solid lubricants in the industry. A solid lubricant is a thin film composed of a single solid or a combination of solids introduced between two rubbing surfaces for the purpose of modifying friction and wear. Solid lubricants are solely used in chiples metal working such as drawing, rolling, stamping, extrusing, spinning and in the machining operation where chips are formed, these solid lubricants are used with fluids or oils as base.

Polymers and polymer-based materials have most of the advantages of other solid lubricants plus they have following additional advantages :

- They absorb vibration and generally noiseless.
- Their conformability is better and they absorb minor dimensional in-accuracy better.
- Polymer components can be easily formed by machining or moulding etc.

From wide range of commercially available polymers, PTFE, polyacetal, polyethylene, nylon and polyamide are commonly used in tribology and PTFE is the outstanding amongst all. However, many reinforcements and fillers are to be incorporated in these materials at PTFE and other polymers have poor creep resistance and poor mechanical stability.

Coefficient of friction of PTFE, sliding on itself, can be as slow as 0.04 but, for other polymers, coefficient of friction varies from 0.15 to 0.4. However, use of polymers is limited by temperature.

The main requirements for solid lubricants are :

(i) Thermal stability

(ii) Low shear strength

(iii) Surface protection and

(iv) Good bonding properties

4. Gases

The principle of operation of a hydrodynamic or hydrostatic bearing using gas as a lubricant is basically same as that of an oil-lubricated bearing. In general, gases have extremely low coefficient of absolute viscosity as compared to oils. Consequently, viscous resistance is very much less. In gas lubrication, the film thickness is thinner than oil-lubricated bearings. **Air** is used as one of the main lubricants and hence the bearings are called as **aerodynamic bearings** or **aerostatic bearings**. To operate an aerodynamic bearing successfully, the following points should be kept in mind.

- The surfaces of journal and bearing should be finished with high accuracy.
- The alignment between the journal and bearing must be very good.
- The dimensions and clearances must be accurate.
- The speed must be high.
- The load should be relatively low.

One of the exclusive advantages of gas bearings is that these can be operated over extremely wide ranges of temperatures. Due to this gas bearings can be used at temperature upto 500°C for prolonged periods. One more important advantage of gas bearings is its low frictional characteristics. For this reason, these bearings are being used in near-static apparatus, such as dynamometers, wind-tunnel balances and other sophisticated mechanical instruments. Low frictional characteristics of externally pressurised gas bearings can be exploited in machine tool slideways. Gas bearings can also be used effectively where it is necessary to keep the environment free from contamination by liquid lubricants, such as in high-purity gas handling plants. However, gas bearings are prone to instability. Therefore, a designer, or a user of gas bearings should be conversant with this characteristics for successful operations.

2.2.2 Properties of Lubricants

The physical and chemical properties of mineral oils affect lubrication. A few properties may have both physical and chemical effects. Recognition of these properties is useful for designing lubrication systems, choosing lubricating oils, diagnosing lubrication, friction and wear problems, and selecting appropriate testing method.

1. Viscosity

The most important single property of a lubricant is its **viscosity.**

'Viscosity' *is defined as the internal frictional resistance offered by a fluid to change its shape or relative motion of its parts.* Sir Isaac Newton was the first person to propose that a force is necessary to shear a fluid film. This force is a measure of internal friction of fluid and it resembles friction between two solid surfaces.

In fluid-film lubrication, load-carrying capacity depends on the viscosity of the lubricant. A lubricant with extremely low viscosity cannot form a layer of fluid film between two sliding surfaces. Again, if the viscosity of a lubricant increases, the frictional resistance and heat generation in the bearing increases. Thus, the choice of a lubricant with proper viscosity is an important consideration in bearing design.

An oil film of thickness h, placed between two parallel plates is shown in Fig. 2.4. The lower plate is stationary while the upper plate is moved with a velocity U by means of a force F.

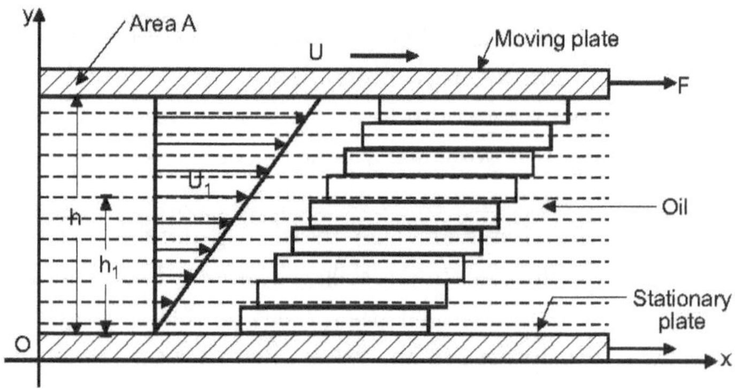

Fig. 2.4 : Newton's law of viscosity

The velocity of oil adhering to the top surface will have the velocity of top surface, whereas the oil particles adhering to the stationary surface (bottom surface) will have zero velocity. Under this condition the velocity of oil at any point in the film is proportional to the distance of the point from stationary surface. It may be considered that the flow is equivalent to a large number of thin layers of fluid sliding over each other. Therefore,

$$\frac{U}{h} = \frac{U_1}{h_1} = \frac{U_2}{h_2} \text{ etc.}$$

This type of orderly movement is called **streamline, laminar** or **viscous flow.**

The tangential force per unit area i.e. $\left(\frac{F}{A}\right)$ is shear stress, while the ratio $\left(\frac{U}{h}\right)$ is the rate of shear. According to Newton's law of viscosity, the shear stress is proportional to the rate of shear at any point in the fluid.

Therefore,

$$\frac{F}{A} \propto \frac{U}{h}$$

or

$$\frac{F}{A} = \mu \frac{U}{h} \quad \text{... (2.1)}$$

The constant of proportionality μ in the above equation is called **absolute (dynamic) viscosity.**

The equation (2.1) can also be written as,

$$F = \mu A \frac{U}{h} \quad \text{... (2.2)}$$

When the velocity distribution is non-linear with respect to h, the term $\left(\frac{U}{h}\right)$ in the above equation is replaced by $\left(\frac{dU}{dh}\right)$ and the equation is rewritten as,

$$F = \mu A \left(\frac{dU}{dh}\right) \quad \text{... (2.3)}$$

Any lubricant which obeys the above law is **Newtonian lubricant.**

The reciprocal of viscosity is called **fluidity.**

$$\frac{1}{\mu} = \frac{\frac{dU}{dh}}{\left(\frac{F}{A}\right)} \quad \text{... (2.4)}$$

The **unit of absolute (dynamic) viscosity** is given by,

From equation (2.1), μ is given by,

$$\mu = \frac{Fh}{AU} \quad \text{... (2.5)}$$

∴ Unit of absolute viscosity is,

$$\Rightarrow \frac{(N)\ (mm)}{(mm^2)\ (mm/s)}$$

$$\Rightarrow N\text{-}s/mm^2$$

In S.I. units, the absolute viscosity is **N-s/m²** or **Pa-s.**

In c.g.s. units, the absolute viscosity is dyne-s/cm². The popular unit of viscosity is **Poise**, which gives **absolute viscosity in dyne-s/cm²**.

Also, $1\ P = 0.1\ Pa\text{-}s$ ... (2.6)

Poise is a large unit and viscosities of most of the lubricating oils are given in terms of **centi-Poise (cP),** which is one-hundredth of a Poise. Therefore, two separate notations are used for viscosity, viz.,

μ = Viscosity in units of (N-s/mm²)

and z = Viscosity in units of (cP)

The relationship between z and μ is as follows :

$$1\ cP = \frac{1}{10^2}\ \text{Poise}$$

$$= \frac{1}{10^2}\left(\frac{\text{dyne-s}}{cm^2}\right)$$

$$= \frac{1}{10^2} \times \left(\frac{N}{10^5}\right)\left[\frac{s}{(10^2\ mm^2)}\right]$$

$$= 10^{-9}\ N\text{-}s/mm^2 \text{ or MPa.s}$$

$$\therefore 1\ N\text{-}s/mm^2 = 10^9\ cP \quad \text{... (2.7)}$$

$$\text{Therefore, } \mu = \frac{z}{10^9} \quad \text{... (2.8)}$$

The viscosity of most lubricant lies in the range of **2 to 400 cP**. To give a quantitative idea of absolute or dynamic viscosity, the water has the viscosity of the order of 1 cP at room temperature, while air is having a viscosity of 0.02 cP.

Another measure of viscosity is **kinematic viscosity (z_k) equals z divided by density (ρ).**

$$z_k = \frac{z}{\rho} \quad \text{... (2.9)}$$

where, z_k = Kinematic viscosity

z = Absolute (dynamic) viscosity

ρ = Mass density of fluid in gm/cm³

In c.g.s. units, the unit of kinematic viscosity is **Stoke (St).**

Also, $1\ St = 1\ cm^2/s$ ... (2.10)

Like poise the stoke is a large unit, so it is used as **Centistoke (cSt)** which is one-hundredth of a stoke.

In S.I. units, the kinematic viscosity is **m²/s.**

Measurement of Viscosity :

Instruments that are used to measure viscosity of a fluid are known as **viscometers.** Most viscometers do not measure the absolute viscosity directly but measure kinematic viscosity z_k.

In practice, it is difficult to carry out an experiment with two parallel plates for measurement of viscosity. The popular method of determining viscosity is to measure the time required for a given volume of oil to pass through a capillary tube of standard dimensions. The oil is kept in a reservoir which is immersed in the constant temperature bath. Based on this principle, there are three commercial viscometers named after Saybolt, Redwood and Engler.

The ASTM standard method recommends the use of a viscometer called the **Saybolt Universal Viscometer.** The method is based on noting the time of emptying a given quantity (60 ml) of the lubricant at a given temperature through a capillary tube of standard dimensions and the time is measured in seconds. The unit of viscosity is called **Saybolt Universal Seconds (SUS)**, which is related to kinematic viscosity by the following relationship.

$$z_k = \left[0.22\,t - \frac{180}{t}\right] \qquad \text{... (2.11)}$$

where, t is viscosity in Saybolt Universal Seconds (SUS) and z_k is the kinematic viscosity in centistokes (cSt). The Saybolt Universal viscometer is widely used in U.S.A.

In SI units, kinematic viscosity $z_k = \left(0.22\,t - \frac{180}{t}\right) \times 10^{-6}$ m^2s

In SI units, dynamic (absolute) viscosity $z = \rho \cdot z_k$ N-s m^2 where ρ-kg/m^3

In Redwood viscometer, 50 ml of lubricating oil is passed through a capillary tube of specific dimensions and the time is measured in terms of **Redwood seconds**. The Redwood viscometer is widely used in U.K.

2. Effect of Temperature on Viscosity

The viscous resistance of a lubricating oil is due to intermolecular forces. As the temperature increases, the oil expands and the molecules move farther apart and the intermolecular forces decrease, with the result, the viscosity decreases. Thus, *viscosity of a lubricating oil decreases with increasing temperature.* The change of viscosity due to a change in temperature is different for different oils. Therefore, the two oils having the same viscosity at some temperature may have different viscosities at another temperature. There is no unique way to represent the variation of viscosity with temperature. A simple equation which gives reasonable accuracy for liquid lubricants is,

$$\log_e \mu = A + \frac{B}{T} \qquad \text{... (2.12)}$$

where, A and B are constants and T is the absolute temperature. The rate of change of viscosity with respect to temperature is indicated by a number called **viscosity index. (V.I.).** To find the V.I. of an oil, its temperature-viscosity relationship must be considered with two standard oils. This is shown graphically in Fig. 2.5.

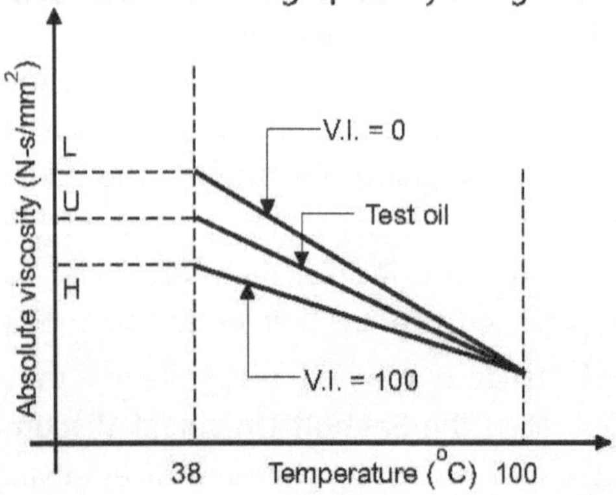

Fig. 2.5

The V.I. can be calculated from the following equation :

$$\text{V.I.} = \frac{L-U}{L-H} \times 100 \qquad \text{... (2.13)}$$

where, U = Viscosity at 38°C of the test oil whose viscosity index is to be calculated

L = Viscosity at 38°C of an oil with V.I. = 0 having same viscosity at 100°C as the test oil whose viscosity index is to be calculated

H = Viscosity at 38°C of an oil with V.I. = 100, having same viscosity at 100°C as the test oil whose viscosity index is to be calculated

In this system, two groups of mineral oils were used for comparison. These are **Pennsylvania** and **Gulf Coast Oils.** The Pennsylvania oils were considered to be superior group and were given V.I. = 100 (least change in viscosity with respect to temperature). The other group, that is Gulf Coast oils were rated V.I. = 0 (maximum change in viscosity with respect to temperature). A V.I. of 90 indicates that the oil with this value thins out less rapidly than oil with a V.I. of 50.

3. Effect of Pressure on Viscosity

As the pressure of lubricating oil is increased, the molecules are forced to come closer thereby, increasing the intermolecular forces. This increases the viscosity.

The viscosity of lubricating oils increases with pressure relatively slowly at low pressure, but the influence of pressure on viscosity at higher pressure is significant. As the pressure is further increased, the rate of change of viscosity increases, until at high pressure the oils become plastic solids. The variation of viscosity with pressure is very important in the case of elastohydrodynamic line and point contacts, such as in cams, gears and rolling-element bearings.

For most mineral oils the following equations give reasonable accuracy.

$$\mu = \mu_0 \exp(\alpha p) \qquad \text{... (2.14)}$$

where, μ_0 = Absolute viscosity at atmospheric pressure

α = Pressure coefficient of viscosity

μ = Absolute viscosity at pressure p

p = Normal pressure

4. Oiliness

The oiliness is the property by virtue of which the lubricant remains attached with the surface. Oiliness is that property of a lubricant that produces low friction under conditions of boundary lubrication. The lower the friction, the greater the oiliness.

5. Pour Point

It is defined as the temperature at which the oil just ceases the flow freely under the specified conditions. Thus, it imposes a lower limit on the oil working temperature and determines the suitability of lubricants for low-temperature installations.

6. Flash Point

It is defined as the lowest temperature at which the vapour given off by oil, when heated in a standard apparatus, ignites momentarily on the application of a flame. It is the measure of the fire hazard. The knowledge of flash point of oils is particularly important when used in high pressure systems of aircraft or in forging process where oil, leaking from a pipe, may come into contact with hot metal leading to fire hazards.

7. Fire Point

It is the temperature at which the lubricating oil gives off sufficient vapour to burn it continuously, when ignited.

8. Oxidation Stability

At room temperatures, the oxidation rate of mineral oils is fairly low but it is greatly accelerated at temperature above 90°C. The oxidation products vary but in general they consist of acidic compounds, sludge and lacquers. The insoluble products may clog oil holes, pipelines, filters and other elements. The soluble products may cause the oil to become corrosive and to increase in viscosity. Thus, the oxidation has an influence on the life of oil as well as the lubricating system.

9. Foaming

Foaming is defined as the production and coalescence of gas bubbles on a lubricant surface. Excessive foaming of oil due to churning may lead to inadequate lubrication and related problems. This is especially true for hydrostatic bearings where the pump used in the hydraulic line cannot deal efficiently with a frothy mixture. Excessive foam can starve bearings and pumps of liquid lubricant (pump cavitation) causing failure, and cause poor performance in hydraulic systems.

10. Thermal Conductivity

Thermal conductivity is the rate of transfer of heat for a given temperature gradient through a material. The rate of transfer of heat from a hot spot to a cooler area in a bearing is a factor in controlling overheating. Therefore, an oil with a high thermal conductivity would be expected to lower the temperature of a bearing. Thermal conductivity of a lubricating oil is used in equations to calculate heat transfer in bearings. Most mineral oils have approximately the same thermal conductivity. Thermal conductivity do not change appreciably with the change in temperature.

11. Demulsibility

This property is usually desired in oils, which may come into contact with water by leakage or by condensation. Demulsibility is the ability of a oil that is insoluble in water to separate from the water with which it may be mixed in the form of an emulsion. The demulsibility number of an oil can be found from a standard test which determines the time in seconds required for a known volume of oil to separate from an equal volume of condensed steam. Emulsifiable oils are used in metal cutting processes where the function of the emulsion is to cool the system rather than to lubricate.

12. Acidity and Alkalinity

Traces of weak organic acids present in lubricating oils do not cause any harm. But acids formed by combustion or oxidation and introduced by contamination may cause harm. Acidity of oil is expressed in terms of **neutralisation number,** which is defined as the weight of potassium hydroxide (KOH) in milligrams required to neutralize one gram of oil.

In some specialised applications, alkalinity is introduced in the lubricants. For example, engine oils are alkaline for the purpose of neutralizing fuel combustion products.

13. Specific Gravity or Relative Density

It is the ratio of density of lubricant to density of water at the same temperature and pressure.

Most mineral oils have relative densities in the range of 0.85 to 0.95.

2.3 VISCOUS FLOW THROUGH RECTANGULAR SLOT [Nov. 15, 17]

The basic theory of viscous flow through rectangular slot has been utilized for design of the load-carrying system of the Hale's Telescope.

Flow of lubricating fluid through rectangular slot will give a fundamental equation in lubrication. The viscous fluid flow through rectangular slot is as shown in Fig. 2.6.

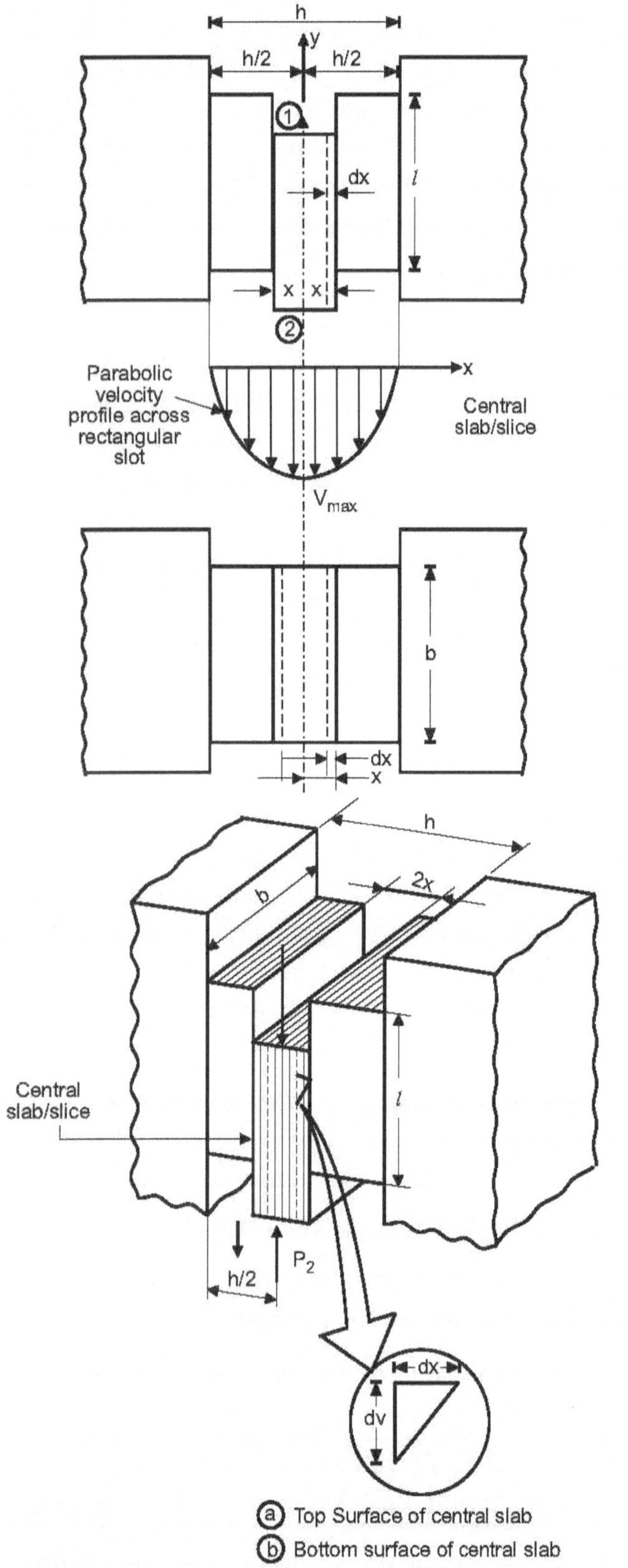

Fig. 2.6 : Viscous flow through rectangular slot

Let, l = Slot length in the direction of fluid flow (mm)

b = Width of slot perpendicular to the direction of flow (mm)

h = Slot thickness or thickness of fluid flow (mm)

ΔP = $(P_1 - P_2)$ – Pressure difference between the two sides of the middle slice (N/mm^2).

μ = Absolute viscosity of the fluid ($N\text{-}s/mm^2$)

2.3.1 Assumptions

- The width of slot (b) is assumed to be very large as compared to the film thickness h, therefore, the losses at the ends of the slot can be neglected.
- The length of the slot (l) is assumed to be large, therefore the losses at entrance and exit can be neglected.
- It is assumed that pressure at end (1) is larger than at end (2), therefore a pressure difference between the two points (1) and (2) causes flow of fluid.
- It is assumed that middle rectangular slab of fluid is being extruded down through the slot due to the downward force created by pressure difference $(P_1 - P_2)$. The thickness of middle fluid slab is '2x', width is 'b' and length is 'l' as shown.

2.3.2 Analysis

The downward force on the middle slab of fluid due to pressure difference, $\Delta P = P_1 - P_2$ is given by,

$$F_y(\downarrow) = (2x) \cdot b \cdot \Delta P$$

Now, due to viscosity of fluid flowing in downward direction, the shear resistance will act on both surfaces of the slab in upward direction.

$\therefore$ The upward resisting force acting on the slab,

$$F_y(\uparrow) = -\mu \cdot A \cdot \frac{dv}{dx}$$

$$F_y(\uparrow) = -\mu\,(2\,b \cdot l)\,\frac{dv}{dx}$$

The negative sign shows that the velocity v decreases as x increases i.e. velocity gradient $\frac{dv}{dx}$ is negative.

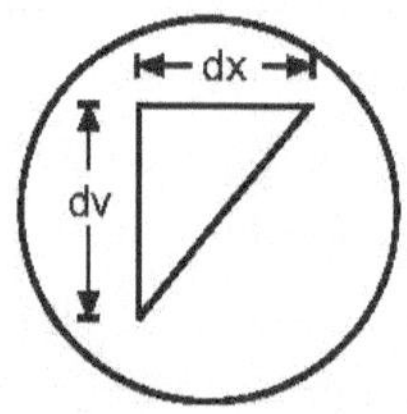

Fig. 2.7

Equating forces in vertical direction for static equilibrium,

$$F_y(\downarrow) = F_y(\uparrow)$$

$$\therefore\quad (2x) \cdot b \cdot \Delta P = -\mu \cdot (2 \cdot b \cdot l) \cdot \frac{dv}{dx}$$

$$dv = -\frac{\Delta P}{\mu l} \cdot x \cdot dx$$

Integrating above equation to get velocity,

$$\therefore \quad \int dv = \int -\frac{\Delta P}{\mu l} \cdot x \cdot dx$$

$$v = -\frac{\Delta P \cdot x^2}{2\mu \cdot l} + C \qquad \text{... (2.15)}$$

where, C is constant of integration and evaluated for the boundary condition.

$$\left(\text{At } x = \pm\frac{h}{2} \Rightarrow v = 0\right)$$

Substituting boundary condition in equation (2.15)

$$0 = -\frac{\Delta P}{2\mu \cdot l}\left(\pm\frac{h}{2}\right)^2 + C$$

$$\therefore \quad C = \frac{\Delta P\, h^2}{8\mu l}$$

Using value of C in equation (2.1) for v, it becomes,

$$v = -\frac{\Delta P\, x^2}{2\,\mu \cdot l} + \frac{\Delta P\, h^2}{8\,\mu \cdot l}$$

$$v = \frac{\Delta P}{2\mu \cdot l}\left[\frac{h^2}{4} - x^2\right] \qquad \text{... (2.16)}$$

This equation shows that the velocity distribution across the rectangular slot is parabolic in nature.

The maximum velocity when x = 0, will be at the centre (shown in Fig. 2.7), and is given by,

$$v_{max} = \frac{\Delta P}{2\mu \cdot l}\left[\frac{h^2}{4} - 0\right]$$

$$v_{max} = \frac{\Delta P \cdot h^2}{8\mu \cdot l} \qquad \text{... (2.17)}$$

Average Velocity of the Lubricating Fluid :

It is equal to two third of the maximum velocity.

$$\text{i.e.} \quad v_{av} = \frac{2}{3} v_{max}$$

$$v_{av} = \frac{2}{3}\left[\frac{\Delta P\, h^2}{8\mu \cdot l}\right]$$

$$v_{av} = \frac{\Delta P \cdot h^2}{12\mu \cdot l} \qquad \text{... (2.18)}$$

Volume Flow Rate of Fluid through Rectangular Slot (Q) :

It is the average velocity multiplied by area of cross-section.

i.e. Q = Average velocity × Area of cross-section of the slot

$$Q = v_{av} \times (b \cdot h)$$

$$Q = \frac{\Delta P\, h^2}{12\,\mu \cdot l}(b \cdot h)$$

$$\therefore \quad Q = \frac{\Delta P \cdot h^3 \cdot b}{12 \cdot \mu \cdot l} \qquad \text{... (2.19)}$$

2.3.3 Applications

It is used for design or evaluation of the load-carrying system of the Hale's telescope.

2.4 HYDROSTATIC STEP BEARING ANALYSIS

[Nov. 16]

It is also called circular step bearing.

Fig. 2.8 illustrates the principle of a hydrostatic step bearing action.

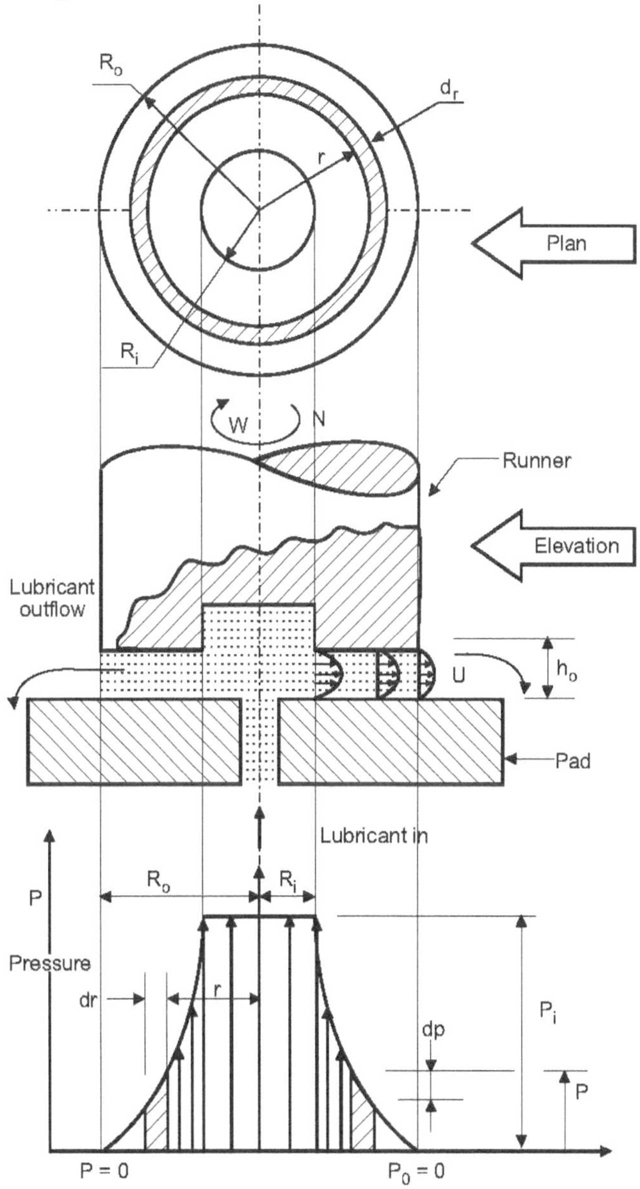

Fig. 2.8 : Pressure distribution in hydrostatic step bearing

Let,
W = Vertical thrust load on runner, (N)
R_0 = Outer radius of shaft, (mm)
R_i = Radius of recess or pocket, (mm)
P_0 = Outlet lubricant pressure, (N/mm^2)
P_i = Inlet lubricant pressure, (N/mm^2) or supply pressure

h_0 = Fluid-film thickness (annulus depth), (mm)

μ = Absolute viscosity of lubricant, (N-s/mm²)

Q = Flow rate of lubricant, (mm³/sec.)

N = Shaft speed, (rpm)

2.4.1 Principle of Working

Lubricant from a reservoir is pumped by a constant displacement pump and is forced into a central circular recess and then flows outwards between the bearing surfaces, developing pressure and separation and returning to a reservoir for recirculation. The thrust load is supported by the fluid within the pocket and land. The outside pressure is usually equal to zero or ambient.

2.4.2 Assumptions

Following assumptions are made in order to derive the load-carrying capacity, flow requirement and frictional power lost.

- The recess depth is quite enough for the pressure in it to be fairly uniform or almost constant.
- For pressure development, the bearing is assumed to have low rotational velocity and its effect is neglected for pressure development.
- The flow can be considered as laminar across the land of the pad.

2.4.3 Analysis

1. Pressure Distribution

For deriving an equation for the total load-carrying capacity of hydrostatic step bearing, it is first necessary to find the pressure distribution on an annular area between the radii R_i and R_0.

We know equation (2.5) for flow rate of lubricant through a slot and is given by,

$$Q = \frac{\Delta P \cdot h^3 \cdot b}{12 \cdot \mu \cdot l}$$

where, ΔP = Pressure difference causing fluid flow, (N/mm²)

b = Width of the slot across the fluid flow, (mm)

l = Length of the slot in the direction of fluid flow, (mm)

h = Thickness of the slot or fluid film, (mm)

μ = Absolute viscosity of the lubricant, (N-s/mm²)

Consider an annular ring at radius r, where ($R_i < r < R_0$) and of width 'dr', along the flow of lubricant. Apply above equation for the flow rate of lubricant across the annular ring.

$$\Delta P = dp; \; b = 2\pi r; \; l = dr; \; h = h_0.$$

Using these quantities in above equation,

$$Q = -\frac{dp \cdot (2\pi r) \cdot h_0^3}{12 \cdot \mu \cdot dr}$$

The negative sign is due to the fact that as radius increases, pressure decreases or $\left(\frac{dp}{dr}\right)$ is negative.

$$dp = -\frac{6\mu \cdot Q}{\pi \cdot h_0^3} \cdot \frac{dr}{r}$$

Integrating above equation, we get,

$$\int dp = \int -\frac{6\mu \cdot Q}{\pi \cdot h_0^3} \cdot \frac{dr}{r}$$

$$P = -\frac{6\mu \cdot Q}{\pi \cdot h_0^3} \cdot \log_e r + C \quad \text{... (2.20)}$$

where, C is constant of integration and can be calculated from,

First Boundary Condition : At $r = R_0$, $P = 0$.

Using this condition in equation (2.19), for P,

$$0 = -\frac{6\mu Q}{\pi h_0^3} \log_e R_0 + C$$

$$\therefore \quad C = \frac{6\mu Q}{\pi h_0^3} \cdot \log_e R_0$$

Substituting value of C in equation (2.19), for pressure,

$$\therefore \quad P = \frac{6\mu Q}{\pi h_0^3} \log_e r + \frac{6\mu Q}{\pi h_0^3} \cdot \log_e R_0$$

$$\therefore \quad P = \frac{6\mu Q}{\pi h_0^3} \log_e \frac{R_0}{r} \quad \text{... (2.21)}$$

2. Flow Rate of Lubricant

Using second boundary condition,

At $r = R_i$, $P = P_i$

Using this condition in above equation,

$$\therefore \quad P_i = \frac{6\mu \cdot Q}{\pi h_0^3} \log_e \frac{R_0}{R_i}$$

$$\therefore \quad Q = \frac{\pi \cdot P_i \cdot h_0^3}{6\mu \cdot \log_e \left(\frac{R_0}{R_i}\right)} \quad \text{... (2.22)}$$

This is an equation for flow rate of lubricant.

Pressure Distribution in Bearing Clearance of Hydrostatic Step Bearing :

Dividing equation for P by equation for P_i,

$$\frac{P}{P_i} = \frac{\log_e \left(\frac{R_0}{r}\right)}{\log_e \left(\frac{R_0}{R_i}\right)}$$

$$\therefore \qquad P = \left[\frac{\log_e\left(\frac{R_0}{r}\right)}{\log_e\left(\frac{R_0}{R_i}\right)}\right] P_i \qquad \ldots (2.23)$$

which gives the pressure distribution on an annular ring from radius R_i to R_0.

3. Load-Carrying Capacity

The total load-carrying capacity of hydrostatic step bearing is equal to summation of two components i.e.

- Load supported by the central recess area (where pressure P_i is constant).
- Load supported by the annular area from R_i to R_0 (where pressure P_i varies).

$\therefore$ Total load,

$$W = W_R + W_A$$

$$\text{i.e.} \quad W = \begin{bmatrix}\text{Load supported by} \\ \text{central recess}\end{bmatrix} + \begin{bmatrix}\text{Load supported by} \\ \text{annular area}\end{bmatrix}$$

$$= (P_i \cdot \pi \cdot R_i^2) + \left(\int_{R_i}^{R_0} P \cdot 2\pi r \cdot dr\right)$$

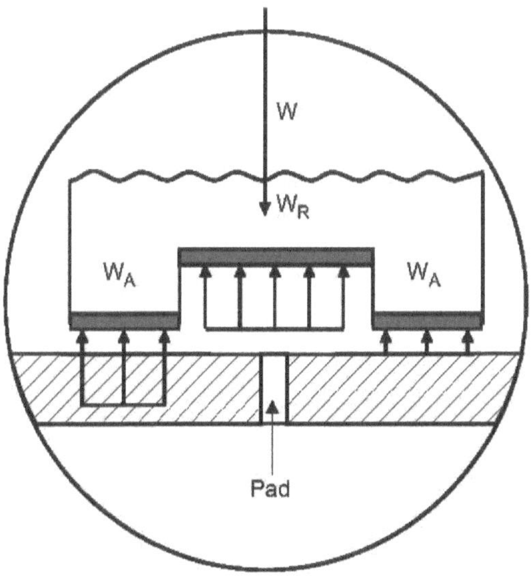

Fig. 2.9

Using equation for pressure distribution in an annular area from radius R_i to R_0,

$$W = P_i \cdot \pi \cdot R_i^2 + \int_{R_i}^{R_0} \frac{P_i \cdot \log_e\left(\frac{R_0}{r}\right)}{\log_e\left(\frac{R_0}{R_i}\right)} \cdot 2\pi r \cdot dr$$

$$W = \pi \cdot P_i \cdot R_i^2 + \frac{2\pi \cdot P_i}{\log_e\left(\frac{R_0}{R_i}\right)} \int_{R_i}^{R_0} \log_e\left(\frac{R_0}{r}\right) \cdot r \cdot dr$$

$$W = \pi \cdot P_i \left[R_i^2 + \frac{2}{\log_e\left(\frac{R_0}{R_i}\right)} \int_{R_i}^{R_0} \log_e\left(\frac{R_0}{r}\right) \cdot r \cdot dr\right]$$

Now, integrating by parts,

$$\left[\because \int AB \cdot dx = A \int B\, dx + \int \int B \cdot \frac{dA}{dx} \cdot dx\right]$$

$$\text{Let,} \qquad I = \int_{R_i}^{R_0} \log_e\left(\frac{R_0}{R_i}\right) \cdot r \cdot dr$$

$$= \left\{\log_e\left(\frac{R_0}{r}\right) \int r\, dr - \int \left[\left(\int r\, dr\right) \cdot \frac{d}{dr} \log_e\left(\frac{R_0}{r}\right)\right] \cdot dr\right\}_{R_i}^{R_0}$$

$$= \left\{\log_e\left(\frac{R_0}{r}\right) \cdot \frac{r^2}{2} - \int \frac{r^2}{2} \cdot \frac{r}{R_0} \cdot \left(-\frac{R_0}{r^2}\right) \cdot dr\right\}_{R_i}^{R_0}$$

$$= \left\{\log_e\left(\frac{R_0}{r}\right) \cdot \frac{r^2}{2} + \frac{r^2}{4}\right\}_{R_i}^{R_0}$$

$$= \left\{\left[\log_e\left(\frac{R_0}{R_0}\right) \cdot \frac{R_0^2}{2} + \frac{R_0^2}{4}\right] - \left[\log_e\left(\frac{R_0}{R_i}\right) \cdot \frac{R_i^2}{2} + \frac{R_i^2}{4}\right]\right\}$$

$$\therefore I = \frac{R_0^2 - R_i^2}{4} - \frac{R_i^2}{2} \cdot \log_e\left(\frac{R_0}{R_i}\right) \qquad \ldots \left(\because \log_e(1) = 0\right)$$

Using I in above integral,

$$W = \pi \cdot P_i \left\{R_i^2 + \frac{2}{\log_e\left(\frac{R_0}{R_i}\right)} \left[\frac{R_0^2 - R_i^2}{4} - \frac{R_i^2}{2} \log_e\left(\frac{R_0}{R_i}\right)\right]\right\}$$

$$= \pi \cdot P_i \left\{R_i^2 + \frac{R_0^2 - R_i^2}{2 \cdot \log_e\left(\frac{R_0}{R_i}\right)} - R_i^2\right\}$$

$$W = \frac{\pi \cdot P_i\,(R_0^2 - R_i^2)}{2 \log_e\left(\frac{R_0}{R_i}\right)} \qquad \ldots (2.24)$$

This equation is used to evaluate load-carrying capacity of hydrostatic step bearing and it is used even if there is no recess, in which case R_i will be the radius of the oil supply-pipe.

Also, we can write equation for P_i in terms of W as below.

$$P_i = \frac{2W \log_e\left(\frac{R_0}{R_i}\right)}{\pi\left(R_0^2 - R_i^2\right)} \qquad \ldots (2.25)$$

4. Dimensionless Form of Flow Rate and Load-Carrying Capacity

The flow rate and load-carrying capacity for a hydrostatic step bearing can often be conveniently expressed in non-dimensional terms as discussed in the following.

(i) Dimensionless form of Flow Rate (Q*) :

We have an equation (2.21), for flow rate of lubricant for circular hydrostatic step bearing.

$$Q = \frac{\pi \cdot P_i \cdot h_0^3}{6\mu \cdot \log_e\left(\frac{R_0}{R_i}\right)}$$

$$\therefore \quad \frac{Q}{\pi \cdot P_i \cdot h_0^3} = \frac{1}{6\mu \cdot \log_e\left(\frac{R_0}{R_i}\right)}$$

$$\therefore \quad Q^* = \frac{1}{6\mu \cdot \log_e\left(\frac{R_0}{R_i}\right)} \quad \text{... (2.26 a)}$$

where, $Q^* = \frac{Q}{\pi \cdot P_i \cdot h_0^3}$ is called normalized or non-dimensional flow rate for circular pad bearing.

Dimensionless form of Load-Carrying Capacity (W*) :

We know the equation for load-carrying capacity of circular pad hydrostatic step bearing,

$$W = \frac{P_i \cdot \pi \cdot (R_0^2 - R_i^2)}{2 \cdot \log_e\left(\frac{R_0}{R_i}\right)}$$

$$\therefore \quad W = \frac{P_i \cdot \pi \cdot R_0^2\left[1 - \left(\frac{R_i}{R_0}\right)^2\right]}{2 \cdot \log_e\left(\frac{R_0}{R_i}\right)}$$

$$\therefore \quad \frac{W}{P_i \cdot \pi \cdot R_0^2} = \frac{1 - \left(\frac{R_i}{R_0}\right)^2}{2 \cdot \log_e\left(\frac{R_0}{R_i}\right)}$$

$$\therefore \quad W^* = \frac{1 - \left(\frac{R_i}{R_0}\right)^2}{2 \cdot \log_e\left(\frac{R_0}{R_i}\right)} \quad \text{... (2.26 b)}$$

where, $W^* = \frac{W}{P_i \cdot \pi \cdot R_0^2}$ is called normalized or non-dimensional load for circular-pad bearing. W* and Q* both depend on shape of the bearing. The variation in these quantities with the ratio of the outer land to inner land radii i.e. $\left(\frac{R_0}{R_i}\right)$ are plotted in Fig. 2.10.

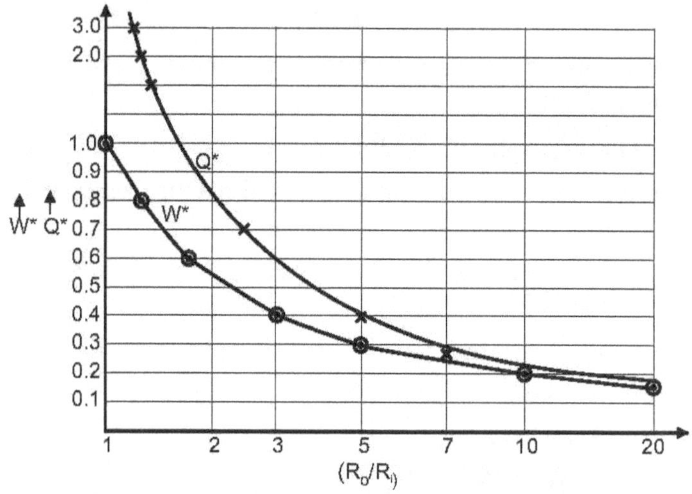

Fig. 2.10 : Variation of non-dimensional load and non-dimensional flow rate with $\left(\frac{R_0}{R_i}\right)$ ratio

2.5 ENERGY LOSSES IN HYDROSTATIC STEP BEARING

During the operation of hydrostatic bearing, the lubricant is continuously pumped from reservoir and supplied in the recess. Thus, there is loss of energy associated with pumping of the lubricant, called pumping power loss and also the energy loss due to viscous friction, called viscous power loss.

Therefore, total power loss is given by,

Total power loss

$$= \begin{bmatrix}\text{Energy loss due} \\ \text{to viscous friction}\end{bmatrix} + \begin{bmatrix}\text{Pumping power} \\ \text{loss}\end{bmatrix}$$

$$\therefore \quad E_T = E_F + E_P$$

Let E_F be Part–I and E_P be Part–II.

2.5.1 Viscous Power Loss (Frictional Power Loss (E_F))

Energy loss is associated in order to overcome the viscous frictional resistance on the step bearing during rotation of shaft. The frictional or viscous power loss can be evaluated as below.

Applying Newton's law of viscosity,

Shear force on area is given by,

$$F = \mu \cdot A \cdot \frac{U}{h}$$

Here, we consider elemental area as annular ring of radius r (where, $R_i < r < R_0$) and width dr across which lubricant is flowing.

The area of elemental ring, $A = 2\pi r \cdot dr$

Velocity at annular ring, $U = v = \dfrac{2\pi r \cdot N}{60}$

and Fluid-film thickness, $h = h_0$.

Using these quantities in above equation for shear force on annular ring (dF), it becomes,

$$dF = \mu \cdot (2\pi r \cdot dr) \frac{v}{h_0}$$

Substituting velocity v in above equation,

$$F = \mu \cdot (2\pi r \cdot dr) \frac{2\pi N \cdot r}{60} \cdot \frac{1}{h_0}$$

$$= \frac{4\pi^2 \cdot \mu \cdot N}{60 \cdot h_0} \cdot r^2 \cdot dr$$

∴ Frictional torque on annular ring is given by,

$$dT = r \cdot dF$$

$$= \frac{4\pi^2 \cdot \mu \cdot N}{60 \cdot h_0} \cdot r^3 \cdot dr$$

∴ Total frictional torque on the step is given by,

$$T = \int_{R_i}^{R_0} dT = \int_{R_i}^{R_0} \frac{4\pi^2 \cdot \mu \cdot N}{60 \cdot h_0} \cdot r^3 \cdot dr$$

$$= \frac{4\pi^2 \cdot \mu \cdot N}{60 \cdot h_0} \left[\frac{1}{4} \cdot r^4\right]_{R_i}^{R_0}$$

$$T = \frac{\pi^2 \cdot \mu \cdot N}{60\, h_0} [R_0^4 - R_i^4] \quad \text{... (2.27)}$$

∴ Frictional or viscous power loss is given by,

$$E_F = \frac{2\pi N \cdot T}{60 \times 10^6}, \text{(kW)}$$

Substituting value of frictional torque in above equation, it becomes

$$= \frac{2\pi \cdot N}{60 \times 10^6} \left[\frac{\pi^2 \cdot \mu \cdot N}{60\, h_0} (R_0^4 - R_i^4)\right] = \frac{2\pi^3 \cdot \mu \cdot N^2 (R_0^4 - R_i^4)}{3600 \times 10^6\, h_0}$$

$$\therefore \quad E_F = \frac{\mu \cdot N^2 (R_0^4 - R_i^4)}{(58.053 \times 10^6)\, h_0}, \text{(kW)} \quad \text{... (2.28)}$$

This is an equation for viscous or frictional power loss and it shows that frictional power loss is inversely proportional to the film thickness (h_0), i.e.

$$E_F \propto \frac{1}{h_0} \quad \text{... (2.29)}$$

where, E_F = Friction or viscous power loss, (kW)

μ = Absolute viscosity of lubricant, ($N\text{-}s/mm^2$)

R_0 = Outer radius of the shaft, (mm)

R_i = Radius of recess or pocket in shaft, (mm)

N = Speed of shaft (rpm)

2.5.2 Pumping Power Loss (E_P)

Power is consumed by the external pump therefore, the energy loss associated with pumping of lubricant is called pumping power loss and is given by,

$$E_P = Q \cdot \Delta P \quad \text{... (2.30)}$$

where, $\Delta P = (P_i - P_0)$

$$= \frac{(Q \times 10^{-9}) \cdot (\Delta P \times 10^6)}{\mu_p \cdot 10^3}, \text{(kW)}$$

$$E_P = \frac{Q(P_i - P_0)}{\mu_p \cdot 10^6}, \text{(kW)} \quad \text{... (2.31)}$$

where, E_P = Pumping power loss, (kW)

P_i = Inlet lubricant pressure, (N/mm^2)

P_0 = Outlet lubricant pressure, (N/mm^2)

μ_p = Pump efficiency

Q = Lubricant flow rate, (mm^3/sec.)

Now, for $P_0 = 0$, $\mu_p = 1$, above equation becomes,

$$E_P = \frac{P_i \cdot Q}{10^6} \quad \text{... (2.32)}$$

Using value of Q in above equation, it becomes,

$$E_P = \frac{\pi \cdot P_i^2 \cdot h_0^3}{6\mu \cdot \log_e\left(\frac{R_0}{R_i}\right)} \cdot \frac{1}{10^6}, \text{(kW)}$$

$$\therefore \quad E_P = \frac{1}{1.91 \times 10^6} \left[\frac{P_i^2 \cdot h_0^3}{\mu \cdot \log_e\left(\frac{R_0}{R_i}\right)}\right], \text{(kW)} \quad \text{... (2.33)}$$

∴ This equation shows that pumping power loss is directly proportional to the cube of film thickness.

$$\text{i.e.} \quad E_P \propto h_0^3 \quad \text{... (2.34)}$$

Adding Part-I and Part-II for total power loss,

∴ Total power loss in hydrostatic step bearing,

$$E_T = \left\{\frac{1}{58.053 \times 10^6}\left[\frac{\mu \cdot N^2(R_0^4 - R_i^4)}{h_0}\right]\right\} + \left\{\frac{1}{1.91 \times 10^6}\left[\frac{P_i^2 \cdot h_0^3}{\mu \cdot \log_e\left(\frac{R_0}{R_i}\right)}\right]\right\} \quad \text{... (2.35)}$$

∴ For most of the bearings, pumping losses are predominent compared to viscous or frictional losses.

∴ Neglecting viscous losses,

Total power loss,

$$E_T = \frac{1}{1.91 \times 10^6} \frac{(P_i^2 \cdot h_0^3)}{\mu \cdot \log_e \left(\frac{R_0}{R_i}\right)}$$

... Neglecting effect of viscous power ... (2.36)

2.6 REYNOLD'S EQUATION FOR HYDRODYNAMIC LUBRICATION

Need : Derivation of the classical Reynolds equation reflects insight into fluid behaviour in bearing lubricant films which then leads to the science of hydrodynamic lubrication. Solution of Reynolds equation can be used to determine the pressure distribution in a bearing with an arbitrary film shape. Once the pressure profile is evaluated, all other bearing performance parameters such as load-carrying capacity (W), friction force (F), flow rates (Q), etc. can easily be determined.

2.6.1 Assumptions in Reynold's Equation

[Aug. 16]

For analysis of hydrodynamic bearing, following basic assumptions are necessary :

- The fluid is considered to be Newtonian (i.e. the shear stress is directly proportional to the rate of shear strain).
- The fluid flow is considered to be viscous and laminar (i.e. no vortices and no turbulence in the fluid flow).
- The fluid is incompressible (i.e. the volume of fluid flowing past any section in unit time will be constant).
- Inertia force $\left(\text{i.e. } F_I = \rho \frac{Du}{Dt}\right)$ due to acceleration of fluid and body forces (i.e. F_B) are small compared with the pressure and viscous force terms i.e. $F_I = 0$ and $F_B = 0$.
- Pressure variation across the film is negligibly small $\left(\text{i.e. } \frac{\partial p}{\partial z} = 0\right)$ but it varies along the length of film (i.e. P = p(x, y)).
- Velocity at any point in the fluid film is a function of both x and y. (Partial derivatives $\frac{\partial u}{\partial z}$ and $\frac{\partial v}{\partial z}$ are large compared with other velocity derivative components). It is due to the fact that u and v are usually greater than w, and z is much smaller dimension.
- Curvature effects are neglected and the thickness of the fluid film is much smaller as compared with length or width of the film.
- Usually, fluid-film thickness = h = 0.0254 mm

And $\frac{h}{l} = \frac{\text{Fluid-film thickness}}{\text{Fluid-film length}} = 10^{-3}$

- There is no slip at the bearing and the fluid-film interface (i.e. the velocity of the lubricant film is identical with the surface velocity at the bearing surface).
- The viscosity of the fluid remains constant as it flows through the bearing.

2.6.2 Reynold's Equation for Two-Dimensional Flow [Nov. 15, May 16, Aug. 16]

Reynold's equation for hydrodynamic lubrication can be derived in following two ways :

- **By Direct Method :** In which we consider equilibrium of forces acting on faces of fluid element, which is dragged into a rapidly narrowing clearance space, converging in the direction of motion.
- **From Navier-Stroke's Equations :** For a Newtonian fluid, the equations of motions for a viscous fluid, called Navier-Stroke's equations are taken into account for deriving Reynold's equation.

1. Reynold's Equation for Two-Dimensional Flow by Direct Method

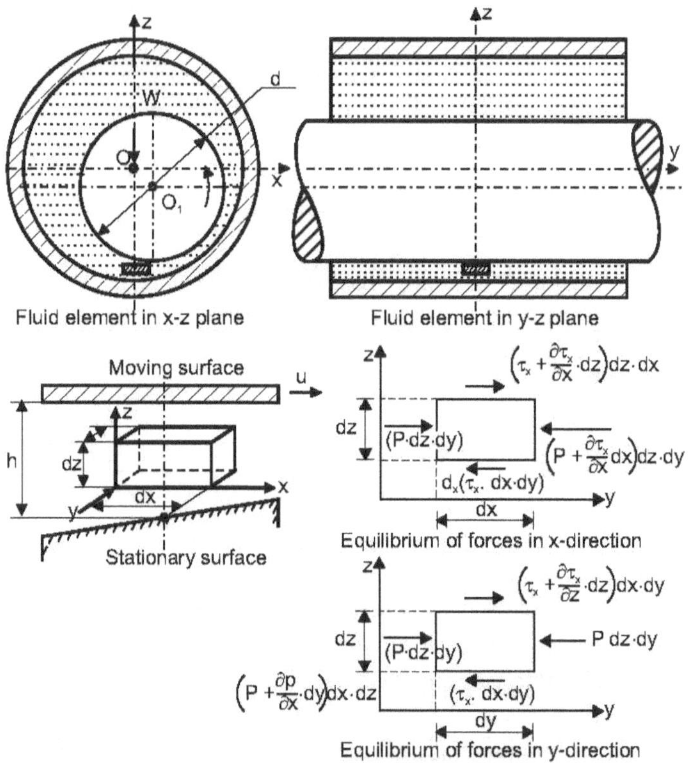

Fig. 2.11

Consider a journal rotating in a bearing with a constant surface velocity u_b. For analysis a fluid element having dimensions dx, dy, dz and x, y, and z directions respectively is taken into account; as shown in Fig. 2.11. Here 'x' axis represents direction of motion of fluid. 'y' axis is parallel to journal axis and 'z' axis is radial direction of journal.

Let u_b, v_b and w_b be the components of velocity of moving surface in X, Y, Z directions respectively.

Velocity Distribution :

Various forces acting on the faces of fluid element in horizontal (i.e. X-direction) are :

- Fluid pressure force ('P' dz dy) in positive x-direction.
- Fluid pressure force $\left(P + \frac{\partial P}{\partial x} dx\right) dy \cdot dz$ in negative x-direction.
- Shear force on bottom surface ($\tau_x \cdot dx \cdot dy$).
- Shear force on top surface $\left(\tau_x + \frac{\partial \tau_x}{\partial z} \cdot dz\right) dx \cdot dy$.

For equilibrium of forces on fluid element (x-z plane) in X-direction, we have,

$$P\, dz \cdot dy - \left(P + \frac{\partial P}{\partial x} \cdot dx\right) dy \cdot dz + \left(\tau_x + \frac{\partial \tau_x}{\partial z} dz\right) dx \cdot dy - (\tau_x \cdot dx\, dy) = 0 \quad \text{... (2.37)}$$

$$P\, dz \cdot dy - P \cdot dy \cdot dz - \frac{\partial P}{\partial x} \cdot dx \cdot dy \cdot dz + \tau_x\, dx \cdot dy + \frac{\partial \tau_x}{\partial z} \cdot dz \cdot dx \cdot dy - \tau_x \cdot dx \cdot dy = 0$$

$$\therefore \quad \frac{\partial P}{\partial x} = \frac{\partial \tau_x}{\partial z}$$

Here,

$$dx \cdot dy \cdot dz \neq 0$$

... (which indicates volume of fluid element)

According to Newton's law of viscosity,

$$\tau_x = \mu \cdot \frac{\partial u}{\partial z}$$

Substituting above equation in equation for $\frac{\partial P}{\partial x}$, we get,

$$\therefore \quad \frac{\partial P}{\partial x} = \frac{\partial}{\partial z}\left(\mu \cdot \frac{\partial u}{\partial z}\right)$$

$$\therefore \quad \frac{\partial P}{\partial x} = \mu \frac{\partial^2 u}{\partial z^2} \quad \text{... (2.38)}$$

Above equation gives pressure gradient in X-direction.

$\therefore$ Pressure gradient in Y-direction and Z-direction are,

$$\frac{\partial P}{\partial y} = \mu \frac{\partial^2 v}{\partial z^2}$$

$$\frac{\partial P}{\partial z} = 0 \quad \text{... (As per assumption number v)}$$

Now, we have an equation

$$\frac{\partial P}{\partial x} = \mu \frac{\partial^2 u}{\partial z^2}$$

$$\frac{\partial^2 u}{\partial z^2} = \frac{1}{\mu} \frac{\partial P}{\partial x}$$

Integrating above equation twice with respect to z,

$$\frac{\partial u}{\partial z} = \frac{1}{\mu} \frac{\partial P}{\partial x} \cdot z + A$$

$$u = \frac{1}{2\mu} \frac{\partial P}{\partial x} \cdot z^2 + A \cdot z + B$$

where, A and B are constants of integration and can be evaluated by using the following boundary conditions.

Boundary Conditions :

At z = 0; $u = u_a = 0$

At z = h; $u = u_b$

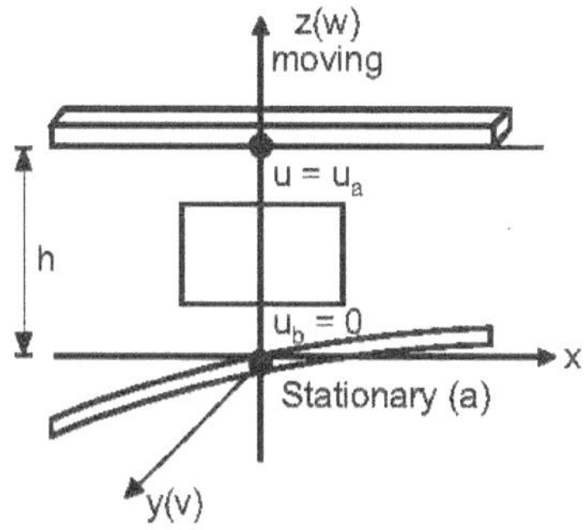

Fig. 2.12

Substituting in above equation for 'u',

$$0 = 0 + 0 + B \Rightarrow B = u_a = 0$$

$$u_b = \frac{1}{2\mu} \frac{\partial P}{\partial x} \cdot h^2 + A \cdot h + B$$

$$\therefore \quad u_b = \frac{1}{2\mu} \cdot \frac{\partial P}{\partial x} \cdot h^2 + A \cdot h + 0$$

$$A = \frac{u_b}{h} - \frac{h}{2\mu} \cdot \frac{\partial P}{\partial x}$$

and $B = 0$

Substituting values of constants A and B in equation for 'u'

$$\therefore \quad u = \frac{1}{2\mu} \frac{\partial P}{\partial x} \cdot z^2 + \left[\frac{u_b}{h} - \frac{h}{2\mu} \frac{\partial P}{\partial x}\right] z + 0$$

$$\therefore \quad u = \frac{1}{2\mu} \frac{\partial P}{\partial x} (z^2 - h \cdot z) + \frac{u_b \cdot z}{h} \quad \text{... (2.39)}$$

'Poiseuille flow term' 'Couette flow term' due to pressure (due to motion of gradient $\frac{\partial P}{\partial x}$ surface b) also called as shear term.

Above equation gives velocity distribution of lubricant as a function of 'z' and pressure gradient $\frac{\partial P}{\partial x}$ (i.e. velocity variation along x-direction for the limits z = 0 to z = h).

Rate of Flow of Fluid :

Rate of flow of fluid (q_x) in x-direction per unit width in y-direction,

$$q_x = \int_0^h u \times 1 \times dz$$

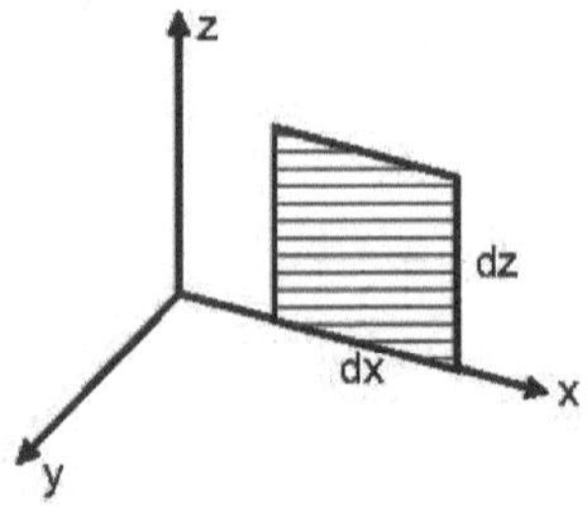

Fig. 2.13

Substituting value of u from equation in equation above,

$$q_x = \int_0^h \left[\frac{1}{2\mu}\frac{\partial P}{\partial x}(z^2 - h \cdot z) + \frac{u_b}{h} \cdot z\right] dz$$

$$= \left[\frac{1}{2\mu} \cdot \frac{\partial P}{\partial x}\left(\frac{z^3}{3} - h\frac{z^2}{2}\right) + \frac{u_b}{h} \cdot \frac{z^2}{2}\right]_0^h$$

$$= \frac{1}{2\mu}\frac{\partial P}{\partial x}\left(\frac{h^3}{3} - \frac{h^3}{2}\right) + \frac{u_b}{h} \cdot \frac{h^2}{2}$$

$$q_x = -\frac{h^3}{12\mu} \cdot \frac{\partial P}{\partial x} + \frac{u_b \cdot h}{2} \quad \text{... (2.40)}$$

Similarly, the rate of flow of fluid in y-direction per unit width in x-direction is,

$$q_y = -\frac{h^3}{12\mu}\frac{\partial P}{\partial y} + \frac{v_b \cdot h}{2} \quad \text{... (2.41)}$$

Consider continuity equation for incompressible flow,

$$\frac{\partial u}{\partial x} + \frac{\partial v}{\partial y} + \frac{\partial w}{\partial z} = 0$$

Now, $\frac{\partial w}{\partial z} = 0$

... (As per assumption (vi) i.e. no flow in z-direction)

$$\therefore \quad \frac{\partial u}{\partial x} + \frac{\partial v}{\partial y} = 0 \quad \text{... (2.42)}$$

Integrating above equation w.r.t. z within the limits z = 0 to z = h,

$$\int_0^h \frac{\partial u}{\partial x} \cdot dz + \int_0^h \frac{\partial v}{\partial y} \cdot dz = 0$$

$$\therefore \quad \frac{\partial}{\partial x}\left[\int_0^h u \cdot dz\right] + \frac{\partial}{\partial y}\left[\int_0^h v\, dz\right] = 0$$

Bracketed quantities indicate the rate of flow of lubricant along x and y-directions respectively.

$$\text{i.e.} \quad q_x = \int_0^h u \cdot dz \quad \text{and} \quad q_y = \int_0^h v \cdot dz$$

Substituting values of q_x from equation (2.40) and q_y from equation (2.41), we get,

Above equation becomes

$$\frac{\partial}{\partial x}[q_x] + \frac{\partial}{\partial y}[q_y] = 0$$

$$\frac{\partial}{\partial x}\left[-\frac{h^3}{12\mu}\frac{\partial P}{\partial x} + \frac{u_b \cdot h}{2}\right] + \frac{\partial}{\partial y}\left[-\frac{h^3}{12\mu} \cdot \frac{\partial P}{\partial y} + \frac{v_b \cdot h}{2}\right] = 0$$

$$\therefore \quad \frac{\partial}{\partial x}\left[\frac{h^3}{12\mu}\frac{\partial P}{\partial x}\right] + \frac{\partial}{\partial y}\left[\frac{h^3}{12\mu} \cdot \frac{\partial P}{\partial y}\right] = \frac{\partial}{\partial x}\left(\frac{u_b \cdot h}{2}\right) + \frac{\partial}{\partial y}\left(\frac{v_b \cdot h}{2}\right)$$

$$\frac{\partial}{\partial x}\left[h^3 \cdot \frac{\partial P}{\partial x}\right] + \frac{\partial}{\partial y}\left[h^3\frac{\partial P}{\partial y}\right] = 6\mu\left[\frac{\partial}{\partial x}(u_b \cdot h) + \frac{\partial}{\partial y}(v_b \cdot h)\right] \quad \text{... (2.43)}$$

which is Reynold's equation for two-dimensional flow.

Simplifying the above equation,

i.e. for constant viscosity and $\frac{\partial h}{\partial y} = 0$

$$\frac{\partial}{\partial x}\left[h^3\frac{\partial P}{\partial x}\right] + \frac{\partial}{\partial y}\left[h^3 \cdot \frac{\partial P}{\partial y}\right] = 6\mu\left[u_b \cdot \frac{\partial h}{\partial x} + 0\right]$$

$$\text{i.e.} \quad \frac{\partial}{\partial x}\left[h^3\frac{\partial P}{\partial x}\right] + \frac{\partial}{\partial y}\left[h^3\frac{\partial P}{\partial y}\right] = 6\mu \cdot u_b \cdot \frac{\partial h}{\partial x} \quad \text{... (2.44)}$$

- Reynold's equation cannot be solved directly. However, solution can be obtained by using numerical methods.

  For infinitely long bearing, as per assumption (vi), there is no flow in y-direction.

  $$\text{i.e.} \quad \frac{\partial P}{\partial y} = 0$$

  Hence, above equation is further reduced to

  $$\frac{\partial}{\partial x}\left[h^3\frac{\partial P}{\partial x}\right] = 6\mu \cdot u_b \cdot \frac{\partial h}{\partial x} \quad \text{... (2.45)}$$

  which is Reynold's equation for one-dimensional flow.

  Exact solution is possible by considering bearing to be infinitely long, or very short.

2.7 RAIMONDI AND BOYD METHOD

- **Objectives :**

  For a hydrodynamic journal bearing of finite length, there is no exact solution, and there is being need to establish the relationship between different variables (i.e. design variables and performance variables) in the design of hydrodynamic journal bearing).

- A. A. Raimondi and John Boyd of Westinghouse Research Laboratories, solved the Reynolds equation on the digital computer by using iteration technique. While analysing it, they used different dimensionless parameters comprising of different design variables (e.g. Dynamic viscosity (μ), Unit bearing pressure (P), Journal speed (N_j), Bearing dimensions (r_j, C, l, β, etc.)) and performance variables (e.g. Coefficient of friction (f), Temperature rise (ΔT), the lubricant flow rate, (Q), Minimum oil-film thickness (h_0)).
- Refer Fig. 2.14 for various bearing performance parameters.

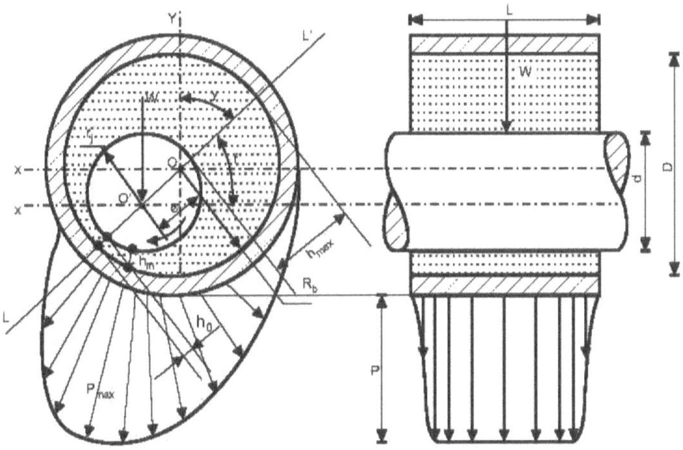

Fig. 2.14

The dimensionless parameters are given for full journal bearing with side leakage in the form of charts and tables by A. A. Raimondi and John Boyd in 1958.

Fig. 2.14 shows the numerical solutions for eccentricity ratio ($\in$) versus Sommerfeld number (S) for a wide range of L/D ratios.

- **Dimensionless Parameters :**

Length to diameter ratio $\left(\frac{l}{d}\right)$

Eccentricity ratio ($\in$)

$$\in = \frac{e}{C} \quad \text{... (By definition)}$$

Referring to Fig. 2.13,

$$R_b = e + r_j + h_0$$

$$R_b - r_j = e + h_0$$

$$e + h_0 = C$$

$$\frac{e}{C} + \frac{h_0}{C} = 1$$

$$\in + \frac{h_0}{C} = 1$$

$$\in = 1 - \frac{h_0}{C}$$

- Minimum film thickness variable $\left(\frac{h_0}{C}\right)$.
- Sommerfeld number (S).

It is an important non-dimensional parameter more generally used for the description of bearing performance. It bears a relationship to journal bearing theory. The equation for Sommerfeld number contains different variables which can be controlled to give better design.

Sommerfeld number can be given by the relation,

$$S = \frac{\mu \cdot N_j}{P}\left(\frac{r_j}{C}\right)^2$$

where, S – Sommerfeld number

μ – Viscosity of the lubricant (N-s/mm²)

N_j – Journal speed (rps)

P – Unit bearing pressure (N/mm²)

Designer can control all the variables in Sommerfeld number.

- Coefficient of friction variable (CFV),

$$CFV = f\left(\frac{r_j}{C}\right) \quad \text{... (2.46)}$$

- Flow variable (FV),

$$FV = \frac{Q}{r_j \cdot C \cdot N_j \cdot l} \quad \text{... (2.47)}$$

where, Q – Total oil flow rate.

- Flow ratio $\left(\frac{Q_s}{\bar{Q}}\right)$
- Maximum film pressure ratio $\left(\frac{P}{P_{max}}\right)$.
- Maximum film pressure angle ($\phi_{P_{max}}$).
- Zero pressure angle (ϕ_{P_0}).
- Minimum film thickness angle (θ).
- Temperature rise variable.

$$\left[\frac{\rho \cdot C_p \cdot \Delta T}{P}\right]$$

where, ρ – Density (kg/m³)

C_p – Specific heat of lubricant (J/kg °C)

P – Bearing pressure (N/m²)

ΔT – Temperature difference (°C)

Fig. 2.15 : Shows eccentricity ($\in$) and $\left(\frac{h_0}{C}\right)$ ratio versus (S) for wide range of $\left(\frac{l}{d}\right)$ ratios.

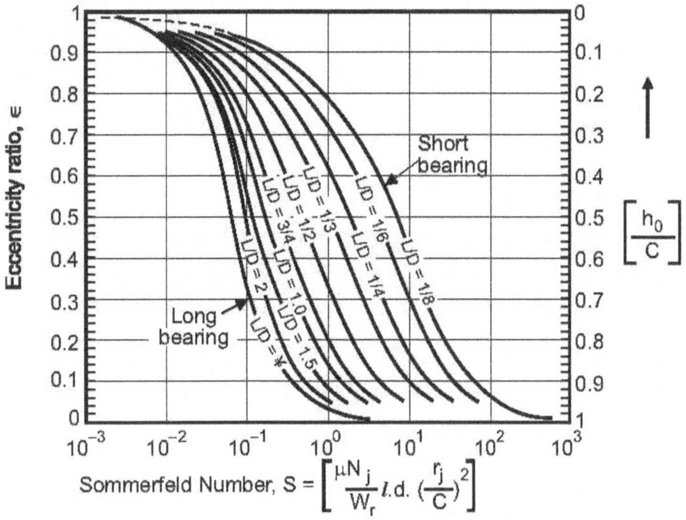

Fig. 2.15 : ∈ and $\left(\frac{h_0}{c}\right)$ versus 'S'

Table 2.1 shows the values of dimensionless parameters for full journal bearings with side leakage.

Table 2.1

$\left(\frac{l}{d}\right)$	∈	$\frac{h_0}{C}$	S	CFV $= f \cdot \left(\frac{r_j}{C}\right)$	FV $= \frac{Q}{r_j \cdot C \cdot N_j \cdot l}$	FR $= \left(\frac{Q_s}{Q}\right)$	$\frac{P}{P_{max}}$	ψ
∞	1.0	0	0	0	0	0	0	0
	0.97	0.03	–	–	–	0	–	–
	0.9	0.1	0.0115	0.756	0.411	0	0.358	31.62
	0.8	0.2	0.021	0.961	0.760	0	0.495	42.22
	0.6	0.4	0.0389	1.20	1.56	0	0.667	54.31
	0.4	0.6	0.0626	1.52	2.26	0	0.764	61.94
	0.2	0.8	0.123	2.57	2.83	0	0.814	67.26
	0.1	0.9	0.240	4.80	3.03	0	0.826	69.10
	0	1.0	∞	∞	3.147	0	–	70.92
∞	1.0	0	0	0	0	1.0	–	0
	0.97	0.03	0.00474	0.514	4.82	0.973	0.152	15.47
	0.9	0.1	0.0188	1.05	4.74	0.913	0.247	26.45
	0.8	0.2	0.0446	1.70	4.62	0.842	0.313	36.24
	0.6	0.4	0.121	3.22	4.33	0.680	0.415	50.58
	0.4	0.6	0.264	5.79	3.99	0.497	0.484	63.10
	0.2	0.8	0.631	12.8	3.59	0.280	0.529	74.02
	0.1	0.9	1.33	26.4	3.37	0.150	0.540	79.5
	0	1.0	∞	∞	3.142	0	–	85
$\left(\frac{1}{2}\right)$	1.0	0	0	0	–	1.0	0	0
	0.97	0.03	0.00609	0.610	5.88	0.980	0.126	13.75
	0.9	0.1	0.0313	1.60	5.69	0.939	0.206	23.66
	0.8	0.2	0.0923	3.26	5.41	0.874	0.267	33.31
	0.6	0.4	0.319	8.10	4.85	0.730	0.365	48.14
	0.4	0.6	0.779	17.0	4.29	0.552	0.441	61.45
	0.2	0.8	2.03	40.9	3.72	0.318	0.506	74.94
	0.1	0.9	4.31	85.6	3.43	0.173	0.523	81.62
	0	1.0	∞	∞	3.142	0	–	88.5
$\left(\frac{1}{4}\right)$	1.0	0	0	0	–	1.0	0	0
	0.97	0.03	0.0101	0.922	6.12	0.984	0.108	12.22
	0.9	0.1	0.0736	3.50	5.91	0.945	0.180	21.85
	0.8	0.2	0.261	8.8	5.60	0.884	0.240	31.04
	0.6	0.4	1.07	26.7	4.99	0.746	0.334	46.72
	0.4	0.6	2.83	61.1	4.37	0.567	0.415	60.86
	0.2	0.8	7.57	153.0	3.76	0.330	0.489	75.18
	0.1	0.9	16.2	322.0	3.45	0.180	0.515	82.31
	0	1.0	∞	∞	3.142		–	89.5

2.8 TEMPERATURE RISE IN HYDRODYNAMIC JOURNAL BEARING

- Generation of heat in hydrodynamic journal bearing is due to the fact that frictional work in bearing is converted to heat which increases the temperature of the lubricating oil. As temperature increases, the viscosity of lubricating oil decreases. The analysis is based on constant viscosity. The viscosity should be determined at average temperature (T_{AV}) of lubricating oil.
- Heat generated in bearing gets dissipated by three basic modes :

  (i) Conduction, (ii) Convection, (iii) Radiation.
- Instead of calculating the heat generated by three basic modes, it is assumed that heat generated is carried away by the flowing lubricating oil in the bearing.

  Temperature rise in the hydrodynamic journal bearings can be calculated as below.

Power Lost in Friction

$$P_F = T_F \cdot \omega_j$$

(a) Friction Power Lost $= \frac{\mu W_r \cdot r_j}{10^3} 2\pi N_j$ (W)

$$P_F = \frac{\mu W_r \cdot r_j \cdot 2\pi N_j}{10^6} \text{ (kW)} \qquad \ldots (2.48)$$

where, T_F – Frictional torque

ω_j – Angular velocity

P_F – Power lost in friction (kW)

W_r – Radial load on journal (N)

N_j – Journal speed (rps)

r_j – Journal radius (mm)

Rate of Heat Generation

$$H_G = P_F$$

(b) Heat Generation $= \dfrac{2\pi N_j \cdot f \cdot W_r}{10^6}$ (kJ/s)

Substituting $CFV = f\left(\dfrac{r_j}{C}\right)$

where, f – Coefficient of friction

$$\therefore \; f = \left(\frac{C}{r_j}\right) CFV$$

$$\text{and } P = \frac{W_r}{2l \cdot r_j}$$

$$\therefore \; W_r = 2l \cdot r_j \cdot P$$

$$\therefore \; H_G = \frac{2\pi N_j \cdot CFV}{10^6}\left(\frac{C}{r_j}\right) 2 \cdot P \cdot l \cdot r_j \quad \text{... (2.49)}$$

Rate of Heat Dissipation

$$H_D = m \cdot C_p \cdot \Delta T$$

Substituting,

Mass flow rate,

$$m = \frac{\rho Q}{10^9}$$

(c) Heat Dissipation

where, ρ – Mass density of ubricating oil (kg/m³)

Q – Total lubricating oil flow rate (mm³/sec.)

$$\text{Now, } FV = \text{Flow variable} = \frac{Q}{r_j \cdot C \cdot N_j \cdot l}$$

$$\therefore \; Q = FV(r_j \cdot C \cdot N_j \cdot l)$$

$$\therefore \; m = \frac{\rho \, FV(r_j \cdot C \cdot N_j \cdot l)}{10^9}$$

$$\therefore \; H_D = \frac{\rho \cdot FV(r_j \cdot C \cdot N_j \cdot l)\, \Delta T}{10^9} \quad \text{... (2.50)}$$

Expressing temperature rise ΔT in dimensionless form,

∴ Equating (b) part and (c) part,

$$H_G = H_D$$

(d) Temperature Rise

$$\therefore \frac{2\pi N_j \cdot CFV\left(\dfrac{C}{r_j}\right) \cdot 2 \cdot P \cdot l \cdot r_j \cdot r_j}{10^6} = \frac{\rho \cdot FV \cdot (C \cdot r_j \cdot N_j \cdot l)\, C_p \cdot \Delta T}{10^3}$$

$$\therefore \quad \Delta T = \frac{4\pi P}{\rho \cdot C_p} 10^3 \left[\frac{CFV}{FV}\right] \quad \text{... (2.51)}$$

Using $\rho = 860$ kg/m³ and $C_p = 1.76$ kJ/kg°C — For most of the lubricating oil

Temperature rise becomes,

$$\Delta T = 8.3023 \cdot P \left[\frac{CFV}{FV}\right] \quad \text{... (2.52)}$$

Average temperature of lubricant is given as,

(e) Average temperature of the lubricant

$$T_{AV} = T_i + \frac{\Delta T}{2} \quad \text{... (2.53)}$$

$$\Delta T = T_0 - T_i$$

where, T_0 – Outlet temperature of lubricating oil

T_i – Inlet temperature of lubricating oil

Total Flow of Lubricating Oil (Q) :

While estimating the temperature rise (ΔT), it is assumed that total heat generated is carried away by the total oil flow (Q) in the bearing which is given by,

Total oil flow = (Circumferential flow) + (Axial flow or side leakage)

$$Q = Q_{CF} + Q_S$$

where, Q_{CF} i.e. circumferential flow is due to the constant circulation of oil round the circumference of journal where, major part of heat is absorbed by lubricating oil and temperature rise takes place.

$$\Delta T = T_0 - T_i$$

where, T_i – Inlet temperature of lubricating oil (°C)

T_0 – Outlet temperature of lubricating oil (°C)

Q_S i.e. Axial flow or side leakage. It is through the bearing ends. Assuming that axial flow takes place at average temperature T_{AV}. If we take into account Q_S, the heat dissipation rate can be changed as follows.

$$H_D = \left[\begin{array}{c}\text{Heat carried due to}\\ \text{circumferential flow}\end{array}\right] + \left[\begin{array}{c}\text{Heat carried due to}\\ \text{side leakage}\end{array}\right]$$

$$= m_c \cdot C_p \cdot \Delta T + m_s \cdot C_p \frac{\Delta T}{2}$$

$$= \frac{\rho \cdot Q_{CF} \cdot C_p \cdot \Delta T}{10^9} + \frac{\rho \cdot Q_S \cdot C_p \cdot \dfrac{\Delta T}{2}}{10^9}$$

$$= \frac{\rho \cdot C_p \cdot \Delta T}{10^9}\left[Q_{CF} + \frac{Q_S}{2}\right]$$

$$= \frac{\rho \cdot C_p \cdot \Delta T}{10^9}\left[(Q - Q_S) + \frac{Q_S}{2}\right]$$

$$= \frac{\rho \cdot C_p \cdot \Delta T}{10^9}\left[Q - \frac{Q_S}{2}\right] = \frac{\rho \cdot C_p \cdot \Delta T \cdot Q}{10^9}\left[1 - \frac{Q_S}{2Q}\right]$$

$$\text{and} \quad H_G = \frac{2\pi N_j \cdot f \cdot W_r \cdot r_j}{10^6}$$

Equating H_D and H_G,

$$\frac{\rho \cdot C_p \cdot \Delta T \cdot Q}{10^9}\left[1 - \frac{Q_S}{2Q}\right] = \frac{2\pi N_j \cdot f \cdot W_r \cdot r_j}{10^6}$$

$$\Delta T = \frac{2\pi \cdot N_j \cdot f \cdot W_r \cdot r_j}{\rho \cdot Q \cdot C_p\left(1 - \dfrac{1}{2} \cdot \dfrac{Q_S}{Q}\right)} \cdot 10^3 \quad \text{... (2.54)}$$

2.9 HYDRODYNAMIC JOURNAL BEARING [Nov. 15]

Journal bearings are designed and most widely used to support a radial load which works on the principle of hydrodynamic action.

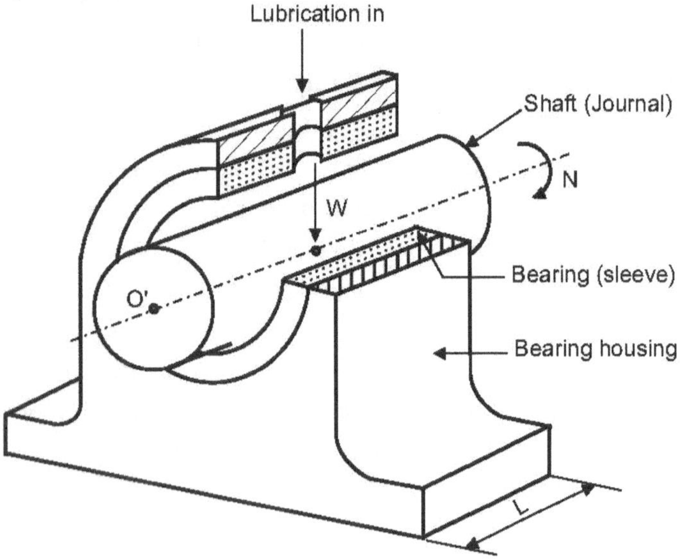

Fig. 2.16

Hydrodynamic Action :

Lubricant is supplied initially in the clearance space between bearing surface and journal, as the journal rotates, the lubricant is pulled into a converging wedge shaped region formed between the moving surfaces. Thus, dynamic pressure built-up, which supports the applied load on journal.

2.9.1 Nomenclature

Fig. 2.17 shows the cross-sectional details of hydrodynamic journal bearing and the terms related with its study. Also the variation of pressure on bearing both circumferentially and along the length of bearing can be illustrated.

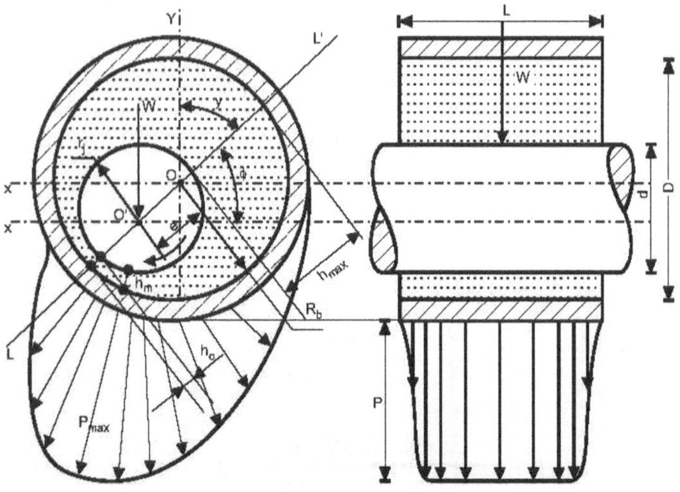

(Radial pressure distribution) **(Axial pressure distribution)**

Fig. 2.17

Notations :

O' – Centre of journal
O – Centre of bearing
W – Radial load on journal
R_b – Radius of bearing
r_j – Radius of journal
h_0 – Minimum oil-film thickness
h_{max} – Maximum oil-film thickness
C – Radial clearance
ω_j – Angular speed of journal
e – Eccentricity
$\in$ – Eccentricity ratio
ψ – Attitude angle
LL' – Line of centres
d – Diameter of journal
D – Diameter of bearing
ϕ – Circumferential co-ordinate

Terms Used :

1. Journal :

It is the portion of circular shaft, supported in a circular bush and take radial load called journal.

2. Bearing (Sleeve) :

Journal is made to rotate in a stationary bush or sleeve called bearing, with a complete 360° arc or various arrangements of a partial arc or arcs in a housing structure.

3. Radial Clearance (C) :

The difference between radii of bearing and that of journal is termed as radial clearance. It can be given as below,

$$C = R_b - r_j \qquad \text{... (2.55)}$$

The bearing and journal operate with a small radial clearance of the order of $\left(\frac{1}{1000}\right)$ of the journal radius.

4. Eccentricity (e) :

It is the distance between the centres of bearing and the journal measured along the line of centres.

$$e = O'O \text{ from figure}$$

5. Eccentricity Ratio ($\in$) :

It is the ratio of the eccentricity to the radial clearance.

$$\text{i.e.} \qquad \in = \frac{e}{C} \qquad \text{... (2.56)}$$

Value of eccentricity ratio varies from 0 to 1.

For concentric bearings,

$$e = 0$$

$$\therefore \qquad \in = 0$$

When journal touches bearing,

$$e = C$$

$$\therefore \quad \epsilon = 1$$

Eccentricity ratio (ϵ) is a function of,

$$\epsilon = f(W_z, \omega_j, \mu, r_j, l, C)$$

where, r_j – Journal radius

l – Length of bearing Geometric variables

C – Radial clearance

W_z – Radial load

ω_j – Angular velocity of journal Operating variables

μ – Dynamic viscosity of lubricant

Also, radial load (W_z) is a function of,

$$W_z = f(e, \omega_j, \mu, r_j, l, C)$$

6. Minimum Oil-Film Thickness (h_0) :

The oil-film thickness measured along the line joining the centres of journal and bearing. It can be given as,

$$H_0 = R_b - (r_j + e)$$

7. Minimum Oil-Film Thickness Ratio $\left(\frac{h_0}{C}\right)$:

It is the ratio of the minimum oil-film thickness to radial clearance.

$$\therefore \quad \text{Minimum oil-film thickness ratio} = \frac{h_0}{C}$$

Now,

$$R_b = e + r_j + h_0$$

$$R_b - r_j = e + h_0$$

$$C = e + h_0$$

$$1 = \frac{e}{C} + \frac{h_0}{C}$$

$$\therefore \quad \frac{h_0}{C} = 1 - \frac{e}{C}$$

$$\therefore \quad \frac{h_0}{C} = 1 - \epsilon \qquad \text{... (2.57)}$$

8. Attitude Angle (ψ) :

It is the angle made by the line of centres with load line (i.e. vertical) such that condition of perfect lubrication occur where film thickness is minimum.

2.9.2 Pressure Distribution

Fig. 2.17 shows the pressure distribution in radial and axial direction.

- **Radial Pressure Distribution :**

It is maximum at particular angle measured from load line.

P_{max} – the maximum pressure developed in the fluid film

- **Axial Pressure Distribution :**
  - Pressure distribution along the length of hydrodynamic journal bearing is almost constant.
  - Pressure suddenly falls at ends of bearing due to atmospheric pressure. It is shown in Fig. 2.17.

2.9.3 Principle of Hydrodynamic Lubrication

Principle of hydrodynamic lubrication in journal bearing can be illustrated by using the following three cases :

Case (i) : Journal at rest.

Case (ii) : Journal starts rotating.

Case (iii) : Journal at full speed.

Case (i) Journal at Rest :

- When the journal is at rest, the weight of the journal squeezes out of the oil film so that the journal rests on the bearing surface i.e. in initial stage, under the radial load, the journal touches the inner bearing surface at point P, and metal-to-metal contact is established at P.
- In this case, radial clearance,

$$C = R_b - r_j$$

And at point P, $h_0 = 0$

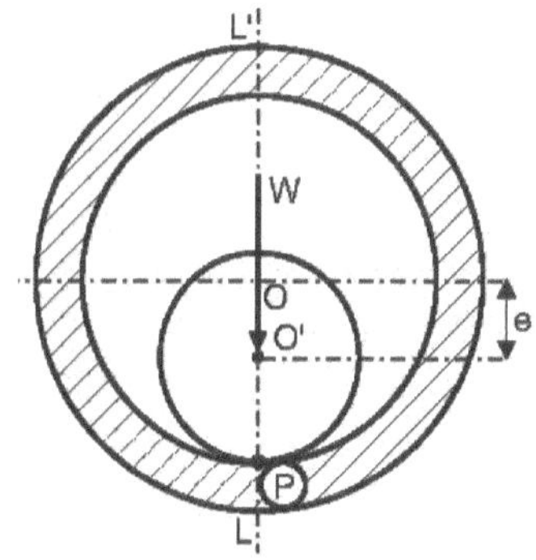

Fig. 2.18

Case (ii) Journal Starts Rotating :

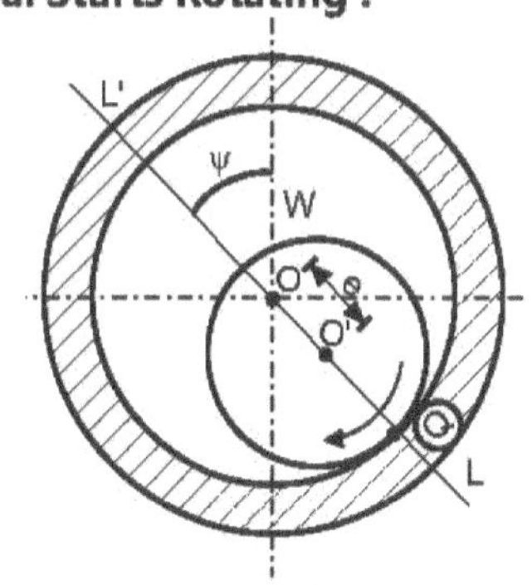

Fig. 2.19

- When journal starts to rotate in clockwise direction inside the bearing, it will climb the bearing wall due to friction up to point Q, such that the line of centres OO' of the bearing and the journal comes to rest at an angle of repose ψ_o.
- At point Q, $h_0 = 0$

Case (iii) Journal at Full Speed :

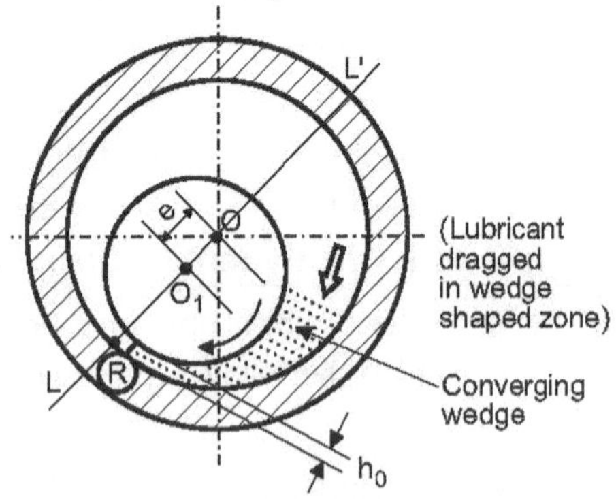

Fig. 2.20

- As the journal speed is further increased, it will drag the fluid into the converging wedge shaped zone; which starts to exert pressure with increasing journal speed.
- At particular journal speed, the pressure becomes sufficient to support the load (W) and the journal is shifted to other side (i.e. to left side) of verticle.
- In this situation, there is minimum clearance and oil-film thickness is minimum (h_{min}) at point R.
- The amount of rotation of the line of centres of journal and bearing from the load line depends on :
  - The magnitude of applied radial load
  - The journal speed and
  - The viscosity of lubricant.
- Thus, in hydrodynamic bearings, it is not necessary to supply the lubricant under pressure. The only requirement is to ensure sufficient and continuous supply of the lubricant.
- The minimum thickness of the fluid film increases with an increase in fluid viscosity and surface speed and decreases with an increase in load.

2.10 BEARING CONSTRUCTIONS

Two basic types of bearing constructions are generally in use, as solid bushing and lined bushing.

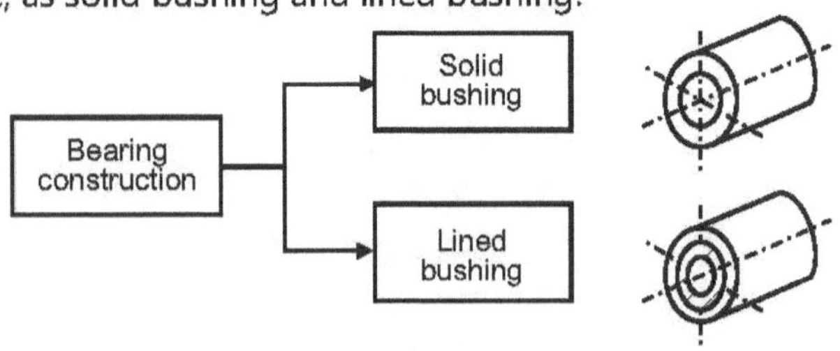

Fig. 2.21

(i) Solid Bushing :

It can be manufactured from a round bar or can be produced by casting process as a single cylindrical unit with central bore and finally, finished to required size by grinding and reaming operations.

Fig. 2.22 shows the solid bushing.

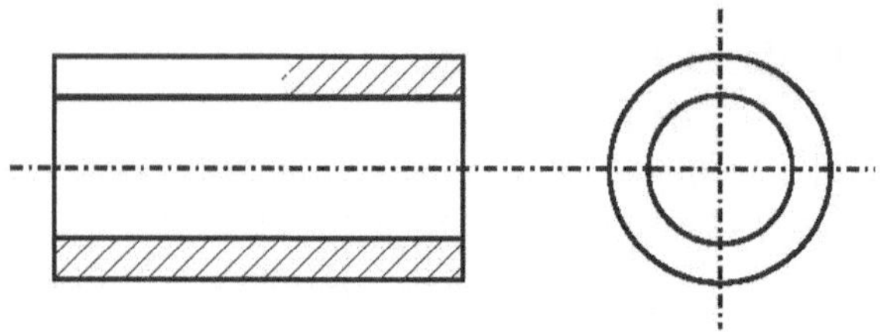

Fig. 2.22 : Solid bushing

e.g. Bronze bearing is a particular example of this type of bushing. When worn out, it can be replaced as a whole element.

(ii) Lined Bushing :

It is similar to solid bushing but has an additional ring inside, generally of babbitt (white metal) material. The outer ring is made up of steel. Thus, babbitt is backed by outer steel bush. Therefore, it is called as lined bushing or simply liners in automotive.

Fig. 2.23 shows the lined bushing.

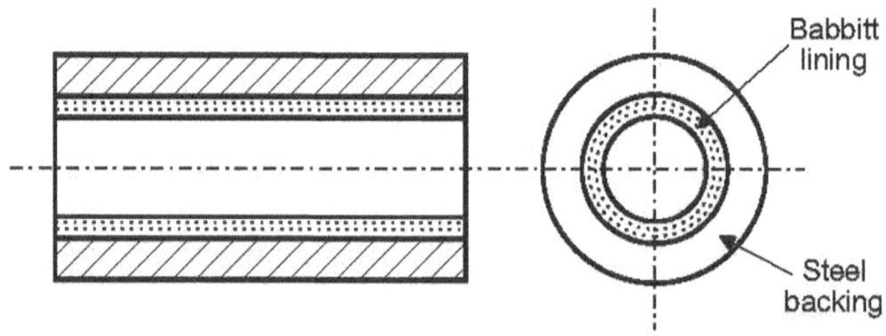

Fig. 2.23 : Lined bushing

Usually, lined bushing is split into two halves and is provided with a locking element which prevents the axial as well as rotational movement of the bearing with respect to the bushing.

Types of Patterns of Oil Grooves :

Suitable positioning and locating an oil groove in the bearing is very much essential to have better performance of the bearing. There are various patterns of oil grooves in the bearings to serve this purpose. Groove have often been placed in bearings with the result that they scrap off the oil and interrupt the film.

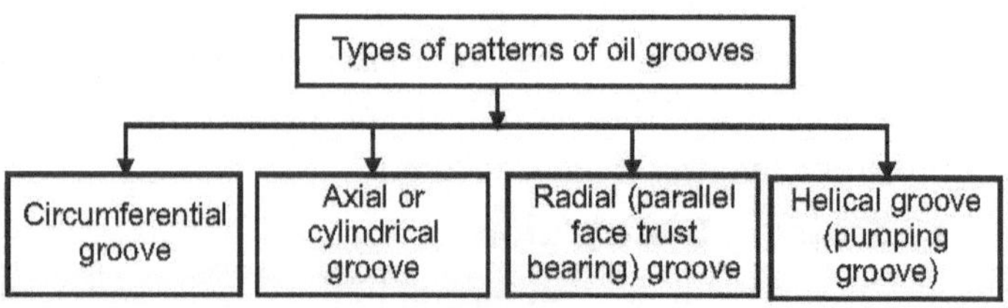

- **Circumferential Oil Groove :**

It distributes lubricant around the shaft at the oil hole when placed at the ends of a bearing. They act as collector rings and salvage the lubricant that is ordinarily forced out at the ends of the bearing. This is shown in Fig. 2.24.

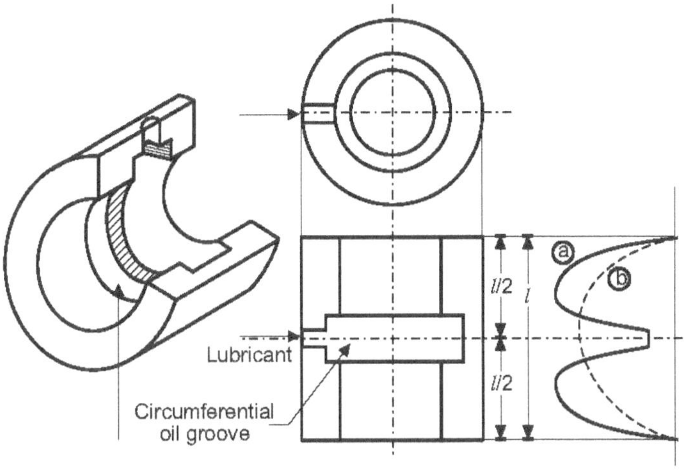

(a) Bearing with groove, (b) Bearing without groove

Fig. 2.24 : Circumferential oil groove

It divides the bearing into two small bearings each of length $\left(\frac{l}{2}\right)$, where l is total length of the bearing. In this type of construction of oil groove pattern, the pressure developed in the bearing along the axis is reduced due to the groove pattern which affects the load-carrying capacity. Thus the load-carrying capacity decreases. But in general, the flow through a journal bearing with a circumferential groove will be greater than the flow through a bearing fed with oil under pressure from a single hole. Approximately, the flow may be four times greater. But sometimes, the centrifugal force acting on the oil in the circumferential groove may build pressure higher than the supply pressure, which may restrict the flow of the lubricant.

Some experimental results show the effect of groove width, location and edge radius on the oil flow rate, such that oil flow rate increases by providing edge radius to a square groove.

Applications :

These types of bearings are most widely used for the crank shaft and connecting rod of automotive engines.

- **Axial Groove or Cylindrical Groove :**

These groove patterns are as shown in Fig. 2.25.

Axial grooves distribute lubricant lengthwise in the bearing or along the axis of the bearing. This type of bearing has an axial or cylindrical groove almost along the total length of the bearing.

As compared to the circumferential groove bearings, the axial groove bearing has higher load-carrying capacity. But such types of bearings are very much prone to shocks and vibrations during operation. Number of similar patterns are obtained by combining axial groove pattern with circumferential groove pattern.

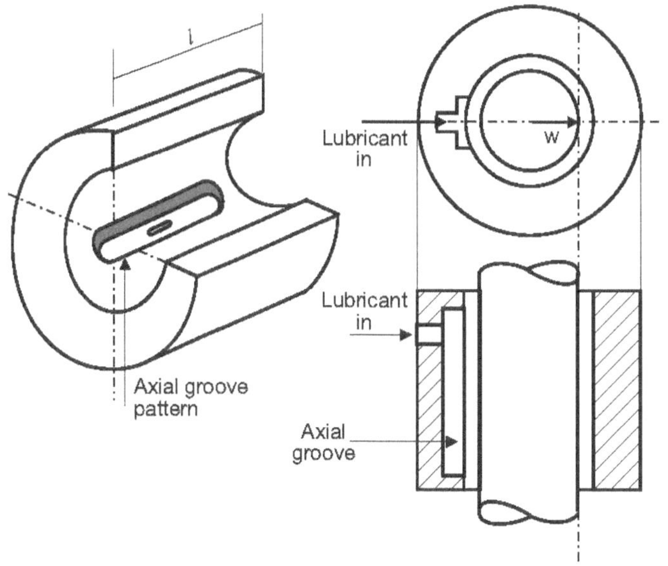

Fig. 2.25

Applications :

These types of grooves in bearings are used for gear boxes and high-speed applications.

- **Radial Oil Groove :**

Sometimes, radial grooves cut in the thrust plate give better results. A parallel face thrust plate as shown in Fig. 2.20 will make a very satisfactory bearing if a radial groove with well rounded or beveled edges are placed in the faces.

Experimental investigation for a flat steel thrust plate of 60 mm outer diameter and 42 mm inner diameter, with a unit load of about 690 kPa gives the values of coefficient of friction for flat plate thrust bearing and for flat plate with radial grooves. (Table 2.2).

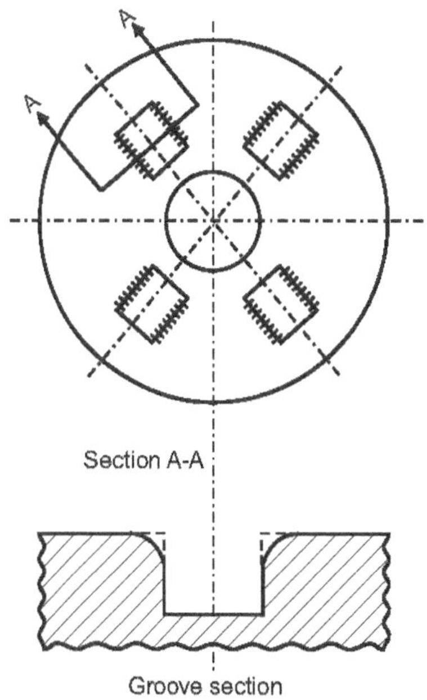

Fig. 2.26 : Parallel face thrust plate with groove section

Table 2.2 : Flat Plate Thrust Bearing with Radial Groove

Speed r.p.m. (N)	Coefficient of Friction for Flat Plate Thrust Bearing	Coefficient of Friction for Flat Plate with Radial Grooves
260	0.011	0.0065
520	0.0099	0.0067
900	0.015	0.0067

- **Helical Groove Pattern :**

It is as shown in Fig. 2.27 where a helical groove is cut inside the bearing. It has a lubricant inside and it flows helically.

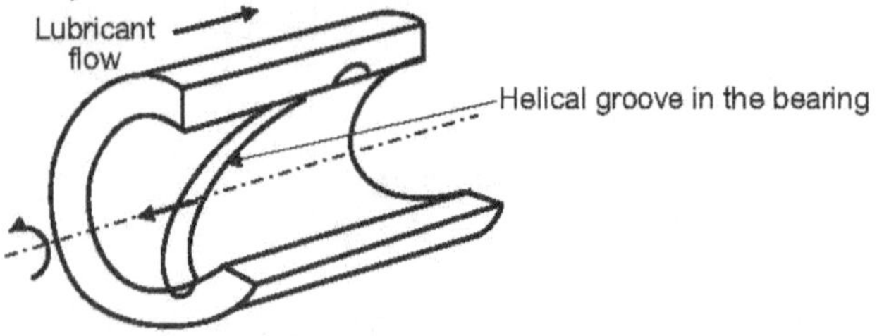

Fig. 2.27 : Helical groove in bearing

As the lubricant flows through the helical groove, the groove provides pumping action for the lubricant to flow. It is used for some of the special applications.

2.11 BEARING MATERIALS

2.11.1 Introduction

There would be no problem in the selection of a bearing material if the operating conditions in a bearing were ideal. If the rubbing surfaces of a bearing were always separated by a film of lubricant, if there were no elastic or thermal distortions, no roughness of surfaces and no grit or other abrasive particles in the lubricant, in that case almost anything would serve as a bearing material. Therefore, the only consideration would be to select a material of sufficient strength. Sometimes, the conditions of operation may be such that complete fluid films are not developed to separate the rubbing surfaces. Some bearings must be operated under conditions that prevent the formation of fluid films. Some important properties need to be taken into account while selecting a suitable material for a particular application of bearing.

These properties are a measure of mechanical, metallurgical and chemical characteristics of bearing materials that have been found essential.

2.11.2 Desirable Properties of Bearing Materials

Following standard requirements are taken into account while selecting a suitable bearing material for a particular application. These are considered as desirable properties of bearing materials.

- Score resistance.
- Mutual solid solubility.
- Compressive strength.
- Fatigue strength.
- Deformability :
  - Conformability
  - Embeddability
  - Bondability
- Corrosion resistance
- Structure
- Thermal properties :
  - Thermal conductivity
  - Thermal expansion
- Cost and availability.

Score Resistance :

- Bearing material should not damage the journal surface when operating under the condition of boundary lubrication. Some materials have a tendency to adhere or weld to the shaft if the fluid-film conditions break down. Thus, bearing materials which operate with low friction, display low score resistance.
- 'Score resistance is defined as the antiweld or antiseizure characteristics of a bearing material and not the action of cutting or grooving of a bearing material by dirt or foreign matter'.

Table 2.3 shows score resistance of some metals.

Table 2.3 : Score Resistance of Some Metals

Score Resistance	Metals
Good	Silver, Cadmium, Tin, Lead
Fair	Carbon, Copper, Selenium
Poor	Magnesium, Aluminium, Copper, Zinc, Barium, Tungsten
Very poor	Beryllium, Silicon, Calcium, Titanium, Chromium, Iron, Cobalt, Nickel, Gold, etc.

- The bearing may score or wipe on the shaft surface if the score resistance is not sufficiently high.
- When score resistance is a primary requirement, white metals are used which includes both the tin-base and lead-base alloys.
- Improved resistance to scoring is obtained at the expense of hardness and strength.
- Sometimes, special bearing materials are used as score resisting materials e.g. silver, trimetal types, gridded bearings, aluminium bronzes.

Mutual Solid Solubility :

- Mutual solid solubility promotes adhesion, surface welding and consequently wear and high friction.
  - Copper, zinc and gold have poor score-resistance against steel and are in the B-subgroup of metals. They have relatively high solubility.
  - Magnesium, calcium and barium have low solubility and have poor score-resistance and are in the A-subgroup.

Compressive Strength :

- It is the ability of the bearing material to carry the imposed load without extrusion or disintegration.
- The bearing material should have high compressive strength to withstand the maximum pressure (which is much greater than the average pressure i.e. load per unit projected area) and to prevent this extrusion or disintegration of the bearing.

Fatigue Strength :

- Sometimes, the load is variable either in magnitude or direction or both the corresponding variables. Stresses imposed upon the bearing material introduces the condition of fatigue. Thus, in certain applications like automotive and stationary engines, bearings are subjected to high fluctuating loads. The bearing materials in that case should possess sufficient endurance strength to avoid the fatigue failure.

  e.g. Fatigue strength of bronze without a steel backing is high.

  Table 2.4 shows order fatigue strength of typical bearing metals.

Table 2.4 : Order Fatigue Strength of Typical Bearing Metals

Metals	Order of Fatigue Strength
Bronzes	1
Copper lead with tin or silver	2
Thin-babbitt	3
Aluminium alloys	4
Copper-lead	5
Cadmium alloys	6
Lead and tin-base babbitts	7

Deformability :

- Some of the operating conditions which have more influence on the imposition of fluctuating load that leads to deformation and in turn total failure are –
  - Time.
  - Temperature : High operating temperatures sharply reduce the strength of many bearing materials.
  - Design of bearing housing.
  - The amount of flexibility in both the shaft and bearing backing : Relatively soft bearing materials when used in thin layers over stronger backing materials, result in increased strength.

  Order of deformability is exactly opposite to that of the order of fatigue strength.

  It can be explained with the following three properties.

Conformability :

- It is the ability of a bearing material to yield to deformation while operating, without causing failure. Thus, conformability is the ability of the bearing material to yield and adjust itself or adopt its shape to that of the journal.
- When loads are applied to the structure the bearings and journals deflect, often leading to edge contact. If the bearing material is conformable, it will adjust itself by wearing or wiping without causing a serious high temperature condition to develop.

  e.g. Tin-base babbitt develops superfacial softening at the points of high friction, wipes locally, relieves itself and settles down to an extended period.

Embeddability :

- 'It is the ability of the bearing material to embed small dust, dirt, hard-metal particles and other foreign material without scoring the journal'.
- When dirt enters the clearance space, it must embed itself in the relatively soft bearing material or jam itself into the clearance space between a hard bearing material and the shaft and cut or gauge a groove in the bearing and shaft. The bearing material should be soft and should have high embeddability characteristics to allow these foreign particles to get embedded in the lining.

Bondability :

- 'It is the ability of the bearing material to form strong bonds with the mating material'.
- For high capacity applications, bearings are made by bonding one or more thin layers of bearing material to a high strength steel shell. Thus, the bondability of the bearing material is an important consideration while selecting suitable bearing material for such type of construction.
- **Corrosion Resistance :**

  At high operating temperatures oxidation of lubricating oil takes place, thereby deteriorating the surface of the bearing and resulting in the formation of acid. This acid attacks vigorously with disastrous results. This deterioration is termed as corrosion. Hence, the bearing material should have sufficient resistance against such kinds of failure due to corrosion.

The danger of bearing corrosion can be greatly diminished by inhibited oils which resist the formation of corrosive acids.

e.g. In applications like engine bearings.

Table 2.5 shows corrosion resistance for different bearing metals.

Table 2.5 : Corrosion Resistance for Metals

Corrosion Resistance	Metals
Non-corrodible	Al-alloys, Tin-base babbitt, lead-base babbitt, cadmium-indium alloys, bronze (low lead).
Intermediate	Bronzes (high lead), copper-lead, Alkali-hardened lead, silver.
Corrodible	Cadmium alloys.

Structure :

- Structure is of great importance. Since a particular structure for each type of bearing material gives the best performance. A good bearing material should possess a duplex structure with hard particles supported in a soft matrix. e.g. in a tin-base babbitt where hard crystals of tin, copper and antimony are supported in a soft, tin-rich matrix.
- The hard particles in the duplex structure of the babbitt contribute very little to the load-supporting capacity of the material. With white metal-alloys, the basic frictional properties are determined essentially by the softer matrix material itself and the hard particles have little influence.

Thermal Properties :

- The bearing material should possess good thermal properties as given below.
  - ➢ **Thermal Conductivity :** The bearing material should possess high thermal conductivity so as to dissipate heat developed in the bearing rapidly.
  - ➢ **Thermal Expansion :** For bearing operating under wide range of temperatures, the bearing material should have low coefficient of thermal expansion. There is no undue change in clearance.

Cost and Availability :

- The selection of a proper bearing material may be influenced by the cost and by the availability of the material. Therefore, the bearing material should have reasonable cost and should be easily available. Table 2.6 gives some of the commercial bearing metals with the approximate order of increasing cost.
- Here, it is difficult to find all the above stated properties to a reasonable level in any single bearing material. Therefore, the bearing material should be selected from the various bearing materials available, depending upon the requirements of the actual service conditions.

Table 2.6 : The Approximate Order of Increasing Cost of Commercial Bearing Metals

Metals	Order of Cost
Bronze bushing	1
Lead-base babbitt	2
Tin-base babbitt	3
Sintered copper-nickel with lead babbitt	4
Aluminum alloys (solid)	5
Cadmium alloys	6
Copper-lead	7
Copper-lead with thin overlay	8
Aluminum-alloy with thin overlay	9
Silver with thin overlay	10

2.11.3 Bearing Materials : Metallic Group

1. Babbitts (white metals)
   (i) Lead-base babbitt
   (ii) Tin-base babbitt
2. Bronze
3. Copper-lead alloys
4. Aluminum alloys
5. Silver
6. Cast-iron
7. Porous Metal Bearings (Sintered Metal Bearings)
8. Hard materials

1. Babbitts (White Metals)

These prove to be excellent bearing materials. They have a silvery appearance and are generally called 'white' metals or babbitt metals. Babbitts are used where score resistance is a primary requirement. Babbitts can be classified into two categories :

- Lead-base babbitt and
- Tin-base babbitt.

Table 2.7 shows the typical composition of these two categories.

Table 2.7 : Composition of Babbitts

Constituent Elements	Types of Babbitts	
	Lead-Base Babbitt	Tin-Base Babbitt
Lead	74	0.5
Tin	9.25 – 10.75	86.0
Antimony	14 – 16	6.0 – 7.5
Copper	0.5	5 – 6.5
Others (Iron, Arsenic, Bismuth, Zinc, Aluminum)	1.0	0.5

(i) The babbitts are to a large extent the standard of desirability for bearing materials in many applications as discussed earlier.

(ii) The babbitts possess excellent bondability, embeddability and conformability but relatively weak fatigue strength, especially at temperature above 121°C.

(iii) Tin-base babbitt is able to wipe locally, relieve any high spots or areas of metallic contact and then reform its contour so as to reestablish fluid-film condition.

(iv) Tin and lead-base babbitts are non-corrodible as tin-base babbitt is one of the most resistant of all the possible bearing materials.

(v) Effects of Lining Thickness on Fatigue Life :

- They are bonded as a thin layer or many thin layers over backing material; like bronze or steel. This results in increased strength with softer bearing material.
- Fig. 2.28 shows a comparison of the life of a bearing under a mean load of 13.79 MPa against varying thickness of white-metal bearing alloys. A graph is plotted from the test results run under accelerated laboratory conditions.

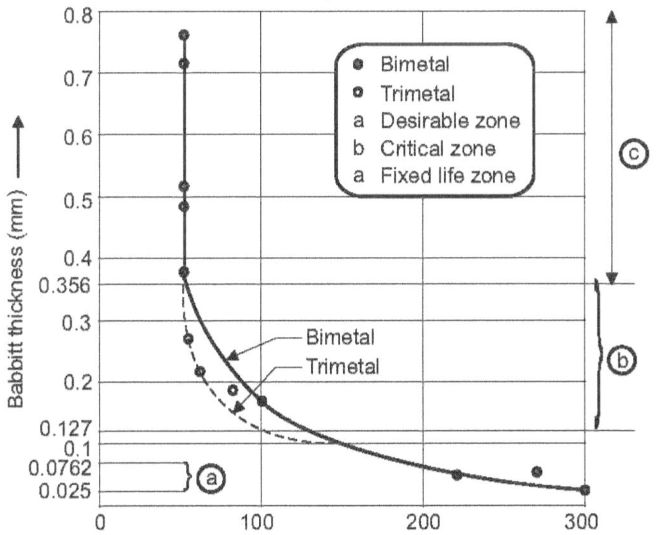

Fig. 2.28 : Bearing life vs. babbitt thickness

- **Fixed-Life Zone :** It is clear that in the thickness range (0.813 mm to 0.356 mm), variation in lining thickness has no effect on the bearing life. It is found to be upto 55 hours.
- **Critical Zone :** The thickness ranging from (0.356 mm to 0.127 mm) there is a very noticeable increase in bearing life. This is called a critical zone. The ultimate gain is reached in the region where the thickness varies from 0.127 mm to 0.0254 mm.
- **Desirable Zone :** For the thickness variation from (0.0762 mm to 0.025 mm) there is ultimate gain in bearing life called as desirable zone.

(vi) Effect of Temperature on Strength and Hardness of Babbitt : High operating temperature sharply reduces the strength of babbitts. Hardness reduces by 50% as temperature rises to 121°C especially in automotive work.

2. Bronzes :

These are most common types of bearings and used where service conditions in terms of load, speed and temperature are moderate. Bronze is an alloy of copper and tin. It is used as machined bush pressed into a shell. The bush may be in one piece or split.

Table 2.8 shows composition of two variables of bronzes, which are commonly used as bearing materials.

Table 2.8 : Composition of Bronzes

Constituent Elements	Composition %	
	Gun Metal	Phosphor Bronze
Copper	87-88	89-90
Tin	10	9-10
Lead	0.3	–
Zinc	2.0	–
Nickel	1.0	–
Phosphorus	–	0.3

Advantages :

- Due to their adequate bearing properties, they are used for most of the applications.
- They have great strength and are helpful if the speeds and temperatures are not so high.
- They can be easily and economically fabricated in special flanged designs as well as in the simple cylindrical forms. Thus, they have excellent casting and machining characteristics. They can be made as a single, solid unit with no bond to fail and no lining to wear through. Thus they are highly reliable.
- The thick walls of bronze provide an adequate reserve for wear which is the basic consideration for industrial machinery exposed to abrasive or contaminating condition.

Disadvantages :

- It has a tendency to adhere to the shaft surface at higher temperatures. When the temperature increases as a result of heavy load, high rubbing speeds and poor lubricating conditions, steel journals get overheated until they turn blue and the bronze gets scored and adheres to the shaft. Therefore, bronze bushings are used for milder applications where mildness is measured in terms of moderate speeds, loads and temperature.

- Compared with babbitts, it has poor conformability and embeddability.

Applications :

They are used in applications like bushings, gears, pump impellers, guides, cams and wire drawing dies where strength is combined with good friction and wear characteristics.

3. Copper-Lead :

These materials are made from a mixture of copper and lead and often with small amounts of other elements like tin, nickel, zinc, iron, etc.

Table 2.9 gives the composition of copper and lead.

Table 2.9

Constituent	Composition %
Copper	55-75%
Lead	25-45%

- They are used where loads are higher than that which can be carried by babbitts.
- They are produced by casting or by sintering the powdered metals, which result in a continuous phase of copper with unalloyed lead dispersed through it.
- They are bonded in thin layers to a backing of steel.

Advantages :

- They have high compressive strength combined with high fatigue strength.
- They are used in heavy duty applications at high temperatures.

Disadvantages :

The conformability of the copper-lead bearing is lower than that of the babbitts. Also, they possess average embeddability. Therefore, they are used for applications with rigid shafts where deflection and end rubbing are less of a problem. The conformability characteristics can be improved by using 'Trimetal construction'.

Trimetal Bearings :

The construction of a bearing obtained by depositing a thin overlay of lead-base babbitt on the copper-lead which improves conformability characteristics of bearing is called as 'Trimetal bearing'. The constructional features of trimetal bearings can be explained with the following Fig. 2.29.

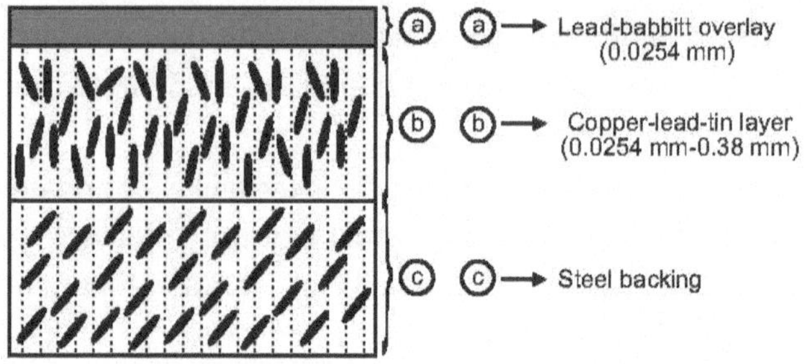

Fig. 2.29 : Trimetal bearing material

- In trimetal bearings, a layer of copper-lead whose thickness varies from 0.254 mm to 0.38 mm is bonded to a thick steel backing.
- Then an overlay of lead babbitt about 0.0254 mm thick is precision plated on the copper-lead.
- The resulting construction produces one of the highest load-carrying plain bearing.
- For thin babbitt overlays, the load varies from limits 13.8 MPa to 27.6 MPa.

  e.g. Moraine - 100 Bearing.

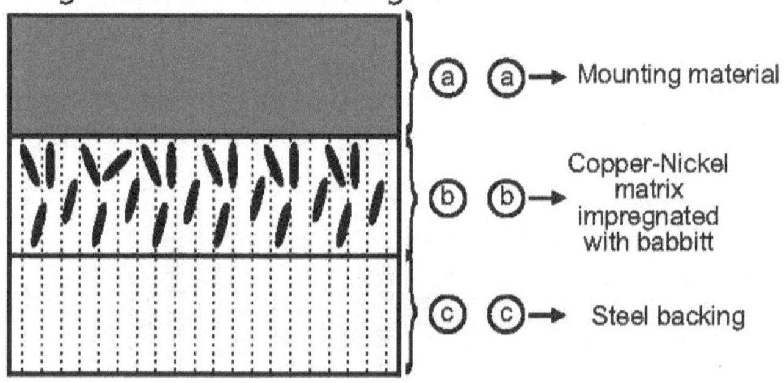

Fig. 2.30 : Moraine - 100 trimetal bearing

- It is manufactured by Moraine Products Division of General Motors.
- A mechanical mixture of pure copper and nickel powder is applied to a steel backing. Then this combination is passed through a furnace. As the copper melts in the reducing atmosphere of the furnace, it brazes the powder to the steel.
- The alloying of the copper and nickel continues till the layer of powder develops a fine porous texture.
- Matrix of copper and nickel is formed on the steel backing and is consequently given a precision rolling.
- Then, it is allowed to pass through a vacuum impregnating machine which fills the pores of the matrix with babbitt.
- The matrix supports the babbitt on its surface and provides a large bond area for both chemical and mechanical bonding.

Applications :

This bearing has proved to be an excellent bearing for engine applications where it provides high strength with good conformability and embeddability.

4. Aluminium Alloy :

Aluminium when alloyed with small amounts of tin, nickel, copper, silicon or cadmium performs very well (Refer composition of type 750). Table 2.10 shows the percentage composition of the three typical aluminium alloys.

Table 2.10 : Percentage Composition of Three Typical Al-alloys

Constituents	Composition % of Types of Al-Alloys		
Al	91.5	89.5	95
Sn	6.5	6.5	0
Ni	1	0.5	0
Cu	1	1	0
Si	0	2.5	4
Cd	0	0	1.0

Advantages :

- **Good Fatigue Strength :** They possess good fatigue strength and yield strength when solid or backed by steel.
- **High Thermal Conductivity :** It has good thermal conductivity which is highly advantageous in conducting heat away from the bearing surface, the source of heat being to friction. This is specifically important for bearings that are not supplied with a sufficient amount of lubricant for cooling purposes. Generally, the bearing operating with mixed or boundary lubrication fall into this category.
- **e.g. Al-Alloy - GM 3889 M :** It is backed by steel and carries a lead-tin copper overlay of about 0.0254 mm thickness.
  - The layer of lead-tin-copper reduces the possibility of corrosion and improves ductility and score resistance.
  - The copper gives the overlay higher load-carrying capacity.
  - It provides excellent service life in engine bearings with unit loads upto 27.5 MPa, average pressure.

Disadvantages :

- **Poor Embeddability :** Al-alloyed bearings have poor embeddability characteristics as compared to that of babbitt.
- **High Thermal Stressing :** For high temperature applications, the expansion of the aluminum alloy may cause high thermal stresses exceeding the elastic limit in compression of the material and resulting in permanent plastic deformation on cooling. The aluminum alloy shrinks and may cause tensile failure cracks. Therefore, Al-alloys should not be used for temperature between 107 to 149°C.
- **Limited Use on Rail Road Axles :** Al-alloyed bearings are being used to a limited extent for rail road axle bearing specially on Italian rail roads.

5. Silver :

They are commonly used in aircraft applications. Pure silver has certain disadvantages.

Disadvantages :

- The performance of silver as a bearing material is not considered to be satisfactory and found to be quite erratic.
- Pure silver has poor score resistance and a low degree of embeddability.
- They have relatively low conformability.
- Cost of silver is high.

To overcome the above problems, the following remedy is suggested.

Overlay of Lead or Lead-Indium on Steel Backing, Improves Embeddability : Performance of silver bearing can be improved by giving an overlay of a soft material such as lead or lead-indium. This makes it more reliable. Most silver bearings are electroplated on a steel backing, annealed and after final machining the bearing surface is given a thin coating of lead or lead-indium of approximately 0.0254 mm thickness. The surface overlays greatly improve the embeddability properties.

Advantages :

- **Increase in Fatigue Resistance :** Under repeated stresses, silver bearings are superior to all other standard materials. Thus, bearing material have high fatigue resistance and they have replaced copper-lead bearings in certain aircraft engine master connecting rod applications with considerable increase in fatigue life.
- **High Thermal Conductivity :** Silver bearings possess excellent thermal conductivity.

6. Cast-Iron :

- It is one of the oldest bearing materials. It is used for relatively light-duty applications.
- It has been suggested that clearances on cast-iron bearings should be a little larger than usual so that if hard particles are torn loose from the cast-iron, they will not tend to jam the clearance space.
- Steel journals have run successfully against cast-iron if the hardness of the journal is in the range from 150 to 250 Brinell.

- The hardness of cast-iron restricts its use to the applications where edge rubbing or shaft deflections are held to a minimum, i.e. the applied load must be light and the alignment between shaft and bearing can be carefully maintained.

7. Porous Metal Bearings (Sintered Metal Bearings) :

- These bearings are used when plain metal bearings are impracticable because of lack of space or inaccessibility for lubrication.
- They are made of powdered metals which are pressed in dies. After compression, they are sintered at a high temperature in a reducing atmosphere. The sintering operation causes the powdered metal bearings to fuse into a strong compact.
- After the sintering process is over, the bearings are submerged in oil for impregnation and finish-sized in a punch press to close tolerances.

The flow diagram of sintering process is given below.

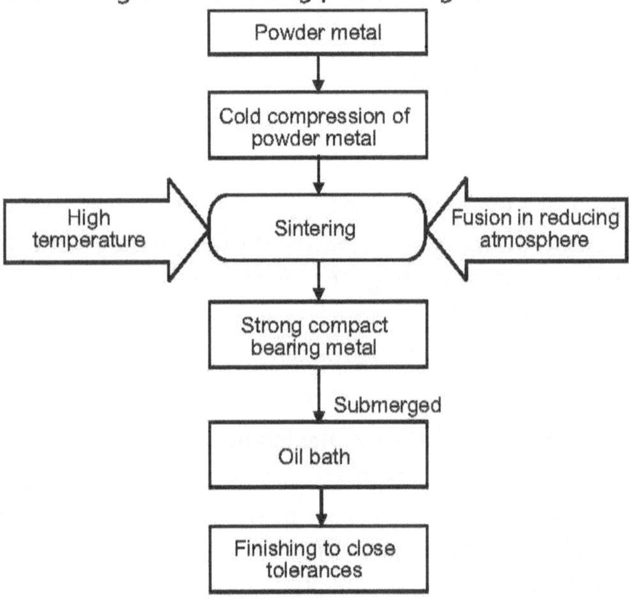

Fig. 2.31

- Following are the varieties of sintered bearings :
  - Iron-base sintered bearings
  - Copper-base sintered bearings
- Sintering temperature for two different powder metals are given below.

Powder Metals	Sintering Temperature
Bronze powder	815°C
Iron (plus copper) powder	1093°C

- Porous bearing is strong and contains voids into which the non-gumming lubricant (which is strongly resistant to oxidation) is drawn. These voids may vary from 16% to 36% of the volume of the bearing.
- For Bronze bearings, lubricant with viscosities 250 to 350 SUS are recommended, and impregnated with graphite, molybdenum disulphide, grease or wax, with 85.5°C maximum suggested temperature. Lubricant may be supplied by wicks or drop-feed oilers to the outside of the bearing housing and by capillary action.
- These bearings are suitable for light loads and moderate speeds. These bearings will tend to build up a thinner hydrodynamic fluid-film than a solid bearing because of the porosity of the material.
- The principle applications for porous metal bearings are probably limited to less than 140 kPa average pressure and moderate speeds, as heating of these bearings is always a problem since there is no circulating oil to remove the heat.
  - **Teflon :** Porous metals impregnated with teflon at light loads and speeds, show that coefficient of friction is 0.05 consistently upto temperatures of approximately 250°C.
  - **Molybdenum Disulphide :** With a sintered copper porous bearing containing molybdenum disulphide, the coefficient of friction is 0.13 to 0.15 upto 300°C.

Applications :

Pedal bushings, water pump bearings, guide bearings on small generators and distributor shafts, vacuum cleaner motors. They are also called less-oil bearings.

8. Hard Materials :

In certain applications, with a high temperature and with the use of lubricants with low viscosity and little natural oiliness perhaps with operations where no lubricant are present. Hard materials are used as bearing material. If the speed is low and wear is an important factor, then both surfaces should be hard. The harder the material, the lesser is the wear.

Minimum total wear takes place when the sliding and stationary surfaces have the same degree of hardness.

e.g.

- Refractory materials and ceramics.
- In case, where there is shortage of high grade cast-iron, surface plates are made of glass, granite and ceramics, or fine grains of crystalline fused alumina bonded with a glass-type bond and containing 10-15% of very fine pores. Also, surface plates are made of glass bonded silicon carbide grains called as 'Crystolon'.

2.12 LUBRICATING OILS

A Lubricating Oil should have Certain Desirable Properties as Listed Below.

- The oil should be available in a wide range of viscosity.
- It should be able to dissipate frictional heat efficiently.
- It should be stable under thermal stresses i.e. the change in viscosity with respect to temperature should be minimum.
- It should be stable under oxidation stresses. i.e. it should be chemically stable and not react with the bearing materials.
- It should have low volatility i.e. it should have a good fire resistance.
- It should have good mechanical stability.
- It should be economical.
- It should have a long life.

The two major categories of lubricating oils are – (i) mineral oils and (ii) natural oils (vegetable or animal oils). The availability, good performance, variety and cheapness of mineral oils have made them the first choice in almost 90% of the applications. These oils are classified by the **Society of Automotive Engineers (SAE)** in terms of grades as discussed below.

2.12.1 SAE Classification

Commonly used viscosity grade has been standardised as **SAE (Society of Automotive Engineers) Viscosity Grades.** Each lubricating oil is designated by a SAE grade or number which is indicative of the viscosity, though it does not directly represent the viscosity. Table 2.11 gives the SAE grades and their viscosity range.

Table 2.11 : SAE Grades and their Viscosities

SAE Grade	Viscosity Range	
	at 0°F (–18°C)	at 210°F (99°C)
	in SUS	in SUS
Motor Oils		
SAE 5 W	Less than 6000	–
SAE 10 W	6000 – 12000	–
SAE 20 W	12000 – 48000	–
SAE 20	–	45 – 58
SAE 30	–	58 – 70
SAE 40	–	70 – 85
SAE 50	–	85 – 110
Gear Oils		
SAE 75 W	Less than 15000	–
SAE 80 W	15000 – 100000	–
SAE 90	–	75 – 120
SAE 140	–	120 – 200
SAE 250	–	200 and above

The viscosities of oils are measured at 210°F (99°C) and viscosities of oils with W suffix are measured at 0°F (–18°C). The W suffix indicates the winter grading of the oils. When SAE number is more, it indicates more viscous oil. e.g. "5W" signifies the oil viscosity when the oil is cold.

Some oils, with polymers added to them have high viscosity indices, and are called **Multigrade Oils.** This is because they are in **One** grade at 0°F and in a **Higher** grade at 210°F. As an example **10W/30** oil may have a viscosity of 8000 SUS at 0°F and 65 SUS at 210°F. It falls into the 10 W range at 0°F and SAE 30 at 210°F. For this reason, it is called **10 W/30**.

2.13 ADDITIVES [Nov. 15, Aug. 16]

In order to have improved characteristics, most modern lubricating oils contain chemical compounds as additives. Thus, additive is a compound that enhances some property, or imparts some new property to the base fluid. In some hydraulic fluid formulations, the additive volume may constitute as much as 20% of the final composition. The most important types of additives include anti-oxidants, anti-wear additives, corrosion inhibitors, viscosity index improvers, and foam suppressants.

- **Viscosity Index Improvers (V.I. Improvers)** are added to an oil to reduce the rate of change of viscosity with temperature and these are usually high-molecular weight polymers.
- **Oiliness Additives :** Number of vegetable, animal and fish oils have a much greater oiliness than mineral oils. To have the benefits of thermal stability and at the same time to retain the benefits of oiliness, the above mentioned oils are added to mineral oils.
- **Pour-Point Depressants** are used to delay the formation of rigid structure from waxy crystals precipitated from cooled mineral oils and this is done by forming a coating on the waxy crystals. Complex polymers are generally used for this purpose.
- **Oxidation Inhibitors** reduce the rate of formation of oxidation products.
- **Detergents or Dispersants** are used mainly in engine oils to keep the engine clear by holding insoluble material in suspension and preventing the formation of

sludge or deposits. Polymer compounds are mostly used as detergents for low-temperature conditions, while for high-temperature conditions, organometallic salts are the usual type of additives.

- **Extreme Pressure Additives (E.P. additives)** allow a mineral oil to operate satisfactorily under conditions of extremely high pressures and temperatures. The most common extreme pressure additives are compounds of sulphur, phosphorous and chlorine. These react with the metal surfaces to form coatings, which exhibit plastic flow and have low-shear resistance.
- **Corrosion Inhibitors :** Alcohols, esters, organic acids, amines or soaps are used as corrosion inhibitors, which are absorbed on or react with the metal surfaces to form protective films.
- **Anti-Foam Additives :** Silicon-based compounds are used as antifoam additives which minimize foaming in situations where this might cause problems.
- **Emulsifiers :** Metallic soaps and petroleum sulphonates are used as emulsifiers to stabilize oil-water emulsions and are particularly used for metal cutting oils, fire resistant hydraulic fluids, etc.
- **Anti-Friction Additives :** Anti-friction, sometimes called lubricity, is defined as the ability of a lubricant to reduce friction, other than by its purely viscous properties. Anti-friction additives reduce friction below that of the base oil alone under conditions of boundary lubrication. The additives are absorbed on, or react with the metal surface or its oxide to form monolayers of low shear strength material. The compounds are long chain (greater than 12 carbon atoms), alcohols, amines, and fatty acids. A classic example is oleic acid reacting with iron oxide to form a film of the iron oleate soap. The low shear strength of the soap film causes the low friction.
- **Anti-Wear Additives :** Anti-wear additives are those which reduce or control wear. They form organic, metallo-organic, or metal salt films on the surface. Sliding or rolling occurs on top of, or within the films, thus reducing metal-to-metal contact. Anti-wear additives only reduce the rate of wear.
- **Anti-Scuff Additives :** Anti-scuff additives are those that prevent scuffing. **Scuffing** is defined as damage caused by solid-phase welding between sliding surfaces. Anti-scuff additives reduce scuffing by forming thick films of high melting point metal salts on the surface which prevent metal to metal contact which, when extensive, may cause scuffing.

There is some overlapping of anti-wear and anti-scuff performance. That is, some additives have good anti-wear properties and can prevent scuffing to a limited degree.

2.14 COMPARISON OF SLIDING AND ROLLING CONTACT BEARINGS

The various factors which influence the selection between the two basic types of bearings are :

Magnitude of load, nature of load, speed, life, frictional loss, space requirement, positional accuracy, noise and cost.

Fig. 2.32 shows the plot of load-carrying capacity against speed for different types of bearings.

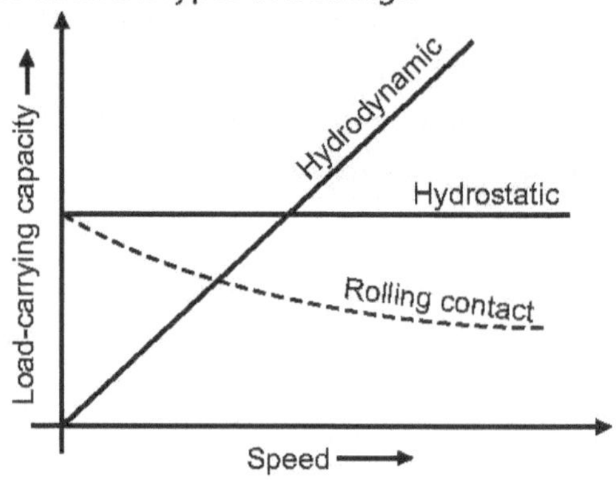

Fig. 2.32 : Load characteristics of bearings

1. Magnitude of Load :

The sliding contact bearings are more suitable for large loads compared to rolling contact bearings.

2. Nature of Load :

The rolling contact bearings are vulnerable to shock loads due to poor damping. Because of the metal-to-metal contact, the rolling elements and races are subjected to plastic deformation under shock and fluctuating loads, which leads to the fatigue failure.

The sliding contact bearings with full-oil film are capable of taking shock and fluctuating loads. The oil film separating the journal and bearing surfaces, absorb the shock. Therefore, the sliding contact bearings are used in heavy I.C. engines as main and connecting rod bearings.

3. Speed :

Because of the centrifugal forces acting on the rolling elements, the rolling contact bearings are not suitable for high speed applications. As shown in Fig. 2.32, the load carrying capacity of the rolling contact bearings, reduces, through marginally, with the speed.

On the other hand, the load-carrying capacity of the hydrodynamic bearings is directly proportional to the speed. As the speed increases, the oil-film thickness as well as the pressure increases. In hydrostatic bearings, the load-carrying capacity is independent of the speed. Therefore, the sliding contact bearings are more suitable for high speed applications.

4. Life :

Due to fluctuating loads, the rolling contact bearings have very limited life compared to sliding contact bearings.

5. Frictional Loss :

The rolling contact bearings have lower starting friction compared to hydrodynamic bearings. In hydrodynamic bearings, during starting metal-to-metal contact in the boundary lubrication region results in a higher starting friction. However, under running conditions, because of the full oil-film the hydrodynamic bearings have lower frictional losses compared to rolling contact bearings. Therefore, the rolling contact bearings are suitable where there are frequent starts and stoppages. On the other hand, the hydrodynamic bearings are suitable where there is comparatively light load at the start and the load increases gradually with speed.

6. Space Requirement :

Rolling contact bearings require more radial space, while sliding contact bearings require more axial space. In addition, sliding contact bearings require an auxiliary equipments like pump, filter, sump, pipings, etc.

7. Positional Accuracy :

The rolling contact bearings give better positional accuracy than the sliding contact bearings. In rolling contact bearings, the relative position of the inner race with respect to the outer race is independent of the operating parameters. On the other hand, in sliding contact bearings, the position of the journal with respect to the bearing varies with the operating parameters.

8. Noise :

Rolling contact bearings generate more noise compared to sliding contact bearings, due to metal-to-metal contact.

9. Cost :

The cost of the sliding contact bearings is much higher compared to the rolling contact bearings, due to auxiliary equipments like pump, filter, sump, pipings, etc. In addition, maintenance cost of sliding contact bearings is also more.

EXERCISE

1. Explain the purpose of lubrication.
2. What do you understand by 'Thick-Film Lubrication' and 'Thin-Film Lubrication ?
3. Explain the basic modes of thick-film lubrication.
4. Explain the following terms :
   (i) Hydrodynamic lubrication
   (ii) Hydrostatic lubrication
   (iii) Boundary lubrication
   (iv) Elasto-hydrodynamic lubrication
   (v) Extreme pressure lubrication
5. Explain the various lubricating oils which are classified according to their functions.
6. State the operating conditions where grease is preferred over mineral oils as a lubricant. State different types of greases along with their applications.
7. Explain the effect of temperature and pressure on the viscosity of lubricating oil.
8. Explain the following properties of lubricants :
   (i) Viscosity
   (ii) Effect of temperature on viscosity
   (iii) Effect of pressure on viscosity
9. Explain the following terms of lubrication :
   (i) SUS (ii) Viscosity index
   (iii) Fluidity (iv) Absolute viscosity
   (v) Kinematic viscosity
10. Explain the various units of absolute viscosity and kinematic viscosity and show the relation between the two viscosities.
11. Explain in detail why viscosity is generally measured in kinematic units rather than absolute units.
12. Write a short note on :
    Measurement of viscosity.
13. What is viscosity index ? Explain the method to find the VI of a test oil.
14. What are additives ? Explain the different types of additives used in lubricating oils.
15. What is the role of additives in lubrication ?
16. State the desirable properties of lubricating oils.
17. Explain 'SAE classification' of lubricating oils.
18. Explain various materials for bearings with their advantages, disadvantages and give atleast two applications of each material.
19. Explain bearing constructions with different oil groove patterns.
20. State assumptions made while deriving Reynold's equation.
21. Derive from basic principles two-dimensional Reynold's equation taking usual notations.

22. Explain Raimondi and Boyd method with reference to each variable involved in it.
23. Write short notes on :
    (i) Sommerfeld Number
    (ii) Design considerations in hydrodynamic journal bearing
    (iii) Temperature rise in hydrodynamic journal bearing
24. Derive expression for flow rate through rectangular slot. What are assumptions made while deriving the equation ?
25. Explain principle of working of hydrostatic step bearing and state assumptions made for the analysis of the hydrostatic step bearing.
26. Derive an expression for oil flow rate and pressure distribution over an annular ring of hydrostatic step bearing.
27. Derive an expression for load-carrying capacity and oil flow rate for hydrostatic step bearing. State the assumptions made.
28. Derive an expression for load-carrying capacity for circular step bearing in terms of supply pressure from fundamental equation of viscous flow and hence determine load-carrying capacity for following data :
    (i) Shaft diameter = 500 mm
    (ii) Recess diameter = 300 mm
    (iii) Supply pressure = 6 MPa
29. State dimensionless form of oil flow rate and load-carrying capacity for hydrostatic step bearing. And plot the variation of these dimensionless parameters with the ratio of the outer land (R_0) to inner land (R_i) i.e. $\left(\frac{R_0}{R_i}\right)$.

◈ ◈ ◈

UNIT II - SPUR GEAR

CHAPTER - 3
SPUR GEAR

3.1 INTRODUCTION

A positive drive, consisting of toothed wheels which by direct contact, transmits power and motion from one shaft to another, is called as gear drive. A gear drive in comparison with belt, rope and chain drives offers the following advantages.

- Velocity ratio is constant.
- It can transmit more power.
- Efficiency is more.
- It can be operated at very slow or very high speeds.
- It is a compact drive.

However there are some disadvantages such as:

- Lubrication is required to be ensured constantly.
- Initial as well as running cost is high.
- Exact centre distance has to be maintained.
- Manufacturing is difficult and hence costly.

3.2 CLASSIFICATION OF GEARS

Gears are classified as follows:

(a) According to the Position of Shaft Axes:

- Parallel
- Intersecting
- Non-parallel and non-intersecting (Skew)

To transmit power between two parallel shafts, spur or helical gears are used. If teeth are cut parallel to the axis of rotation, it is known as spur gear [Refer Fig. 3.1 (a)]. If teeth are cut inclined with the axis of rotation it is known as helical gear [Refer Fig. 3.1 (b)].

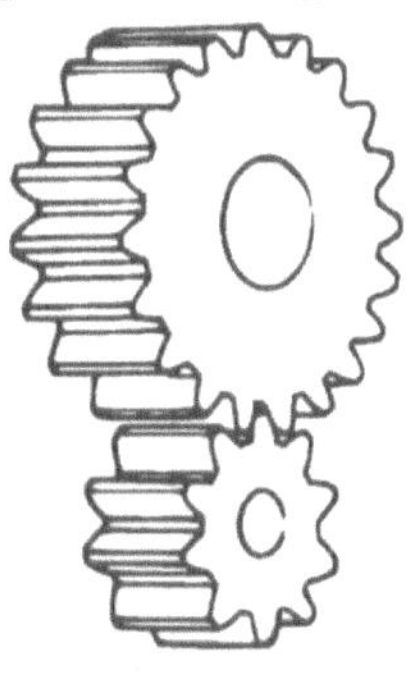

(a) Spur Gear

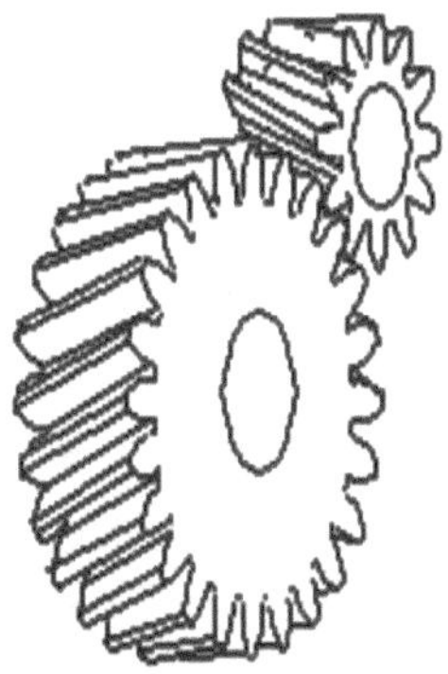

(b) Helical Gear

Fig. 3.1

If power is to be transmitted between two intersecting shafts, bevel gear is used. In bevel gears, the teeth are cut along the generator of frustum of cone. Such bevel gear is called as straight teeth bevel [Refer Fig. 3.1 (c)]. If teeth are cut with a curvature on the face of bevel, it is called spiral bevel gear.

If power is to dbe transmitted between two non-parallel non-intersecting (skew) shafts, spiral gears are used. In this, two helical gears with same hand of helix or opposite hand with unequal helix angles are meshed giving skew arrangement of shafts. The angle between shaft axes is dependent on the hand of helix and value of the helix angle. If the power is to be transmitted at right angles with skew arrangement of shaft axes, we use worm and worm gear. In this drive, worm is a screw thread meshed with a helical gear. [Refer Fig. 3.1 (e)].

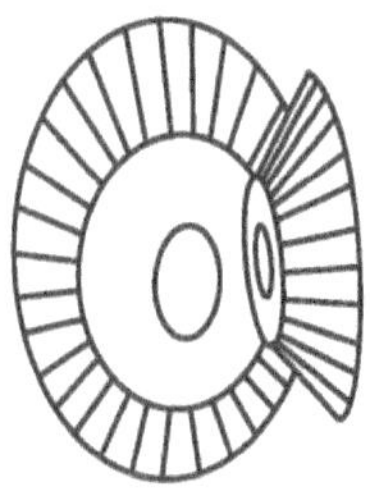

(c) Bevel Gears

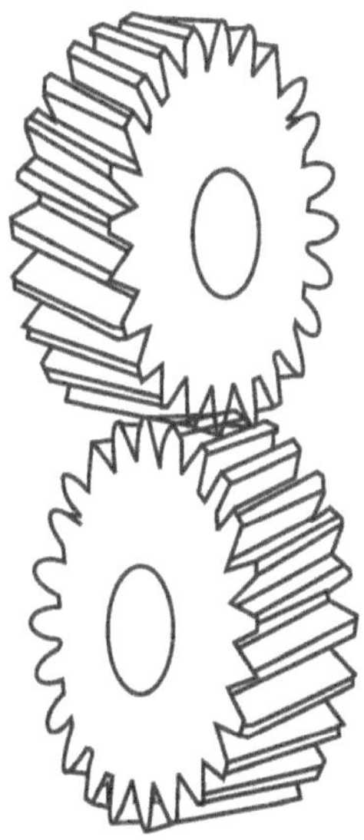

(d) Crossed Helical Gears

(e) Worm and Worm Gear

Fig. 3.1

(b) According to Pitch Line Velocity:

- Gears are classified with this criterion as
- Low speed gears- pitch line velocity less than 3 m/s.
- Medium speed gears- pitch line velocity between 3 m/s to 15 m/s.
- High speed gears- pith line velocity more than 15 m/s.

(c) According to Type of Gearing:

- External gear
- Internal gear
- Rack and pinion

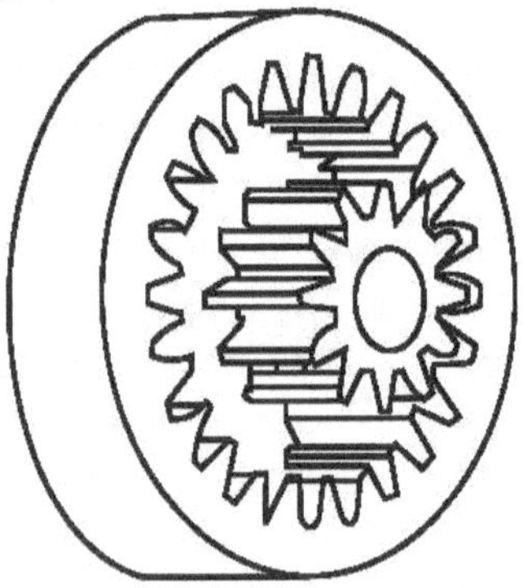

(f) External - Internal Gear

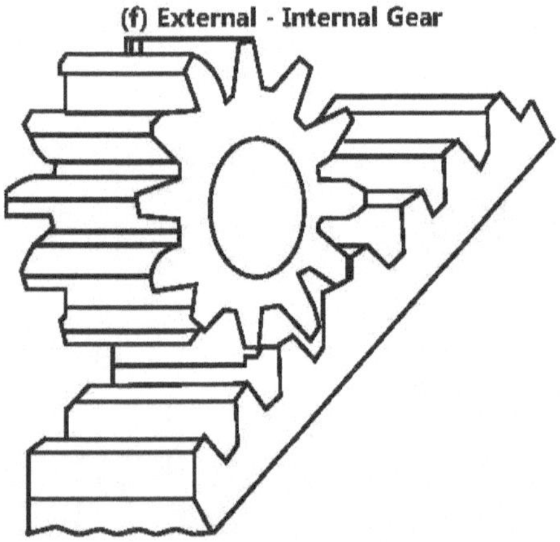

(g) Rack and Pinion

Fig. 3.1

External gears are more popular because of ease of manufacture and ease of achieving the required precision. In this case, the direction of rotation of meshing gears is opposite to each other while for internal gears it is same. Rack and pinion is used to covert rotary motion into sliding motion.

3.3 LAW OF GEARING

- It states that the common normal drawn at any point of contact must intersect the line joining centres of rotation giving always the same pitch points for all points of contact to get constant angular velocity ratio.
- In Fig. 3.2, Two profiles 1 and 2 are making contact at point C. A common normal drawn from this point C intersects line O_1O_2 at point P.
- For any other point of contact on these two profiles, the same intersection point between O_1O_2 and common normal (point P) should be obtained. CA and CB are the velocities of point C on gears 1 and 2 respectively. Their component CD along the common normal at C must be same if there is to be no slip.
- It can be shown mathematically that, the ratio of the angular velocities of the two profiles is inversely proportional to the ratio of distance between the pitch point and the centre of rotation.

$$\frac{\omega_2}{\omega_1} = \frac{O_1P}{O_2P} = \text{Constant}$$

- If for any point of contact, the pitch point (point of intersection between common normal drawn at point of contact and the line joining centres of rotation) remains same, the ratio O_1P/O_2P will be constant or the ratio of angular velocities will be same. This is the law of gearing.

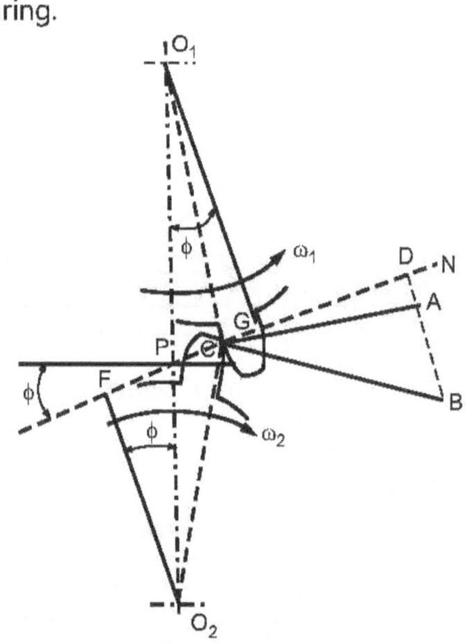

Fig. 3.2: Law of Gearing

3.4 GEAR TOOTH TERMINOLOGY

Following are the important terms to be used in this unit (Refer Fig. 3.3).

- **Pitch Circle:** It is the imaginary circle which by pure rolling action would give the same motion as that of actual gear. Diameter of this circle is known as pitch circle diameter.

  This is the most important term which defines the profile of a gear tooth.
- **Pitch Point:** It is point of tangency between two pitch circles of meshing gears.
- **Pitch Surface:** The rolling surface of two discs which has the diameter equal to pitch circle.

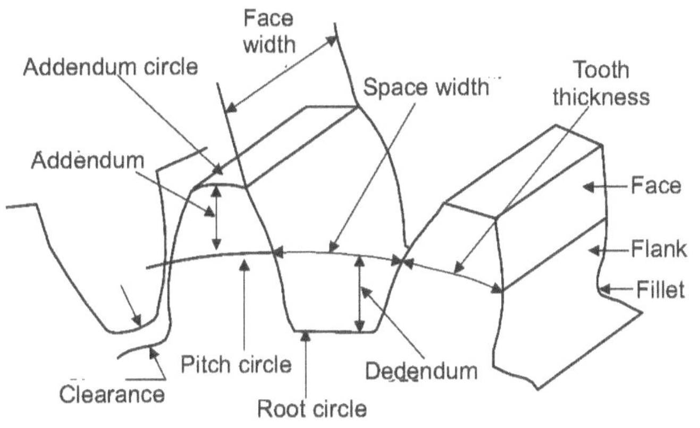

Fig. 3.3: Gear Tooth Terminology

- **Root Circle:** It is the circle concentric with pitch circle above which the tooth profile starts. Base circle is the circle above which involute tooth profile starts. Hence, the root circle may or may not be same as the base circle. In case of involute profiles, the root and base circle are different and for cycloidal profiles they are same.
- **Pressure Angle (ϕ):** It is the angle between common normal drawn at point of contact and common tangent to pitch circles. This angle determines the torque transmission.
- **Face:** The portion of gear tooth above the pitch surface is called as face of a gear tooth.
- **Flank:** The portion of the gear tooth below the pitch surface is called as flank.
- **Addendum (h_a):** The radial height of the gear tooth above the pitch circle is called as addendum, and circle (concentric with pitch circle) passing through it is called addendum circle. It is the largest circle of the gear.
- **Dedendum (h_f):** The radial depth of the gear tooth below pitch circle is called as Dedendum and the circle (concentric with pitch circle) passing through it is called dedendum circle. It is the smallest circle passing through the gear tooth profile.
- **Circular Pitch (p_c):** It is the distance between the similar points of the successive teeth measured along the pitch circle. Mathematically, it is ratio of circumference of the pitch circle and the number of teeth on the gear. $p_c = \frac{\pi d}{z}$ where d is the diameter of the pitch circle and Z is the number of teeth.
- **Diametral Pitch (p_d):** It is the ratio of the number of teeth to the pitch circle diameter.
- **Module (m):** It is the ratio of pitch circle diameter to the number of teeth. It is reciprocal of diametral pitch. This is the most important term as it decides the proportions of gear tooth profile and dimensions of gears. In the gear design once the module is known all other major dimensions can be known. Hence once the module is estimated by calculations, the list of standard modules is referred to finalize the value. Closest value of the calculated module will be chosen from the list. The standard modules list gives the information of available cutters. There are two series of standard modules namely module of first choice and second choice. As far as possible the module of first choice shall be chosen. Refer Table 3.1.
- **Clearance (C):** It is the radial distance between the top of tooth to the bottom of tooth, in a meshing gear.
- **Total Depth:** It is sum of the addendum and dedendum.
- **Working Depth:** It is the sum of addendum of two meshing teeth.
- **Tooth Thickness:** It is the width of the gear tooth measured along pitch circle.
- **Tooth Space:** It is the width of space between two successive teeth measured along the pitch circle.
- **Backlash:** It is the difference between the tooth space and tooth thickness measured along the pitch circle. Small backlash is desirable as it avoids jamming by accommodating the inaccuracies and also compensates for thermal expansion of the profile. The backlash may be obtained, either by setting the tool deeper into the gear blank or slight increase in the centre distance.
- **Face Width (b):** It is the width of gear tooth measured parallel to the axis of rotation.
- **Fillet Radius:** It is the radius connecting the tooth profile with root or base circle.

Table 3.1: List of Standard Modules

Module of First Choice (mm)	1, 1.25, 1.5, 2, 2.5, 3, 4, 5, 6 , 8, 10, 12, 16, 20, 25, 32, 40, 50
Module of Second Choice (mm)	1.125, 1.375, 1.75, 2.25, 2.75, 3.5, 4.5, 5.5, 7, 9, 11, 14, 18, 22, 28, 36, 45

3.5 STANDARD SYSTEM OF GEAR TOOTH PROFILE

The profiles which satisfy this condition are called conjugate profiles. For the standardization two profiles are generally used namely:

1. Involute profiles.
2. Cycloidal profiles.

The comparison of these two profiles is given below:

Table 3.2 : Comparison between Involute Profiles and Cycloidal Profiles

Involute Profiles	Cycloidal Profiles
It is defined as the locus of the end of a thread which is tightly wound on a cylinder and released. It gives a single curve known as involute.	It is a combination of two curves obtained by rolling a circle inside and outside of a pitch circle for one complete revolution without slipping. The two opposite curvatures obtained this way are known as hypo-cycloid and picycloids respectively.
Path of contact is a straight line and lies along the common normal	Path of contact is same as the arc of contact and it lies along the rolling circle.
The pressure angle remains constant as all points of contact lie along same common normal.	Common normal changes the inclination with the common tangent to the pitch circles for different points of contact.
For little variation in centre distance the velocity ratio is unaffected.	Exact centre distance has to be maintained else the velocity ratio will differ.
Easy to manufacture and inspect.	Difficult to manufacture.
As pressure angle remains constant, they are suitable for power transmission.	As the pressure angle varies as the point of contact, used for motion transmission.
Interference occurs .	No interference as the exactly opposite curvatures makes contact.
Due to interference there are limitations for minimum number of teeth.	No such limitations.
Weaker due to radial flank.	Stronger because of wider base.

3.6 STANDARD SYSTEMS OF GEAR TOOTH

There are four different standard gear tooth systems. These systems define the profile of the rack cutter (Refer Fig. 3.4) which is to be used for manufacturing the gears on large scale.

1. 14.5° composite system (consists of two curvatures one for cutting involute portion and other curve for cutting fillet).
2. 14.5° full depth involute system.
3. 20° full depth involute system.
4. 20° stub involute system.

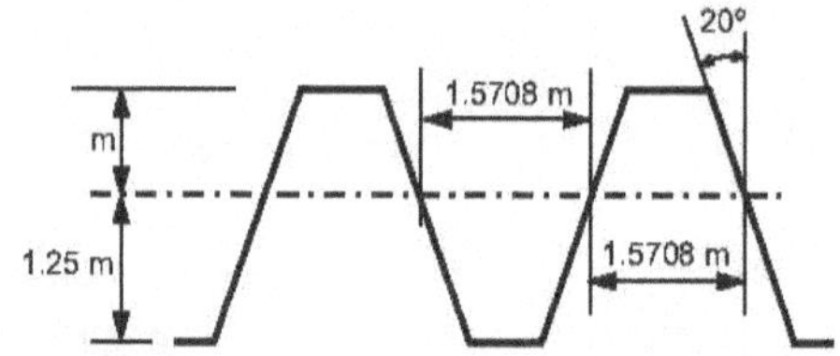

Fig. 3.4 (a): 20° Full Depth Involute Rack

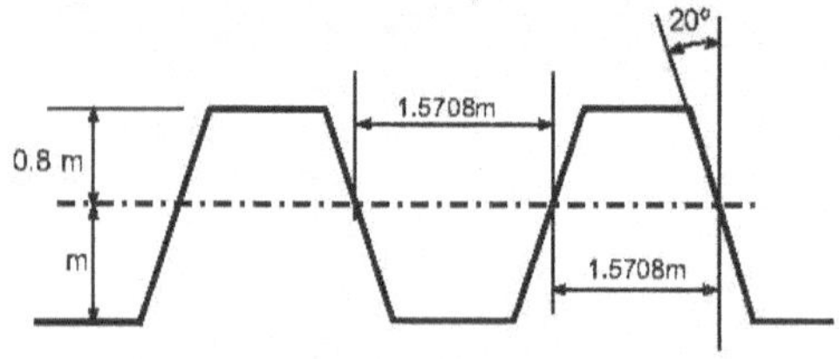

Fig. 3.4 (b): 20° Stub Involute Rack

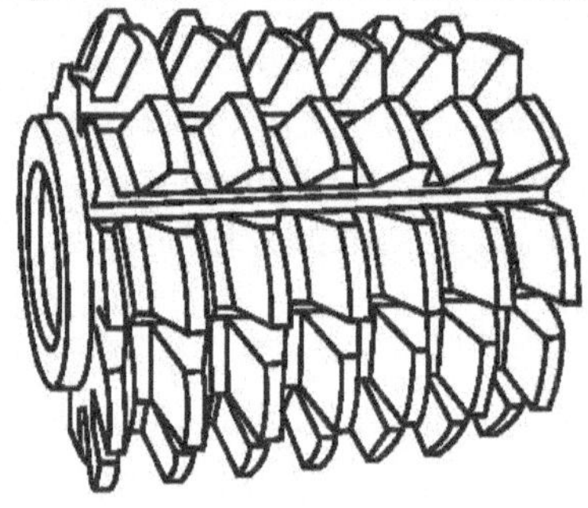

Fig. 3.4 (c): Hob cutter

The gears manufactured using the standard cutters are known as standard gears and have proportions of the profile in terms of module as given in Table 3.3.

Table 3.3: Proportions of Geometry for Various Standard Gear Tooth Systems

Modules	14.5° Full Depth Involute System	20° Full Depth Involute System	20° Stub Involute System
Addendum	1 m	1 m	0.8 m
Dedendum	1.157 m	1.25 m	1 m
Working depth	2 m	2 m	1.60 m
Minimum total depth	2.157 m	2.25 m	1.8 m
Tooth thickness	1.5708 m	1.5708 m	1.5708 m
Minimum clearance	0.25 m	0.25 m	0.2 m
Fillet radius at root	0.4 m	0.4 m	0.4 m

Generally, 20° full depth involute system is preferred as it gives wider base in turn more stronger teeth. Also it helps in reduction of interference and longer path of contact.

3.7 SPUR GEARS - NUMBER OF TEETH AND FACE WIDTH

3.7.1 Number of Teeth

- In the design of gears, the important initial step is choosing number of teeth. As the number of teeth decides the operating speed and the torques on various shafts in the gear train or gear box, it becomes very much important to choose the appropriate number of teeth to satisfy the functional requirement of the gear train or gear box under design. Minimum number of teeth required to avoid interference for a standard full depth involute system is given by $Z_{min} = \frac{2}{\sin^2 \phi}$

  where, ϕ is the pressure angle.

- Although, the primary step for selecting the number of teeth is based on the functional requirement, the second step which would be applied after this is inclusion of hunting tooth in the number of teeth finalized in first step.

Hunting Tooth:

- Consider a gear pair in mesh, in which pinion has 'x' number of teeth and gear has '2x'. It means that for one complete revolution of gear, the pinion will rotate twice and same pair of teeth would mesh with each other always.
- If the number of teeth on gear is increased to '2x + 1', different teeth of pinion and gears would mesh in successive rotations and the uniform wear of all teeth will take place.

3.7.2 Face Width

- In order to obtain uniform distribution of tangential tooth load over the face width of the gear, it is important to maintain optimum face width. If the face width is more the force distribution will not be uniform as there will be more load in the middle of the face width compared to the ends and if the face width is very small, more wear would take place and the shock absorption capacity will be reduced.
- The face width is expressed in terms of module and usually face width is taken in the range 8 m to 12 m.

$$8\text{ m} < b < 12\text{ m}$$

3.8 DESIRABLE PROPERTIES AND SELECTION OF GEAR MATERIALS

Following properties should be possessed by the gear materials:

- The gear material should have more yield strength in order to withstand heavy static and dynamic loading along with softer core to absorb the shocks.
- The coefficient of friction should be less to avoid wear and heat generation.
- The coefficient of thermal expansion should be more.
- It should be heat treatable as case hardening is required in order to avoid pitting failure.
- It should have better machinality as this would decide the precise profile cutting.
- It should possess fairly good corrosion resistance.
- For selection of gear materials please refer Table 3.3.

3.9 CONSTRUCTIONAL DETAILS OF GEAR WHEEL

The gear construction is done on the basis of size and material of the gear. Based on size and method of manufacture the following gear constructions are usually used.

1. **Integral Gear:** The gear is made integral with the shaft if the root circle of the gear is slightly greater than the diameter of shaft with a difference in root circle diameter and the shaft diameter almost equal to the radius of shaft [Fig. 3.5 (a)]. In this case, key is not required. If gear fails, the shaft also needs to be replaced.

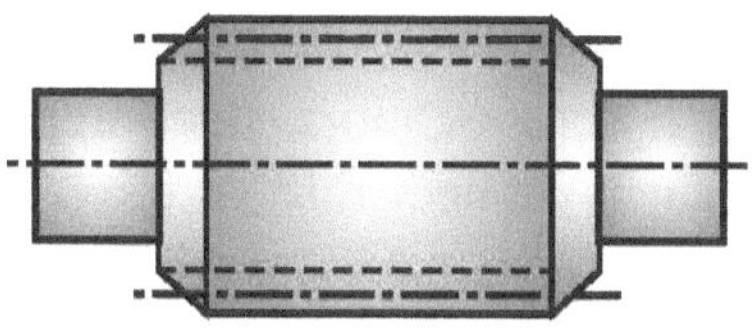

(a) Integral Gear

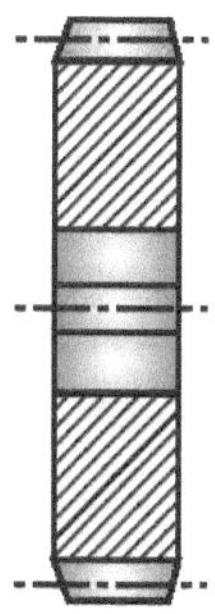

(b) Plate or Disc Gear

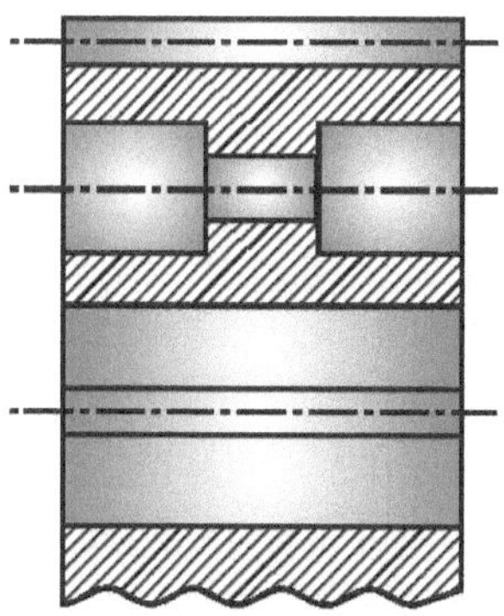

(c) Forged Gear

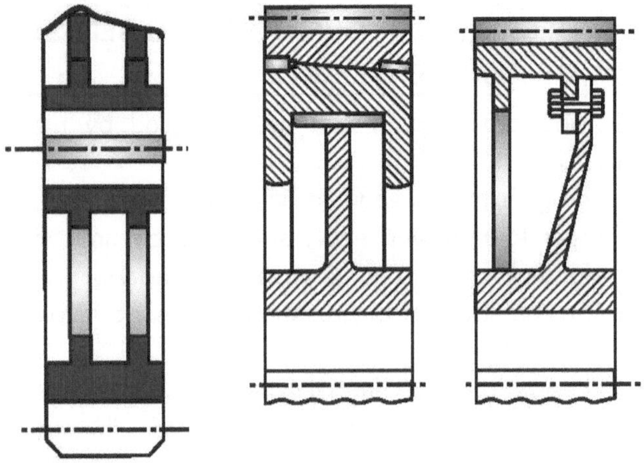

(d) Cast Gears with Two Webs (e) Rimmed Gears

Fig. 3.5

2. **Plate or Disc Type:** If the gear blank radius is up to 75 mm gears are machined on rolled bars. The details of the construction are shown in Fig. 3.5 (b).
3. **Forged Gear:** The gears with a radius of gear blank ranging from 75 mm to 200 mm are forged. The details of the construction are given in the Fig. 3.5 (c). The forged gear is strong and light in weight and low cost in case of mass scale manufacturing.
4. **Cast Gear:** There are two different construction methods for the gears of this type. The cast gears will be solid cast with a web if a gear blank diameter is up to 900 mm and will be a rimmed gear with two webs will be preferred if the blank diameter is more than 1000 mm. [Refer Fig. 3.5 (d)].
5. **Cast Iron Hub Connected with Steel Rim:** Rimmed construction with arms or webs connecting the rim with hub is used if the gear blank diameter is more than 1000 mm. In such construction a steel ring on which gear teeth are cut is connected to cast iron rim either by using bolts or grub screws. The thickness of the rim is decided considering the cutting of teeth. [Refer Fig. 3.5 (e)]

3.10 FORCE ANALYSIS

- When two gears mesh with each other and driving gear starts rotating, the driving force is exerted on the meshing driven gear along the line of action (common normal drawn at point of contact). Thus, the force transmission line is inclined with the line of motion transmission by an angle equal to pressure angle.
- This driving force (P_n) will be resolved along the line of motion (P_t) (i.e. common tangent to pitch circles) and along the line joining centers of rotation (P_r). The component along the line of motion is called as tangential force which is a desirable force producing the output torque and the component along the line joining centres of rotation is called as radial force which is undesirable as it tries to separate two meshing gears.

Torque supplied by the driving shaft is given by $T = \frac{60P \times 10^6}{2\pi N}$ N-mm

Torque transmitted is given by $T = P_t \times \frac{d_p}{2}$ N-mm.

where, P = Power supplied in kW

N = Speed of driver in r.p.m.

P_t = Tangential tooth force in N

d_p = Pitch circle diameter of the driver in mm

- Direction of rotation of the driver as shown is anticlockwise. The normal force P_n is transmitted along the line of action, which is further resolved into two components. Refer Fig. 3.6.

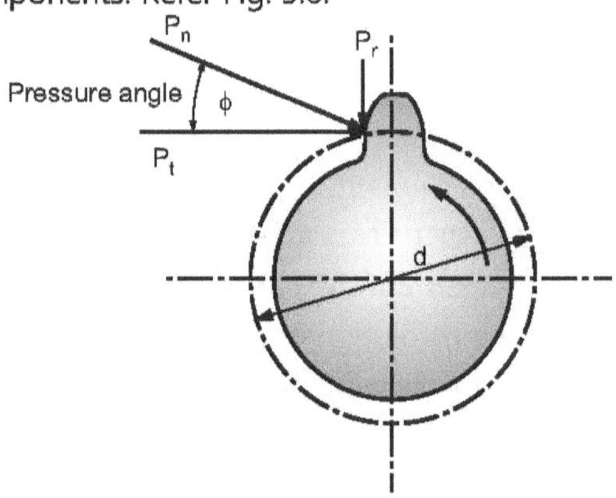

Fig. 3.6: Force Analysis

Magnitude of tangential component in terms of normal force will be,

$$P_t = P_n \cos \phi$$

and the radial tooth force is given by

$$P_r = P_n \sin \phi$$

- The tangential force P_t can be determined from the known values of power, speed and p.c.d. of driver. Knowing P_t one can find the normal and radial force magnitudes using following equations

$$P_r = P_t \tan \phi$$

and $$P_n = \frac{P_t}{\cos \phi}$$

- This is an approximate method of determining the tooth forces. The exact method is much complex and is out of scope for this book. For the above analysis the following assumptions are considered.

- Only one pair of teeth takes the total load (while in actual case more than one pair shares the normal force transferred from driver to driven).
- The magnitude of the resultant load is considered to be constant.
- Static loading conditions are considered.

3.11 BASIC MODES OF GEAR TOOTH FAILURES

Basically, the gear tooth in mesh may break in bending due to excessive loading and as the profiles are in continuous contact it may fail by destruction of surfaces in contact.

(a) Bending or Tooth Breakage Failure:

During meshing, the gear tooth is subjected to static and dynamic loadings. The normal force is resolved in different directions and if the dominant force component (tangential force) is considered, the tooth acts as a cantilever beam which starts bending due to this tangential force, and for a certain load it may break. In order to avoid this kind of failure, one may estimate the module that avoids this failure or select a higher grade material.

(b) Surface Destruction:

There are various ways to cause the destruction of the profile surfaces as described below.

1. **Abrasive Wear:** In a gear box, lubricating oil is required in order to limit the temperature of gear box in desired range and reduce the friction. The foreign particles, wear debris or dirt present in the lubricant oil, comes in contact with the interacting profiles (as the oil is splashed) which destructs the interacting profiles. To avoid this kind of failure, oil filter should be used at the inlet port of the lubricating oil. Also the oil should be drained at a regular frequency such that the wear debris are driven out and then fresh oil should be poured.
2. **Corrosive Wear:** This type of wear is characterized by chemical action of ingredients in the lubricant like high pressure additives. Presence of moisture, foreign materials in the lubricant may corrode the gear tooth profile. In order to avoid this kind of failure, proper lubricant should be chosen, entry to the foreign materials should be restricted and oil be replaced at regular frequency.
3. **Initial (Corrosive) Pitting:** The high spots in the profiles due to inaccuracies get worn out progressively and load is redistributed. Small pits are formed at these high spots after the wear. This is known as initial pitting. The initial pitting can be avoided by machining the gears accurately and ensuring positional accuracy such that the gear tooth is loaded uniformly across the face width.
4. **Destructive Pitting:** This failure occurs when the stress induced in the surface exceeds the surface endurance strength. When gears are meshed, the radial load puts pressure at the interface and the stress is induced which is called as Hertz or contact stress in the interface. If this contact stress is more than the surface endurance strength, the material removal will start and small pits are formed which progressively develops a crack which continues propagating and fatigue failure occurs. Actually, gears require soft core for damping the shock loading but to improve the surface endurance strength gears need case hardening. In the subsurface region, another failure may occur called as spalling in which due to low subsurface strength, cracks are initiated and progressively develop and appear on the surface. To avoid this kind of failure the case depth of hardening should be increased. But if the case depth is increased, the gear will become more brittle. This failure can be avoided by improving the surface endurance strength by improving surface hardness.
5. **Scoring:** This failure occurs if lubrication system fails. If sufficient quantity of lubricating oil doesn't reach at the contacting interface, the temperature of the gear tooth goes on increasing. At some moment even gear teeth may get welded. Due to supplied power, these welds get broken or teeth may break. This kind of failure is most likely in case the operating speed of gears is high, or the inaccurate profiles are meshed or heavy tooth load is applied. Remedies for this failure are:

- Ensure the presence of the lubricant with sufficient quantity. Using an oil level indicator or dip stick one can ensure the sufficient quantity of lubricant and also to limit the operating speed. In some cases additional fins are required to cool the lubricant. If natural convection is not giving required temperature a fan is required additionally to cool the fins and in turn the lubricant.
- Surface finish of the gear tooth profiles should be good.
- Limit the sliding velocity.
- Apply safe load on the gear tooth.

3.12 SELECTION OF TYPES OF GEARS

The Table 3.4 gives broad guidelines for the selection of type of gear to be used for a given application.

Table 3.4 : Guidelines for Selection of Gear Type

Type	Features	Application	Remarks
Spur	Parallel shafts, High speeds and loads, Maximum efficiency.	Applicable to all types of trains and wide range of velocity ratios.	Simplest tooth elements offering maximum precision First choice recommended for all gear meshes, except where very high speed or loads occur.
Helical	Parallel shafts, Very high load and speed, Efficiency slightly less than spur.	Most applicable for very high speeds and high loads, Also applicable where spur is used.	Equivalent quality to spur gears except the complication of helix angle. Recommended for all very high speed and load cases. Only the axial thrust should be accommodated.
Crossed helical	Skew shafts, Point contact, High sliding, low speeds, Light loads.	Relatively low velocity ratio, low speeds and light loads only, Any angle skew shafts	To be avoided for precision meshes, Point contact limits capacity and precision, Suitable for right angle drives with less loads, Less expensive substitute for bevel gears, Good lubrication required due to point contact and high sliding.
Internal spur	Parallel shafts, high speed, high load.	In case of high speed and high load, offers low sliding and high stress loading, good for high capacity and long life in planetary gear trains to produce large reduction ratios.	Not recommended for precision meshes, Because of design, fabrication and inspection limitations, should only be used if internal feature is necessary.
Bevel	Intersecting shafts, High speeds, high loads.	Right angle shafts, intersecting shafts, high velocity ratios, high speeds and heavy loads, Low efficiency.	Good choice for right angle ratios, particularly low ratios, however complicated tooth form and lubrication limits achievement of precision, should be located at one of the less critical meshes in the train.
Worm	Right-angles, Skew shafts, High velocity ratio, High speed and loads, low efficiency, Most designs non reversible.	High velocity ratios, High loads, Angular meshes.	Worm can be made with high precision, but worm gear has inherent limitations. To be considered for average precision cases but can be of high precision with care. Best choice for combination of right angle with higher velocity ratios. High sliding requires excellent lubrication.

3.12.1 Selection of Materials For Gears

Table 3.4 gives comparison of ferrous materials used for the gears based on strength and hardness.

Table 3.5: Gear Materials

Material	Minimum Tensile Strength, N/mm²	IS Designations	BHN
Malleable Cast iron	280-320	WM280, BM320	217 core
Cast iron	200- 350	FG200	300 core
Cast steel	550	CS550	145 core
Plain carbon steel	500-800	50C4, 45C8, 50C8	143-248 core
Alloy steel	550-1550	55Cr70, 40Cr1, 27Mn2, 37Mn2, 45Mn2 Mo28, 35Mn2 Mo45, 40 Cr1 Mo28, 40Cr1 Mo60, 40Ni3, 30 Ni4 Cr1	200-500 core

3.13 ADDENDUM MODIFICATION OF GEARS

- In case of gears with involute teeth, interference will occur when the number of teeth selected is less than minimum number of teeth required to avoid interference. In this phenomenon, the addendum part of the gear tooth will make contact with non-involute part of other meshing tooth in the region between the base circle and the dedendum circle. To avoid this contact, the addendum of one of the gears is reduced or modified.
- Practically, while manufacturing the gear, the rack cutter will be shifted by some distance away from the gear blank and for cutting the other meshing gear the rack cutter is shifted towards the centre of the gear blank by same distance. This process is also known as rack shift (Fig 3.7). Variation in addendum modification coefficient is shown in Fig. 3.7 as 'x'.

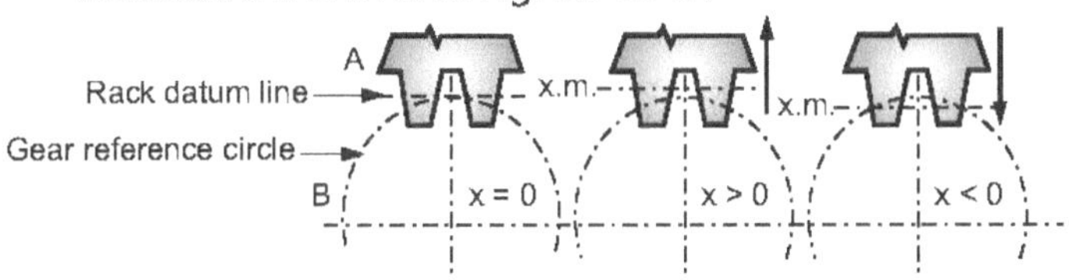

Fig. 3.7: Rack Shift Method

3.14 DESIGN CONSIDERATIONS

For design of gears the following important points should be considered:

- The gear tooth should be strong to withstand static and dynamic loading.
- The gear tooth must resist deflection, accelerations and stress concentration.
- The tooth should have good wear resistance ensuring long life.
- Economy of material and space constraint shall be considered.
- Interference shall be avoided using appropriate length of path of contact.
- Perfect alignment and elastic deformation of gears, shafts, supporting members shall be considered.
- Appropriate lubrication system to ensure better performance.

3.15 BEAM STRENGTH (LEWIS EQUATION)

- In 1892 Wilfred Lewis developed an equation stating the bending strength of a gear tooth. Beam strength of a gear tooth is defined as the maximum tangential load that tooth can sustain without bending or tooth breakage. The analysis done by Lewis was based on the following assumptions.
  - ➢ The tangential force is uniformly distributed over the face width of the gear.
  - ➢ The effect of separating force (radial force) is neglected.
  - ➢ Stress concentration effect is neglected.
  - ➢ Only one pair of teeth is in contact which takes the total load at any time.
- In the assumptions stated above the tangential load is considered to be uniformly distributed over the face width of the gear. The face width is expressed in terms of module and usually face width is taken in the range 8 m to 12 m.
- Refer Fig. 3.8, which shows a gear tooth profile and an uniform strength parabola inscribed within the tooth profile. Except at base, the gear tooth profile is stronger than the parabola. At section 'AA' the profile and parabola have same strengths. Hence at this section,

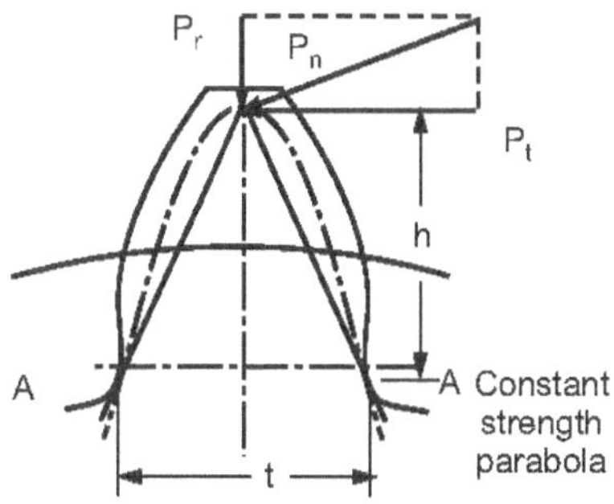

Fig. 3.8: Gear Tooth as Cantilever

Bending moment, $M = P_t \times h$

Moment of inertia of the section at AA, $I = bt^3/12$

and $Y = t/2$

Hence, bending stress at section AA is given by

$$\sigma_b = \frac{M}{\left(\frac{I}{Y}\right)} = \frac{P_t h \times 12 \times \left(\frac{t}{2}\right)}{bt^3} = P_t\left(\frac{6h}{bt^2}\right) \quad \ldots (3.1)$$

Or $$P_t = mb\sigma_b\left(\frac{t^2}{6hm}\right)$$

… (Multiplying and dividing by module m)

Let, $$Y = \left(\frac{t^2}{6hm}\right)$$

$$P_t = \sigma_b \cdot b \cdot Y \cdot m$$

where, P_t = Tangential tooth load in N

m = Module of the gear in mm

b = Face width in mm (to be chosen between 8 m < b < 12 m)

σ_b = Permissible bending stress in N/mm^2 (To be taken as $S_{ut}/3$)

t = Tooth thickness at root in mm

h = Height of gear tooth which is length of the cantilever in mm and

Y = Lewis form factor

- This equation gives a relationship between tangential tooth load and bending stress. If the tangential tooth load is varied, the induced bending stress also varies. Beam strength is the ultimate or maximum value of tangential load a gear tooth can sustain without breaking. Hence to get the beam strength of a given gear tooth the permissible value of the bending stress should be used. The tangential load thus obtained using permissible bending stress is considered as beam strength and will be denoted by S_b hence fourth.

$$S_b = \sigma_b \cdot b \cdot Y \cdot m \qquad \ldots (3.2)$$

- The permissible value of bending stress for any gear material is taken as one third of ultimate tensile strength. This is done to consider the endurance strength of a gear tooth, as gear tooth is not subjected to static loading. The bending stress on an individual gear tooth is varying. Hence to find the corrected endurance strength considering all the factors such as size factor, shape factor, reliability factor, fatigue stress concentration factor and the approximate relation for the endurance strength of a standard specimen, the corrected endurance strength can be obtained as one third of ultimate tensile strength.
- In the beam strength equation, Y is a Lewis form factor which can be obtained by empirical relations given below or the tables given in data books and standards may be used. In both the cases the value of Lewis form factor is based on the number of teeth. If the number of teeth is more, the Lewis form factor will be higher. Following empirical equations may be used for determining the Lewis form factor.

For 14.5° full depth involute and composite systems,

$$Y = 0.389 - \frac{2.148}{Z}$$

For 20° full depth involute systems,

$$Y = 0.484 - \frac{2.87}{Z}$$

For 20° stub involute systems,

$$Y = 0.55 - \frac{2.64}{Z}$$

3.16 VELOCITY FACTOR

The tangential load is a static load. In addition to this, because of the following reasons, dynamic load is induced between the meshing teeth.

- In accurate tooth profile
- Uneven tooth space
- Bearing misalignment
- Inertia of moving parts.

This dynamic load can be estimated by either approximate or exact analysis. While designing a gear pair in the initial stages it is recommended to use approximate method and once the module is known the exact method can be used to check the design.

The approximate analysis was developed by Barth and he came out with empirical relations to find the velocity factor (also known as Barth factor) to consider the effect of dynamic load. This factor is used to find the magnitude of the maximum tangential load for design of gears.

- For ordinary and commercially cut gears made with form cutters and pitch line velocity less than 10 m/s.

$$C_v = \frac{3}{3 + v}$$

- For generated teeth with pitch line velocity less than 20 m/s.

$$C_v = \frac{6}{6 + v}$$

- For precise teeth shaved, ground and lapped and pitch line velocity more than 20 m/s.

$$C_v = \frac{5.6}{5.6 + \sqrt{v}}$$

In all the above relations, v is the pitch line velocity in m/s units.

3.17 SERVICE FACTOR

- It is a load modifying factor used to find the maximum tangential load. While the forces on meshing gear teeth are analyzed, the effect of inertia and other resisting forces for acceleration or in the instant when the gears start rotating from rest condition are not considered. Hence to incorporate all such conditions in the analysis a simple way is to modify the magnitude of tangential load obtained using the mean torque. Hence, the service factor may be given by ratio of maximum tangential load actually applied to the tangential load obtained using mean torque value.
- The service factor depends on type of prime mover and the type of driven machinery and can be obtained

using the recommendations given in the Table 3.6. The service factor for a same application with different prime movers is different. (e.g.: If prime mover is a motor, the service factor would be smaller and for same driven machine if prime mover is I.C. Engine the service factor is higher.) If the type of prime mover is same and the driven machine is different, the service factor will be different.

- Depending on whether an electric motor starts on no load or full load, the starting torque of the motor is higher compared to its rated torque. This ratio also represents service factor.

  The range of recommended service factor is from 1.10 to 2.25.

  Let T be the mean torque obtained from equation

$$T = \frac{60P \times 10^6}{2\pi N}$$

  The tangential load would be found using equation

$$P_t = \frac{2T}{d}$$

  The tangential load will get modified using equation $= C_s P_t$.

Table 3.6: Cs (Considering 8 hrs/day operation)

Power Source	Load on Driven Machine		
	Uniform	Moderate Shock	Heavy Shock
Uniform - Electric motor	1.00	1.25	1.75
Light shock - Multi cylinder engine	1.25	1.50	2.00
Heavy shock - Single cylinder engine	1.50	1.75	2.25

3.18 LOAD CONCENTRATION FACTOR (C_L)

- In the Lewis beam strength equation it was assumed that the tangential tooth load is uniformly distributed over the face width. But practically because of misalignments in gears, bearings or errors, elastic deformation of gears or supporting elements like bearings, housings or casing the tangential tooth load gets concentrated at any position along the face width.
- To consider this effect load concentration factor is used to modify the tangential tooth load. The value of the load concentration factor can be chosen from the Table 3.5. The range of this factor varies from 1.3 to 2.2. The lower value is recommended in the precise mounting and the higher value corresponds to average accuracy in mounting the bearings.

3.19 EFFECTIVE LOAD ON GEAR (P_{eff})

For the design of the spur gear based on the beam strength, it is already discussed that the tangential load will be considered. Further the modifying factors for this tangential load were discussed. Considering all these modifying factors the effective load is obtained. It is assumed that the gear tooth is subjected to this effective load and the module will be obtained.

$$P_{eff} = \frac{C_s C_L P_t}{C_v}$$

[To be used in the initial stages of the design] ... (3.3)

$$P_{eff} = C_s C_L P_t + P_d$$

[To be used after knowing the module] ... (3.4)

Here, P_{eff} is the effective tooth load in N.

C_s is the service factor

C_L is the load concentration factor

C_v is the velocity factor

P_t is the tangential tooth load in N

P_d is the dynamic tooth load obtained by Buckingham's equation in N

Buckingham's Incremental Dynamic Tooth Load Equation

- The effective load obtained using velocity factor is only an approximate estimate of the dynamic load. In practice it was observed that the dynamic load depends greatly on accuracy with which the gears are cut and gears are mounted. Elaborate experimentation led Buckingham to arrive at an equation which can estimate the incremental dynamic load. The incremental load is to be added to the maximum tangential load to obtain effective load.

$$P_d = \frac{21v\,(C \cdot b + C_s C_L P_t)}{21v + \sqrt{(C \cdot b + C_s C_L P_t)}} \quad \text{... (3.5)}$$

This equation is also written as

$$P_d = \frac{21v\,(C \cdot b + P_{tmax})}{21v + \sqrt{(C_b + P_{tmax})}}$$

where, $P_{tmax} = C_s \cdot C_L \cdot P_t$

v = Pitch line velocity in m/s

b = The face width of gear in mm

C = Deformation factor in N/mm

$$C = e \cdot \frac{K}{\left[\frac{1}{E_p} + \frac{1}{E_g}\right]}$$

K = Factor depending upon tooth form

K = 0.107 for 14.5° full depth involute and composite systems.
= 0.111 for 20° full depth involute teeth.
= 0.115 for 20° stub involute systems.

E_p, E_g are the moduli of elasticity for pinon and gear materials respectively in N/mm^2.

e is the total tooth error in the pinion and the gear, mm.

With standard values of E_p, E_g for steel materials = 2.07 × 10^5 N/mm^2 and K for 20° pressure angle

C = 11500*e (N/mm) for Steel Pinion with Steel gear
C = 10000*e (N/mm) for steel Pinion on Cast Iron Gear
C = 8900*e (N/mm) for Cast Iron Pinion with Cast Iron gear

- To find the tooth error, standards give one to twelve grades of the accuracy levels of gear cutting obtained by considering various manufacturing processes and the error equations related to them where e = 16.0 + 1.25ϕ, and ϕ is called tolerance factor and to be found using equation, $\phi = m + 0.25\sqrt{d}$. Refer Table 3.7 for error equations for various grades.

  E.g. For a gear of grade 8, to find the error in pinion in microns, use $e_p = 16.0 + 1.25\,(m + 0.25\sqrt{d_p})$ with values of d_p and module m for the pinion. For error on gear use d_g in place of d_p. (Values of d_p or d_g should be in mm)
- To find the total error, add the tooth error of pinion and gear. These equations give value of error in microns to be converted in mm.

  Total tooth error: $e = e_p + e_g$

Table 3.7: Tooth Profile Error

Grade	Error in Tooth Profile (in micron)
1.	e = 0.80 + 0.06 ϕ
2.	e = 1.25 + 0.10 ϕ
3.	e = 2.00 + 0.16 ϕ
4.	e = 3.20 + 0.25 ϕ
5.	e = 5.00 + 0.40 ϕ
6.	e = 8.00 + 0.63 ϕ
7.	e = 11.0 + 0.90 ϕ
8.	e = 16.0 + 1.25 ϕ
9.	e = 22.0 + 1.80 ϕ
10.	e = 32.00 + 2.50 ϕ
11.	e = 45 + 3.55 ϕ
12.	e = 63.00 + 5.00 ϕ

3.20 WEAR STRENGTH (BUCKINGHAM'S EQUATION)

- Gear tooth fails in pitting i.e. surface destruction if the induced contact stress at the surface exceeds the surface endurance strength of the material. To avoid this failure, the size and the surface hardness should be so chosen that the gear becomes capable of taking the effective tooth load. This theory was invented by Buckingham.
- Refer Fig. 3.9 (a) in which two cylinders of diameters d_1 and d_2 and unit length (length = l) are making contact with each other. These cylinders are pressed against each other. The contact or Hertz stress is induced at the interface. It is given by

$$\sigma_c = \frac{2P}{\pi h l} \quad \ldots (3.6)$$

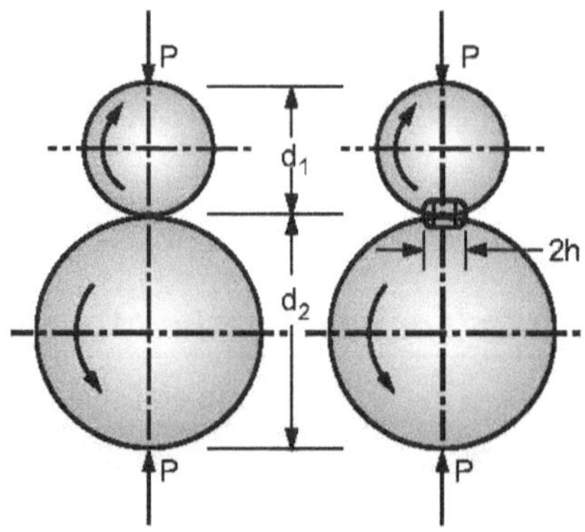

Fig. 3.9 (a): Cylinders Rolling on Each Other

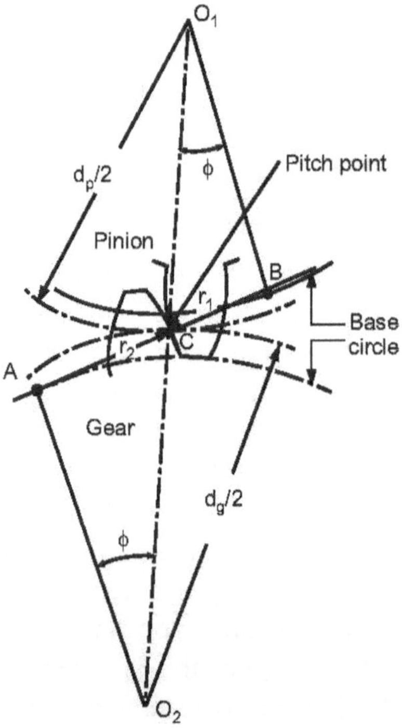

Fig. 3.9 (b): Gears in Mesh

Here,
$$h = \left[\frac{2P(1-\nu)^2\left[\frac{1}{E_1}+\frac{1}{E_2}\right]}{\pi l\left(\frac{1}{d_1}+\frac{1}{d_2}\right)}\right]^{\frac{1}{2}}$$

Where, σ_c = The maximum compressive stress

P = Compressive force applied to press the cylinders

h = Half width of deformation

l = Axial length of cylinder

d_1, d_2 = Diameters of the cylinders

E_1, E_2 = Modulii of elasticity for cylinder materials

υ = Poission's ratio.

The effect of compressive force P is to deform the cylinder in approximate elliptical shape with 2h as width and l as length. Substituting h in σ_c equation and squaring both sides,

$$\sigma_c^2 = \frac{1}{\pi(1-\upsilon^2)}\left(\frac{P}{l}\right)\left[\frac{\left(\frac{1}{r_1}+\frac{1}{r_2}\right)}{\left(\left[\frac{1}{E_1}+\frac{1}{E_2}\right]\right)}\right] \quad \ldots (3.7)$$

Where, r_1 and r_2 are the radii of the two cylinders rolling with a radial pressure. Considering Poisson's ratio as 0.3,

$$\sigma_c^2 = 0.35\left(\frac{P}{l}\right)\left[\frac{\left(\frac{1}{r_1}+\frac{1}{r_2}\right)}{\left(\left[\frac{1}{E_1}+\frac{1}{E_2}\right]\right)}\right]$$

The above equation gives the relation for the maximum compressive stress or Hertz stress for the cylinders rolling with radial pressure.

When only one pair of teeth takes the load, contact takes place at pitch point. Refer Fig. 3.10 (b). The analysis presented here is considering the contact takes place at the pitch point. Hence, the radius of curvature of the gear tooth is selected at pitch point. Also the dynamic load is imposed on the tooth near pitch line area and the wear on gear tooth usually occurs near the pitch line.

If same theory is to be applied to the gears, following substitutions shall be made.

$$r_1 = O_1C \cdot \sin\phi = \frac{d_p \sin\phi}{2} \text{ and } r_2 = O_2C \cdot \sin\phi = \frac{d_g \sin\phi}{2}$$

which are the radii of curvature of the tooth profiles. Length of cylinder would be replaced by the face width l = b.

$$\left(\frac{1}{r_r}+\frac{1}{r_2}\right) = \frac{2}{\sin\phi}\left[\frac{1}{d_p}+\frac{1}{d_2}\right] \quad \ldots (3.8)$$

If Q is considered as ratio factor given by $\frac{2Z_g}{Z_p+Z_g}$

Or
$$Q = \frac{2d_g}{d_p+d_g} \ldots \text{ [as d = mZ]}$$

$$\left(\frac{1}{d_p}+\frac{1}{d_g}\right) = \frac{d_p+d_g}{d_p d_g} = \frac{2}{Qd_p} \quad \ldots (3.9)$$

From equations (a) and (b),

$$\left(\frac{1}{r_1}+\frac{1}{r_2}\right) = \frac{4}{Qd_p \sin\phi}$$

The force along the pitch line is P_n which is $P_n = \frac{P_t}{\cos\phi}$.

Substituting all these values in the contact stress equation (3.7),

$$\sigma_c^2 = \frac{1.4P_t}{bQd_p \sin\phi\cos\phi\left(\frac{1}{E_p}+\frac{1}{E_g}\right)} \quad \ldots (3.10)$$

Defining a load stress factor K as,

$$K = \frac{\sigma_c^2 \sin\phi\cos\phi\left(\frac{1}{E_p}+\frac{1}{E_g}\right)}{1.4}$$

$$P_t = bQd_pK$$

This equation gives the relation between the contact stress and the applied tangential load. If load increases the induced contact stress also increases. If the induced stress exceeds the endurance strength of gear surface, pitting failure occurs. Hence in order to avoid this failure, the maximum safe tangential force known as wear strength (S_w) can be estimated based on the permissible values of the endurance limit.

$$\therefore \quad S_w = bQd_pK \quad \ldots (3.11)$$

Load Stress Factor K can be Calculated as follows:

For 20° pressure angle, steel as material for gears, the modulus of elasticity is 207000 N/mm^2 , and the compressive stress according to Gustav Niemen,

$$\sigma_c = 0.27 \text{ BHN kgf/mm}^2$$
$$= (0.27 \times 9.81) \text{ BHN N/mm}^2$$

Considering the material of pinion and gear as steel, and substituting the value of contact stress, the load stress factor will be approximately

$$K = 0.16\left(\frac{BHN}{100}\right)^2$$

... (for pinion and gear made of steel)

$$K = 0.18\left(\frac{BHN}{100}\right)^2$$

... (for steel pinion and cast iron gear)

$$K = 0.21\left(\frac{BHN}{100}\right)^2$$

... (for pinion and gear made of cast iron)

With the following substitutions wear strength of a gear tooth in terms of module is estimated.

- b, the face width taken between 8 m to 10 m.
- Q, the ratio factor is evaluated by equation $\frac{2Z_g}{Z_p + Z_g}$
- d_p, the pitch circle diameter can be taken as $m \times Z_p$
- K, the load stress factor $K = 0.16\left(\frac{BHN}{100}\right)^2$

3.21 GEAR DESIGN

Gear designer is mainly confronted with two situations,

1. In the first situation, he is required to find the required module and thus design the gear pair for a given load and required speed.
2. In the second situation, for a given gear drive designer is required to find factor of safety for transmitting certain power. He may face a situation where a given gear pair is reporting failure and is interested in knowing the theoretical factor of safety for transmitting that power.

The following procedure may be adopted for the estimation of the module.

Estimate Load in Terms of Module:

- Find pitch line velocity $v = \frac{\pi \cdot d \cdot N_p}{60 \times 1000}$ ... m/sec. where ($d = Z_p \times m$) is in mm, and Z_p is known.
- Find tangential load, $P_t = \frac{\text{Power (kW)} \times 1000}{v}$ ... (N) in terms of module.
- Find velocity factor C_v using appropriate relationship in terms of module.
- Knowing factor C_s, C_L and C_v find, $P_{eff} = \frac{C_s C_L P_t}{C_v}$

Estimate Beam and Wear Strength in Terms of Module:

- For checking the beam strength, initially the weak element out of pinion and gear is identified using equation $S_b = \sigma_b \cdot b \cdot Y \cdot m$. If pinion and gear materials are same, pinion will always be weaker of the two since 'Y' for smaller value of Z will be smaller and other quantities are same for both. For different materials for pinion and gear, first calculate the product of permissible bending stress and the Lewis form factor $(\sigma_b Y)_{pinion}$, $(\sigma_b Y)_{gear}$ for pinion as well as gear. The weaker element will have this product smaller of the two. Considering b = 10 m, beam strength will be calculated in terms of module which is unknown at this stage. Smaller S_b controls the design for beam strength.
- Wear strength (S_w) is estimated in terms of module and is estimated for a pair. The hardness estimated is applicable for both pinion and gear. However, since pinion will rotate more revolutions compared to gear, it will wear out more. Hence in practice gear is provided with the calculated value of hardness and pinion hardness may be increased by 25 to 30 BHN.
- The module for the weaker element is found out based on either Beam strength or Wear strength. Determine the module based on the smallest value of the Beam or Wear strength.
- Effective load and the strength can now be equated considering factor of safety as below.

  Smaller of (S_b) or (S_w) = $P_{eff} \times$ (FS)
- Unknown module can now be found out.

Procedure to Find Factor of Safety/Transmission Capacity for a given Gear Drive (Module Known):

- Find pitch line velocity, $v = \frac{\pi \cdot d \cdot N_p}{60 \times 1000}$ ... m/sec. where, d is in mm
- Find tangential load, $\frac{\text{Power (kW)} \times 1000}{v}$ ... N
- Knowing factor C_s and C_L find, $P_{t\,max} = C_s \cdot C_L \cdot P_t$
- Designer decides the grade of gear cutting accuracy based on the technical requirements and its cost impact. Higher accuracies will result into lesser gear cutting errors. This will reduce dynamic loads and improve tangential load capacity. This however also increases the cost. Designer has to strike a balance. Once the accuracy grade is selected, permissible tooth errors on pinion (e_p) and on gears (e_g) can be estimated. Knowing pinion/gear materials and permissible error ($e = e_p + e_g$), deformation factor C can be found out.
- Find dynamic load P_d using: $P_d = \frac{21\,v\,(C \cdot b + P_{tmax})}{21\,v + \sqrt{(C_b + P_{tmax})}}$
- Find $P_{eff} = P_{t\,max} + P_d$
- Find the beam strength S_b and Wear Strength S_w as discussed above.
- Effective load and the strength can now be equated considering factor of safety as below.

  Smaller of (S_b) or (S_w) = $P_{eff} \times$ (FS)
- The unknown either FS or $P_{t\,max}$ can be found out from above equation.

3.22 CROWNING OF GEAR TOOTH

Crowning is a method of strengthening a gear tooth in which uniform distribution of the load along the face width is ensured. When gears mesh with each other the force is transferred from driver to the driven gear along the common normal. The distribution of these forces on the gear tooth is assumed to be uniform. But due to the following reasons this distribution does not remain uniform.

- There are inaccuracies in the tooth profile due to manufacturing limitations and distortion of the profile may occur during heat treatment. These inaccuracies will redistribute the loads resulting in non-uniform loading of gear tooth.
- The elastic deformation of the shaft may also vary the load distribution.
- Also if positional accuracy of bearings and shafts is not up to the mark, the distribution of load will not be uniform.

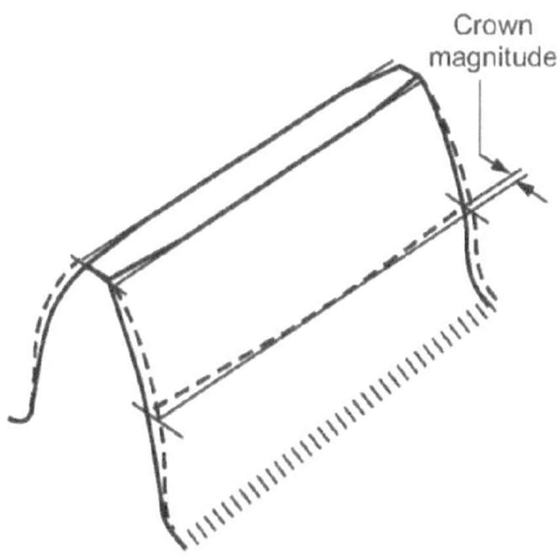

(a)

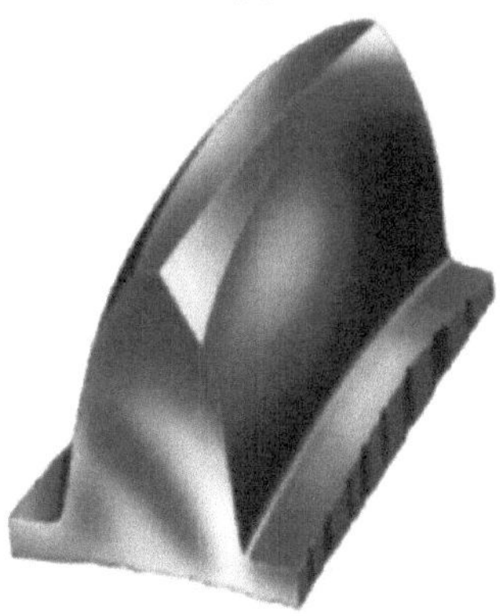

(b)

Fig. 3.10: Crowning of Tooth Profile

As a result of this, the maximum pressure gets shifted to the end of tooth along the face width. If at the end of gear tooth extra curvature is added, the maximum pressure will shift towards the central region. Practically, the crowning is done using gear shaving method. (Refer Fig. 3.10).

3.23 GEAR LUBRICATION METHODS

- Gear lubrication plays a very important role in successful working of a gear drive. Even a properly designed and mounted gear drive can fail with inadequate lubrication. Gears operating at lower pitch line velocities (in case of hand driven mechanisms) are grease lubricated. For medium range of pitch line velocities the gears are placed in an enclosure and dipped in oil. Rotation of the gears induces a splash which will reach the contacting teeth and bearings. The lubrication method in this case is splash lubrication. The type of lubricant used can be straight mineral oil or extreme-pressure lubricants.
- In some cases to improve the heat transfer a jet of oil is sprayed on the contacting surfaces using an oil pump. In both of these cases a special care has to be taken in order to avoid leakage of oil out of the gear box. Gasket at the mating parts of the gear box casing, oil seals on the shafts which are extruding out of the casing, oil inlet and drain plugs and oil level indicators have to be provided. Also an air breather which ensures the atmospheric pressure inside the casing is required.

SOLVED EXAMPLES

Example 3.1: *Design a spur gear pair with 20° involute teeth based on beam strength. The velocity factor is to be used to account for dynamic load. The pinion shaft is connected to a 12 kW, 1440 rpm motor. The reduction ratio is 4 : 1 and pinion has 18 teeth. Pinion as well as gear are made of 40 C8 (S_{ult} = 610 MPa). The factor of safety and service factors both can be taken as 1.5.*

Specify the dimensions of gears and suggest suitable hardness for gear.

Given data: Input power = 12 kW, N_p = 1440 rpm, $\frac{Z_g}{Z_p} = 4$, $Z_p = 18$, $(S_{ult}) = 610\ N/mm^2$, C_s = FS = 1.5.

Solution: Effective load (Based on P_t, C_s, C_v). We first find tangential load on the pinion.

Torque on Pinion:

$$T_p = \frac{60 \times P}{2\pi \cdot N} \times 10^6\ N{\cdot}mm$$

$$= \frac{60 \times 12}{2\pi \times 1440} \times 10^6 = 79577.5\ N{\cdot}mm$$

Tangential Load P_t:

$$P_t = \frac{2 \cdot T}{m \cdot Z_p} = \frac{2 \times 79577.5}{m \times 18} = \frac{8842}{m}$$

Velocity Factor C_v: (Assuming velocity to be less than 10 m/sec.)

$$C_v = \frac{3}{3 + v}$$

where, $$v = \frac{\pi\, d_p \cdot N_p}{60 \times 1000} = \frac{\pi \times 18 \times m \times 1440}{60 \times 1000}$$

$$= 1.36\ m$$

$$C_v = \frac{3}{3 + 1.36\ m}$$

Effective Load, P_{eff}:

$$P_{eff} = \frac{C_s}{C_v} \cdot P_t = \frac{1.5}{\left[\frac{3}{(3 + 1.36\ m)}\right]} \times \frac{8842}{m}$$

$$P_{eff} = \frac{13263\ (1 + 0.453\ m)}{m}$$

Beam Strength:

Since pinion and gear are made from same material, pinion is weaker. Hence, here, we find beam strength for pinion.

$$S_b = \sigma_b \cdot b \cdot Y \cdot m$$

where, $$\sigma_b = \frac{S_{ult}}{3} = \frac{610}{3}$$

$$b = 10 \cdot m$$

$$Y = 0.484 - \frac{2.87}{Z_p} = 0.484 - \frac{2.87}{18} = 0.324$$

$$S_b = \frac{610}{3} \times (10\ m) \times 0.324 \times m$$

$$= 658.8\ m^2$$

S_w cannot be calculated as hardness is not known.

∴ Equating bending strength with P_{eff} considering FS.

$$S_b = FS \times P_{eff}$$

$$658.8\ m^2 = 1.5 \times \left[\frac{13263\ (1 + 0.453\ m)}{m}\right]$$

$$658.8\ m^3 = 19894.5 + 9012.2\ m$$

$$m^3 - 13.67\ m - 30.2 = 0$$

Solving, $m \simeq 4.5$ mm

Next preferred module value is 5 mm.

Hence, we select m = 5 mm (for both pinion and gear).

Pinion width = 10 m = 50 mm

Pinion diameter = $m \cdot Z_p$ = 90 mm

Gear diameter = $m \cdot Z_g$ = 360 mm

Exact value of effective load

$$P_{eff} = \frac{13263\ (1 + 0.453\ m)}{m}, \text{ with } m = 5 \text{ mm}$$

$$= 8661\ N$$

Wear Strength : (In terms of BHN)

$$S_w = b \cdot Q \cdot d_p \cdot K$$

where, $$Q = \frac{2Z_g}{Z_g + Z_p} = \frac{2 \times 72}{18 + 72} = 1.6$$

$$K = 0.16 \left(\frac{BHN}{100}\right)^2$$

$$S_w = (10 \times 5)\ (1.6)\ (90) \times 0.16 \left(\frac{BHN}{100}\right)^2$$

$$= 1152 \left(\frac{BHN}{100}\right)^2$$

Equating wear strength with effective load considering FS,

$$S_w = FS\ [P_{eff}]$$

$$1152 \left(\frac{BHN}{100}\right)^2 = 1.5\ [8661]$$

$$BHN = 335.8 \simeq 336$$

Example 3.2: *A pair of spur gears with a 20° pressure angle consists of a 25 teeth pinion with 60 teeth gear. The module is 5 mm while its face width is 50 mm. The pinion rotates at 600 rpm. The gears are made of steel and heat treated to a surface hardness of 250 BHN. Assume that the dynamic load is accounted by means of velocity factor. The service factor and factor of safety are 1.75 and 2.1 respectively. Calculate wear strength of the gear, the static load that gears can transmit without pitting and rated power that can be transmitted.*

Assume $C_v = \frac{3}{3 + v}$

Given data: Z_p = 25, Z_g = 60, m = 5 mm, b = 50 mm, N_p = 600, BHN = 250, C_s = 1.75, FS = 2.1.

Solution: Design in this case is based only on wear strength.

(i) Wear Strength: (b = 50, $d_p = m \times Z_p$ = 125)

$$S_w = b \cdot K \cdot Q \cdot d_p$$

$$K = 0.16 \left(\frac{BHN}{100}\right)^2 \text{ for steel on steel}$$

$$= 0.16\ (2.5)^2 = 1.0$$

$$Q = \frac{2Z_g}{Z_g + Z_p} = 1.4117$$

$$S_w = 50 \times 1.0 \times 1.41 \times 125 = 8823\ N$$

(ii) Effective Load:

$$P_{eff} = \frac{C_s}{C_v} \cdot P_t$$

Pitch line velocity v

$$v = \frac{\pi d_p \cdot N_p}{60 \times 1000}$$

$$= \frac{\pi \times 125 \times 600}{60 \times 1000} = 3.927\ m/sec$$

$$C_v = \frac{3}{3 + v} = \frac{3}{3 + 3.927} = 0.433$$

$$\therefore \quad P_{eff} = \frac{1.75}{0.433} \cdot P_t = 4.041 \cdot P_t$$

Equating $P_{eff.}$ with wear strength considering factor of safety

$$S_w = FS \cdot (P_{eff})$$

$$8823 = 2.1 \times 4.041\ P_t$$

$\therefore$ Static load capacity,

$$P_t = 1040 \text{ N}$$

(iii) Rated Power Capacity

$$\left(\textbf{Torque} = \mathbf{P_t \times \frac{d_p}{2}} = \textbf{65000 N·mm}\right)$$

$$P = \frac{2\pi NT}{60 \times 10^6} = \frac{2\pi \times 600 \times 65000}{60 \times 10^6}$$

$$= 4.08 \text{ kW}$$

Example 3.3: *A pair of 20° full involute teeth spur gear is to transmit 35 kW at 750 rpm. The velocity ratio is ratio 3.5 : 1. The pinion is made of steel having allowable stress of 110 N/mm². The gear is made of cast iron having allowable stress of 60 N/mm².*

The pinion has 16 teeth and the face width is 12 times the module. Determine the module, face width and pitch diameter of both pinion and gear from the standpoint of strength taking into consideration the velocity factor.

Tooth form factor is $Y = 0.484 - \dfrac{2.85}{Z}$

Velocity factor $C_v = \dfrac{5}{5 + v}$

Assume service factor as 1.5 and load concentration factor as 1.3. Consider factor of safety as 1.5.

Given data: 35 kW, N_p = 750 rpm, $\frac{Z_2}{Z_1} = 3.5$, $(\sigma_{all})_{pinion} = 110$ N/mm², $(\sigma_{all})_{gear} = 60$ N/mm², $Z_p = 16$, $Z_g = 56$, $b = 12\ m$, $C_s = 1.5$, $C_L = 1.3$.

Solution:

Effective Load:

$$\text{Tangential load} = \frac{\text{Power}}{\text{Pitch line velocity}} = \frac{35 \times 10^6}{\left(\dfrac{\pi \times 16\ m \times 750}{60}\right)}$$

$$P_t = \frac{35 \times 10^3}{0.628\ m} = \frac{55704}{m}$$

$$C_v = \frac{5}{5 + v} = \frac{5}{(5 + 0.628\ m)}$$

$$\therefore \quad P_{eff} = \left(\frac{C_s C_L}{C_v}\right) P_t$$

$$= \frac{1.5 \times 1.3\ (5 + 0.628\ m)}{5} \times \frac{55704}{m}$$

$$P_{eff} = \frac{108622.8\ (5 + 0.628\ m)}{5\ m}$$

Beam Strength:

$$S_b = \sigma_b \cdot b \cdot Y \cdot m$$

Since pinion and gear materials are different, we need to find which one is weaker.

$$Y_p = 0.484 - \frac{2.85}{Z_p} = 0.484 - \frac{2.85}{16} = 0.306$$

$$Y_g = 0.484 - \frac{2.85}{Z_g} = 0.484 - \frac{2.85}{56} = 0.4331$$

$$(\sigma_b \cdot Y)_{pinion} = 110 \times 0.306 = 33.66$$

$$(\sigma_b \cdot Y)_{gear} = 60 \times 0.4331 = 25.986$$

$$\text{Since, } (S_b \cdot Y)_{gear} < (S_b \cdot Y)_{pinion}$$

Gear is weaker. Hence, beam strength will be calculated based on gear strength.

$$\therefore \quad S_b = (\sigma_b \cdot Y)_{gear} \times b \times m$$

$$= 25.986 \times 12.m \times m = 311.84\ m^2$$

Equating beam strength with P_{eff} considering factor of safety

$$S_b = (P_{eff})\ FS$$

$$311.84\ m^2 = \left[\frac{108622.8\ (5 + 0.628\ m)}{5m}\right] \times 1.5$$

$$m^3 = 104.49\ (5 + 0.628\ m)$$

$$m^3 - 65.625\ m - 522.45 = 0$$

By trial and error method

$$m \simeq 10.7 \text{ mm}$$

Select next preferred module size = 12 mm

(a) Module ≃ 12 mm

(b) Face width = 12 m = 144 mm

(c) Pinion pitch circle diameter = $m \times Z_p$ = 192 mm

Gear pitch circle diameter = $m \times Z_g$ = 672 mm

Example 3.4: *A pair of spur gear with 20° full depth involute consists of 20 teeth pinion meshing with 60 teeth on gear. The pinion rotates at 1000 rpm and receives 20 kW power through its shaft. The pinion as well as gear are made up of steel and the permissible bending stress is 260 N/mm². The gears are machined to grade 7. For this grade e = 11 + 0.9 ϕ where $\phi = (m + 0.25\sqrt{d})$ where, m and d are the module and pcd respectively. Use Lewis form factor.*

$$Y = 0.484 - \frac{2.85}{Z} \text{ and } C_v = \frac{6}{6 + v}$$

Design the gears, specify their dimensions and find out the required value of surface hardness of gear. Take load concentration factor as 1.3, service factor as 1.25. Consider dynamic load by Buckingham's equation. Required factor of safety is 1.5. Deformation factor 'C' for gear spur is 11500 · e N/mm.

Given data: $Z_p = 20$, $Z_g = 60$, $N_p = 1000$ rpm, Power = 20 kW, $(\sigma_b)_{all} = 260$ N/mm^2, $C_s = 1.25$, $C_L = 1.3$, FS = 1.5.

Solution:

Pitch line velocity,

$$v = \frac{\pi d_p N_p}{60 \times 10^3} = \frac{\pi \times 20\ m \times 1000}{60 \times 10^3} = 1.047\ m$$

Tangential force,

$$P_t = \frac{\text{Power}}{\text{Velocity}} = \frac{20 \times 10^3}{1.047\ m} = \frac{19102.2}{m}\ (N)$$

Velocity factor,

$$C_v = \frac{6}{6 + v} = \frac{6}{6 + 1.047\ m}$$

Effective load,

$$P_{eff} = \left(\frac{C_s \cdot C_L}{C_v}\right) P_t$$

$$= \frac{1.25 \times 1.3}{\left(\dfrac{6}{6 + 1.047\ m}\right)} \times \frac{19102.2}{m}\ N$$

$$P_{eff} = \frac{5173.5\ (6 + 1.047\ m)}{m}$$

Beam Strength: ($\sigma_b = 260$ N/mm^2, b = 10 m)

$$S_b = \sigma_b \cdot b \cdot Y \cdot m$$

Since pinion and gear materials are same, pinion is weaker. Hence, we obtain S_b for pinion.

Lewis form factor,

$$Y_p = 0.484 - \frac{2.85}{20} = 0.3415$$

$$S_b = 260 \times 10\ m \times 0.3415 \times m$$

$$S_b = 888\ m^2$$

Finding Module:

Equating bending strength to P_{eff} considering factor of safety

$$S_b = (P_{eff})\ (FS)$$

$$888\ m^2 = \frac{5173.5\ (6 + 1.047\ m)}{m} \times 1.5$$

$$888\ m^2 = \frac{7760.25\ (6 + 1.047\ m)}{m}$$

$$m^3 - 9.149\ m - 52.4 = 0$$

Solving $m \simeq 4.8$

Considering next preferred module and checking against dynamic load we select m = 6 mm.

Checking design for dynamic load by Buckingham's equation.

Dynamic Load:

$$P_d = \frac{21v\ (b \cdot C + C_s \cdot C_L \cdot P_t)}{21\ v + \sqrt{(b \cdot C + C_s \cdot C_L \cdot P_t)}}$$

With b = 60 mm, $d_p = 20 \times 6 = 120$ mm, $d_g = 60 \times 6 = 360$ mm.

Total Tooth Error (e):

Error on pinion e_p

$$= 11 + 0.9\ (m + 0.25\sqrt{d_p})$$

$$= 11 + 0.9\ (6 + 0.25\sqrt{120}) = 18.86\ \mu m$$

Error on gear e_g

$$= 11 + 0.9\ (6 + 0.25\sqrt{360}) = 20.67\ \mu m$$

Total tooth error

$$e = e_p + e_g = 39.53\ \mu m = 0.03953\ mm$$

$$v = 1.05 \times m = 6.3\ m/sec.$$

$$P_t = \frac{19102.2}{m} = 3184\ N$$

Deformation factor,

$$C = 11500 \times e = 454.6\ N/mm$$

$$\therefore \quad P_d = \frac{21 \times 6.3\ (60 \times 454.6 + 1.25 \times 1.3 \times 3184)}{21 \times 6.3 + \sqrt{(60 \times 454.6 + 1.25 \times 1.3 \times 3184)}}$$

$$= \frac{132.3\ (32450)}{132.3 + \sqrt{32450}}$$

$$= \frac{132.3 \times 32450}{132.3 + 180.2} = 13738\ N$$

Precise Value of (P_{eff}):

$$P_{eff} = (C_s \cdot C_L \cdot P_t + P_d)$$

$$= (1.25 \times 1.3 \times 3184 + 13738)$$

$$= 5174 + 13738 = 18912\ N$$

Factor of Safety:

$$FS = \frac{\text{Beam strength}}{P_{eff}} = \frac{888\ m^2}{18912} = 1.69$$

Since this is higher than required value of 1.5, design is safe.

$$d_p = m \times Z_p = 6 \times 20 = 120\ mm$$

$$d_g = m \times Z_g = 6 \times 60 = 360\ mm$$

$$b = 10\ .\ m = 60\ mm$$

Centre diameter,

$$a = \frac{d_p + d_g}{2} = 240\ mm$$

Addendum, h_a = 1 m = 6 mm

Dedendum, h_f = 1.25 m = 7.5 mm

Required Surface Hardness:

Wear strength= $b \cdot Q \cdot K \cdot d_p$

$$Q = \frac{2Z_g}{Z_g + Z_p} = \frac{2 \times 60}{60 + 20} = \frac{120}{80} = 1.5.$$

To avoid pitting failure

$$b \cdot Q \cdot K \cdot d_p > (FS) \cdot P_{eff}$$

$$60 \times 1.5 \times K \times 120 > 1.5 \times 18912$$

$$K > 2.62$$

$$0.16 \left(\frac{BHN}{100}\right)^2 > 2.62$$

$$BHN > 404.6$$

Minimum required BHN is 405.

Example 3.5: *Following data is given for a pair of spur gear with 20° fuel depth involute teeth. Number of teeth on pinion = 20, Number of teeth on gear = 50. Speed of pinion = 1200 rpm. Module = 4 mm. Service factor = 1.5. Face width = 32 mm. Both gears are made of steel with an ultimate strength of 600 N/mm². Using velocity factor to account for dynamic load, calculate beam strength and rated power that gears can transmit without bending failure. Factor of safety is 1.5. Use* $C_v = \frac{3}{3+v}$

Given data: $Z_p = 20$, $Z_g = 50$, $N_p = 1200$ rpm, m = 4 mm, $C_s = 1.5$, b = 32 mm, $(\sigma_b)_{all} = 600$ N/mm², FS = 1.5.

Solution: Since both pinion and gear are of same material, pinion is weaker. Hence, beam strength will be calculated based on pinion strength.

$$S_b = \sigma_b \cdot b \cdot Y \cdot m$$

$$\left(\sigma_b = \frac{600}{3}, b = 32 \text{ mm}, m = 4 \text{ mm}\right)$$

$$Y = 0.484 - \frac{2.87}{Z_p} = 0.484 - \frac{2.87}{20} = 0.3405$$

$$\therefore \quad S_b = \frac{600}{3} \times 32 \times 0.3405 \times 4 = 8717 \text{ N}$$

$$\text{Velocity,} \quad v = \frac{\pi \cdot d_p N_p}{60 \times 1000} = \frac{\pi \times (4 \times 20) \times 1200}{60 \times 1000}$$

$$= 5.026 \text{ m/sec.}$$

$$C_v = \frac{3}{3+v} = \frac{3}{3+5.026} = 0.374$$

Effective load on gear tooth,

$$P_{eff} = \left[\frac{C_s \times C_L}{C_v}\right] P_t = \left(\frac{1.5 \times 1}{0.374}\right) P_t = 4.013 \, P_t$$

Equating beam stress with P_{eff}. Considering factor of safety,

$$S_b = (P_{eff}) \text{ FS}$$

$$8717 = [4.013 \times P_t] \, 1.5$$

$$P_t = 1448.2 \text{ N}$$

Torque transmitted

$$= \frac{P_t \times d_p}{2} = \frac{1448.2 \times 80}{2} = 57925 \text{ N·mm}$$

Rated power transmitted,

$$P = \frac{2\pi NT}{60 \times 10^6} = \frac{2\pi \times 1200 \times 57925}{60 \times 10^6}$$

$$= 7.28 \text{ kW}$$

Example 3.6: *The pinion shaft of a spur gear pair is connected to a 12 kW electric motion running at 1440 rpm. Gear tooth system is 20° full depth involute. Pinion has 18 teeth and the gear has 72 teeth. Module is 6 mm. Application factor is 1.25 and service factor 1.4. Face width of the gear is 60 mm. Gears are machined to grade 8. Deformation factor is to be obtained from the relation* $C = \frac{K \cdot e}{\left[\frac{1}{E_p} + \frac{1}{E_g}\right]}$ *with K = 0.11. Pinion and gear materials are alloy steel with* $S_{ult} = 800$ *MPa and E = 2.1 × 10⁵ N/mm². Gears are heat treated to 360 BHN. Assuming the dynamic load to be accounted by Buckingham equation.*

Find: (a) Factor of safety against bending failure.

(b) Factor of safety against pitting failure.

Given data: 12 kW, 1440 rpm, $Z_p = 18$, $Z_g = 72$, m = 6 mm, $C_L = 12.5$, $C_s = 1.4$, b = 60 mm.

Solution:

Tangential Load on Pinion:

Linear velocity at contact point,

$$v = \frac{\pi d_p \cdot N_p}{60 \times 1000} = \frac{\pi (18 \times 6) \, 1440}{60 \times 1000}$$

$$= 8.14 \text{ m/sec}$$

$$\text{Tooth load,} \quad P_t = \frac{\text{Power}}{\text{Velocity}} = \frac{12 \times 10^3}{8.14} = 1474.2 \text{ N}$$

Tooth error for accuracy grade 8 is given by,

$$e = 16 + 1.25 \, [m = 0.25 \sqrt{d}]$$

$$e_p = 16 + 1.25 \, [6 + 0.25 \sqrt{6 \times 18}] = 26.75$$

$$e_g = 16 + 1.25 \, [6 + 0.25 \sqrt{6 \times 72}] = 29.99$$

$$\therefore \quad e = e_p + e_g = 56.74 \, \mu\text{m} = 0.05674 \text{ mm}$$

Deformation factor (E for steel: 2.1 × 10⁵ N/mm²)

$$C = \frac{K \cdot e}{\left[\frac{1}{E_p} + \frac{1}{E_g}\right]} = \frac{0.11 \times 0.05674}{\frac{1}{10^5}\left[\frac{1}{2.1} + \frac{1}{2.1}\right]}$$

$$= 655.34 \text{ N/mm}$$

$$P_d = \frac{21 \cdot v \, (b \cdot C + C_s \cdot C_L \cdot P_t)}{21v + \left[\sqrt{b \cdot C + C_s \cdot C_L \cdot P_t}\right]}$$

$$= \frac{21 \times 8.14 \, (60 \times 655.34 + 1.4 \times 1.25 \times 1474.2)}{21 \times 8.14 + \left[\sqrt{60 \times 655.34 + 1.4 \times 1.25 \times 1474.2}\right]}$$

$$P_d = \frac{21 \times 8.14 \, (41900.25)}{21 \times 8.14 + \sqrt{41900 \cdot 25}}$$

$$= \frac{7162428}{375.63} = 19068 \text{ N}$$

$$P_{eff} = C_s \cdot C_L \cdot P_t + P_d$$

$$= 1.4 \times 1.25 \times 1474.2 + 19068$$

$$= 2580 + 19068 = 21648 \text{ N}$$

Beam Strength:

Since pinion and gear are of same material, beam strength will be calculated based on pinion which is weaker.

$$S_b = \sigma_b \cdot b \cdot Y \cdot m$$

$$\left(b = 60 \text{ mm}, m = 6 \text{ mm}, \sigma_b = \frac{800}{3} \text{ N/mm}^2\right)$$

$$Y = 0.484 - \frac{2.87}{Z_p} = 0.484 - \frac{2.87}{18} = 0.3245$$

$$S_b = \frac{800}{3} \times 60 \times 0.3245 \times 6 = 31152 \text{ N}$$

Wear Strength:

$$S_w = b \cdot Q \cdot d_p \cdot K$$

$$\left[b = 60, K = 0.16\left(\frac{BHN}{100}\right)^2, d_p = 108\right]$$

$$Q = \frac{2Z_g}{Z_g + Z_p} = \frac{2 \times 72}{72 + 18} = 1.6$$

$$K = 0.16\left(\frac{360}{100}\right)^2 = 2.07$$

$$S_w = 60 \times 1.6 \times 108 \times 2.074 = 21499 \text{ N}$$

Factor of Safety:

$$S_b = [P_{eff}] \text{ FS}$$

$$\therefore \quad \text{FS} = \frac{31152}{21648} = 1.44 \text{ (In bending)}$$

$$S_w = [P_{eff}] \text{ FS} \quad \therefore \text{ FS} = \frac{21499}{21648} = 0.99 \text{ (In wear)}$$

Example 3.7: *Three gears with idler arrangement are connected as shown below. Gear P is the driver with 18 teeth, Q is the idler with 54 teeth and R is the driven with 36 teeth. All gears are 20° full depth involute profiles and module is 4 mm. The driver P is transmitting 5 kW power at 1500 rpm. Assuming all gear are centrally mounted on their respective shafts. Find:*

(i) Torque transmitted by each shaft.

(ii) Forces acting on each gear and show their directions.

(iii) Radial force acting on each bearing holding shaft 'P'.

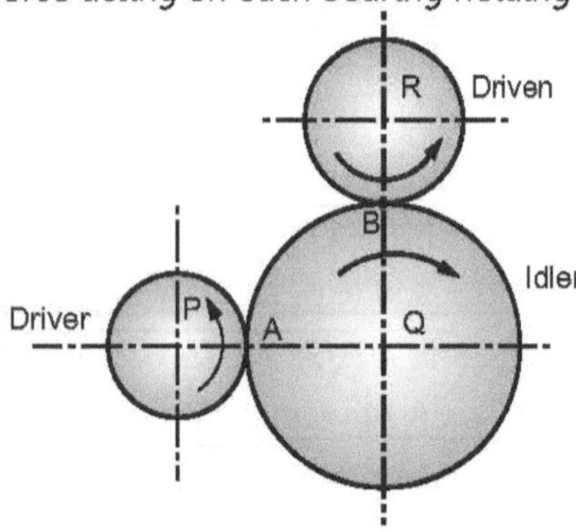

Fig. 3.11

Given data: Input power P = 5 kW at 1500 rpm.

Solution:

(i) Torque on Shaft P:

$$T_p = \frac{\text{Power} \times 60 \times 10^6}{2\pi N}$$

$$= \frac{5 \times 60 \times 10^6}{2\pi \times 1500} = 31831 \text{ N·mm}$$

Gear R rotates at $\left(\frac{18}{36}\right)$ speed as 'P': 750 rpm.

Torque on Shaft R:

$$T_R = \frac{\text{Power} \times 60 \times 10^6}{2\pi N} \times \frac{5 \times 60 \times 10^6}{2\pi \times 750}$$

$$= 63662 \text{ N·mm}$$

Idler does not transmit any torque.

$\therefore$ Torque on shaft Q is zero.

(ii) Forces on Gears 'P':

Tangential Force on Driver 'P':

$$(P_{tp}) = \frac{\text{Torque}}{d_p/2} = \frac{31831}{(m \cdot Z_p/2)} = \frac{31831}{(5 \times 18/2)} = 707.4 \text{ N}$$

Radial Force on Driver Gear 'P':

$$(P_{rp}) = (P_{tp}) \times \tan\phi = 707.4 \tan 20° = 257.5 \text{ N}$$

Forces on Gear 'Q' and 'R':

Tangential and radial forces on Q acting at contact point, 'A' will be same as above but opposite in direction.

Thus, $(P_{tQ}) = (P_{tP})$ and $(P_{rQ}) = (P_{rP})$

Tangential forces between Q and R will again be of same magnitude since whatever torque is received at point A is transmitted further at point B. Further same will be the magnitude of forces transmitted on gear 'R' at point B. Forces with their directions are shown in the Fig. 3.12.

Thus, $P_{tQ} = P_{tQ} = P_{tR} = 707.4$ N

and $P_{rp} = P_{rQ} = P_{rR} = 257.5$ N

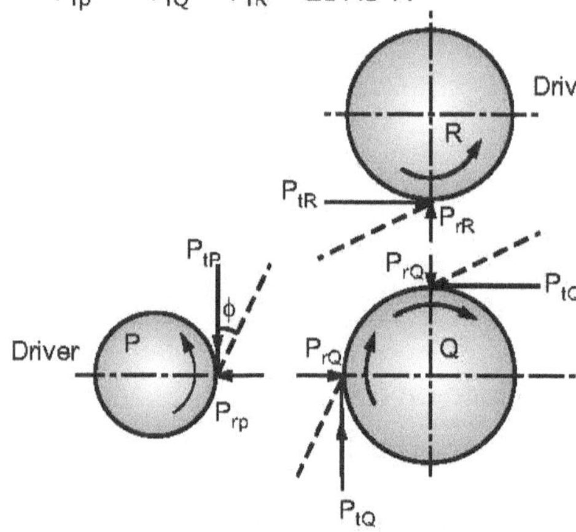

Fig. 3.12

Force Directions:

(a) Between P and Q: P is the driver and Q is the driven. On the driver, tangential force direction is opposite to direction of motion. Hence, at point A force on 'P' is in the downward direction. On the driven, tangential force direction is in the same direction of motion. Hence, at point A force on 'Q' is in the upward direction.

(b) Between Q and R: Q is the driver and R is driven. Force directions between Q and R can be decibel on above lines.

(iii) Resultant Radial Force on Shaft 'P':

$$= \sqrt{P_{tP}^2 + P_{rP}^2} = \sqrt{(707.4)^2 + (257.5)^2}$$

$$= 752.8 \text{ N}$$

Since gears are centrally mounted.

$$\text{Radial force on each bearing} = \frac{752.8}{2} = 376.4 \text{ N}$$

Example 3.8: *A pair of spur gears with 20° full depth involute teeth system consists of pinion with 21 teeth running at 720 rpm meshing with gear running at 378 rpm. Pinion is made of alloy steel with* S_{ult} *= 620 N/mm*^{2} *and gear with plain carbon steel having* S_{ult} *= 540 N/mm*^{2}*. Gear pair is heat treated to 300 BHN. Face width is 10 times module. Service factor and factor of safety are 1.5 and 1.75 respectively. Gear pair is manufactured by hobbing process. Centre distance between pinion and gear shaft is 152.5 mm. Assuming velocity factor accounts for dynamic load, calculate.*

(i) Beam strength

(ii) Wear strength

(iii) Maximum static load that gear pair can transmit.

(iv) Rated power that gear pair transmit. Take velocity factor, $C = \frac{6}{6+v}$

Given data: $Z_p = 21$, $N_p = 720$ rpm, $N_g = 378$ rpm, $(S_{ult})_{pinion} = 620$ N/mm^2, $(S_{ult})_{gear} = 540$ N/mm^2, BHN = 300, $C_s = 1.5$, FS = 1.85.

Solution: Centre distance

$$= \frac{d_p + d_g}{2} = 152.5 \text{ mm},$$

$$Z_g = 21 \times \frac{720}{378} = 40$$

$$d_p = m \cdot Z_p = 5 \times 21 = 105 \text{ mm}$$

$$d_g = m \cdot Z_g = 5 \times 40 = 200 \text{ mm}$$

Thus, $d_g = 200$ mm, $d_p = 105$ mm,

Module = 5 mm

(i) Beam Strength:

Pinion and gear are of different materials. Hence, we must find which one is weaker.

$$(\sigma_b \cdot Y)_{Pinion} = \frac{S_{ult}}{3} \times \left(0.484 - \frac{2.87}{Z_p}\right)$$

$$= \frac{620}{3} \times \left(0.484 - \frac{2.87}{21}\right) = 71.78$$

$$(\sigma_b \cdot Y)_{gear} = \frac{S_{ult}}{3} \times \left(0.484 - \frac{2.87}{Z_g}\right)$$

$$= \frac{540}{3} \times \left(0.484 - \frac{2.87}{40}\right) = 74.2$$

Since,

$(\sigma_b \cdot Y)_{Pinion} < (\sigma_b, Y)_{gear}$

Pinion is weaker.

Beam strength is to be calculated based on pinion strength.

Taking width = 10 · m

$$\text{Beam strength} = \sigma_b \cdot b \cdot Y \cdot m$$

$$= (71.78)\ (10\ \ m)$$

$$S_b = 717.8\ m^2 = 17945 \text{ N}$$

(ii) Wear Strength:

$$= b \cdot Q \cdot d_p \cdot K$$

[With b = 50 mm, d_p = 105 mm]

$$Q = \frac{2Z_g}{Z_g + Z_p} = \frac{2 \times 40}{40 + 21} = 1.3115$$

$$K = 0.16 \left(\frac{BHN}{100}\right)^2 = 0.16\ (3)^2 = 1.44$$

$$S_w = 50 \times 1.311 \times 105 \times 1.44$$

$$S_w = 9911 \text{ N}$$

(iii) Static Load Capacity (P_t):

Velocity factor C_v

$$= \frac{6}{6+v} = \frac{6}{\left(6 + \frac{\pi \times 105 \times 720}{60 \times 1000}\right)}$$

$$= \frac{6}{6 + 3.958} = 0.602$$

$$P_{eff} = \left(\frac{C_s \cdot C_L}{C_v}\right) P_t = \left(\frac{1.5 \times 1}{0.602}\right) P_t$$

$$= 2.492\ P_t$$

Since wear strength < Beam strength,

Equating wear strength with effective load and considering factor of safety

$$S_w = [P_{eff}]\ FS$$

$$9911 = [2.492\ P_t] \times 1.75$$

$$P_t = 2272 \text{ N}$$

Static load capacity is 2272 N

(iv) Rated Power Transmitted:

$$\text{Power} = P_t \times v = \frac{2272 \times 3.958}{1000} = 8.99 \text{ kW}$$

Example 3.9: *A spur gear pair is to transmit 8 kW power at 750 rpm to another machine at 300 rpm. Pinion and gear, both are made of same steel having ultimate tensile strength of 620 N/mm*^{2}*. Surface hardness is 330 BHN. Pinion has 20 teeth, the module is 5 mm and face width is 60 mm. Service factor is 1.5 and load concentration factor is 1.4.*

Determine the factor of safety in bending and wear.

(i) By considering that velocity factor accounts for dynamic load.

(ii) By considering dynamic load using Buckingham equation.

Use the following data:

Barth factor, $C_v = \dfrac{3}{3+v}$

Tooth profile error (e) = 0.05 mm

Dynamic factor, $C = K \cdot e \left[\dfrac{E_p \cdot E_g}{E_p + E_g}\right]$

Tooth form factor, K = 0.111

$E = 2.1 \times 10^5$ *N/mm² for both materials.*

Given data:

Power = 8 kW, N_p = 750 rpm, N_g = 300 rpm, 330 BHN, (S_{ult}) = 620 N/mm², Z_p = 20, m = 5, b = 60, C_s = 1.5, C_L = 1.4.

Solution: (i) Factor of Safety in Bending and Wear:

Velocity at pitch point

$$= v = \frac{\pi d_p \cdot N_p}{60 \times 1000} = \frac{\pi (5 \times 20)\, 750}{60 \times 1000}$$

$$= 3.93 \text{ m/sec.}$$

Tangential force,

$$P_t = \frac{\text{Power}}{\text{Velocity}} = \frac{8 \times 1000}{3.93} = 2037.2 \text{ N}$$

Velocity factor,

$$C_v = \frac{3}{3+v} = \frac{3}{3+3.93} = 0.433$$

$$P_{eff} = \left[\frac{C_s \cdot C_L}{C_v}\right] P_t$$

$$= \left[\frac{1.5 \times 1.4}{0.433}\right] \times 2037.2 = 9880 \text{ N}$$

Beam Strength (S_b):

Since pinion and gear have same material, pinion is weaker.

$$(\sigma_b Y)_{pinion} = \left(\frac{S_{ult}}{3}\right)_{pinion} \times Y_p$$

$$= \left(\frac{620}{3}\right) \times \left(0.484 - \frac{2.87}{20}\right) = 70.37$$

$$S_b = \sigma_b \cdot b \cdot Y \cdot m = 70.37 \times 60 \times 5$$

$$= 21111 \text{ N}$$

Wear Strength (S_w): b = 60 mm, d_p = 100 mm,

$$S_w = b \cdot Q \cdot d_p \cdot K$$

$$Q = \frac{2Z_g}{Z_g + Z_p} = \frac{2 \times 50}{50 + 20} = 1.428$$

$$k = 0.16 \left(\frac{BHN}{100}\right)^2 = 0.16 \left(\frac{330}{100}\right)^2 = 1.742$$

$$S_w = 60 \times 1.428 \times 100 \times 1.742 = 14929 \text{ N}$$

Factor of safety in bending (based on velocity factor)

$$= \frac{21111}{7057.3} = 2.99$$

Factor of safety in wear (based on velocity factor)

$$= \frac{14929}{7057.3} = 2.11$$

(ii) Factor of Safety Based on Buckingham Equation:

Dynamic load,

$$P_b = \frac{21 \cdot v\, (b \cdot C + P_{tmax})}{21 \cdot v + \sqrt{b \cdot C + P_{tmax}}}$$

$$P_{tmax} = C_s \cdot C_L \cdot P_t$$

$$= 1.5 \times 1.4 \times 2037.2 = 4278 \text{ N}$$

$$C = K \cdot e \left[\frac{E_p \cdot E_g}{E_p + E_g}\right]$$

$$= 0.111 \times 0.05 \left[\frac{(2.1)^2 \times 10^{10}}{2 \times 2.1 \times 10^5}\right]$$

$$= 582.75 \text{ N/mm}$$

$$P_d = \frac{21 \times 3.93\, (60 \times 582.75 + 4278)}{21 \times 3.93 + \sqrt{(60 \times 582.75 + 4278)}}$$

$$= \frac{21 \times 3.39 \times 39243}{21 \times 3.93 + 198.1}$$

$$= 11541 \text{ N}$$

$$P_{eff} = C_s \cdot C_L \cdot P_t + P_d$$

$$= 4278 + 11541 = 15819 \text{ N}$$

Factor of safety in bending $= \dfrac{S_b}{P_{eff}} = \dfrac{21111}{15819} = 1.33$

Factor of safety in wear $= \dfrac{S_b}{S_w} = \dfrac{14929}{15819} = 0.94$

(Design in inadequate against wear strength)

As a corrective action wear strength has to be increased or tooth pitch error should be reduced by going for more accurate tooth cutting and finishing.

Example 3.10: *A spur gear pinion having 20 teeth made of plain carbon steel with* S_{ult} *= 540 N/mm² is to mesh with gear having 85 teeth made of gray cast iron FC 260. Pinion shaft is connected to 10 kW. 1440 rpm electric motor. Starting torque is twice rated torque. Tooth system is 20° full-depth involute. Face width is 10 times module for which load distribution factor is 1.4. Gears are machined to meet specifications of grade 7; for which deformation factor is 11500 e.*

(a) If factor of safety against bending is 1.5, design gear pair by using velocity factor and Buckingham's equation for dynamic load.

(b) If factor of safety against pitting is 2.0 specify hardness.

Take *Velocity factor,* $C_v = \dfrac{3}{3+v}$

Deformation factor, C = 11500 e

$$\textit{Lewis form factor, } \gamma = 0.484 - \frac{2.87}{Z}$$

$$\textit{Load stress factor, } K = 0.18\left[\frac{BHN}{100}\right]^2$$

For grade 6:

$$e = 8 + 0.63\,[m + 0.25\sqrt{d}]$$

Given data: $Z_p = 20$, $Z_g = 85$, $(S_{ult})_{pinion} = 540\ N/mm^2$, $(S_{ult})_{gear} = 260\ N/mm^2$, Power = 10 kW, $N_p = 1440$ rpm, $C_s = 2$, b = 10.m, $C_L = 1.4$, FS = 1.5 in bending.

Solution: Effective Load considering Velocity Factor:

Pitch line velocity,

$$v = \frac{\pi d_p \cdot N_p}{60 \times 1000} = \frac{\pi\,(20 \cdot m)\,1440}{60 \times 1000}$$

$$= (1.508.m)\ m/sec.$$

Tangential force,

$$P_t = \frac{Power}{v} = \frac{10 \times 10^3}{1.508} = \left(\frac{6631.3}{m}\right) N$$

Velocity factor,

$$C_v = \frac{3}{3 + v} = \frac{3}{(3 + 1.508\ m)}$$

$$P_{eff} = \left(\frac{C_s \cdot C_L}{C_v}\right) P_t$$

$$= \frac{2 \times 1.4\,(3 + 1.508\ m)}{3} \times \left(\frac{6631.3}{m}\right)$$

$$P_{eff} = \frac{6189.2\,(3 + 1.508\ m)}{m}$$

Beam Strength:

Since pinion and gear materials are different

$$(\sigma_b \cdot Y)_p = \frac{540}{3} \times \left(0.484 - \frac{2.87}{Z_p}\right)$$

$$= \frac{540}{3} \times \left(0.484 - \frac{2.87}{20}\right) = 61.29$$

$$(\sigma_b \cdot Y)_g = \frac{260}{3} \times \left(0.484 - \frac{2.87}{Z_g}\right)$$

$$= \frac{260}{3} \times \left(0.484 - \frac{2.87}{85}\right) = 39.02$$

Thus, gear is weaker and beam strength is calculated based on gear strength.

∴ Beam strength

$$S_b = (\sigma_b \cdot Y)\, b \cdot m = (39.02) \times 10\ m \times m$$

$$= 390.2\ m^2$$

Equating beam strength with effective load considering factor of safety

$$S_b = [P_{eff}]\ FS$$

$$390.2\ m^2 = \left[6189.2\,\frac{(3 + 1.508\ m)}{m}\right] \times 1.5$$

$$390.2\ m^3 - 13999.97\ m - 27851.4 = 0$$

$$m^3 - 35.88\ m - 71.38 = 0$$

Solving by trial and error, m = 6.9 mm

We select next preferred size of module = 8 mm.

Checking by Considering Buckingham's Equation:

Dynamic load,

$$P_d = \frac{21\,v\,(b \cdot C + F_t)}{21 \cdot v + \sqrt{b \cdot C + F_t}}$$

where, $(P_{t\,max}) = F_t = C_s \cdot C_L\,(P_t) = 2 \times 1.4\left(\frac{6631.3}{8}\right)$

$$= 2321\ N$$

Deformation factor,

$$C = 11500 \cdot e$$

$$e_p = 8 + 0.63\,(m + 0.25\sqrt{d_p})$$

$$= 8 + 0.63\,(8 + 0.25\sqrt{20 \times 8})$$

$$= 15.03\ \mu m$$

$$e_g = 8 + 0.63\,(8 + 0.25\sqrt{85 \times 8})$$

$$= 17.14\ \mu m$$

$$e = e_p + e_g = 32.17\ \mu m = 0.03217\ mm$$

Deformation constant,

$$C = 11500 \times 0.0321 = 369.96$$

$$P_d = \frac{21 \times (1.508 \times 8)\,(80 \times 369.96 + 2321)}{21 \times (1.508 \times 8) + \sqrt{80 \times 369.96 + 2321}}$$

$$P_d = \frac{21 \times 12.064\,(31917.8)}{21 \times 12.064 + (178.6)} = 18718\ N$$

$$P_{eff} = P_{tmax} + P_d = 2321 + 18717$$

$$= 21039\ N$$

$$S_b = 390.2\ m^2 = 390.2 \times 64 = 24972.8\ N$$

$$FS = \frac{24972.8}{21039} = 1.186$$

Since, the factor of safety < 1.5.

We need to redesign.

Trial 1: Consider module = 10 · mm

Tangential force,

$$P_t = \left(\frac{6631.3}{m}\right) = \frac{6631.3}{m}$$

Velocity factor,

$$v = \frac{\pi d_p \cdot N_p}{60 \times 1000} = (1.508 \cdot m)$$

$$P_{eff} = \frac{6189.2\,(3 + 1.508\ m)}{m} = 11190\ N$$

Checking for dynamic load

$$e_p = 8 + 0.63\,(m + 0.25\sqrt{d_p})$$

$$= 8 + 0.63\,(10 + 0.25\sqrt{20 \times 10})$$

$$= 16.52\ \mu m$$

$$e_g = 8 + 0.63\,(m + 0.25\sqrt{d_g})$$

$$= 8 + 0.63\,(10 + 0.25\sqrt{85 \times 10})$$

$$= 18.89\ \mu m$$

$$e = e_p + e_g = 35.41\ \mu m = 0.03541\ mm$$

$$C = 11500 \times e = 407.2 \text{ N/mm}$$

$$P_d = \frac{21 \cdot v\,(b \cdot C + F_t)}{21 \cdot v + \sqrt{b \cdot C + F_t}}$$

With $(P_{tmax}) = F_t = C_s \cdot C_L \cdot P_t$

$$= 2 \times 1.4 \times \left(\frac{6631.3}{10}\right) = 1856.77$$

$$P_d = \frac{21 \times 1.508 \times 10\,(100 \times 407.2 + 1856.77)}{21 \times 1.508 \times 10 + \sqrt{100 \times 407.2 + 1856.77}}$$

$$= \frac{316.68\,(40720 + 1856.77)}{316.68 + 206.3}$$

$$= \frac{316.68\,(42576.77)}{523.00} = 25780 \text{ N}$$

$$P_{eff} = P_{tmax} + P_d = 1856.77 + 25780$$

$$= 27637 \text{ N}$$

$$S_{bmax} = 390.2\,m^2 = 39020 \text{ N}$$

$$FS = \frac{39020}{27637} = 1.41$$

Since FS < 1.5.

We need to take trial with m = 12 mm

Trial 2: $e_p = 8 + 0.63\,(m + 0.25\sqrt{d_p})$

$$= 8 + 0.63\,(12 + 0.25\sqrt{20 \times 12}$$

$$= 18.00 \text{ μm}$$

$$e_g = 8 + 0.63\,(12 + 0.25\sqrt{85 \times 12})$$

$$= 20.59 \text{ μm}$$

$$e = e_p + e_g = 38.59 \text{ μm} = 0.03859 \text{ mm}$$

$$C = 11500 \times e = 443.785 \text{ N/mm}$$

$$P_{t\,max} = F_t = 2 \times 1.4 \times \left(\frac{6631.3}{m}\right) = 1547.3 \text{ N}$$

$$P_d = \frac{21 \times 1.508 \times 12\,(120 \times 443.785 + 1547.3)}{21 \times (1.508 \times 12) + \sqrt{120 \times 443.785 + 1547.3}}$$

$$= \frac{20825446.8}{614.113} = 33911 \text{ N}$$

$$P_{eff} = P_{tmax} + P_d = 1547.3 + 33911$$

$$P_{eff} = 35458.3 \text{ N}$$

$$S_b = 390.2\,m^2 = 390.2\,(12)^2 = 56188.8 \text{ N}$$

$$\text{Factor of safety} = \frac{S_b}{P_{eff}}$$

$$\frac{56188.8}{35458.3} = 1.58$$

∴ We select module = 12 mm

Hardness Required for Factor of Safety = 2.

With module m = 12 mm.

$$P_{eff} = 35448.3 \text{ N}$$

For $S_w = 35458 \times 2 = 70916$ N

$$S_w = b \cdot Q \cdot d_p \cdot K\; b = 10\,m = 120 \text{ mm}$$

$$d_p = m \cdot Z_p = 12 \times 20 = 240 \text{ mm}$$

$$Q = \frac{2Z_g}{Z_g + Z_p} = \frac{2 \times 85}{85 + 20} = 1.69$$

$$\therefore \quad K_{required} = \frac{70916}{120 \times 240 \times 1.69} = 1.456$$

$$K = 1.456 = 0.18\left(\frac{BHN}{100}\right)^2$$

$$BHN = 285$$

Material hardness required is 285 BHN.

EXERCISE

1. State and explain different types of gear tooth failures, their causes and remedies.
2. Explain the following terms: (a) Hunting tooth, (b) Crowning of gear tooth.
3. What is dynamic load on gears? State the factors on which the dynamic load depends and explain the methods of estimation of dynamic load on gear tooth.
4. Discuss the standard systems of gear tooth.
5. Explain why Involute profile is preferred over cycloidal profile for gear tooth.
6. Discuss lubrication of gears.
7. Write a note on gear lubrication.
8. What are the effects of decreasing the pressure angle in gear design?
9. What is addendum modification in gears? State the advantages and limitations of it.
10. Explain the selection of gear type for a given application.
11. What are various materials suitable for the gears?
12. Explain design considerations for the gears.
13. Discuss the desirable properties of gear material.
14. Explain in brief selection of number of teeth on the gears.
15. Explain force analysis for spur gear pair.
16. Explain the various assumptions in Lewis beam strength equation.
17. What are different types of gear tooth failures and there remedies?
18. Derive an expression for the beam strength of the gear.
19. Why the pinion is always weaker in pitting?
20. How the surface endurance strength is improved?

◈◈◈

CHAPTER - 4

HELICAL GEARS

4.1 INTRODUCTION

- In case of spur gear, teeth are cut parallel to the axis of the shaft. Due to this the contact between teeth in mesh occurs completely along the face width. This results in sudden impact of load from driving tooth onto the driven tooth which results in generation of noise, more so when gears rotate at high speed.
- To overcome this limitation helical gears were introduced. In case of helical gears teeth are cut on a cylindrical blank at some angle with the axis of the shaft. Because of this, the contact between meshing teeth starts with a point contact on the edge and develops into line contact whose length of contact gradually increases. Thus, impact is less, resulting in a smooth and silent transmission of power. Another advantage of helical gears is since teeth are cut at an angle, contact ratio (i.e. number of pairs of teeth in contact) is more as compared to the spur gear with same number of teeth and gear width.
- Thus, helical gear will transmit more power compared to spur gear. Contact ratio being higher, helical gears can also permit lesser number of teeth compared to spur gear of same diameter and module. Helical gears would however introduce axial thrusts which is not present in spur gears and would demand higher shaft alignment accuracies. (Refer Fig. 4.1)
- Axial thrust in helical gears can be eliminated by introducing double helical gears. These are called Herringbone gears. Teeth are cut along the width in two parts. One half is with LH helix and the other half with RH helix. The axial thrust produced on two parts opposes each other and the resultant is zero. Thus, these gears are able to eliminate axial thrust. Cost of manufacture and alignment requirements are more stringent thus increasing the cost.
- The proper selection of type of gear is a combination of different requirements and their technical cost influence.
- To combine the advantages of low noise of helical gears and low cost of spur gears, in many applications involving multistage gear boxes, a helical gear pair is used in the first stage where speed are high and torque (and thus the tangential force and axial thrust) will be less. Helical gears can also be used to transmit power between two non-parallel and non-intersecting shafts. For transmitting power between two parallel shafts, the helix angle of the meshing gears has to be same and hand of helix must be opposite. In this chapter, we limit our discussions only to single helical gears transmitting power between two parallel shafts.

(a) Single Helical Gear (b) Double Helical Gear

Fig. 4.1

4.2 IMPORTANT TERMS FOR HELICAL GEARS

- **Helix Angle (ψ):** It is the angle made by the axis of the tooth with axis of rotation.
- **Transverse Circular Pitch (P_c):** It is the distance between two successive teeth measured along pitch circle in transverse plane.

$$P_c = \frac{\pi \cdot d}{Z}$$

where, Z is the number of teeth.

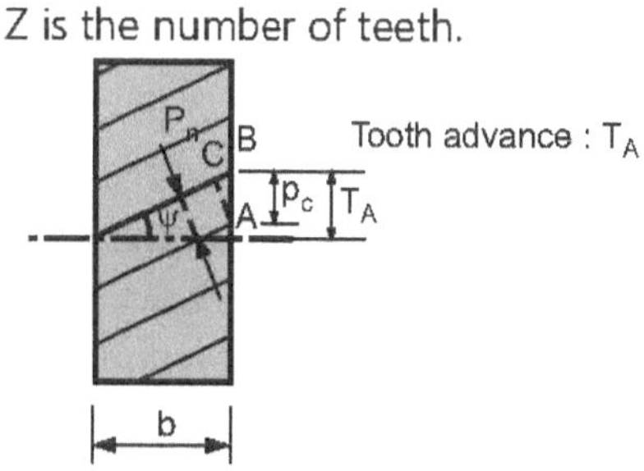

Fig. 4.2: Helical Gear

- **Normal Circular Pitch (P_n):** It is the distance between two successive teeth measured along pitch circle in the plane perpendicular to the axis of teeth.
- **Axial Pitch (P_a):** It is the distance between two successive teeth measured along the axis of rotation.

From triangle ABC,

$$\frac{P_n}{P_c} = \frac{AC}{AB} = \cos \psi$$

$$P_n = P_c \cdot \cos \psi$$

Similarly, $m_n = m \cdot \cos \psi$

where, m is the transverse module

m_n is the normal module.

In helical gears, normal module is treated as standard since the tooling is made accordingly.

Table 4.1 gives the values of normal module of first and second preference.

Table 4.1: List of Standard Modules

Module of First Choice (mm)	1, 1.25, 1.5, 2, 2.5, 3, 4, 5, 6, 8, 10, 12, 16, 20, 25, 32, 40, 50
Module of Second Choice (mm)	1.125, 1.375, 1.75, 2.25, 2.75, 3.5, 4.5, 5.5, 7, 9, 11, 14, 18, 22, 28, 36, 45

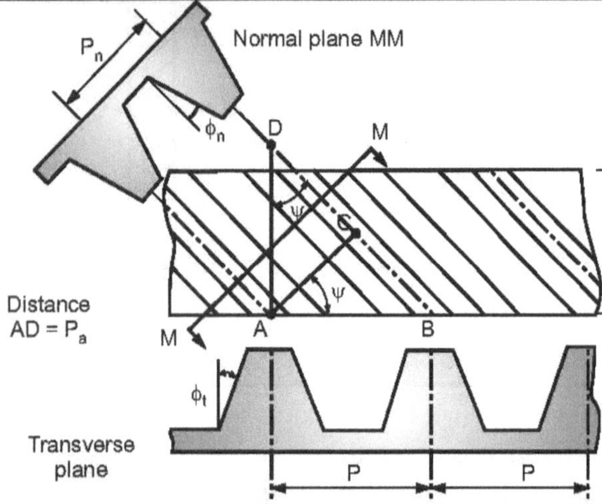

Fig. 4.3: Terminology of Helical Gears

From triangle ABD (Refer Fig. 4.3),

$$P_a = \frac{P_c}{\tan \psi}$$

- **Transverse Pressure Angle (ϕ_t):** Pressure angle is the angle between tangent to the pitch circles and common normal drawn at point of contact measured in transverse (plane of rotation) plane.
- **Normal Pressure Angle (ϕ_n):** This is the pressure angle measured in the normal plane (Plane perpendicular to the teeth). Standard normal pressure angles are 14.5°, 20°, 25°.

  It can be seen

$$\tan \phi_t = \frac{\tan \phi_n}{\cos \psi}$$

4.3 FORMATIVE NUMBER OF TEETH (VIRTUAL MEMBER OF TEETH)

- In the previous chapter, we have analysed spur gear. In order to simplify the analysis of helical gear, an imaginary spur gear (also called as formative spur gear or equivalent spur gear) is considered and on this spur gear number of teeth are found, which are called virtual number of teeth or formative number of teeth.
- Consider a helical gear cut by a plane A-A, normal to the tooth axis and passing through its center. (Refer Fig. 4.4).

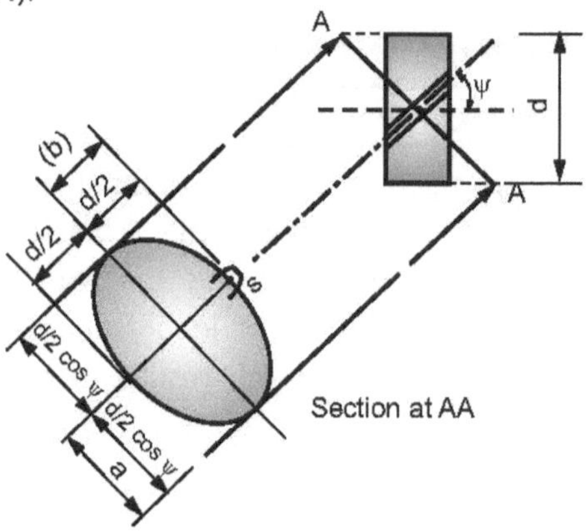

Fig. 4.4

- The cross section along plane A-A will appear as an ellipse with its semi major axis (a) as $\left(\frac{d}{2 \cos \psi}\right)$ and semi minor axis (b) as $\left(\frac{d}{2}\right)$.

  The radius of curvature r' at point 'S' is given by coordinate geometry as

$$r' = \frac{a^2}{b} = \frac{\left(\frac{d}{2 \cos \psi}\right)^2}{(d/2)} = \frac{d}{2 \cos^2 \psi}$$

- So, the imaginary spur gear, in the plane AA has pitch circle radius r' (diameter d') and module m_n. Formative number of teeth for a helical gear are the number of teeth that can be generated on the surface of a cylinder having a radius equal to the radius of curvature at a point at the tip of the minor axis of an ellipse (radius at point 'S' in the Fig. 4.4) obtained as above.

  The pitch circle diameter of the equivalent spur gear will be d'.

Thus, $$d' = 2r' = \frac{d}{\cos^2 \psi}$$

Number of teeth on this imaginary spur gear are

$$z' = \frac{d'}{m_n} = \frac{d}{m_n \cos^2 \psi}$$

Since, $$d = Z \times m = \frac{Z m_n}{\cos \psi}$$

$\therefore$ $$Z' = \frac{Z}{\cos^3 \psi}$$

where, Z' = Virtual or formative number of teeth

Z = Actual number of teeth

- The virtual number of teeth being larger than the actual number of teeth, Lewis form factor gives higher values and the gear has better strength compared to spur gears with same number of teeth.

4.4 TOOTH PROPORTIONS FOR HELICAL GEAR

Addendum: $h_a = m_n$

Dedeundum: $h_f = 1.25\ m_n$

Clearance: $0.25\ m_n$

Addendum circle diameter: d_a

$$d_a = d + 2h_a = \frac{Z \cdot m_n}{\cos\psi} + 2m_n$$

$$d_a = m_n\left[\frac{Z}{\cos\psi} + 2\right]$$

Dedendum circle diameter:

$$d_f = m_n\left[\frac{Z}{\cos\psi} - 2.5\right]$$

Helix angle normally varies from 15 to 25°.

Face Width (b):

For smooth operation of helical gears and to ensure contact ratio more than one, the distance between the leading edge of the tooth and the trailing edge of the same tooth (T_A) must be more than the circular pitch.

Thus, the width of a gear and its helix angle are interrelated.

Contact ratio = Tooth Advance (T_A)/ Circular pitch (P_c)

... should be greater than 1.

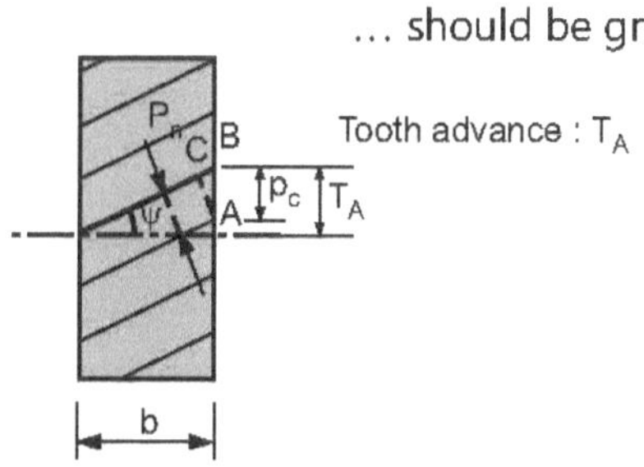

Fig. 4.5

From Fig. 4.5,

$$T_A = b \cdot \tan\psi$$

Therefore, $$b > \frac{P_c}{\tan\psi} = \frac{\pi \cdot m}{\tan\psi}$$

$$= \frac{\pi \cdot m_n}{\cos\psi \cdot \tan\psi} = \frac{\pi \cdot m_n}{\sin\psi}$$

Thus, $$b_{min} = \frac{\pi \cdot m_n}{\sin\psi}$$

In practice 'b' should be at least = $1.15\ b_{min}$

Satisfying this condition usual values selected are between (9 to 12) m_n.

4.5 FORCE ANALYSIS

Fig. 4.6 shows pitch cylinder of a driven gear. The direction of rotation of the gear is clockwise as shown. Therefore, the direction of tangential force will be in the same direction as the direction of rotation i.e. P_t will assist the motion at the contact point. Let P_n be the normal force transferred from the driving tooth on the driven gear tooth along the common normal drawn at point of contact. This force is inclined to the axis of rotation by helix angle ψ. The force P_n is resolved in three perpendicular directions which align with,

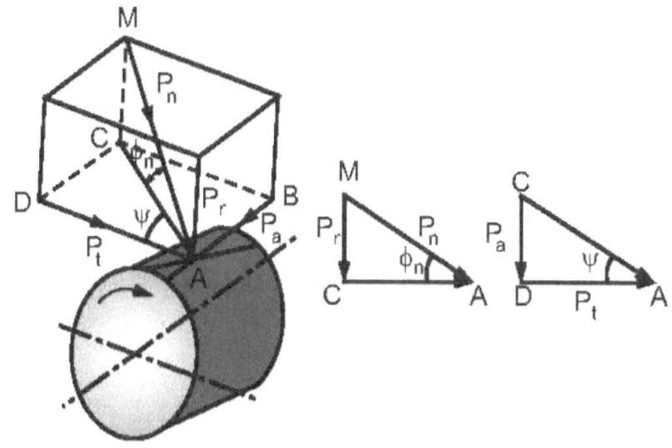

Fig. 4.6

- Tangential force: P_t
- Radial force: P_r and
- Axial force: P_a

Referring Fig 4.6

Radial component:

$$P_r = P_n \cdot \sin\phi_n \qquad \text{...(4.1)}$$

Component along AC (X):

$$X = P_n \cdot \cos\phi_n \qquad \text{... (4.2)}$$

Further $$P_a = X \cdot \sin\psi = P_n \cdot \cos\phi_n \cdot \sin\psi \qquad \text{... (4.3)}$$

$$P_t = X \cdot \cos\psi = P_n \cdot \cos\phi_n \cdot \cos\psi \qquad \text{... (4.4)}$$

From equations (4.3) and (4.4),

$$\mathbf{P_a = P_t \cdot \tan\psi} \qquad \text{... (4.5)}$$

From equations (4.1) and (4.4),

$$\mathbf{P_r = P_t\left(\frac{\tan\phi_n}{\cos\psi}\right)} \qquad \text{... (4.6)}$$

In practice, P_t can be found out by knowing the power to be transmitted.

and $$P_t = \frac{P}{v} = \frac{60 \times 10^6\ P}{\pi d \times N}$$

where, P = Power in kW

v = Pitch line velocity in m/s

d = Pitch circle diameter in mm

N = Speed r.p.m.

Once the tangential force P_t is found out, P_r and P_a can be estimated by using the equations (4.5) and (4.6) above. These forces will ultimately be transferred to the shaft and the bearings. The directions of these forces on driving and driven gears can be found using following guidelines.

- The direction of tangential force on the driven gear tooth will be in the same direction as the motion at the point contact.
- Radial force is always a separating force and acts in a direction such that it will try to increase the center distance.

- The direction of axial force can be decided as below.
- Identify which one is a driving gear and find hand of helix on it.

(a) For this RH/ LH thumb rule can be applied. Care should be taken to ensure that thumb is along the axis of rotation.

(b) Alternately for a horizontal shaft helical gear, trace a downward arrow along the tooth axis (Refer Fig. 4.7). If the arrow points towards left side, gear helix is LH or if it points to right side, gear is a RH gear.

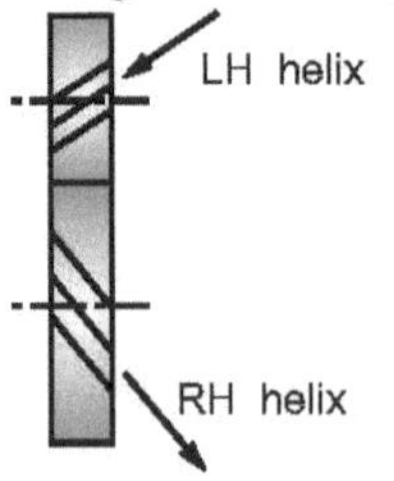

Fig. 4.7

To find the direction of axial component of load.

- On the driving gear use left hand for LH helix and right hand for RH helix.
- Align fingers in direction of rotation. The thumb indicates direction of axial component for driving gear. For driven gear the thrust direction will be opposite to that on driving gear.

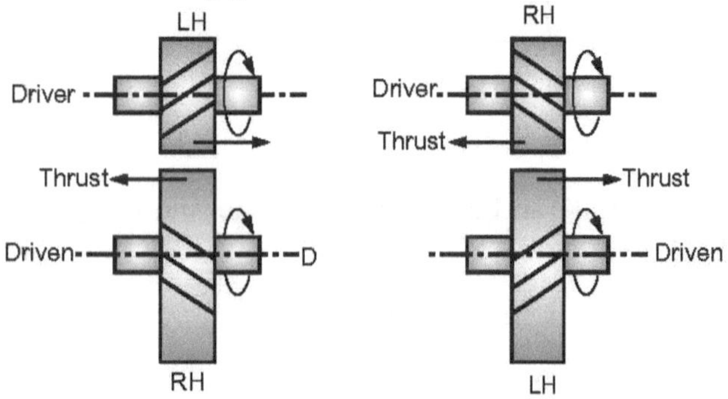

Fig. 4.8: Helical Gear: Axial Thrust Direction

Axial thrust directions for two configurations are shown in Fig 4.8.

4.6 BEAM STRENGTH OF HELICAL GEAR (LEWIS EQUATION)

The analysis of beam strength for a helical gear is performed on the similar lines as that of spur gear by considering equivalent spur gear for a given helical gear. This equivalent spur gear has pitch circle diameter as d', module as m_n and number of teeth as Z'.

Beam strength for a spur gear is: $S_b = \sigma_b \cdot b \cdot Y \cdot m$

When applied to the equivalent spur gear equation will be

$$S_{bn} = \sigma_b \cdot b' \cdot Y' \cdot m_n$$

Thus, $$S_{bn} = \sigma_b \cdot \left(\frac{b}{\cos\psi}\right) \cdot Y' \cdot m_n$$

'S_b' for the spur gear gets converted to S_{bn}. (Fig 4.9). This is because, the force considered for a spur gear beam strength is its capacity to withstand a force perpendicular to tooth axis. S_{bn} is perpendicular to tooth axis. However its component S_b (= $S_{bn} \cdot \cos\psi$) only is useful for torque transmission.

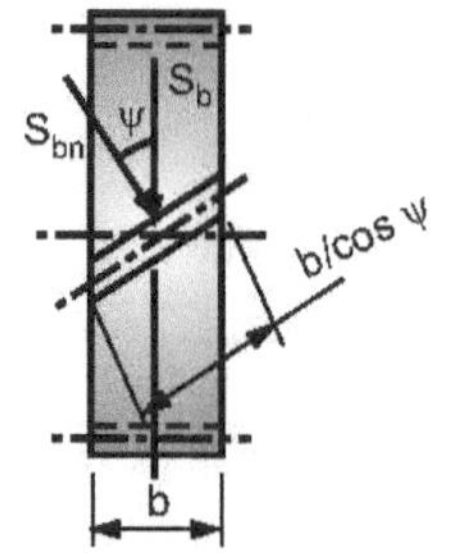

Fig. 4.9: Beam Strength for Helical Gear

$$S_b = S_{bn} \times \cos\psi$$

Therefore, $$S_b = \frac{\sigma_b \cdot b \cdot Y' \cdot m_n}{\cos\psi} \times \cos\psi$$

Thus, $$S_b = \sigma_b \cdot b \cdot Y' \cdot m_n \qquad \ldots (4.7)$$

Where, Y' is Lewis form factor based on formative number of teeth.

For 14.5° full depth involute and composite systems,

$$Y' = 0.389 - \frac{2.148}{Z'}$$

For 20° full depth involute systems,

$$Y' = 0.484 - \frac{2.87}{Z'}$$

Equation (4.7) is the beam strength equation for helical gear. In order to avoid bending of tooth beam strength has to be more than effective load on gear tooth.

Denoting factor of safety as FS

$$S_b > P_{eff}$$
$$S_b = (FS)\, P_{eff}$$

4.7 WEAR STRENGTH OF HELICAL GEAR

Wear strength of helical gear can be found out by modifying the wear strength equation of the spur gear. To do this an equivalent (formative or imaginary) spur gear is considered having pitch circle diameter as d', module as m_n and number of teeth as Z'.

Wear strength of spur gear is given by

$$\therefore \quad S_w = bQd_pK$$

Wear strength of equivalent spur gear will be

$$S_{wn} = \left(\frac{b}{\cos\psi}\right) \cdot Q' \left(\frac{d_p}{\cos^2\psi}\right) K$$

Since, $$Q' = \frac{2Z_g'}{Z_g' + Z_p'} = \frac{2Z_g}{Z_g + Z_p} = Q$$

$$\therefore \quad S_{wn} = \frac{b \cdot Q \cdot d_p \cdot K}{\cos^3 \psi}$$

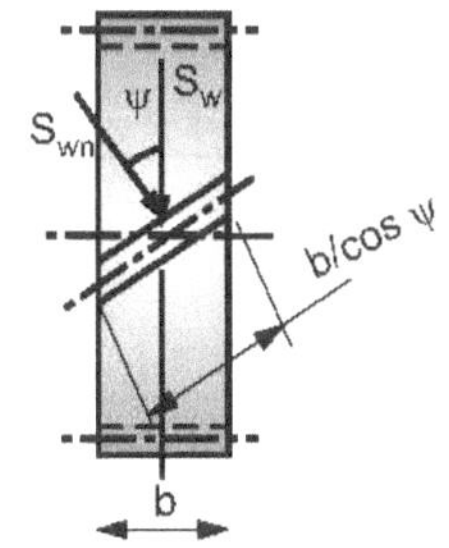

Fig. 4.10: Wear Strength for Helical Gear

With the same comments applicable as for the beam strength equation (Refer Fig. 4.10), wear strength equation reduces to

$$S_w = S_{wn} \cdot \cos \psi = \frac{b \cdot Q \cdot d_p \cdot K}{\cos^2 \psi} \quad \dots (4.8)$$

Values of K can be similarly obtained as in case of spur gear.

$$K = 0.16 \left(\frac{BHN}{100}\right)^2$$

... (for pinion and gear made of steel)

$$K = 0.18 \left(\frac{BHN}{100}\right)^2$$

... (for steel pinion and cast iron gear)

$$K = 0.21 \left(\frac{BHN}{100}\right)^2$$

... (for pinion and gear made of cast iron)

Equation (4.8) is the wear strength equation. To avoid pitting failure wear strength should be greater than the effective load on meshing teeth.

$$S_w > P_{eff}$$

Denoting factor of safety as FS

$$S_w = (FS)\, P_{eff}$$

4.8 EFFECTIVE LOAD

Similar to spur gear, in the preliminary stage, approximate approach is selected to account for dynamic load.

$$P_{eff} = \frac{C_L\, C_s\, P_t}{C_v}$$

where, C_L = Load concentration factor

C_s = Service factor

C_v = Velocity factor

- The velocity factor for helical gears depends on pitch line velocity, gear cutting accuracy and accuracy of gear mounting. Typical value could be $C_v = 6/(6 + V)$.
- Values of load concentration factor and service factors are based on experience on previous equipments. In absence of such data, guide line values can be referred from design data handbooks.
- Once the value of module is estimated based on preliminary calculations, gear dimensions are known, manufacturing methods can be finalized and then Buckingham's equation can be used to have more precise estimation of the dynamic load. Thus, dynamic load is calculated from Buckingham's incremental dynamic tooth load equation.

Incremental dynamic tooth load is given by

$$P_d = \frac{21 \cdot V\,(b \cdot C \cos^2 \psi + P_{tmax}) \cos \psi}{21 \cdot V + \sqrt{b \cdot C \cdot \cos^2 \psi + P_{tmax}}}$$

where, $P_{tmax} = (C_s \cdot C_l) \cdot P_t$

C is the deformation constant (N/mm)

With standard values of E_p, E_g for steel materials = 2.07×10^5 N/mm^2 and K for 20° pressure angle.

C = 11500 × e (N/mm) for Steel pinion with steel gear

C = 10000 × e (N/mm) for Steel pinion on Cast Iron gear

C = 8900 × e (N/mm) for Cast Iron pinion with Cast Iron gear

Tooth error for grade 8 pinion and gear is given by

$$e_p = 16.0 + 1.25\left(m + 0.25\sqrt{d_p}\right)$$

and $$e_g = 16.0 + 1.25\left(m + 0.25\sqrt{d_g}\right)$$

Table 4.2: Tooth Profile Error

Grade	Error in Tooth Profile (in micron)
1	e = 0.80 + 0.06φ
2	e = 1.25 + 0.10φ
3	e = 2.00 + 0.16φ
4	e = 3.20 + 0.25φ
5	e = 5.00 + 0.40φ
6	e = 8.00 + 0.63φ
7	e = 11.00 + 0.90φ
8	e = 16.00 + 1.25φ
9	e = 22.00 + 1.80φ
10	e = 32.00 + 2.50φ
11	e = 45.00 + 3.55φ
12	e = 63.00 + 5.00φ

Precise estimation of effective load can now be obtained as,

$$P_{eff} = C_s \cdot C_L \cdot P_t + P_d$$

4.9 HELICAL GEAR DESIGN PROCEDURE

Design procedure for helical gears is similar to that for spur gears. However, it is important to note following points.

- Center distance: (a) can be obtained as below:

$$a = \frac{d_p}{2} + \frac{d_g}{2} = \frac{m \cdot Z_p}{2} + \frac{m \cdot Z_g}{2}$$

$$= m \cdot \left(\frac{Z_p + Z_g}{2}\right)$$

Thus, $$a = m\left(\frac{Z_p + Z_g}{2}\right) = \frac{m_n}{\cos \lambda}\left(\frac{Z_p + Z_g}{2}\right)$$

where, m_n is the standard value of module to be selected from preferred values.

- Beam strength equation and wear strength equations get modified as explained above.
- In case of spur gears, only radial loads are passed on to the shaft because of (tangential + radial) load on the gear. For helical gears forces on gears will be tangential + radial + axial. Loads transferred to the shaft will be radial + axial.

SOLVED EXAMPLES

Example 4.1: *A pair of helical gear consists of 24 teeth pinion meshing with 120 teeth gear. The pinion rotates at 720 rpm. The normal pressure angle is 20° while helix angle is 25°. The face width is 40 mm and the normal module is 4 mm. The pinion as well as gear are made of steel 40 C_8 (S_{ult} = 600 N/mm²) with surface hardness of 350 BHN. The service factor and the factor of safety are 1.5 and 2 respectively. Assuming that velocity factor accounts for the dynamic load, calculate the power transmitting capacity of gears. Use* $C_v = \frac{5.6}{5.6 + \sqrt{v}}$.

Given data: $Z_p = 24$, $Z_g = 120$, $\phi_n = 20°$, $\psi = 25°$, $N_p = 720$ rpm, $b = 40$ mm, $m_n = 4$ mm, $S_{ult} = 600$ N/mm², $C_s = 1.5$, FS = 2.

Solution: Virtual number of teeth, $Z_p' = \frac{Z_p}{\cos^3 \psi} = \frac{24}{\cos^3 25} = 32.239$

As material for pinion and gear is same, pinion is weaker. Hence beam strength is found for the pinion. Lewis form factor for virtual number of teeth on pinion.

$$Y_p' = 0.484 - \frac{2.87}{Z_p'} = 0.484 - \frac{2.87}{32.239}$$

$$= 0.3949$$

Beam Strength of Pinion Material:

$$S_b = \sigma_b \cdot b \cdot Y' \cdot m_n$$

$$= \frac{600}{3} \times 40 \times 0.3949 \times 4 = 12636.8 \text{ N}$$

Wear strength is given by

$$S_w = \frac{b \cdot Q \cdot d_p \cdot K}{\cos^2 \psi}$$

where, $$d_p = \frac{m_n \cdot Z_p}{\cos \psi} = \frac{4 \times 24}{\cos 25} = 105.924 \text{ mm}$$

and Ratio factor,

$$Q = \frac{2 \cdot Z_g}{Z_g + Z_p} = \frac{2 \times 120}{120 + 24} = 1.667$$

$$K = 0.16\left(\frac{BHN}{100}\right)^2 = 1.96$$

$$\therefore \quad S_w = \frac{40 \times 1.667 \times 105.924 \times 1.96}{\cos^2 25}$$

$$S_w = 16853 \text{ N}$$

Since, $S_b < S_w$, load capacity will be based on beam strength (S_b).

Velocity factor C_v is given as

$$C_v = \frac{5.6}{5.6 + \sqrt{v}}$$

where, $$v = \frac{\pi d_p \times N_p}{60 \times 1000}$$

$$= \frac{\pi \times 105.924 \times 720}{60 \times 1000} = 3.993 \text{ m/sec.}$$

$$C_v = \frac{5.6}{5.6 + \sqrt{3.993}} = 0.737$$

Effective load, $$P_{eff} = \left(\frac{C_s \times C_L}{C_v}\right) P_t$$

$$= \left(\frac{1.5 \times 1}{0.737}\right) P_t = 2.035\ P_t$$

Equating S_b with P_{eff} considering factor of safety

$$S_b = [P_{eff}]\ FS$$

$$12636.8 = [2.035 \cdot P_t]\ 2$$

$$\therefore \quad P_t = 3104.86 \text{ N}$$

Power transmitting capacity

$$\text{Power} = P_t \times v = 3104.86 \times 3.993$$

$$= 12397.7 \text{ Watts}$$

$$\text{Power} = 12.4 \text{ kW}$$

Example 4.2: *A pair of helical gears consists of 18 teeth pinion with 45 teeth gear having 20° normal pressure angle. Pinion receives 7.5 kW power at 2000 rpm. Determine the module and face width of gear to avoid bending failure. Assume the following data.*

(a) Helix angle = 23°, (b) Safe stress for pinion and gears: 150 N/mm², (c) Factor of safety = 2 and service factor = 1.2. Determine the tangential, axial and radial components of the resultant tooth force.

Use $C_v = \frac{5.6}{5.6 + \sqrt{v}}$.

Given data: $\phi_n = 20°$, $\psi = 23°$, $Z_p = 18$, $Z_g = 45$, Power = 7.5 kW, $N_p = 2000$ rpm, $(\sigma_b)_{all} = 150 \text{ N/mm}^2$, FS = 2, $C_s = 1.25$.

Solution: Effective Force Based on Velocity Factor:

Pitch line velocity,

$$v = \frac{\pi d_p \cdot N_p}{60 \times 1000} = \pi \cdot \left(\frac{m_n \cdot Z_p}{\cos \psi}\right) \cdot \frac{N_p}{60 \times 1000}$$

$$= \frac{\pi \times m_n \cdot 18}{0.9205} \times \frac{2000}{60 \times 1000}$$

$$= (2.0478 \cdot m_n) \text{ m/sec}$$

Tangential force,

$$P_t = \frac{\text{Power}}{\text{Velocity}} = \frac{7.5 \times 10^3}{2.0478 \times m_n} = \left(\frac{3662.46}{m_n}\right) \text{N}$$

$$C_v = \frac{5.6}{5.6 + \sqrt{v}} = \frac{5.6}{5.6 + \sqrt{2.0478\, m_n}}$$

Effective force, (Assuming $C_L = 1$, since not given)

$$P_{eff} = \left(\frac{C_s \cdot C_L}{C_v}\right) \cdot P_t$$

$$= (1.25) \times \left(\frac{3662.46}{m_n}\right) \times \frac{5.6 + \sqrt{2.0478\, m_n}}{5.6}$$

$$P_{eff} = \frac{817.5}{m_n}\left(5.6 + \sqrt{2.0478\, m_n}\right)$$

Beam Strength:

Since pinion and gear materials are same, pinion is weaker. Hence beam strength is found for pinion.

$$Z'_p = \frac{Z_p}{\cos^3 \psi} = \frac{18}{\cos^3 23} = 23.078$$

Lewis form factor

$$Y'_p = 0.484 - \frac{2.87}{Z'_p} = 0.484 - \frac{2.87}{23.078}$$

$$= 0.3596$$

Pinion Beam Strength:

$$S_b = \sigma_b \cdot b \cdot Y'_p \cdot m_n$$

$$= 150 \times 10 \cdot m_n \times 0.3596 \times m_n$$

$$= 539.4\, m_n^2$$

To avoid bending failure, equating beam strength to P_{eff}, considering factor of safety.

$$S_b = (P_{eff}) \cdot FS$$

$$539.4\, m_n^2 = \left[\frac{817.5}{m_n}\left(5.6 + \sqrt{2.0478\, m_n}\right)\right] \times 2$$

Simplifying,

$$m_n^3 - 4.335\sqrt{m_n} - 16.97 = 0$$

Solving by trial and error, $m_n = 2.9$ mm

We select next higher standard module, $m_n = 3$.

Gear face width will be $b = 10 \times m_n = 30$ mm.

$$S_b = 539.4\, m_n^2 = 4854.6 \text{ N}$$

$$P_{eff} = 2201.0 \text{ N}$$

Forces on Gears:

(1) Tangential force,

$$P_t = \frac{3662.46}{m_n} = 1220.8 \text{ N}$$

(2) Radial force,

$$P_r = \left(\frac{\tan \phi_n}{\cos \psi}\right) \cdot P_t = 618.4 \text{ N}$$

(3) Axial force,

$$P_a = (\tan \psi) \cdot P_t = 518.2 \text{ N}$$

Assuming pinion having LH helix and rotating in clockwise direction as viewed left side, forces are as shown in Fig. 4.11.

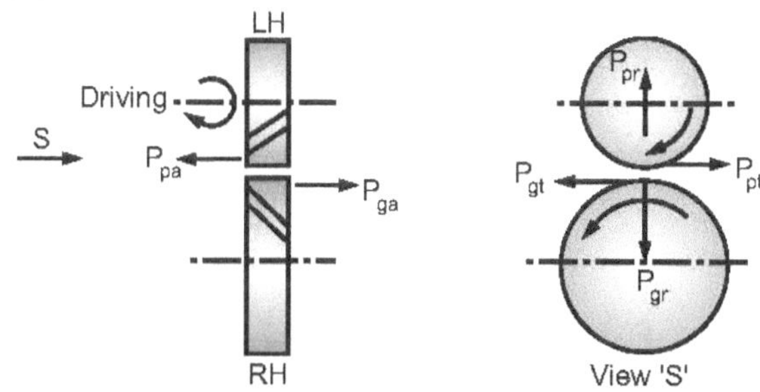

Fig. 4.11

Example 4.3: *Following data is given for a helical gear pair hardened to 350 BHN (20° full depth involute) transmitting 15 kW power from an electric motor at 720 rpm to a machine running at 360 rpm.*

Number of teeth on pinion = 18

Centre distance = 180 mm

Helix angle = 26°

Helix angle = 12 × Normal module

Permissible bending stress = 150 N/mm² (for both)

Service factor = 1.1.

Combined tooth error = 40 microns

Deformation factor = (11600 × Error) N/mm

Find FS in bending and pitting. Use

$$F_d = \frac{21 \cdot V\,(b \cdot C \cos^2 \psi + F_{tmax}) \cos \alpha}{21 \cdot V + \sqrt{b \cdot C \cdot \cos^2 \psi + P_{tmax}}},$$

$$Y = 0.484 - \frac{2.87}{Z'}$$

Given data: $\phi_n = 20°$, Power = 15 kW, $N_p = 720$ rpm, $N_g = 360$ rpm, $Z_p = 18$, C.D. = 180 mm, $\psi = 26°$, $b = 12 \times m_n$, $(\sigma_b)_{all} = 150 \text{ N/mm}^2$, $C_s = 1.1$, $e = 40\ \mu$m, $C = 11600 \times e$.

Solution: $Z_p = 18$, $d_p = \frac{m_n \cdot Z_p}{\cos\psi}$, $d_g = \frac{m_n \cdot Z_g}{\cos\psi}$

Centre distance,

$$a = 180 = \frac{d_p + d_g}{2} = \frac{m_n}{2\cos\psi}(Z_p + Z_g)$$

$$\therefore \quad 180 = \frac{m_n(18+36)}{2\times\cos 26}, \quad m_n = 5.992 \simeq 6 \text{ mm}$$

Thus, gears have normal module $m_n = 6$ mm.

Pitch line velocity,

$$v = \frac{\pi \cdot d_p \cdot N_p}{60\times1000} = \pi\left(\frac{m_n \cdot Z_p}{\cos\psi}\right)\frac{720}{60\times1000}$$

$$= 4.53 \text{ m/sec}$$

Tangential force,

$$P_t = \frac{\text{Power}}{\text{Velocity}} = \frac{15\times10^3}{4.53} = 3311.3 \text{ N}$$

$$P_{tmax} = P_t \times C_s \times C_L = 3311.3\times1.1\times1$$

$$= 3643 \text{ N}$$

Deformation factor,

$$C = 11600\times e = 11600\times0.04 = 464 \text{ N/mm}$$

Dynamic load,

$$P_d = \frac{21\cdot V(b\cdot C\cos^2\psi + P_{tmax})\cos\psi}{21\cdot V + \sqrt{b\cdot C\cdot\cos^2\psi + P_{tmax}}}$$

$$= \frac{21\times4.53(72\times464\times\cos^2 26 + 3643)\cos 26}{21\times4.53 + \sqrt{(72\times464\times\cos^2 26 + 3643)}}$$

$$P_d = 9694.63 \text{ N}$$

Equivalent load considering dynamic load by Buckingham's equation

$$P_{eff} = P_{tmax} + P_d$$

$$P_{eff} = 3643 + 9694.63 = 13338 \text{ N}$$

Bending Strength:

Since pinion and gear have same material, pinion is weaker.

∴ Beam strength is found for pinion

$$S_b = \sigma_b\, b\cdot Y'_p \cdot m_n$$

$$Z'_p = \frac{Z_p}{\cos^3\psi} = \frac{18}{\cos^3 26} = 24.79$$

$$Y'_p = 0.484 - \frac{2.87}{Z'_p} = 0.484 - \frac{2.87}{24.79} = 0.368$$

$$S_b = (150)\times72\times0.368\times6 = 23846.4 \text{ N}$$

Wear Strength:

$$S_w = \frac{b\cdot Q\cdot d_p\cdot K}{\cos^2\psi}$$

$$d_p = \frac{m_n\cdot Z_p}{\cos\psi} = \frac{6\times18}{\cos 26} = 120.161 \text{ mm}$$

$$K = 0.16\left(\frac{350}{100}\right)^2 = 1.96$$

$$Q = \frac{2Z_g}{Z_g + Z_p} = \frac{2\times36}{18+36} = 1.333$$

$$S_w = \frac{72\times1.333\times120.161\times1.96}{\cos^2 26} = 27981 \text{ N}$$

$$(FS)_{bending} = \frac{S_b}{P_{eff}} = \frac{23846.4}{13338} = 1.788$$

$$(FS)_{wear} = \frac{S_w}{P_{eff}} = \frac{27981}{13338} = 2.1$$

Example 4.4: *A pair of parallel helical gears consists of 18 teeth pinion rotating at 6000 rpm and supplying 2.5 kW power to gear. The speed reduction is 4 : 1. Normal pressure angle and helix angles are 20° and 23° respectively. Both gears are made of hardened steel (S_{ult} = 600 N/mm²). Motor starting torque is 150% of the rated torque and required factor of safety is 1.8. Gears are manufactured to grade 6 for which error may be taken as e = 8 + 0.63 ϕ where, $\phi = m_n + 0.25\sqrt{d}$. Gears are hardened to 250 BHN. Estimate the normal module by considering velocity factor. $C_v = \frac{5.6}{5.6+\sqrt{v}}$. Determine dynamic load by considering Buckingham's equation. Take deformation factor as $C = 11500\cdot e$.*

Given data: $Z_p = 18$, $N_p = 6000$ rpm, Speed ratio = 4, $\phi_n = 20°$, $\psi = 23°$, $S_{ult} = 600$ N/mm², $C_s = 1.5$, FS = 1.8, BHN = 250.

Solution: Effective Load Based on Velocity Factor:

$$d_p = \frac{m_n\cdot Z_p}{\cos\psi} = \frac{18\times m_n}{\cos 23} = 19.555\, m_n$$

$$v = \text{Pitch line velocity}$$

$$= \frac{\pi\cdot d_p\cdot n_p}{60\times1000}$$

$$= \frac{\pi\times19.555\times m_n\times6000}{60\times1000} = 6.143\, m_n$$

Tangential force,

$$P_t = \frac{\text{Power}}{\text{Velocity}} = \frac{2.5\times10^3}{6.143\, m_n} = \frac{407}{m_n}$$

Velocity factor,

$$C_v = \frac{5.6}{5.6+\sqrt{v}} = \frac{5.6}{5.6+\sqrt{6.143\, m_n}}$$

Effective force,

$$P_{eff} = \left(\frac{C_s\cdot C_L}{C_v}\right)P_t$$

$$= 1.5\times1\times\left(\frac{5.6+\sqrt{6.143\, m_n}}{5.6}\right)\frac{407}{m_n}$$

$$= \frac{109}{m_n}\left(5.6+\sqrt{6.143\, m_n}\right)$$

Beam Strength:

Pinion is weaker since material is same for pinion and gear. Hence beam strength is calculated for pinion.

$$S_b = \sigma_b\cdot b\cdot Y'\cdot m_n$$

$$Z_p' = \frac{Z_p}{\cos^3\psi} = \frac{18}{\cos^3 23} = 23.08$$

$$Y_p' = 0.484 - \frac{2.87}{Z_p'} = 0.3596$$

$$S_b = \left(\frac{600}{3}\right) \times 10 \cdot m_n \times 0.3596 \times m_n$$

$$S_b = 719.2\ m_n^2$$

Wear Strength:

$$S_w = \frac{b \cdot Q \cdot d_p \cdot K}{\cos^2\psi}$$

where, $Q = \frac{2Z_g}{Z_g + Z_p} = \frac{2 \times 72}{18 + 72} = 1.6$

$$K = 0.16\left(\frac{BHN}{100}\right)^2 = 0.16\left(\frac{250}{100}\right)^2 = 1$$

$$S_w = \frac{10 \times m_n \times 1.6 \times 19.555\ m_n \times 1}{\cos^2(23)}$$

$$= 369.25\ m_n^2$$

As $S_w < S_b$, wear strength will govern the design.

Equating S_w with P_{eff} considering factor of safety.

$$S_w = [P_{eff}]\ FS$$

$$369.25\ m_n^2 = \frac{109}{m_n}\left(5.6 + \sqrt{6.143\ m_n}\right)$$

$$m_n^3 - 0.732\sqrt{m_n} - 1.65$$

$$m_n \simeq 1.4 \text{ mm}$$

We select $m_n = 2$ mm

For checking the design by considering dynamic load using Buckingham equation.

Dynamic Load:

$$d_p = 19.555 \times m_n,\ d_g = \frac{72 \times m_n}{\cos 23} = 156.43 \text{ mm}$$

Tooth error,

$$e = 8 + 0.63\,(m_n + 0.25\sqrt{d})$$

$$e_p = 8 + 0.63\,(2 + 0.25\sqrt{19.555 \times 2})$$

$$= 10.166\ \mu m$$

$$e_g = 8 + 0.63\,(2 + 0.25\sqrt{156.43})$$

$$= 11.23\ \mu m$$

$$e = e_p + e_g = 21.396\ \mu m = 0.021396 \text{ mm}$$

$$C = 11500 \times e = 246 \text{ N/mm}$$

$$P_d = \frac{21 \cdot v\,(b \cdot C \cdot \cos^2\psi + P_{tmax})\cos\psi}{21 \cdot v + \sqrt{b \cdot C\cos^2\psi + P_{tmax}}}$$

With $v = 6.143 \times m_n = 12.286$ m/sec.

and $P_{tmax} = (C_s \cdot C_L)\,P_t = (1.5 \times 1)\frac{407}{2}$

$$= 305.3 \text{ N}$$

$$P_d = \frac{21 \times 12.286\,(20 \times 246\cos^2 23 + 305.3)\cos 23}{21 \times 12.286 + \sqrt{20 \times 246\cos^2 23 + 305.3}}$$

$$P_d = 3270.5 \text{ N}$$

$$P_{eff} = P_{tmax} + P_d = 3270.5 + 305.3$$

$$= 3575.8 \text{ N}$$

Beam strength

$$S_b = 719.2\ m_n^2 = 2877 \text{ N}$$

$$\therefore \quad FS = \left(\frac{2877}{3575}\right) = 0.8$$

Wear strength

$$S_w = 369.25\ m_n^2 = 1476 \text{ N}$$

$$\therefore \quad FS = \left(\frac{1476}{3575}\right) = 0.4$$

Trial 1: Since FS is even less than 1 and we require FS = 2. Let us select $m_n = 4$.

$$e_p = 8 + 0.63\,(4 + 0.25\sqrt{19.555 \cdot m_n})$$

$$= 11.91\ \mu m$$

$$e_g = 8 + 0.63\,[4 + 0.25\sqrt{(19.555\ m_n \times 4)}]$$

$$= 13.3\ \mu m$$

$$e = e_p + e_g = 25.21\ \mu m = 0.02521 \text{ mm}$$

$\therefore$ Deformation constant,

$$C = 11500 \times e = 289.9 \text{ N/mm}$$

$$b = 10 \cdot m_n = 40 \text{ mm}$$

$$v = 6.143 \times m_n = 24.572 \text{ m/sec.}$$

$$P_t = \frac{407}{m_n} = \frac{407}{4} = 101.75$$

$$P_{tmax} = \left(\frac{C_s \cdot C_L}{1}\right) P_t = 1.5 \times 1.0 \times 101.75$$

$$= 156.62 \text{ N}$$

By Buckingham equation

$$P_d = \frac{21 \times 24.572\,(40 \times 289.9 \times \cos^2 23 + 156.62)\cos 23}{21 \times 24.572 + \sqrt{40 \times 289.9\cos^2 23 + 156.72}}$$

$$= 7847.6 \text{ N}$$

$$P_{eff} = P_{tmax} + P_d = 152.62 + 7695 = 7847.6 \text{ N}$$

Beam strength,

$$S_b = 719.2\ m_n^2 = 11508 \text{ N} \qquad \therefore FS = 1.47$$

Wear strength,

$$S_w = 369.25\ m_n^2 = 5907 \text{ N} \qquad \therefore FS = 0.75$$

Thus, we see that results are better than original but still inadequate since factor of safety required in bending as well as wear should be two.

Trial 2:

We select module = 6 mm. Further, since $S_w << S_b$. We increase material hardness to 350 BHN so that S_w will be improved.

We conduct above steps again and also find new value of S_w.

$$e_p = 8 + 0.633\,(4 + 0.25\sqrt{19.555 \times 6})$$
$$= 12.23\ \mu m$$
$$e_g = 8 + 0.63\,(4 + 0.25\sqrt{19.555 \times 4 \times 6})$$
$$= 13.93\ \mu m$$
$$e = e_p + e_g = 26.16\ \mu m = 0.02616\ mm$$

Thus, $C = 11500 \times e = 300.8$ N/mm

Width, $b = 10 \times m_n = 60$ mm

$$v = 6.143 \times m_n = 36.858\ m/sec$$
$$P_t = \frac{407}{m_n} = 67.8\ N$$
$$P_{tmax} = 1.5 \times 1 \times 67.8 = 101.7\ N$$

By Buckingham Equation

$$P_d = \frac{21 \times 36.858\,(60 \times 300.8 \times \cos^2 23 + 101.7)\cos 23}{21 \times 36.858 + \sqrt{60 \times 300.8 \cos^2 23 + 101.7}}$$
$$P_d = 12931\ N$$
$$P_{tmax} = 101.7\ N$$
$$P_{eff} = P_{tmax} + P_d = 13033\ N$$

Beam strength = $719.2\ m_n^2 = 25893$ N

$$\text{Wear strength} = \frac{b \cdot Q \cdot d_p \cdot K}{\cos^2 \psi}$$
$$= \frac{60 \times 1.6 \times (19.555 \times 6) \times 0.16 \left(\frac{350}{100}\right)^2}{\cos^2 23}$$
$$= 26054\ N$$
$$\text{FS in bending} = \frac{S_b}{P_{eff}} = \frac{25893}{13033} = 1.987$$
$$\text{FS in wear} = \frac{S_w}{P_{eff}} = \frac{26054}{13033} = 1.999$$

Thus, the requirements are almost satisfied and we can select module = 6 mm with increased hardness for the surfaces to be 350 BHN.

Example 4.5: *A steel helical pinion (S_{ult} = 800 N/mm²) having 16 teeth meshes with a steel gear (S_{ult} = 720 N/mm²). The pair transmits 20 kW power from 720 rpm motor to a 144 rpm machine. The application factor is 1.2 and load concentration factor is 1.1. Required factor of safety is 1.5. The face width is 10 times the normal module. The tooth system is 20° full depth involute while the helix angle is 25°. The gear is machined to grade. Deformation factor for the gear pair is 11000 · e N/mm. Design the gear pair by using the velocity factor and Buckingham's equation for dynamic load. Suggest the surface hardness for gear pair. Use the following data:*

$$Y = 0.484 - \frac{2.87}{Z'}$$
$$C_v = \frac{5.6}{5.6 + \sqrt{v}} \qquad e = 11 + 0.9\,(m_n + 0.25\sqrt{d})$$

Load stress factor,

$$K = 0.16\left(\frac{BHN}{100}\right)^2 \quad P_d$$
$$= \frac{21 \cdot v\,(b \cdot C \cdot \cos^2 \psi + P_{tmax})\cos \psi}{21 \cdot v + \sqrt{b \cdot C \cos^2 \psi + P_{tmax}}}$$

Given data: Z_p = 16, Z_g = 80, Power = 20 kW, $(S_{ult})_{pinion}$ = 800 N/mm², $(S_{ult})_{gear}$ = 720 N/mm², C_s = 1.2, C_L = 1.1, FS = 1.5, b = 10 · m_n, ϕ_n = 20°, ψ = 25°.

Solution:

Effective Force:

Pinion diameter d_p

$$= \frac{m_n \cdot Z_p}{\cos \psi} = \frac{m_n\,16}{\cos 25} = 17.65\ m_n$$

Pitch line velocity,

$$v = \frac{\pi \cdot d_p \cdot N_p}{60 \times 1000} = \frac{\pi \times 17.65 \cdot m_n \times 720}{60 \times 1000}$$
$$= 0.665\ m_n$$

Tangential force,

$$P_t = \frac{\text{Power}}{v} = \frac{20 \times 10^3}{0.665\ m_n} = \frac{30075.2}{m_n}$$

Effective force,

$$P_{eff} = \left(\frac{C_s \cdot C_L}{C_v}\right) \cdot P_t$$
$$= \frac{1.2 \times 1.1\,(5.6 + \sqrt{0.665\ m_n})}{(5.6)} \times \frac{30075}{m_n}$$
$$P_{eff} = \frac{7089.1}{m_n}\,[5.6 + \sqrt{0.665\ m_n}]$$

Beam Strength:

$$S_b = \sigma_b \cdot b \cdot Y'\ m_n$$
$$Y'_p = 0.484 - \frac{2.87}{(Z_p/\cos^3 \psi)} = 0.35$$
$$Y'_g = 0.484 - \frac{2.87}{(80/\cos^3 \psi)} = 0.457$$
$$(\sigma_b \cdot Y'_p) = \frac{800}{3} \times 0.35$$
$$= 93.3 \qquad (\sigma_b\,Y'_g) = \frac{720}{3} \times 0.457 = 109.68$$

Since $(\sigma_b\,Y'_p)$ is smaller than $(\sigma_b\,Y'_g)$

Pinion is weaker. Hence beam strength is found for the pinion.

Beam Strength:

$$S_b = \sigma_b \cdot b \cdot Y' \cdot m_n$$
$$= 93.3 \times 10 \cdot m_n \times m_n = 933\ m_n^2$$

Equating beam strength with P_{eff} considering factor of safety

$$S_b = FS\,(P_{eff})$$
$$933\ m_n^2 = 1.5\left[\frac{7089.1}{m} \cdot \left(5.6 + \sqrt{0.665\ m_n}\right)\right]$$

$$m_n^3 - 9.29\sqrt{m_n} - 63.8 = 0$$

Solving by trial and error, $m_n = 4.4$

We select module = 5 mm.

Dynamic Load using Buckingham Equation:

$V = 0.665\ m_n = 3.325$ m/sec, $b = 10 \times m_n = 50$ mm

$$P_{tmax} = P_t \times C_s \cdot C_L = \frac{30075}{m_n} \times 1.2 \times 1.1 = 7939 \text{ N}$$

$$d_p = \frac{m_n \cdot Z_p}{\cos\psi} = \frac{5 \times 16}{\cos 25}$$

$$= 88.27 \text{ mm } d_g = \frac{m_n \cdot Z_g}{\cos\psi} = \frac{5 \times 80}{\cos\psi}$$

$$= 441.35 \text{ mm}$$

$$e_p = 11 + 0.9\,(m_n + 0.25\sqrt{d_p}) = 17.61\ \mu\text{m}$$

$$e_g = 11 + 0.9\,(m_n + 0.25\sqrt{441.35})$$

$$= 20.22\ \mu\text{m}$$

$$e = e_p + e_g$$

$$= 37.83\ \mu\text{m} = 0.03783 \text{ mm C}$$

$$= 11000 \times 0.03783 = 416 \text{ N/mm}$$

$$P_d = \frac{21 \cdot v\,(b \cdot C\cos^2\psi + P_{tmax})\cos\psi}{21 \cdot v + \sqrt{b \cdot C\cos^2\psi + P_{tmax}}}$$

$$= 6948 \text{ N}$$

$$P_{eff} = P_{tmax} + P_d = 7939 + 6948 = 14887 \text{ N}$$

Beam strength,

$$S_b = 933\ m_n^2 = 23325 \text{ N}$$

$$FS = \frac{S_b}{P_{eff}} = \frac{23325}{14887} = 1.56$$

Surface Hardness for Gear:

Wear strength is given by

$$F_w = \frac{b \cdot Q \cdot d_p \cdot K}{\cos^2\psi}$$

$b = 50$ mm, $d_p = 88.27$ mm;

$$Q = \frac{2Z_g}{Z_g + Z_p} = \frac{2 \times 80}{16 + 80} = 1.667$$

$$K = 0.16\left(\frac{BHN}{100}\right)^2$$

$$F_w = \frac{50 \times 1.667 \times 88.27 \times 0.16}{\cos^2\psi}\left(\frac{BHN}{100}\right)^2$$

$$= 1433.13\left(\frac{BHN}{100}\right)^2$$

Equating wear strength with P_{eff} considering factor of safety

$$F_w = FS\,(P_{eff})$$

$$1433.13\left(\frac{BHN}{100}\right)^2 = 1.5\,(14887)$$

Solving required value of hardness = 395 BHN.

Example 4.6: *A pair of helical gears consists of 26 teeth pinion meshing with a 52 teeth gear. The normal module is 4 mm and normal pressure angle is 20°. Find the required value of helix angle if centre distance is to be maintained exactly 170 mm. The pinion transmits 5 kW power at 1500 rpm. Pinion has right hand helix and rotates clockwise when seen from the right side. Show the forces on pinion and gear.*

Solution: $Z_p = 26$, $Z_g = 52$, $m_n = 4$ mm

$$\text{Centre distance, CD} = \frac{d_p}{2} + \frac{d_g}{2}$$

$$170 = \frac{m_n \cdot Z_p}{2 \cdot \cos\psi} + \frac{m_n \cdot Z_g}{2 \cdot \cos\psi}$$

$$= \frac{1}{\cos\psi}\left(\frac{4 \times 26}{2} + \frac{4 \times 52}{2}\right)$$

$$\cos\psi = 0.9176$$

Helix angle, $\psi = 23.42°$

Torque on pinion 'A'

$$= \frac{(\text{Power})_{kW} \times 60 \times 10^6}{2\pi N} = \frac{5 \times 60 \times 10^6}{2\pi \times 1500}$$

$$= 31830 \text{ N·mm}$$

Tangential force on pinion

$$= \frac{31830}{\left(\frac{d_p}{2}\right)} = \frac{31830}{\left(\frac{m_n \times Z_p}{2 \times \cos\psi}\right)} = 561 \text{ N} = P_t$$

Axial force on pinion

$$= P_a = P_t \times \tan\psi = 243 \text{ N} = P_a$$

Radial force on pinion

$$= P_t \times \frac{\tan\phi_n}{\cos\psi} = 561 \times \frac{\tan 20°}{\cos 23.41°}$$

$$= 223 \text{ N} = P_r$$

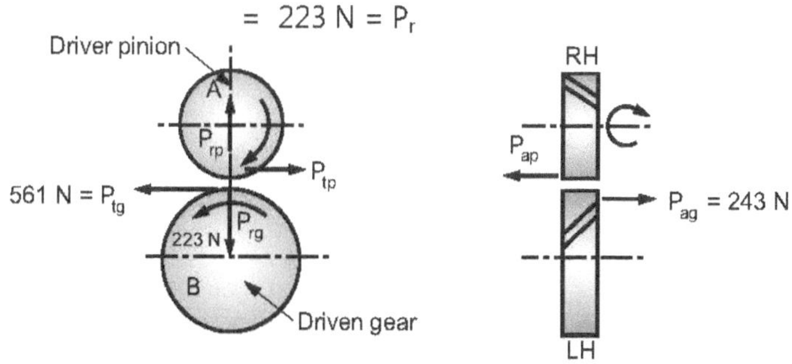

Fig. 4.12

Example 4.7: *A pair of helical gears with 18 teeth on pinion and 54 teeth on gear transmits 10 kW power at 1500 rpm. Both gears are made of case hardened steel (S_{ult} = 720 N/mm²). Normal pressure angle is 20° and the helix angle is 24°. Motor starting torque is double the rated torque and load concentration factor is 1.1. Gears are finished to grade 6 and desired factor of safety is 1.6.*

1. *Estimate normal module considering dynamic load by velocity factor. Select standard module.*
2. *Determine dynamic load using Buckingham's equation and factor of safety in bending.*
3. *Recommend suitable hardness for gears.*

For grade 6: $e = 8 + 0.63\,(m + 0.25\sqrt{d})$

Deformation factor, $C = 11500 \times e$

$$C_v = \frac{5.6}{5.6 + \sqrt{v}}\,P_d$$

$$= \frac{21 \cdot v\,(b \cdot C \cdot \cos^2\psi + P_{tmax})\cos\psi}{21 \cdot v + \sqrt{b \cdot C\cos^2\psi + P_{tmax}}}$$

Given data: $Z_p = 18$, $Z_g = 54$, Power = 10 kW, N_p = 1500 rpm, $S_{ult} = 720\ N/mm^2$, $\phi_n = 20°$, $\psi = 24°$, $C_s = 2.0$, $C_L = 1.1$.

Solution: Effective Force:

$$d_p = \frac{m_n \cdot Z_p}{\cos\psi} = \frac{18 \cdot m_n}{\cos 24} = 19.7\ m_n$$

$$v = \text{Pitch line velocity} = \frac{\pi \cdot d_p \cdot N_p}{60 \times 1000}$$

$$= \frac{\pi \times 19.7\ m_n \times 1500}{60 \times 1000} = 1.547\ m_n$$

$$\text{Tangential force, } P_t = \frac{\text{Power}}{\text{Velocity}} = \frac{10.0 \times 10^3}{1.547\ m_n} = \frac{6464.12}{m_n}\ N$$

$$\text{Velocity factor, } C_v = \frac{5.6}{5.6 + \sqrt{1.547\ m_n}}$$

Effective force,

$$P_{eff} = \left(\frac{C_s \cdot C_L}{C_v}\right)P_t$$

$$= \frac{2.0 \times 1.1 \times (5.6 + \sqrt{1.547\ m_n})}{5.6} \times \frac{6464.12}{m_n}$$

$$P_{eff} = \frac{2540\,(5.6 + \sqrt{1547\ m_n})}{m_n}$$

Beam Strength:

Pinion is weaker since material is same for pinion and gear.

∴ Beam strength for pinion:

$$S_b = \sigma_b \cdot b \cdot Y'\ m_n$$

Lewis form factor,

$$Y'_p = 0.484 - \frac{2.87}{Z'_p}$$

where, $$Z'_p = \frac{Z_p}{\cos^3\psi} = \frac{18}{\cos^3 24} = 23.61$$

$$Y'_p = 0.484 - \frac{2.87}{23.61} = 0.3624$$

Beam Strength:

$$S_b = \sigma_b \cdot b \cdot Y' \cdot m_n = \frac{720}{3} \cdot (10\ m_n) \times 0.3624 \times m_n$$

$$S_b = 869.76\ m_n^2$$

Equating beam strength with effective load considering factor of safety

$$S_b = [P_{eff}]\ FS$$

$$869.76\ m_n^2 = \left[2540\,\frac{(5.6 + \sqrt{1.547\ m_n})}{m_n}\right]1.6$$

$$m_n^3 - 5.58\sqrt{m_n} - 26.16 = 0$$

Solving by trial and error, m_n = 3.4 mm. We select m_n = 4 mm.

Check for Dynamic Load:

$$e_p = 8 + 0.63\,(m_n + 0.25\sqrt{d_p})$$

$$= 8 + 0.63\,(4 + 0.25\sqrt{78.8}) = 11.92\ \mu m$$

$$e_g = 8 + 0.63\,(m_n + 0.25\sqrt{d_g})$$

$$= 8 + 0.63\,(4 + 0.25\sqrt{236.4}) = 12.94\ \mu m$$

$$e = e_p + e_g = 24.86\ \mu m$$

Deformation factor,

$$C = 11500 \times e = 11500 \times 0.02486 = 285.89\ N/mm$$

$$v = 1.547\ m_n = 6.188\ m/sec.$$

$$P_{tmax} = \frac{6464.12}{m_n} = 1616\ N$$

$$P_{tmax} = (C_s \cdot C_L)\,P_t = 2 \times 1.1 \times 1616 = 3555\ N$$

$$P_d = \frac{21 \cdot v\,(b \cdot C \cdot \cos^2\psi + P_{tmax})\cos\psi}{21 \cdot v + \sqrt{b \cdot C\cos\psi + P_{tmax}}}$$

$$= \frac{21 \times 6.188\,(40 \times 285.89 \times \cos^2 24 + 3555)\cos 24}{21 \times 6.188 + \sqrt{(40 \times 285.89\cos^2 24 + 3555)}}$$

$$P_d = 6363.2\ N$$

$$P_{eff} = P_{tmax} + P_d = 3555 + 6363.2 = 9918.2\ N$$

$$\text{Factor of safety in bending} = \frac{S_b}{P_{eff}} = \frac{869.76\ m_n^2}{P_{eff}} = 1.4$$

Deciding Hardness:

Wear strength should be S_b = 13917 N

$$S_w = \frac{b \cdot Q \cdot d_p \cdot K}{\cos^2\psi}$$

where, $$Q = \frac{2Z_g}{Z_g + Z_p} = \frac{2 \times 54}{18 + 54} = 1.5.$$

$$\therefore\quad K_{reqd.} = \frac{13917 \times \cos^2 24}{40 \times 1.5 \times (19.7 \times 4)} = 2.456$$

$$K = 2.456 = 0.16\left(\frac{BHN}{100}\right)^2$$

∴ Required hardness, BHN = 392

EXERCISE

1. What is formative number of teeth in helical gears? Derive the expression for formative number of teeth in helical gears.
2. Derive a relation for virtual number of teeth for a helical gear.
3. What is the relationship between transverse and normal pressure angle and helix angle?
4. State two advantages and disadvantages of herringbone and double helical gear, also draw the sketches.
5. Obtain an expression for the formative number of teeth.

CHAPTER - 5
BEVEL GEAR

5.1 INTRODUCTION

- When power is to be transmitted between two intersecting shafts, bevel gears are used. In bevel gears, teeth are cut on a frustum of a cone.

Types of Bevel Gears :

- Bevel gears are of two types, straight teeth bevel and curved or spiral teeth bevel. In case of straight teeth bevel, the teeth are cut along the generators of the cone and if the teeth are extended they intersect with the axis of rotation [refer Fig. 5.1 (a)]. If curved teeth are cut on the frustum of cone the gear is called spiral teeth bevel [refer Fig. 5.1 (b)].

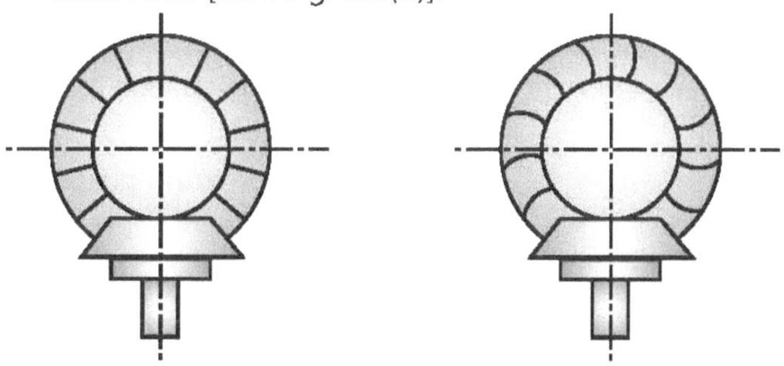

(a) Straight Teeth Bevel **(b) Spiral Teeth Bevel**

Fig. 5.1

- The point of intersection is known as apex point which is common for both, driving and driven bevel gears. Straight teeth bevel gears are easy to manufacture, easy to maintain while the spiral bevel, because of complicated shape of tooth requires special purpose machines. Because of this they are costlier compared with straight tooth bevel. The straight bevel gears create noise if operated at higher speeds while because of curved teeth, the contact between two meshing spiral gears take place gradually and results in quiet operation. Hence, they are suitable for high speed applications.
- The bevel gear and pinion are designed in pairs. Hence, they are replaced together in pair. Although bevel gears can be used for almost any angle between shaft axes, they are most commonly used to connect shafts at right angles which are intersecting. In this chapter, only straight tooth bevel gears transmitting power at right angles will be discussed.

5.2 TERMINOLOGIES OF BEVEL GEARS

The terminologies of bevel gears are given below. (Refer Fig. 5.2)

- **Pitch Cone:** The tooth axes (pitch line) if extended, intersect the centre of rotation at apex point 'O'. The cone formed by extending the tooth axes is called as pitch cone.
- **Pitch Cone Angle (γ):** The angle contained in the pitch line and the axis of rotation is called as pitch cone angle.
- **Cone Distance (A_0):** It is the slant height of the pitch cone.
- **Pitch Circle Diameter (d):** It is the diameter of the pitch circle at the larger end of bevel gear.
- **Addendum (h_a):** It is the height of gear tooth above pitch circle measured at larger end of the gear.
- **Dedendum (h_f):** It is the depth of gear tooth below the pitch circle measured at larger end of the gear.

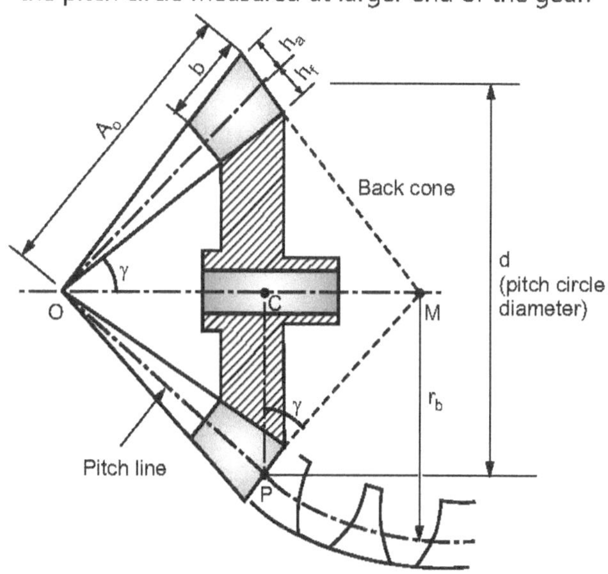

Fig. 5.2 : Bevel Gear Terminology

- **Back Cone:** It is the imaginary cone formed by drawing perpendiculars to the tooth axes (pitch line) on the opposite side of the pitch cone.
- **Back Cone Radius (r_b):** It is the slant height of the back cone.

5.3 VIRTUAL NUMBER OF TEETH (Z')

For simplifying the analysis of bevel gears, an imaginary spur gear is considered which has pitch circle radius as back cone radius of the bevel gear and number of teeth as Z'. These teeth on imaginary spur gear are known as formative or virtual number of teeth. Referring to the Fig. 5.3.

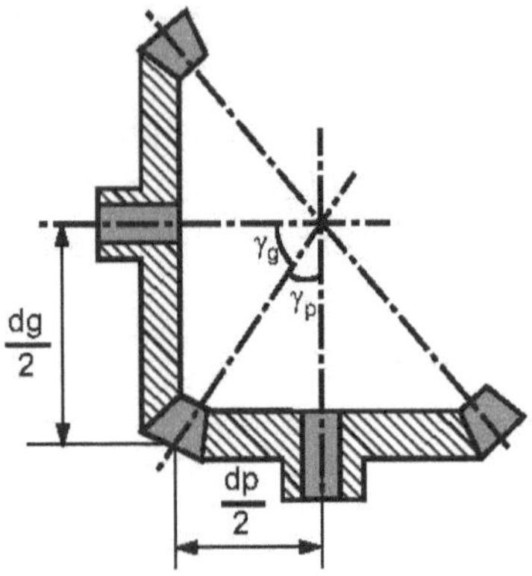

Fig. 5.3 : Bevel Gears in Mesh

For the back cone,

$$\text{Diameter} = 2r_b = Z' \times m$$

$$Z' = \frac{2r_b}{m} \qquad \text{... (5.1)}$$

Where m is the module at large end.

For the pitch cone,

$$Z = \frac{d}{m} \qquad \text{... (5.2)}$$

Dividing equation (5.9) by (5.10),

$$\frac{Z'}{Z} = \left(\frac{2r_b/m}{d/m}\right) = \frac{2r_b}{d}$$

In triangle PCM,

$$\sin(90^\circ - \gamma) = \frac{d/2}{r_b}$$

$$\text{or} \quad \cos\gamma = \frac{d}{2r_b}$$

$$\text{Hence,} \quad \frac{Z'}{Z} = \frac{1}{\cos\gamma}$$

$$Z' = \frac{Z}{\cos\gamma} \qquad \text{... (5.3)}$$

Equation (5.11) is very important which shall be used in the design of bevel gears.

Geometric Relationships and Cone Distance (A_0):

Refer Fig. 5.3 which shows bevel gears in mesh.

Let γ_p, γ_g = Pitch cone angles of pinion and gear respectively.

d_p, d_g = Pitch circle diameters of pinion and gear measured at the large end.

$$\tan\gamma_p = \frac{d_p/2}{d_g/2} = \frac{d_p}{d_g} = \frac{mZ_p}{mZ_g} = \frac{Z_p}{Z_g} \qquad \text{... (5.4)}$$

$$\tan\gamma_g = \frac{d_g/2}{d_p/2} = \frac{d_g}{d_p} = \frac{mZ_g}{mZ_p} = \frac{Z_g}{Z_p} \qquad \text{... (5.5)}$$

Considering that the two shafts are at right angles

$$\gamma_p + \gamma_g = 90^\circ$$

Cone distance (slant height) is given by

$$A_o = \sqrt{\left(\frac{d_p}{2}\right)^2 + \left(\frac{d_g}{2}\right)^2} \qquad \text{...(5.6)}$$

5.4 FORCE ANALYSIS

The Fig. 5.4 shows the various forces on the pinion and gear teeth. For the purpose of force analysis these forces are assumed to act at the midpoint of the face width along the pitch line.

Let,

P_n - Resultant tooth force acting along common normal drawn at point of contact.

r_{mp}, r_{mg} - Radius of bevel pinion and gear at midpoint of face width

P_{tp}, P_{tg} - Tangential tooth forces on pinion and gear

P_{rp}, P_{rg} -Radial tooth forces on pinion and gear

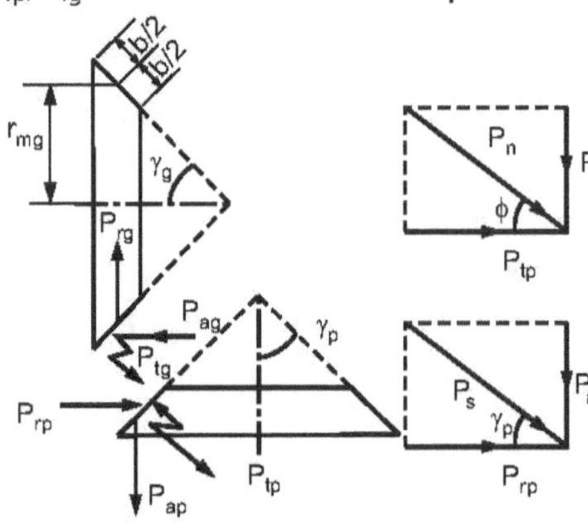

Fig. 5.4: Bevel Gear Pair in Contact

P_{ap}, P_{ag} - Axial tooth forces on pinion and gear

ϕ - Pressure angle

From Fig. 5.4.

$$r_{mp} = \left[\frac{d_p}{2} - \frac{b \sin\gamma_p}{2}\right] \qquad \text{... (5.7)}$$

$$r_{mg} = \left[\frac{d_p}{2} - \frac{b \sin\gamma_p}{2}\right] = \left[\frac{d_g}{2} - \frac{b \sin(90 - \gamma_p)}{2}\right]$$

$$= \left[\frac{d_g}{2} - \frac{b \cos\gamma_p}{2}\right] \qquad \text{... (5.8)}$$

Refer Fig. 5.4.

At the midpoint of the face width, the resultant force (P_n) is resolved into two components, tangential (P_t) and the other separating force (P_s). The separating force (P_s) is further resolved into two components as radial (P_{rp}) and axial force (P_{ap}). The magnitudes of the forces on the pinion teeth can be obtained as below.

If T_p is the torque on the pinion, tangential force (P_{tp}) on the pinion is given by

$$P_{tp} = \frac{T}{r_{mp}} \quad \text{... (5.9)}$$

Knowing (P_{tp}), normal and separating force on the pinion is obtained as

$$P_n = P_{tp} \cos \phi \quad \text{... (5.10)}$$

$$P_s = P_{tp} \tan \phi \quad \text{... (5.11)}$$

Separating forces are further resolved into radial and axial (thrust) forces on the pinion.

$$P_{rp} = P_s \cos \gamma_p = P_{tp} \tan \phi \cos \gamma_p, \text{ and} \quad \text{... (5.12)}$$

$$P_{ap} = P_s \sin \gamma_p = P_{tp} \tan \phi \sin \gamma_p \quad \text{... (5.13)}$$

Once the forces on the pinion are found using above equations, the forces on gear tooth can be found out using following relations (Refer Fig. 5.4).

$$P_{tp} = P_{tg}, P_{rg} = P_{ap} \text{ and } P_{ag} = P_{rp} \quad \text{... (5.14)}$$

Force Directions on Pinion and Gear:

- The direction of tangential force on pinion and gear is found by same method adopted for spur gear. On the driving gear direction of tangential force is opposite to motion and on the driven gear, the direction of force is same as direction of motion at the contact point.
- Radial and axial forces on pinion and gear are separating in nature. Hence, radial and axial force directions are easy to establish since both will act in such a direction as to separate pinion and gear.

5.5 BEAM STRENGTH OF BEVEL GEAR

To find the beam strength of bevel gear an imaginary spur gear is considered having number of teeth equal to virtual number of teeth. Considering an element of thickness dx situated at a distance x from the apex O and radius as r_x. Beam strength for this element with virtual number of teeth will be

$$S_{bx} = \sigma_b \cdot b_x \cdot Y' \cdot m_x \quad \text{... (5.15)}$$

where,

S_{bx} Beam strength of the element

b_x Face width of the element

σ_b Allowable bending stress

Y' Lewis form factor based on the virtual number of teeth

m_x Module for the element (Module for the bevel gear will not be constant along the width since the gear radius varies along its width)

From Fig. 5.5,

$$\frac{r_x}{R} = \frac{x}{A_0} \quad \text{OR} \quad r_x = \frac{xR}{A_0}$$

Considering the elemental section

$$m_x = \frac{2r_x}{Z} = \frac{2xR}{ZA_0}$$

Since, $$m = \frac{2R}{Z}$$

$$m_x = m\left(\frac{x}{A_0}\right) \quad \text{... (5.16)}$$

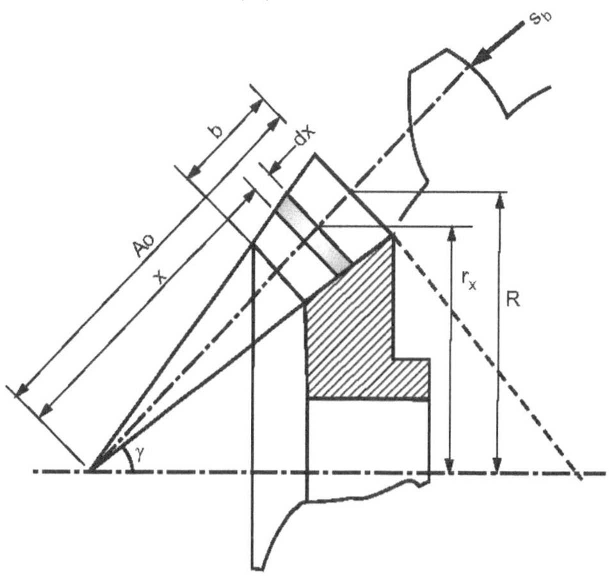

Fig. 5.5 : Beam Strength of Bevel Gear

$$S = \sigma_b dx Y'\left(\frac{mx}{A_0}\right)$$

$$= \frac{\sigma_b \cdot Y' \cdot m \cdot mdx}{A_0} \quad \text{... (5.17)}$$

Torque on pinion shaft due to tangential force S_b on this element will be

$$T_x = S_{bx} \times r_x = \frac{\sigma_b \cdot Y' \cdot m \cdot R \cdot x^2\, dx}{A_0^2} \quad \text{... (5.18)}$$

Total torque on pinion shaft will be obtained by integrating the torque due to element within the range (A_0-b) to A_0

$$T = \int T_x dx = \int_{A_0-b}^{A_0} \frac{\sigma_b \cdot Y'\, m \cdot x^2\, R}{A_0^2} dx$$

$$= \left(\frac{\sigma_b\, Y'\, mR}{A_0^2}\right) \int_{A_0-b}^{A_0} x^2\, dx$$

$$= \left(\frac{\sigma_b Y' mR}{A_0^2}\right)\left[\frac{x^3}{3}\right]_{A_0-b}^{A_0}$$

$$= \sigma_b bY' mR\left[1 - \frac{b}{A_0} + \frac{b^2}{3A_0}\right] \quad \dots (5.19)$$

Assuming S_b (tangential force) to be acting at the large end,

$$T = S_b \times R \quad \dots (5.20)$$

Equating equations (5.27) and (5.28)

$$S_b = \sigma_b bY'm\left[1 - \frac{b}{A_0} + \frac{b^2}{3A_0}\right] \quad \dots (5.21)$$

In the above equation since $\leq \frac{1}{3} A_0$, the term $\left(\frac{b^2}{3A_0^2}\right)$ is neglected being very small.

Hence, beam strength of a bevel gear is given by

$$S_b = \sigma_b bY'm\left[1 - \frac{b}{A_0}\right] \quad \dots (5.22)$$

where,

- S_b – Beam strength in N
- m – Module in mm
- b – Face width 10 m or $\frac{A_0}{3}$ (whichever is small)
- σ_b – Permissible bending stress $\sigma_b = \frac{S_{ut}}{3}$
- Y' – Lewis form factor based on virtual number of teeth

$$Y' = 0.484 - \frac{2.87}{Z'} \text{ for } 20° \text{ full depth involute system}$$

- A_0 – Cone distance

To avoid bending failure, the beam strength should be greater than the effective load.

5.6 WEAR STRENGTH

For the straight tooth bevel, the contact between the meshing teeth is line contact. This is similar to the spur gear pair in mesh. Hence for the equivalent spur gear with number of teeth equal to the virtual number of teeth, the wear strength equation can be written as:

$$S_w = b \cdot Q' \cdot K \cdot d_p' \quad \dots (5.23)$$

where,

- b – Face width of bevel gear (10 m or $A_0/3$ whichever smaller)
- Q' – Ratio factor given by equation Q'

$$= \frac{2Z_g'}{Z_g' + Z_p'} \quad \dots (5.24)$$

Q' is the ratio factor for the equivalent spur gear pair.

For 90° angle between pinion and gear shaft we have

$$Z_p' = \frac{Z_p}{\cos \gamma_p} \quad \text{and} \quad Z_g' = \frac{Z_g}{\sin \gamma_p}$$

Substituting in the equation (5.24)

$$Q' = \frac{2Z_g}{Z_g + Z_p \cdot \tan \gamma_p}$$

K – Load stress factor

$$K = 0.16\left(\frac{BHN}{100}\right)^2$$

... for steel pinion meshing with steel gear

$$K = 0.18\left(\frac{BHN}{100}\right)^2$$

... for steel pinion meshing with cast iron gear

$$K = 0.21\left(\frac{BHN}{100}\right)^2$$

... for cast iron pinion meshing with cast iron gear

d_p' - Pitch circle diameter of equivalent spur gear. $d_p' = \frac{d_p}{\cos \gamma_p}$

$$S_w = \frac{bQ' Kd_p}{\cos \gamma_p} \quad \dots (5.25)$$

In usual practice one of the bevel gears is fitted on a shaft with overhang (i.e. at the end of shaft). Due to this there may be excessive deflection of the shaft. The load between pinion and gear teeth is therefore not uniformly distributed along the width and gear pair does not transmits the forces to its capacity. Hence, the force transmission is considered to be 25% less than its full load capacities. **Buckingham's Wear Strength Equation is Expressed as,**

$$S_w = \frac{0.75\, bQ' Kd_p}{\cos \gamma_p} \quad \dots (5.26)$$

5.7 EFFECTIVE LOAD

- For the design of the gears based on the beam or wear strength, it is already discussed that the effective tangential load is obtained by considering the tangential load with all the modifying factors such as service factor, load concentration factor and velocity factor. It is assumed that the gear tooth is subjected to this effective load. Equating effective load with the beam and wear strength considering appropriate factor of safety, the module will be obtained.

- Like spur gears, in the initial stages of gear design, the dynamic load effect for the bevel gears is considered using approximate approach by using Barth or velocity factor.

$$P_{eff} = \frac{C_s C_L P_t}{C_v} \quad \text{... (5.27)}$$

[To be used in the initial stages of the design when module is unknown]

- To determine the actual (available) factor of safety when module is known, the exact dynamic load using Buckingham's equation should be found. The effective load will be found based on exact dynamic load.

$\therefore \quad P_{eff} = C_s\, C_L\, P_t + P_d$ [To be used after knowing the module]

Here, P_{eff} - Effective tooth load in N

C_s - Service factor

C_L - Load concentration factor

C_v - Velocity factor

P_t - Tangential tooth load in N

P_d - Dynamic tooth load obtained by Buckingham's equation in N

Velocity or Barth factor in case of bevel gears may be considered using either of the two equations given below:

$$C_v = \frac{6}{6 + v} \quad \text{or} \quad C_v = \frac{5.6}{5.6 + \sqrt{v}}$$

Buckingham's incremental dynamic tooth load equation is used to determine dynamic load.

$$P_d = \frac{21v\,(bC + C_s C_L P_t)}{21v + \sqrt{(bC + C_s C_L P_t)}} \quad \text{... (5.28)}$$

Here, V - Pitch line velocity in m/s

b - Face width of gear in mm

C - Deformation factor in N/mm

$$C = \frac{Ke}{\left[\frac{1}{E_p} + \frac{1}{E_g}\right]}$$

K is factor depending upon tooth form

K = 0.107 for 14.5° full depth involute and composite systems.

= 0.111 for 20° full depth involute teeth.

= 0.115 for 20° stub involute systems.

E_p, E_g are the moduli of elasticity for pinon and gear materials respectively in N/mm^2.

e is the total tooth error in the pinion and the gear, (= $e_p + e_g$) mm.

Procedure for obtaining tooth error 'e' is same as in case of spur gear and helical gears.

5.8 DESIGN OF STRAIGHT TEETH BEVEL GEARS

The design of straight teeth bevel gears is almost similar to the spur gears.

Following is the procedure for the design:

- Calculate the pitch line velocity in terms of module from the given data.
- Find the tangential load in terms of module using relation $P_t = \frac{P}{v}$. In this relation, the power should be considered in watts and the pitch line velocity in m/s.
- Calculate C_v in terms of module.
- Calculate P_{eff} in terms of module.
- Determine the beam strength of the weaker element (pinion or gear) in terms of module using equation (5.22).
- Determine the pitch cone angles for the pinion and gear using equations (5.6) and (5.5).
- Determine the virtual (formative) number of teeth for pinion and gear using equation (5.3).
- Determine the Lewis form factor based on virtual number of teeth on pinion and gear.
- Obtain the product of bending stress and Lewis form factor for pinion and gear. Select the smaller of the two to obtain the beam strength.
- Determine A_0 using equation (5.6) and $A_0/3$ and decide the face width. (10 m or $A_0/3$, smaller of the two).
- Determine the wear strength of the pair (considering smaller of the hardness values for pinion and gear) in terms of module using equation (5.26).
- Determine the ratio factor Q' based on the virtual number of teeth on pinion and gear.
- Calculate load stress factor K.
- Compare the beam strength and wear strength and identify the smaller of the two.
- Equate the smaller of beam and wear strength with effective load considering factor of safety to obtain the module.
- Select the standard module and calculate, face width, pitch circle diameters of pinion and gear.
- Calculate pitch line velocity, tangential load, tooth error and deformation factor.
- Calculate the exact dynamic load using Buckingham's equation and new effective load based on exact dynamic load.

- Calculate the beam and wear strength values and obtain the actual (available) factor of safety.
- If the actual factor of safety is more than the initially taken, the gear design is safe. Else, go for a new iteration with higher module and repeat the steps from 10 to 13 till the actual factor of safety is larger than the initially taken.

SOLVED EXAMPLES

Example 5.1: *A pair of straight bevel gears, intersecting at right angles, consists of 20 teeth pinion meshing with 44 teeth gear. Pinion and gear are made of case hardened steel (UTS = 680 N/mm²) and gear (UTS = 600 N/mm²). Pinion transmits 12 kW power at 1500 rpm. Service factor is 1.75. Assuming that velocity factor accounts for dynamic load, design the gear pair for a factor of safety of 1.5. Assume surface hardness of 380 BHN for the gear pair. Use* $C_v = \dfrac{5.6}{5.6 + \sqrt{v}}$.

Given data : $Z_p = 20$, $Z_g = 44$, $(S_{ult})_{pinion} = 680$ N/mm², $(S_{ult})_{gear} = 600$ N/mm², Power = 12 kW, $N_p = 1500$ rpm, $C_s = 1.75$, FS = 1.5, BHN = 380.

Solution: Effective Load:

$$d_p = m \cdot Z_p = 20 \cdot m$$

$$v = \text{Pitch line velocity}$$

$$= \frac{\pi \cdot d_p \cdot N_p}{60 \times 1000} = \frac{\pi \times 20\, m \times 1500}{60 \times 1000}$$

$$= 1.5708\, m \text{ (m/min)}$$

Tangential force,

$$P_t = \frac{\text{Power}}{\text{Velocity}} = \frac{12 \times 10^3}{1.5708}$$

$$= \frac{7639.4}{m}$$

$$\text{Velocity factor} = \frac{5.6}{5.6 + \sqrt{v}} = \frac{5.6}{5.6 + \sqrt{1.5708\, m}}$$

$$P_{eff} = \left(\frac{C_s \cdot C_L}{C_v}\right) P_t$$

$$= \frac{1.75 \times 1 \times (5.6 + \sqrt{1.5708\, m})}{5.6} \times \frac{7639.4}{m}$$

$$P_{eff} = \frac{2387.3\, (5.6 + \sqrt{1.5708\, m})}{m}$$

Beam Strength:

$$S_b = \sigma_b \cdot b \cdot Y' \cdot m \left[1 - \frac{b}{A_o}\right]$$

$$\tan \gamma_p = \frac{Z_p}{Z_g} = \frac{20}{44} = 0.4545,\ \gamma_p = 24.44,\ \gamma_g = 65.56^\circ$$

$$Z_p' = \frac{Z_p}{\cos \gamma_p} = \frac{20}{0.910} = 21.97$$

$$Y_p' = 0.484 - \frac{2.87}{Z_p'} = 0.353$$

$$Z_g' = \frac{Z_g}{\cos \gamma_g} = \frac{44}{0.413} = 106.35\ Y_g'$$

$$= 0.484 - \frac{2.87}{Z_g'} = 0.457$$

$$(\sigma_b \cdot Y_p')_p = \frac{680}{3} \times 0.353 = 80.01$$

$$(\sigma_b \cdot Y_g')_g = \frac{600}{3} \times 0.457 = 91.4$$

Since $(\sigma_b\, Y_p') < (\sigma_b \cdot Y_g')$, pinion is weaker. Beam strength will be based on pinion strength.

$$d_p = m \cdot Z_p = 20 \cdot m, \quad d_g = m\, Z_g = 44 \cdot m$$

Cone distance,

$$A_o = \sqrt{\left(\frac{d_p}{2}\right)^2 + \left(\frac{d_g}{2}\right)^2} = \sqrt{100\, m^2 + 484\, m^2}$$

$$= 24.18\ (m)$$

Now, $$b = \frac{1}{3} A_o$$

$$= 8.055\, m \text{ or } 10\, m \text{ (whichever is smaller)}$$

$$\therefore \text{Width, } b = 8.055 = 8.1\, m$$

Beam Strength: S_b

$$S_b = (\sigma_b \cdot Y_p')_p \cdot b \cdot m \left[1 - \frac{b}{A_o}\right]$$

$$= 80.01 \times 8.1\, m \times m \left[1 - \frac{8.1}{24.17}\right]$$

$$= 430.9\, m^2$$

Wear Strength S_w:

$$\text{Ratio factor } Q' = \frac{2 \cdot Z_g'}{Z_g' + Z_p'} = \frac{2 \times 106.35}{106.35 + 21.97} = 1.658$$

$$K = 0.16 \left(\frac{BHN}{100}\right)^2 = 0.16 \left(\frac{380}{100}\right)^2 = 2.31$$

$$S_w = \frac{0.75\, b \cdot Q' \cdot d_p \cdot K}{\cos \gamma_p}$$

$$S_w = \frac{0.75 \times 8.1\, m \times 1.658 \times 20\, m \times 2.31}{0.91}$$

$$= 511.4\, m^2$$

Since, $S_b < S_w$, the gear pair is weaker in beam strength, hence check should be against S_b.

Equating S_b with effective force P_{eff} considering desired factor of safety.

$$S_b = (P_{eff})\, FS$$

$$430.9\ m^2 = \frac{2387.3\ (5.6 + \sqrt{1.5708\ m})}{m} \times 1.5$$

Simplifying,

$$m^3 - 10.41\sqrt{m} - 46.5 = 0$$

Solving trial and error,

$$m \simeq 4.1$$

We select next higher standard module, m = 5.

Hence, $m = 5$ mm

$d_p = 20\ m = 100$ mm

$d_g = 44$ mm $= 220$ mm

Addendum=1 m = 5 mm

Example 5.2: *Straight bevel gears with 20 and 32 teeth are in mesh with their axes intersecting at right angles. Electric motor developing 10 kW power at 1500 rpm is connected to the pinion shaft. The load distribution factor is 1.2 and starting torque on motor is 150% of the rated torque. Both pinion and gear made of alloy steel having ultimate tensile strength (UTS) of 1000 N/mm² and are heat treated to a surface hardness of 400 BHN. Module and face width are 5 mm and 45 mm respectively. Gears are machined to grade 7. Take deformation factor as 11600 · e N/mm. Assuming that dynamic load is accounted by Buckingham's equation. Find:*

1. *Factor of safety in bending.*
2. *Factor of safety against pitting.*

Tooth system used is 20° full depth involute.

$$e = 11 + 0.9\ (m + 0.25\sqrt{d_m})$$

Given data: $Z_p = 20$, $Z_g = 32$, Power = 10 kW, $N_p = 1500$ rpm, $C_l = 1.2$, $C_s = 1.5$, $S_{ult} = 1000$ N/mm², BHN = 400, m = 5 mm, b = 45 mm.

Solution: $d_p = m \cdot Z_p = 100$ mm

$$\tan\gamma_p = \frac{Z_p}{Z_g} = 0.625,\ \gamma_p = 32°,\ \gamma_g = 58°,$$

$D_g = m \cdot Z_g = 160$ mm.

Pitch line velocity,

$$v = \frac{\pi \cdot d_p \cdot N_p}{60 \times 1000} = 7.854\ \text{m/sec}$$

Tangential force,

$$P_t = \frac{\text{Power}}{\text{Velocity}} = \frac{10 \times 10^3}{7.854} = 1273.2\ \text{N}$$

Maximum tangential force,

$$P_{tmax} = (C_s \cdot C_l)\ P_t = 1.5 \times 1.2 \times P_t = 2291.8\ \text{N}$$

Tooth Error (e):

$$d_{mp} = d_p - b \sin\gamma_p = 100 - 45 \sin 32$$
$$= 76.15\ \text{mm}$$

$$d_{mg} = d_g - b \sin\gamma_g = 160 - 45 \sin 58$$
$$= 121.84\ \text{mm}$$

$$e_p = 11 + 0.9\ (5 + 0.25\sqrt{76.15})$$
$$= 17.46\ \mu \cdot m$$

$$e_g = 11 + 0.9\ (5 + 0.25\sqrt{121.84})$$
$$= 17.98\ \mu \cdot m$$

$$e = 35.44\ \mu m = 0.03544\ \text{mm}$$

$$C = 11600 \times 0.03544 = 411.18\ \text{N/mm}$$

Incremental dynamic load

$$P_d = \frac{21 \cdot v\ (b \cdot C + P_{tmax})}{21 \cdot v + \sqrt{b \cdot C + P_{tmax}}}$$

$$= \frac{21 \times 7.854\ (45 \times 411.7 + 2291.8)}{21 \times 7.854 + \sqrt{45 \times 411.18 + 2291.8}}$$

$$= 11094.8\ \text{N}$$

$$P_{eff} = P_{tmax} + P_d$$
$$= 2291.8 + 11094.8 = 13386.6\ \text{N}$$

Beam Strength:

Since pinion and gear materials are same, pinion is weaker. Therefore, beam strength is calculated based on pinion strength.

$$S_b = \sigma_b \cdot b \cdot Y' \cdot m\left[1 - \frac{b}{A_o}\right]$$

$$\sigma_b = \frac{1}{3} \times 1000 = 333.3\ \text{N/mm}^2,\quad b = 45\ \text{mm}$$

$$Y' = 0.484 - \frac{2.87}{Z'_p} = 0.484 - \frac{2.87}{(Z_p/\cos\gamma_p)}$$

$$= 0.484 - \frac{2.87}{23.58} = 0.362$$

$$A_o = \sqrt{\left(\frac{d_p}{2}\right)^2 + \left(\frac{d_g}{2}\right)^2} = \frac{1}{2}\sqrt{(100)^2 + (160)^2}$$

$$= 94.34\ \text{mm}$$

$$S_b = 333.33 \times 45 \times 0.362 \times 5\left[1 - \frac{45}{94.34}\right]$$

$$= 14200\ \text{N}$$

Wear Strength:

$$S_w = \frac{0.75\ b \cdot Q'\ d_p \cdot K}{\cos\gamma_p};$$

$$Z'_p = \frac{Z_p}{\cos\gamma_p} = 23.58,\quad Z'_g = \frac{Z_g}{\cos\gamma_g} = 60.38$$

$$Q' = \frac{2Z'_g}{Z'_g + Z'_p} = \frac{2 \times 60.38}{23.58 + 60.38} = 1.438$$

$$K = 0.16\left(\frac{BHN}{100}\right)^2 = 0.16\left(\frac{400}{100}\right)^2 = 2.56$$

$$S_w = \frac{0.75 \times 45 \times 1.438 \times 100 \times 2.56}{0.848}$$

$$= 14653.4 \text{ N}$$

Factor of safety in bending

$$= \frac{\text{Beam strength}}{P_{eff}} = \frac{14200}{13386.6} = 1.06$$

Factor of safety in wear

$$= \frac{\text{Wear strength}}{P_{eff}} = \frac{14653.4}{13386.6} = 1.095$$

Example 5.3: *A straight bevel pinion with 24 number of teeth meshes with a 56 teeth gear. The pinion transmits 10 kW power while rotating at 1500 rpm. Motor starting torque is 40% more than the rated torque and load concentration factor is 1.3. Pinion and gear both are made of alloy steel with their UTS values 820 and 680 N/mm² respectively. Hardness for pinion and gear is 400 and 380 respectively. Deformation factor may be taken as 11500 · e N/mm with error 'e' given by* $e = 8 + 0.63\left[m + 0.25\sqrt{d_m}\right]$. *Design the gear for a factor of safety 1.2.*

Use the following relation.

$$C_v = \frac{6}{6 + v}; \; Y = 0.484 - \frac{2.87}{Z'}$$

$$F_d = \frac{21 \cdot v\,(b \cdot C + F_{tmax})}{21 \cdot v + \sqrt{b \cdot C + F_{tmax}}}$$

Given data: $Z_p = 24$, $Z_g = 56$, Power = 10 kW, $N_p = 1500$ rpm, $C_s = 1.4$, $C_l = 1.3$, $(S_{ult})_{pinion} = 820$ N/mm², $(S_{ult})_{gear} = 680$ N/mm², BHN: 400/380, FS = 1.2.

Solution: Effective Load: $d_p = m \cdot Z_p = 24 \cdot m$

$$v = \frac{\pi\, d_p\, N_p}{60 \times 1000} = 1.885 \text{ m}$$

Tangential force,

$$P_t = \frac{\text{Power}}{v} = \frac{10 \times 10^3}{1.885 \text{ m}} = \frac{5305}{m}$$

$$C_v = \frac{6}{6 + v} = \frac{6}{6 + 1.885 \text{ m}}$$

$$P_{eff} = \frac{C_s \cdot C_l}{C_v} \times P_t$$

$$= \frac{1.4 \times 1.3\,(6 + 1.885 \text{ m})}{6} \times \frac{5305}{m}$$

$$= \frac{1609\,(6 + 1.885 \text{ m})}{m}$$

Beam Strength:

$$S_b = \sigma_b \cdot b \cdot Y' \cdot m\left[1 - \frac{b}{A_o}\right]$$

$$\tan \gamma_p = \frac{Z_p}{Z_g} = \frac{24}{56} = 0.428; \; \gamma_p = 23.2°,$$

$$\gamma_g = 66.8°$$

$$Z_p' = \frac{Z_p}{\cos \gamma_p} = \frac{24}{\cos 23.2} = 26.11;$$

$$Y_p' = 0.484 - \frac{2.87}{Z_p'} = 0.374$$

$$Z_g' = \frac{Z_g}{\cos \gamma_g} = \frac{56}{\cos 66.8} = 142.15;$$

$$Y_g' = 0.484 - \frac{2.87}{Z_g'} = 0.464$$

$$(\sigma_b Y_p')_p = \frac{820}{3} \times 0.374 = 102.2$$

$$(\sigma_b Y_g')_g = \frac{680}{3} \times 0.464 = 105.63$$

Since $(\sigma_b Y_p') < (\sigma_b Y_g')$, pinion is weaker.

Beam strength will be calculated based on pinion.

$$d_p = m \cdot Z_p = 24 \text{ m}$$

$$d_g = m \cdot Z_g = 56 \cdot m$$

Cone distance,

$$A_o = \sqrt{\left(\frac{d_p}{2}\right)^2 + \left(\frac{d_g}{2}\right)^2} = \sqrt{144 \text{ m}^2 + 784 \text{ m}^2}$$

$$= 30.46 \text{ m}$$

Now, $b = \frac{1}{3} A_o = 10.15$ m or $10 \cdot m$

whichever is smaller.

∴ $b = 10 \cdot m$

Beam Strength

$$S_b = (\sigma_b \cdot Y_p') \times b \cdot m\left[1 - \frac{b}{A_o}\right]$$

$$= 102.2 \times 10 \cdot m \times m\left(1 - \frac{10.0}{30.46}\right)$$

$$= 686.4 \text{ m}^2$$

Wear Strength:

Ratio factor,

$$Q' = \frac{2 \cdot Z_g'}{Z_g' + Z_p'} = \frac{2 \times 142.15}{142.15 + 26.11} = 1.69$$

$$K = 0.16\left[\frac{BHN}{100}\right]^2 = 0.16\,(3.8)^2 = 2.31$$

$$S_w = \frac{0.75 \cdot b \cdot Q'\, d_p \cdot K}{\cos \gamma_p}$$

$$= \frac{0.75 \times 10 \cdot m \times 1.69 \times 24 \text{ m} \times 2.31}{\cos 23.2}$$

$$= 764.6 \text{ m}^2$$

Since $S_b < S_w$, design will be based on beam strength. Equating beam strength with effective load considering factor of safety

$$686.4\ m^2 = [P_{eff}]\ FS$$

$$686.4\ m^2 = \frac{1609\ (6 + 1.885\ m)}{m}$$

$$m^3 - 4.419\ m - 14 = 0$$

Solving, $m \simeq 3.1$

We select m = 4. Further, we check for dynamic load.

Check with m = 4:

Tangential load,

$$P_t = \frac{5305}{m} = \frac{3505}{4} = 1326.3\ N$$

$$(P_t)_{max} = (C_s \cdot C_l)\ P_t = 1.4 \times 1.3 \times 1326.3 = 2413.7\ N$$

Dynamic Load P_d:

$$d_{mp} = d_p - b \cdot \sin\theta_p = 4 \times 24 - 40 \times \sin 23.2 = 80.24\ mm$$

$$d_{mg} = d_g - b \sin\theta_g = 4 \times 56 - 40 \times \sin 66.8 = 187.23\ mm$$

$$e_p = 8 + 0.63\left(m + 0.25\sqrt{d_m}\right) = 8 + 0.63\left(4 + 0.25\sqrt{80.24}\right) = 11.93\ \mu m$$

$$e_g = 8 + 0.63\left(4 + 0.25\sqrt{187.23}\right) = 12.67\ \mu m$$

$$e = e_p + e_g = 0.0246\ mm$$

$$\therefore \quad C = 11500 \times 0.0246 = 282.9\ N/mm$$

$$v = 1.885\ m = 7.54\ m/sec.$$

$$P_d = \frac{21 \cdot v\ (b \cdot C + P_{tmax})}{21 \cdot v + \sqrt{b \cdot C + P_{tmax}}} = \frac{21 \times 7.54\ (40 \times 282.9 + 2413.7)}{158.34 + 117.174}$$

$$P_d = 7890\ N$$

$$\therefore \quad P_{eff} = P_{tmax} + P_d = 2413 + 7890 = 10303.7\ N$$

$$\text{Beam strength} = 686.4\ m^2 = 10982\ N\ FS$$

$$= \frac{S_b}{P_{eff}} = \frac{10982}{10303.7} = 1.06$$

$$\text{Wear strength} = 764.6\ m^2 = 12236\ N$$

$$FS = \frac{S_w}{P_{eff}} = \frac{12233.6}{10303.7} = 1.18$$

This is smaller than required factor of safety of 1.2. Hence, we take Trial 2 with m = 5 mm.

Trial 2:

Tangential load,

$$P_t = \frac{5305}{5} = 1061\ N$$

$$P_{tmax} = (C_s \cdot C_l)\ P_t = 1.4 \times 1.3 \times 1061 = 1931\ N$$

Tooth Error:

$$d_{mp} = d_p - b \sin\theta_p = 5 \times 24 - 50 \sin 23.2 = 100.3\ mm$$

$$d_{mg} = d_g - b \sin\theta_g = 5 \times 56 - 50 \sin 66.8 = 234.04\ mm$$

$$e_p = 8 + 0.63\left(5 + 0.25\sqrt{104.24}\right) = 12.73\ \mu m$$

$$e_g = 8 + 0.63\left(5 + 0.25\sqrt{243.23}\right) = 13.56\ \mu m$$

$$e = e_p + e_g = 0.02628\ mm$$

$$\therefore \quad C = 11500 \times e = 303\ N/mm$$

$$v = 1.885\ m = 9.425\ m/sec$$

$$P_d = \frac{21 \cdot v\ (b \cdot C + P_{tmax})}{21 \cdot v + \sqrt{b \cdot C + P_{tmax}}} = \frac{21 \times 9.425\ (50 \times 303 + 1931)}{21 \times 9.425 + \sqrt{50 \times 303 + 1931}} = 10287\ N$$

$$P_{eff} = 1931 + 10287 = 12218\ N$$

$$\text{Beam strength} = 686.4\ m^2 = 17160\ N\ FS = \frac{17160}{12218} = 1.4$$

$$\text{Wear strength} = 764.6\ m^2 = 19115\ NFS = \frac{19115}{12218} = 1.56$$

$\therefore$ Safe.

Example 5.4: *A straight bevel gear pair has 24 teeth pinion meshing with 60 teeth gear. Teeth are cut to 20° pressure angle full involute and the module is 4 mm. Two kW power is transmitted at 1000 rpm. The arrangement is as shown in Fig. 5.6.*

Find: (a) Tangential, (b) Axial and (c) Radial forces acting on pinion and gear. Assuming suitable direction of rotation, indicate directions of forces.

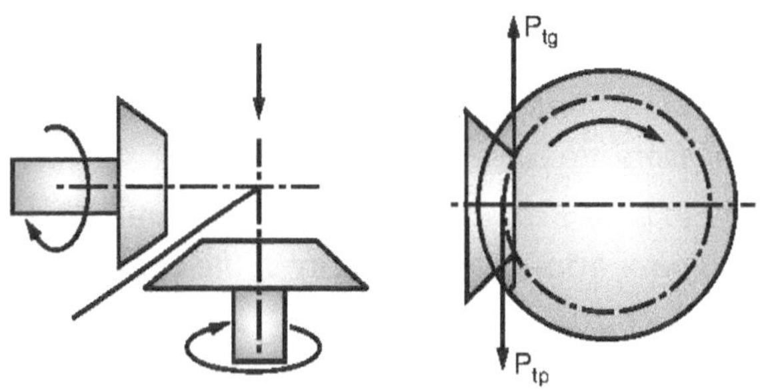

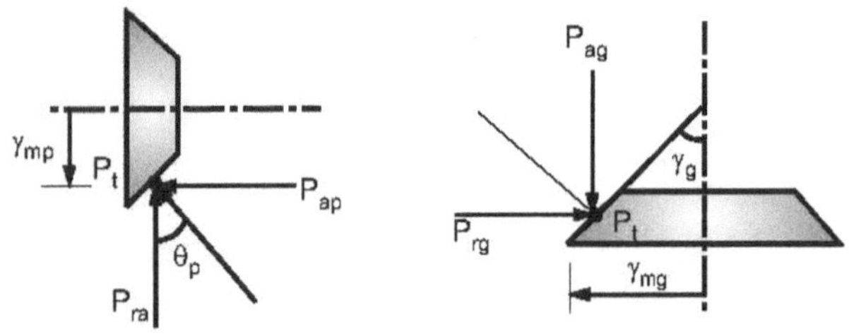

Fig. 5.6

Solution: $\tan\gamma_p = \frac{Z_p}{Z_g} = \frac{24}{60} = 0.4,\ \gamma_p = 21.8°$

Torque on pinion,

$$T_p = \frac{\text{Power} \times 60 \times 1000}{2\pi \cdot N_p} = 19.1 \text{ Nm}$$

Tangential force on pinion,

$$P_{tp} = \frac{T_p}{\left(\frac{d_{mp}}{2}\right)} = \frac{19100}{\left(\frac{d_{mp}}{2}\right)}$$

Taking gear width = 10 m = 40 mm

$$d_{mp} = d_p - b \cdot \sin\theta_p$$
$$= 96 - 40 \cdot \sin 21.8 = 81.14 \text{ mm}$$
$$d_{mg} = d_g - b \cdot \sin\theta_g$$
$$= 240 - 40 \sin 68.2 = 202.86 \text{ mm}$$

$$\therefore \quad P_{tp} = \frac{T_p}{\left(\frac{d_{mp}}{2}\right)} = \frac{19100}{\left(\frac{81.14}{2}\right)} = 470.8 \text{ N}$$

$$P_{tp} = P_{tg} = 470.8 \text{ N}$$
$$P_{ap} = P_t \cdot \tan\phi \cdot \sin\gamma_p$$
$$= 470.8 \times \tan 20 \cdot \sin 21.8 = 63.6 \text{ N}$$
$$P_{rp} = P_t \cdot \tan\phi \cdot \cos\gamma_p$$
$$= 470.8 \times \tan 20 \cos 21.8$$
$$= 159.1 \text{ N}$$
$$P_{ag} = P_{r\,pinion} = 63.6 \text{ N}$$
$$P_{rg} = P_{ap} = 159.1 \text{ N}$$

Example 5.5: *A pair of bevel gears having 16 and 72 teeth has a module of 6 mm at the outer diameter. Pinion is made of alloy steel with a UTS of 700 N/mm² and surface hardness 260 BHN. Gears are made of cast iron with a UTS of 280 N/mm² and hardness 210 BHN. Assume velocity factor accounts for dynamic loads. The pinion rotates at 720 rpm and gears are 20° full involute.*

1. *Find: (a) Beam strength, (b) Wear strength and (c) Power transmission capacity of the gears.*
2. *Find forces on pinion.*

Consider service factor of 1.5 and factor of safety 2.

Use the following relations: $b = \frac{1}{3} A_o$ *or* $10 \cdot m$ *whichever is smaller.*

$$Y' = 0.484 - \frac{2.87}{Z'};\ C_v = \frac{5.6}{5.6 + \sqrt{v}};\ K = 0.18\left(\frac{BHN}{100}\right)^2$$

Given data: $Z_p = 16$, $Z_g = 72$, $m = 6$ mm, $(S_{ult})_p = 700$ N/mm², $(S_{ult})_{gear} = 280$ N/mm², BHN: 260/210, $C_s = 1.5$, FS = 2, $N_p = 720$ rpm, $\phi = 20°$.

Solution: Effective Load: $d_p = m \cdot Z_p = 6 \times 16 = 96$ mm

$$v = \text{Pitch line velocity}$$
$$= \frac{\pi \cdot d_p \cdot N_p}{60 \times 1000}$$
$$= \frac{\pi \times 96 \times 720}{60 \times 1000} = 3.619 \text{ m/sec}$$

Tangential force= P_t

$$\text{Velocity factor} = \frac{5.6}{5.6 + \sqrt{v}} = 0.746$$

Effective load,

$$P_{eff} = \left(\frac{C_s \cdot C_l}{C_v}\right) P_t = \frac{1.5 \times 1}{0.746} \cdot P_t = 2.01\, P_t$$

1. Beam Strength:

$$S_b = \sigma_b \cdot b \cdot Y' m \left[1 - \frac{b}{A_o}\right]$$

$$\tan\gamma_p = \frac{Z_p}{Z_g} = \frac{16}{72} = 0.222,\ \gamma_p = 12.53°,\ \gamma_g = 77.47°$$

$$Z'_p = \frac{Z_p}{\cos\gamma_p} = \frac{16}{\cos 12.53} = 16.39$$

$$Y'_p = 0.484 - \frac{2.87}{Z'_p} = 0.3088$$

$$Z'_g = \frac{Z_p}{\cos\gamma_g} = \frac{72}{\cos 77.47} = 331.9$$

$$Y'_g = 0.484 - \frac{2.87}{Z'_g} = 0.475$$

$$(\sigma_b \cdot Y')_p = \frac{700}{3} \times 0.3088 = 72.05$$

$$(\sigma_b Y')_g = \frac{280}{3} \times 0.475 = 44.33$$

Since $(\sigma_b Y')_g < (\sigma_b Y')_p$, gear is weaker and beam strength will be calculated for gear.

$$d_p = m \cdot Z_p$$
$$= 6 \times 16 = 96 \text{ mm}$$
$$d_g = 6 \times 72 = 432 \text{ mm}$$

Cone distance,

$$A_o = \sqrt{\left(\frac{d_p}{2}\right)^2 + \left(\frac{d_g}{2}\right)^2} = 221.26$$

$b = \frac{1}{3} A_o = 73.75$ or $10\ m = 60$ mm whichever is smaller. Hence, b = 60 mm

(a) Beam Strength:

$$S_b = (\sigma_b \cdot Y)_g\, b \cdot m \left[1 - \frac{b}{A_o}\right]$$

$$= 44.36 \times 60 \times 6 \left[1 - \frac{60}{221.26}\right] = 11631 \text{ N}$$

(b) Wear Strength:

$$S_w = \frac{0.75\, d_p \cdot b\, Q' \cdot K}{\cos \gamma_p}$$

$$Q' = \frac{2 \cdot Z'_g}{Z'_g + Z'_p} = \frac{2 \times 331.9}{331.9 + 16.39} = 1.906$$

$$K = 0.18 \left(\frac{BHN}{100}\right)^2 = 0.18 \left(\frac{210}{100}\right)^2 = 0.794$$

$$S_w = \frac{0.75 \times 96 \times 60 \times 1.906 \times 0.794}{0.9762}$$

$$= 6697.1 \text{ N}$$

(c) Power Transmission Capacity:

Since $S_w < S_b$, wear strength will govern the capacity. Equating wear strength with effective load considering factor of safety,

$$S_w = (P_{eff})\, FS$$

$$6697.1 = (2.01\, P_t)\, 2$$

$$\therefore \quad P_t = 1666 \text{ N (Tangential load)}$$

Power that can be transmitted will be

$$\text{Power} = \frac{2\pi\, N_p \times \left(P_t \times \frac{d_p}{2}\right)}{60 \times 1000 \times 10^3}$$

$$= \frac{2\pi \times 720 \times (1666 \times 48)}{60 \times 1000 \times 10^3}$$

$$= 6.03 \text{ kW}$$

2. Forces on Pinion: (At rated torque/power)

$$P_t = 1666 \text{ N}$$

$$P_a = P_t \cdot \cos \phi \cdot \sin \gamma_p$$

$$= 1666 \cos 20 \sin 12.53 = 339.6 \text{ N}$$

$$P_r = P_t \cdot \cos \phi \cdot \cos \gamma_p$$

$$= 1666 \cos 20 \cos 12.53 = 1528.2 \text{ N}$$

Example 5.6: *The pinion of a bevel gear pair is connected to 5 kW electric motor running at 1440 rpm. Reduction in speed is 3 : 1 from pinion to gear. Pinion has 18 teeth and is made from steel having UTS of 520 N/mm² and surface hardness of 300 BHN. Gear is made from cast iron having UTS of 300 N/mm² and hardness 300 BHN. Pressure angles for the gears is 20° and service factor is 1.2. Deformation factor 'C' may be taken as 116 N/mm.*

Design the gears for a factor of safety of 1.25.

Take $C_v = \frac{6}{6 + v}$, $K = 0.18 \left(\frac{BHN}{100}\right)^2$.

Given data: Reduction 3 : 1, $Z_p = 18$, $Z_g = 54$, Power = 5 kW, $N_p = 1440$ rpm, $(S_{ult})_p = 520$ N/mm², $(S_{ult})_g = 300$ N/mm², $C_s = 1.2$, FS = 1.25, BHN = 300, C = 116 N/mm.

Solution: Effective Load: $d_p = m \cdot Z_p = 18\ m$

$$v = \frac{\pi\,(18\ m) \times 1440}{60 \times 1000} = (1.357\ m) \text{ m/sec}$$

Tangential load,

$$P_t = \frac{\text{Power}}{\text{Velocity}} = \frac{5 \times 10^3}{1.357\ m} = \frac{3684.6}{m} \cdot \text{N}$$

$$P_{eff} = \left(\frac{C_s \cdot C_l}{C_v}\right) P_t = \frac{1.2 \times 1}{\left(\frac{6}{6 + 1.357\ m}\right)} \times \frac{3684.6}{m}$$

$$P_{eff} = \frac{736.92 \times (6 + 1.357\ m)}{m}$$

Beam Strength:

$$\tan \gamma_p = \frac{Z_p}{Z_g} = \frac{1}{3},\ \gamma_p = 18.435°,\ \gamma_g = 71.565°$$

$$Z'_p = \frac{Z_p}{\cos \gamma_p} = \frac{18}{\cos 18.435} = 18.974,$$

$$Z'_g = \frac{Z_g}{\cos \gamma_g} = 170.76$$

$$Y'_p = 0.484 - \frac{2.87}{Z'_p} = 0.3327$$

$$Y'_g = 0.484 - \frac{2.87}{Z'_g} = 0.4672$$

$$(\sigma_b \cdot Y')_p = \frac{520}{3} \times 0.3327 = 57.67$$

$$(\sigma_b \cdot Y')_g = \frac{300}{3} \times 0.4672 = 46.72$$

Since $(\sigma_b Y')_g < (\sigma_b \cdot Y')_p$, gear is weaker.

Beam strength will be based on gear capacity.

Beam Strength:

$$S_b = (\sigma_b \cdot Y)_g \, b \cdot m\left[1 - \frac{b}{A_o}\right]$$

Cone distance,

$$A_o = \sqrt{\left(\frac{d_p}{2}\right)^2 + \left(\frac{d_g}{2}\right)^2}$$

$$= m\sqrt{\left(\frac{18}{2}\right)^2 + \left(\frac{54}{2}\right)^2} = 28.46 \text{ m}$$

$$b = \frac{1}{3}A_o = 9.5 \text{ m or } 10 \text{ m whichever is smaller}$$

$$\therefore \quad b = 9.5 \text{ m}$$

$$S_b = (46.72)\, 9.5 \text{ m} \cdot \text{m}\left[1 - \frac{1}{3}\right] = 295.5 \text{ m}^2$$

$$S_b = 295.5 \text{ m}^2$$

Wear Strength:

$$S_w = \frac{0.75\, d_p \cdot b \cdot Q' \cdot K}{\cos \gamma_p}$$

$$Q' = \frac{2Z_g'}{Z_g' + Z_p'} = \frac{2 \times 170.76}{170.76 + 18.974} = 1.799$$

$$K = 0.18\left(\frac{BHN}{100}\right)^2 = 1.62$$

$$S_w = \frac{0.75 \times 18 \text{ m} \times 9.5 \text{ m} \times 1.799 \times 1.62}{\cos 18.435}$$

$$= 394 \text{ m}^2$$

Since $S_b < S_w$

Beam strength will govern the design.

Equating beam strength with P_{eff}, considering factor of safety.

$$S_b = [P_{eff}]\, FS$$

$$295.5 \text{ m}^2 = \frac{736.92\,(6 + 1.357 \text{ m})}{\text{m}} \times 1.25$$

Simplifying,

$$m^3 - 4.23 \text{ m} - 18.7 = 0$$

Solving by the trial and error $m \simeq 3.2$

We select next higher standard module = 4 mm.

Checking dynamic load by Buckingham's equation

$$\text{Velocity,} \quad v = 1.357 \text{ m} = 5.428 \text{ m/sec}$$

$$C = 116 \text{ N/m}$$

$$b = 9.5 \text{ m} = 38 \text{ mm}$$

$$P_{tmax} = P_t\,(C_s \cdot C_l) = 1.2 \times \frac{3684.6}{\text{m}}$$

$$= 1105.4 \text{ N}$$

Check for Dynamic Load: (Buckingham's Equation)

$$P_d = \frac{21 \cdot v\,(b \cdot C + P_{tmax})}{21 \cdot v + \sqrt{b \cdot C + P_{tmax}}}$$

$$= \frac{21 \times 5.428\,(38 \times 116 + 1105.4)}{21 \times 5.428 + \sqrt{38 \times 116 + 1105.4}}$$

$$= 3339.3 \text{ N}$$

$$\therefore \quad P_{eff} = P_{tmax} + P_d$$

$$= 1105.4 + 3339.3 = 4444.7 \text{ N}$$

Factor of safety based on

Beam strength FS:

$$\frac{S_b}{P_{eff}} = \frac{295.5 \text{ m}^2}{4444.7} = \frac{4728}{4444.7} = 1.06$$

Based on wear strength FS $= \frac{S_w}{P_{eff}} = \frac{394 \text{ m}^2}{4444.7} = \frac{6304}{4444.7} = 1.42$

Since factor of safety based on beam strength is < 1.25 design is not safe.

Hence, we need to increase module to next standard size.

Trial 1: We select module m = 5 mm.

$$D_p = m \cdot Z_p = 90 \text{ mm}$$

$$D_g = m \cdot Z_g = 270 \text{ mm}$$

$$v = 1.357 \times m = 6.785 \text{ m/sec}$$

$$A_o = 28.46 \times m = 142.3 \text{ mm}$$

$$b = 9.5 \text{ m} = 47.5 \text{ mm} \simeq 48 \text{ mm}$$

$$P_t = \frac{3684.6}{m} = \frac{3684.6}{5} = 737 \text{ N}$$

$$P_{tmax} = (C_s \cdot C_l)\, P_t = (1.2 \times 1)\, 737 = 884.4 \text{ N}$$

Dynamic Load:

$$P_d = \frac{21 \cdot v\,(b \cdot C + P_{tmax})}{21 \cdot v + \sqrt{b \cdot C + P_{tmax}}}$$

$$= \frac{21 \times 6.785\,(48 \times 116 + 884.4)}{21 \times 6.785 + \sqrt{48 \times 116 + 884.4}}$$

$$P_d = 4126.1 \text{ N}$$

$$\therefore \quad P_{eff} = P_{tmax} + P_d$$

$$= 884.4 + 4126.1 = 5010.5 \text{ N}$$

Factor of safety based on

Beam strength FS

$$= \frac{S_b}{P_{eff}} = \frac{295.5 \text{ m}^2}{5010.5} = \frac{7387.5}{5010.5} = 1.47$$

Wear strength FS

$$= \frac{S_w}{P_{eff}} = \frac{394 \text{ m}^2}{5010.5} = \frac{9850}{5010.5} = 1.96$$

This satisfies the requirements.

Hence, we select module m = 5 mm.

Example 5.7 : *A pair of straight bevel gears mounted on shafts that are intersecting at right angles, consists of a 20 teeth pinion meshing with a 32 teeth gear. The pinion shaft is connected to an electric motor developing 8.0 kW rated power at 1440 rpm. The starting torque of the motor is 150% of the rated torque. The pressure angle is 20°. Both gears are made of case hardened steel (σ_{ut} = 680 N/mm^2). The teeth on gears are generated and finished by grinding and lapping processes to meet the requirements of class-3 grade. The factor of safety in preliminary stages of gear design is 2.*

(i) In the initial stages of gear design, assume that velocity factor accounts for the dynamic load and that the pitch line velocity is 7.5 m/s. Estimate the module based on beam strength. Select the first preference value of module and calculate the main dimensions of the gears.

(ii) Determine the dynamic load using Buckingham's equation and find out the effective load for above dimensions. What is the correct factor of safety for bending?

(iii) Specify the surface hardness for the gears assuming a factor of safety of 2.

Use the following data:

- *Velocity factor,* $C_v = \dfrac{5.6}{5.6 + \sqrt{v}}$
- *Determination factor, C = 11400 × e N/mm*
- *Buckingham's equation,* $F_d = \dfrac{21 \cdot v\,(C \cdot b + F_t)}{21 \cdot v + \sqrt{C \cdot b + F_1}}$
- *Maximum expected error between two meshing teeth (mm)*

Module (mm)	upto 4	5	6	7	8
Class – 3 (e – mm)	0.0125	0.0125	0.0150	0.0170	0.0190

Given data: Z_p = 20, Z_g = 32, Power = 8 kW at 1440 rpm, C_s = 1.5, S_{ult} = 680 N/mm^2, FS = 2, v = 7.5 m/sec.

Solution:

(i) Effective Force:

$$\text{Velocity} = \frac{\pi\,(m \cdot Z_p)\,1440}{60 \times 1000} = 1.508\text{ m}$$

$$P_t = \frac{\text{Power}}{\text{Velocity}} \times \frac{8 \times 10^3}{1.508\text{ m}} = \frac{5305.2}{m}\text{ m/sec}$$

Velocity factor,

$$C_v = \frac{5.6}{5.6 + \sqrt{7.5}} = 0.6716$$

$$P_{eff} = \left(\frac{C_s \cdot C_i}{C_v}\right)$$

$$= \frac{5305.2}{m} = \left(\frac{1.5 \times 1}{0.6716}\right)\frac{5305.2}{m}$$

$$= \frac{11849}{m}\text{ N}$$

$$\tan\gamma_p = \frac{Z_p}{Z_g} = \frac{20}{32} = 0.625, \quad \gamma_p = 32°, \quad \gamma_g = 58°$$

Beam Strength:

$$S_b = \sigma_b \cdot Y' \cdot b \cdot m\left[1 - \frac{b}{A_o}\right]$$

Since material is same for pinion and gear, pinion is weaker.

$$Z_p' = \frac{Z_p}{\cos\gamma_p} = \frac{20}{0.848} = 23.58$$

$$Z_g' = \frac{32}{\cos\gamma_g} = \frac{32}{0.5299} = 60.386$$

$$Y_p' = 0.484 - \frac{2.87}{Z_p'} = 0.3623$$

$$A_o = \sqrt{\left(\frac{d_p}{2}\right)^2 + \left(\frac{d_g}{2}\right)^2}$$

$$= m\sqrt{\left(\frac{Z_p}{2}\right)^2 + \left(\frac{Z_g}{2}\right)^2}$$

$$= m\sqrt{10^2 + 16^2}$$

$$= 18.87\text{ m}$$

$$b = \frac{1}{3}A_o$$

= 6.3 m or 10 m whichever is smaller

∴ We take b = 7 m

$$S_b = \left(\frac{680}{3}\right) \times 0.3623 \times 7m \cdot m\left[1 - \frac{7\text{ m}}{18.87\text{ m}}\right]$$

$$= 361.6\text{ m}^2$$

Equating beam strength with effective force considering factor of safety

$$S_b = (P_{eff})\text{ FS}$$

$$361.6\text{ m}^2 = \frac{11849}{m} \times 2$$

$$m^3 = 65.53$$

$$m = 4.03\text{ mm}$$

∴ We select next higher standard module = 5 mm

Main gear dimensions will be

$$d_p = 20 \times 5 = 100\text{ mm}$$

$$d_g = 32 \times 5 = 160\text{ mm}$$

$$b = 7 \cdot m = 35 \text{ mm}$$

$$h_a = 1 \cdot m = 5 \text{ mm}$$

$$h_f = 1.25 \cdot m = 6.25 \text{ mm}$$

(ii) Dynamic Load by Buckingham's Equation:

$$P_d = \frac{21 \cdot v\,(C \cdot b + P_{tmax})}{21 \cdot v + \sqrt{C \cdot b + P_{t\,max}}}$$

$$v = \frac{\pi \times d_p \cdot N_p}{60 \times 1000}$$

$$= \frac{\pi \times 100 \times 1440}{60 \times 1000}$$

$$= 7.54 \text{ m/sec}$$

$$P_t = \frac{5305.2}{m} = 1061 \text{ N}$$

$$P_{tmax} = (C_s \cdot C_l)\,P_t$$

$$= 1.5 \times 1061 = 1591.5 \text{ N}$$

$$b = 7\,m = 35 \text{ mm}$$

$$C = 11400 \times e \quad (e = 0.0125 \text{ mm from table})$$

$$= 11400 \times 0.0125$$

$$= 142.5 \text{ N/mm}$$

$$P_d = \frac{21 \times 7.54\,(142.5 \times 35 + 1591.5)}{21 \times 7.54 + \sqrt{142.5 \times 35 + 1591.5}}$$

$$= 4350.5 \text{ N}$$

$$P_{eff} = P_{tmax} + P_d$$

$$= 1591.5 + 4350.5$$

$$= 5942 \text{ N}$$

Beam strength = $361.6\ m^2 = 9040$ N

$\therefore$ Factor of safety = $\frac{9040}{5942} = 1.52$

(iii) Surface Hardness:

Wear strength required considering factor of safety = 2.

$$S_w = 2\,(P_{eff}) = 2 \times 5942 = 11884 \text{ N}$$

$$S_w = \frac{0.75\,d_p \cdot b\,Q'\,K}{\cos \gamma_p}$$

$$Q' = \frac{2Z_g'}{Z_g' + Z_p'}$$

$$= \frac{2 \times 60.386}{60.386 + 23.58} = 1.439$$

$$K = 0.16\left(\frac{BHN}{100}\right)^2$$

Thus, $11884 = \frac{0.75 \times 100 \times 35 \times 1.439 \times 0.16}{\cos 32°}\left(\frac{BHN}{100}\right)^2$

$$BHN = 408$$

Example 5.8: *A pair of cast iron bevel gears connects two shafts at right angles. Pinion and gear diameters are required to be 80 and 100 mm respectively. The tooth profiles of the gears are $14^1/_2°$ composite form and both materials have tensile strength of 240 N/mm². Pinion transmits 3 kW power at 1440 rpm. Considering that velocity factor accounts for dynamic load and taking gear with $b = \frac{1}{3}$ cone distance (slanting height). Service factor is 1.2. Take Lewis form factor $Y' = 0.039 - \frac{2.15}{Z'}$, $K = 0.21\left(\frac{BHN}{100}\right)^2$. Use $C_v = \frac{6}{6 + v}$. Take surface hardness as 280 to pinion and gear. Required minimum factor of safety is 1.2.*

(a) Find the module and number of teeth on each gear considering bending failure.

(b) Check the design for wear.

Given data: $d_p = 80$ mm, $d_g = 100$ mm, $S_{ult} = 240$ N/mm², Power = 3 kW, $N_p = 1440$ rpm, $b = \frac{1}{3} A_o$

Solution: $\tan \gamma_p = \frac{Z_p}{Z_g} = \frac{d_p}{d_g}$

$$= \frac{80}{100};\ \gamma_p = 38.66°,\ \gamma_g = 51.34°$$

(a) Equivalent Tangential Force:

Pitch line velocity,

$$v = \frac{\pi\,d_p\,N_p}{60 \times 1000}$$

$$= 6.0318 \text{ m/sec}$$

Tangential load,

$$P_t = \frac{\text{Power}}{\text{Velocity}} = \frac{3 \times 10^3}{6.0318}$$

$$= 497.4 \text{ N}$$

Velocity factor,

$$C_v = \frac{6}{6 + v} = \frac{6}{6 + 6.0318}$$

$$= 0.4987$$

Effective load,

$$P_{eff} = \left(\frac{C_s \cdot C_l}{C_v}\right) P_t$$

$$P_{eff} = \left(\frac{1.2 \times 1}{0.4987}\right) 497.4$$

$$= 1196.9 \text{ N}$$

Beam Strength:

Since pinion and gear materials are same, pinion is weaker.

Beam strength of pinion is the deciding factor.

$$Z'_p = \frac{Z_p}{\cos \gamma_p}$$

$$= \frac{\left(\frac{d_p}{m}\right)}{\cos \gamma_p}$$

$$= \frac{80}{m \cos \gamma_p} = \frac{102.4}{m}$$

$$Z'_g = \frac{Z_g}{\cos \gamma_g}$$

$$= \frac{\frac{d_g}{m}}{\cos \gamma_g} = \frac{100}{m \cos \gamma_g} = \frac{160}{m}$$

Lewis form factor,

$$Y'_p = 0.39 - \frac{2.12}{Z'_p}$$

$$= 0.39 - \frac{(2.12)\, m}{102.4}$$

$$= (0.39 - 0.0207\, m)$$

Slanting height,

$$A_o = \sqrt{\left(\frac{d_p}{2}\right)^2 + \left(\frac{d_g}{2}\right)^2}$$

$$= \sqrt{40^2 + 50^2} = 64 \text{ mm}$$

Beam Strength: S_b

Considering width $b = \frac{1}{3} A_o$

$$S_b = \sigma_b \cdot b\, Y' \cdot m \cdot \left[1 - \frac{b}{A_o}\right]$$

$$= \left(\frac{240}{3}\right)\left(\frac{64}{3}\right)(0.39 - 0.0207\, m) \cdot m\left[1 - \frac{1}{3}\right]$$

$$= 1137.78\,(0.39 - 0.0207\, m)\, m$$

$$= 443.7\, m - 23.55\, m^2$$

Equating effective loads and beam strength, considering factor of safety

$$S_b = (P_{eff})\, FS$$

$$(443.7\, m - 23.55\, m^2) = (1196.9) \times 1.2 = 1436.3$$

Simplifying,

$$m^2 - 18.8\, m + 61 = 0$$

Solving $m = 4.2$

We select next higher preferred size m = 5 mm.

$$Z_p = \frac{d_p}{m} = \frac{30}{5} = 16$$

$$Z_g = \frac{100}{m} = 20$$

(b) Checking for Wear Strength:

$$S_w = \frac{d_p \cdot b \cdot Q' K}{\cos \gamma_p}$$

$$Z'_p = \frac{102.4}{m} = 20.48$$

$$Z'_g = \frac{160}{m} = 32$$

$$Q' = \frac{2Z'_g}{Z'_g + Z'_p}$$

$$= \frac{2 \times 32}{32 + 20.48} = 1.2195$$

$$K = 0.21\left(\frac{BHN}{100}\right)^2$$

$$= 1.646$$

$$S_w = \frac{80 \times 22 \times 1.2195 \times 1.646}{\cos 38.66}$$

$$= 4524 \text{ N}$$

With this factor of safety available is

$$FS = \frac{\text{Wear strength } (S_w)}{P_{eff}}$$

$$= \frac{4524}{1196.2} = 3.78$$

Thus, design is safe in wear also.

EXERCISE

1. What are the advantages and disadvantages of straight bevel gears and spiral bevel gears.
2. Compare straight bevel, spiral bevel, hypoid bevel gear with sketch.
3. Derive an expression for beam strength of straight bevel gears.
4. With neat sketch discuss force analysis of bevel gears.

5. Obtain an expression for beam strength of bevel gear in the following form.

$$S_b = mb\sigma_b Y\left[1 - \frac{b}{Ao} + \frac{b^2}{3A_o^2}\right]$$

6. Derive an expression for beam strength of straight bevel gear tooth.
7. Explain formative number of teeth for bevel gears.
8. With neat sketch explain components of forces acting on straight bevel gear.
9. What are the different methods of mounting of bevel gear? Explain any one with sketch.
10. Compare straight bevel, spiral bevel hypoid bevel gear with sketch.
11. Explain the bevel gear terminology.
12. Obtain an expression for the formative number of teeth.
13. Derive an expression for the beam strength of a bevel gear
14. Compare straight teeth and spiral teeth bevel.
19. Why the wear strength of the bevel gear is 25% less than its full load capacity?

CHAPTER - 6
WORM GEARS

6.1 WORM / WORM GEARS

- Simplest way to obtain large speed reduction with high torque in a compact space is with worm gear drives.
- A worm drive is a cylindrical worm with a shallow spiral thread that engages with the worm gear in a non-intersecting perpendicular axes configuration. Worm and gear have same hand of helix. Worm is like a screw and can have single or multiple starts. Meshing action is combined rolling and sliding.

(a) (b)

Fig. 6.1: Typical Worm and Worm Wheel Set in Mesh

Advantages of Worm/Worm Gear Pair:

- Worm gear drives are used for higher speed ratios ranging from 10 : 1 to 100 : 1.
- Attains high reduction ratio in single pair. This gives a compact arrangement. For this reason worm gear drives are very common in industrial applications where compactness is important.
- It can be designed for irreversibility. This makes it specially important for applications where irreversibility is a critical requirement.
- Operation is smooth and silent.

Limitations:

- The efficiency of worm – worm gear is however low and therefore they are not usually used for transmitting very large powers. (usually < 75 kW).
- Low efficiency results into more heat generation. The temperature rise of the lubricant limits the power transmission capacity of the gear pair.
- Gear is usually made of Phosphor Bronze which is a costly material.

Worm-Worm Gear Materials:

- Since sliding is predominant in case of worm gear drive, different materials are used for worm and worm wheel. Worm material is hardened and made of better grade material. Worm wheel teeth are difficult to manufacture and therefore are made of soft material.
- Worm undergoes more number of contact stress cycles than the worm gear and therefore a stronger material is usually selected for worm.

6.2 WORM/WORM GEARS – DIMENSIONS AND PROPORTIONS

Worm:

- **Axial Pitch (p_x):** It is the distance from any point on one thread to the corresponding point on the adjacent thread, measured parallel to the worm axis.

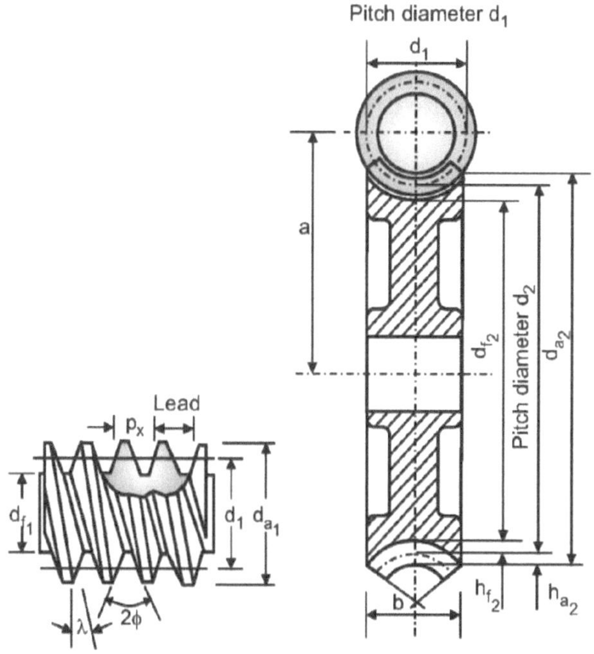

Fig. 6.2: Worm and Worm Wheel - Major Dimensions

- **Lead (L):** It is the linear distance travelled by a thread when one complete revolution is given to the worm.

  Then lead, $L = Z_1 \times p_x$ ... (6.1)

  where Z_1 is the number of starts on the worm.

  For a single start worm, lead is equal to the axial pitch and for double start worm thread, lead is twice the axial pitch and so on. Axial pitch of the worm is equal to the circular pitch of the worm gear.

- **Lead Angle (λ):** It is the angle between the tangent to the thread helix on the pitch cylinder and the plane normal to the axis of the worm.

  Lead angles (λ) are usually selected in the range 9° to 45°, since angles smaller than 9° result in rapid wear of worm tooth and very low efficiency.

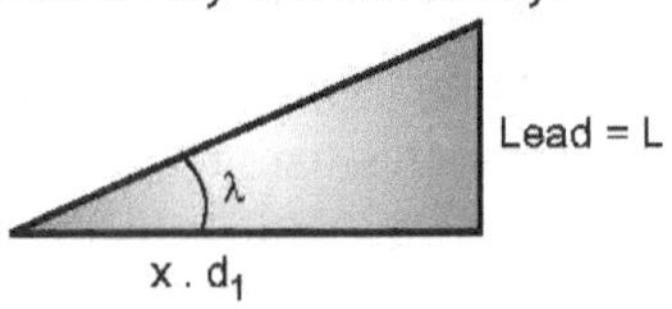

Fig. 6.3: Thread Helix on Worm

- **Helix Angle of Worm (ψ_w):** It is the angle between the tangent to the thread helix on the pitch cylinder and the axis of the worm. Thus,

  Helix angle, $\psi = 90 - \lambda$

- **Velocity Ratio:**

  Let, N_1 = Worm speed (rpm)

  N_2 = Worm gear speed (rpm)

  D = Pitch circle diameter of worm gear

  Linear Velocity of the Worm:

$$V_1 = \frac{L \times N_1}{60}$$

- **Linear Velocity of the Worm Gear:**

$$V_2 = \frac{\pi \times d_2 \times N_2}{60}$$

Since $V_1 = V_2$

$$L \times N_1 = \pi \times d_2 \times N_2$$

Therefore, Velocity Ratio:

$$V_R = \frac{N_1}{N_2} = \frac{\pi \times d_2}{L}$$

Since, $\pi \times d_2 = Z_2 \times p_c$ and $L = Z_1 \times p_x$

$$V_R = \frac{Z_2 \times p_c}{Z_1 \times p_x} = \frac{Z_2}{Z_1} \text{ (Since } p_c = p_x) \qquad \dots (6.2)$$

$$V_R = \frac{\text{Number of teeth in gear}}{\text{Number of starts on worm}}$$

- **Other Relations:**

Axial pitch, $p_x = \pi \times m = \frac{\text{Lead}}{Z_1}$

where, m is the module in mm.

With multi start threads, the velocity ratio reduces but the efficiency increases. For velocity ratios above 30, usually single start worm will be preferred.

Table 6.1: Selection of Velocity Ratio Vs No. of Starts on Worm

Number of starts on Worm (Z_1)	1	3	4
Velocity Ratio (i)	12-36	8-12	6-10

- **Diametral Quotient (q):** It is defined as the ratio of pitch circle diameter of worm to its module.

$$q = \frac{d_1}{m} \qquad \dots (6.3)$$

where, d_1 = Pitch circle diameter of the worm

As can be seen, for a given value of module, the worm diameter 'd_1' increases as 'q' increases. Worm diameter serves as a shaft diameter to carry bending moment as well as torque. A larger 'q' will therefore result in a stronger worm but at the same time size as well as the cost goes up.

Values of standard module are: 1, 1.25, 1.5, 2, 2.5, 3, 4, 5, 6, 8, 10, 12, 16, 20 mm.

Values of q range from 8 to 18. Higher values used usually for low modules.

Standard values are 8, 10, 12.5, 16, 20, 25.

- **Transverse Pressure Angle (ϕ_t):** It is the pressure angle measured in plane of rotation (transverse plane) of gear.
- **Normal Pressure Angle (ϕ_n):** It is the pressure angle measured in the plane perpendicular to the teeth (normal plane) of gear.

$$\tan \phi_t = (\tan \phi_n / \cos \lambda) \text{ OR}$$

$$\tan \phi_t = (\tan \phi_n / \cos \phi_g)$$

ϕ_n value is standard value selected from: 14.5°, 20°, 25°, 30°.

Efficiency of worm/worm gear drive depends on values of coefficient of friction (μ), lead angle (λ) and pressure angle (ϕ_n). Greater normal pressure angles are selected for larger lead angles.

Normally selected values are:

Lead Angle (λ)	4-15°	15-30°
Pressure Angle (ϕ_n)	14.5°	20°

- **Face Width of Gear Tooth (b):** Referring to the Fig. 6.4, showing worm/worm gear pair, tangent to the pitch circle of the worm intersects its addendum circle at B and B_1. BB_1 is the minimum width that should be provided on the worm gear.

Let b = Width of the gear tooth (mm)

d_{a1} = Addendum circle diameter of the worm

d_1 = Pitch Circle Diameter of the worm

h_{a1} = Addendum of the worm

Width 'b' should be ≥ BB_1

$$BB_1 = 2 \cdot BC$$

$$= 2 \cdot \sqrt{OB^2 - OC^2}$$

$$= 2 \cdot \sqrt{\left(\frac{d_{a1}}{2}\right)^2 - \left(\frac{d_1}{2}\right)^2} = \sqrt{d_{a1}^2 - d_1^2}$$

Since $\quad d_{a1} = d_1 + 2h_{a1}$

$$b \geq \sqrt{4 \cdot d_1 \cdot h_{a1} + 4 \cdot h_{a1}^2}$$

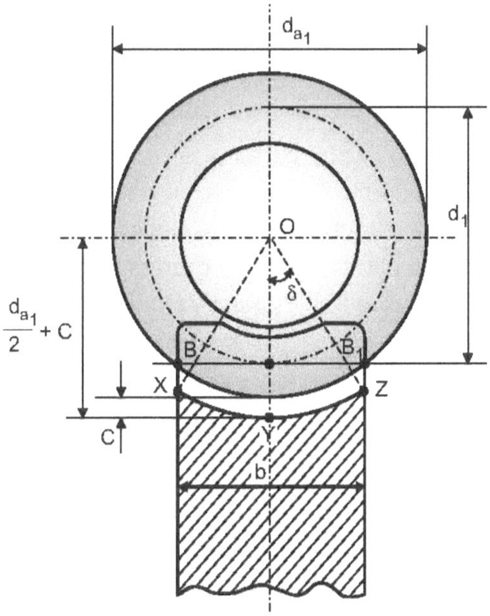

Fig. 6.4: Gear Width

$$b \geq 2\sqrt{q \cdot m\,(m) + m^2} \geq 2 \cdot m\sqrt{q + 1}$$

Thus $\quad b \geq 2 \cdot m\sqrt{q + 1} \quad \ldots (6.4)$

A minimum width of $0.73 \times d_1$ is recommended for the worm gear.

Thus, $\quad b \geq 2 \cdot m\sqrt{q + 1} \quad$ OR $\quad 0.73 \times d_1$

whichever is larger.

Table 6.2 shows proportions and major dimensions for a worm and worm wheel.

Table 6.2: Standard Proportions of Worm and Worm Wheel

Worm	
Addendum of Worm	$h_{a1} = m$
Dedenum of Worm	$h_{f1} = (2.2 \cos \lambda - 1) \cdot mm$
Outer Diameter	$d_{a1} = d_1 + 2.h_{a1}$
Root Diameter	$d_{f1} = d_1 - 2.h_{f1}$
Length of Worm	$L_w = \pi \cdot m \cdot (4.5 + Z_g/50)$
Worm Wheel	
Addendum of Wheel	$h_{a2} = 2(\cos \lambda - 1) \cdot m$
Dedendum of Worm	$h_{f2} = (1 + 0.2 \cos \lambda) \cdot m$
Outer Diameter	$d_{a2} = d_2 + 2 \cdot h_{a2}$
Root Diameter	$d_{f2} = d_2 - 2 \cdot h_{f2}$
Clearance	$C = 0.2 \cos \lambda$
Face Width	$b = 0.73\, d_1$

6.3 GEAR ENVELOPING AND RIM CONSTRUCTIONS

Single Enveloping: In a single enveloping worm set width of the worm gear is cut into concave surface. This partially encloses the worm. As can be seen from the Fig. 6.5 (a), only the middle thread on the worm will have full engagement and the side threads have partial engagement. This is used in high speed and low load condition.

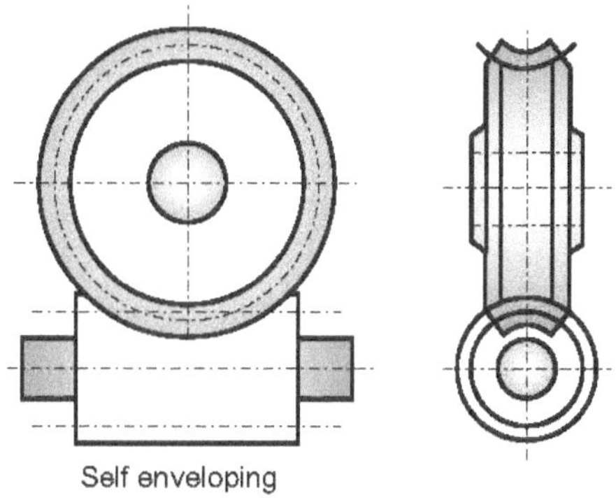

(a) Single Enveloping

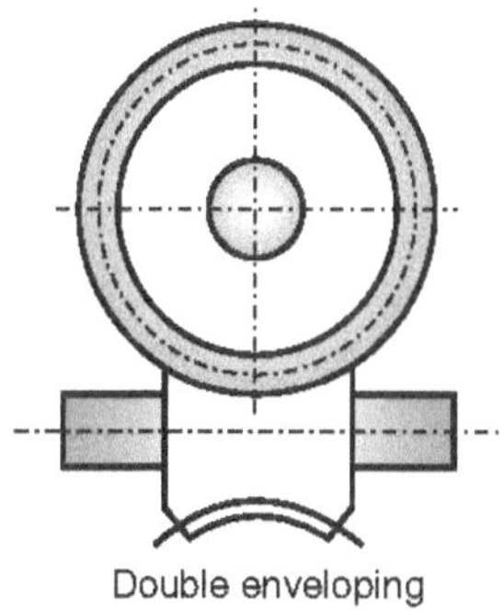

(b) Double Enveloping

Fig. 6.5: Worm Gears

Double Enveloping:

- In this case both, the width of the gear and the worm along its length, are cut with a concave shape to get better engagement as shown in Fig. 6.5 (b). With this both the worm and worm gear partially enclose each other. There are more teeth in contact and there is area contact rather than line contact. This permits higher load transmission.
- Double enveloping gears are difficult to manufacture and mount compared with the single enveloping gears because of critical alignment requirements. They are used for higher load transmissions.

Worm Wheel Rim Construction:

- The following Fig. 6.6 shows different rim constructions for a worm wheel.
- First arrangement shows straight teeth cut on a rim with a form cutter. This along with a straight worm will

give no enveloping condition. This will give low efficiency and is used for very light service involving low load and intermittent service.

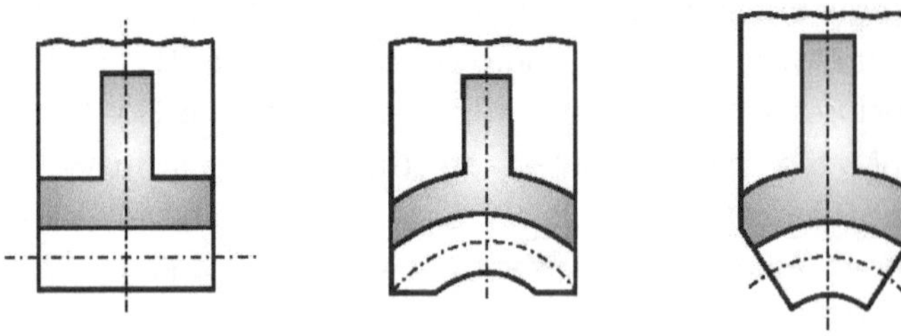

(i) (ii) (iii)

Fig. 6.6: Wheel Rims

- In second arrangement, teeth are cut with a hob and ends of the gear teeth are cut parallel to the gear axis. This arrangement along with a straight worm will give single enveloping arrangement.
- The third arrangement, in which worm gear teeth are cut radially towards the worm axis, characterises worm envelopment. This arrangement is used for heavy services.
- First arrangement is easiest to manufacture but will give least efficiency on the other hand third arrangement is most difficult will give maximum efficiency. Alignment during mounting is also requires closer tolerances and hence is costly.

6.4 STRENGTH OF WORM AND WORM GEAR

Lewis Equation for Strength of Gear Tooth (Beam Strength), S_b:

- The approach used to find beam strength is similar to that of spur gears.
- Since the material of the worm possesses higher mechanical properties than those of the worm wheel, the calculations are done to determine the strength of the worm wheel teeth only. The maximum load (S_b), which a worm wheel tooth can transmit, considering the tooth as a cantilever beam, is given as

$$S_b = \sigma_b \cdot b \cdot Y \cdot m_n = \sigma_b \cdot b \cdot Y \cdot m \cdot \cos\lambda \quad \ldots (6.5)$$

where σ_b is the strength of the gear tooth material (N/mm²) which is its maximum permissible bending stress.

This could be taken as $= \frac{1}{3} \cdot S_{ult}$

b: Width of the gear wheel (mm) : Minimum of $b = 0.73 \times d_1$ or $b = 2m\sqrt{q+1}$

Y: Lewis form factor = 0.314 for 14.5° pressure angle
= 0.393 for 20° pressure angle

Buckingham's Wear Load Capacity of Gear Tooth (Wear Strength) S_w:

- Wear load capacity is calculated for worm and worm wheel pair as such and not for individual worm or worm wheel. Buckingham suggested following equation to find wear load rating in a worm gearing.

$$S_w = d_g \cdot b \cdot K_w \quad \ldots (6.6)$$

where, K_W (N/mm²) is the worm gear load stress factor (or wear constant). The factor depends on combination of material pair for worm and pressure/ lead angle.

Table 6.3: Typical Values of K_w

Worm	Worm Gear	K_w (N/mm²)
Steel	Phosphor Bronze	0.51
Hardened Steel	Cast Iron	0.42
Hardened Steel	Phosphor Bronze	0.68
Hardened Steel	Chilled Phosphor Bronze	1.0

6.5 FORCE ANALYSIS OF WORM GEAR DRIVE

- Consider a worm – worm wheel drive shown in Fig. 6.7 with worm rotating in anticlockwise direction. The worm shown has right hand threads and the worm is considered driver, as this will be the normal case.

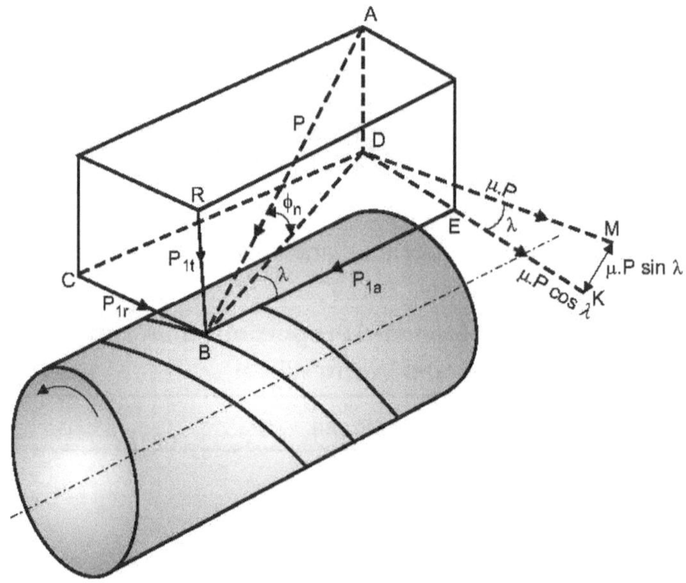

Fig. 6.7: Force Analysis : Forces on Worm

Let

P : be the normal resultant force acting between worm and worm wheel teeth

ϕ_n : Normal pressure angle

λ : Lead angle on worm (Helix angle on the worm gear)

P acts along AB which is perpendicular to BC and acts at an angle ϕ_n.

Components of P acting on the worm are as shown in the Fig. 6.7. Resolving P in the plane ABD.

$$\text{Radial force } P_{1r} = RB = AD = P \cdot \sin\phi_n \quad \ldots (6.7)$$

Normal force, BD = AB cos ϕ_n = P · cos ϕ_n ... (6.8)

Normal force can be further resolved into

P_{1t} = CB = DE = BD · sin λ = P. cos ϕ_n . sin λ ... (6.9)

P_{1a} = BE = BD · cos λ = P · cos ϕ_n . cos λ ... (6.10)

- Friction cannot be neglected between worm and worm wheel since there is sliding plus rolling action. Considering friction between and worm wheel, frictional force (μ.P) acts perpendicular to P and is considered to be acting in the plane of sliding. Frictional force (μ.P) as shown in Fig. 6.7 acts along DM which is perpendicular to BD. The direction of μ.P is such that it is opposing anticlockwise motion of worm at point B. Thus direction will be form D to M.

  Friction force DM is further resolved as below.

  Along, DK = μ.P. cos λ (in the same direction as tangential force P_{1t}) and

  Along, KM = μ.P. sin λ (in the direction opposite to axial force P_{1a})

  Superimposing frictional component on P_{1t} and P_{1a},

  Total forces will be,

$$P_{1a} = P \cdot \cos\phi_n \cdot \cos\lambda - \mu \cdot P \cdot \sin\lambda$$

$$= P(\cos\phi_n \cdot \cos\lambda - \mu \cdot \sin\lambda) \quad \ldots (6.11)$$

$$P_{1t} = P \cdot \cos\phi_n \cdot \sin\lambda + \mu \cdot P \cdot \cos\lambda$$

$$= P(\cos\phi \sin\lambda + \mu \cdot \cos\lambda) \quad \ldots (6.12)$$

$$P_{1r} = P \cdot \sin\phi = P_{2r} \quad \ldots (6.13)$$

Prom this, P_{1a} and P_{1r} can be expressed in terms oP P_{1t} as

$$P_{1a} = P_{1t}\frac{(\cos\phi_n \cdot \cos\lambda - \mu \cdot \sin\lambda)}{(\cos\phi_n \cdot \sin\lambda + \mu \cdot \cos\lambda)} \quad \ldots (6.14)$$

$$P_{1r} = P_{1t}\frac{\sin\phi_n}{(\cos\phi_n \cdot \sin\lambda + \mu \cdot \cos\lambda)} \quad \ldots (6.15)$$

- In practice knowing power transmission from the worm, values of P_{1t} will be known. $P_{1t} = \left(\frac{M_{1t}}{\frac{d_1}{2}}\right)$. Once value of P_{1t} is known, values of P_{1a} and P_{1r} can be found out using above relationships.

  Fig. 6.8 shows the forces P_{1r}, P_{1t}, P_{1a} which are the radial, tangential and axial forces on the worm.

  Equal and opposite forces will be acting on the worm wheel

$$P_{2t} = P_{1a}$$

$$P_{2a} = P_{1t} \quad \ldots (6.16)$$

$$P_{2r} = P_{1r}$$

and these are shown in the Fig. 6.8.

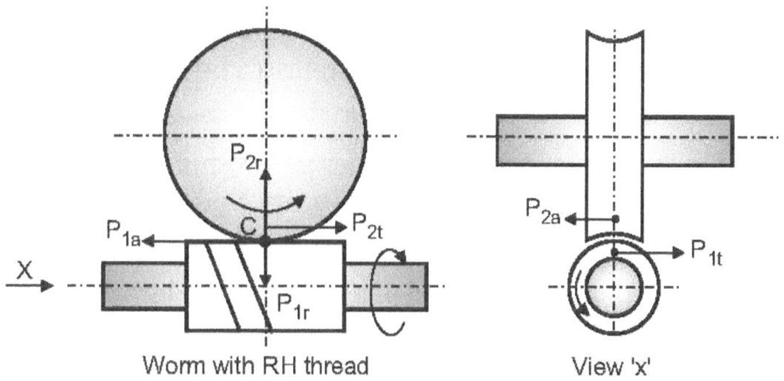

(a) Worm with RH Threads **(b) View - X**

Fig. 6.8: Forces on Worm Gear

Thus, the tangential, radial and axial components of forces on worm and worm wheel can be found out.

- In practice designer will be required to establish the force and directions without going into force analysis every time. Before deciding the direction of forces of the worm and the worm wheel, it is necessary to first establish the direction of rotations of wheel.

Direction of Rotation for Worm and Gear:

- In Fig. 6.9, consider worm to be rotating in anticlockwise direction as seen from left side and let the worm be having LH threads. Worm is considered similar to a thread and the worm wheel as a nut.
- Applying LH thumb rule, with fingers curling in the direction of rotation, worm will tend to move to right. The worm is not allowed to move axially and therefore will push the gear towards left side and thus the wheel will rotate clockwise. These directions of movements can be imagined about the contact point 'C' in Fig. 6.9.
- If the hand of helix on worm was RH instead of LH, then we apply RH thumb rule and the directions of P_{2t} and P_{1a} would have been exactly opposite.

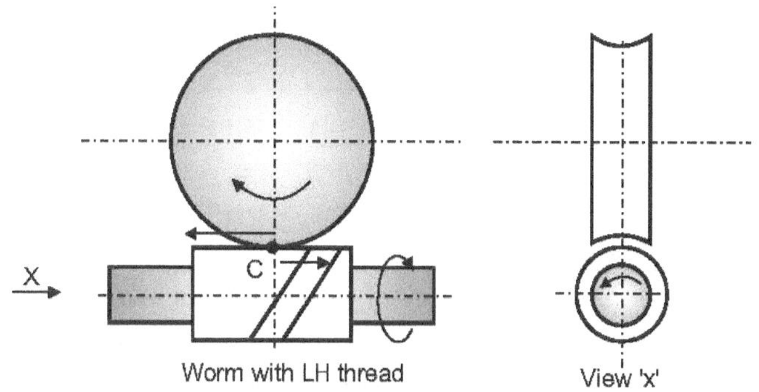

Fig. 6.9: Worm and Gear Rotation Directions

Direction of Forces for Worm and Gear:

- Once the directions of rotation are established, force directions can now be decided.
- Force on the driver member is opposite to direction of motion.

- Force on the driven member is in the direction of motion. The direction of motion is to be determined at the contact point 'C' shown in the Fig. 6.10.
- Fig. 6.10 shows the worm gear pair with its direction of rotation as obtained in Fig. 6.9. The worm gear linear direction of motion is toward left as seen at point 'C'. Hence tangential force P_{2t} is also towards left. The axial force P_{1a} is in opposite to P_{2t} (to the right.).

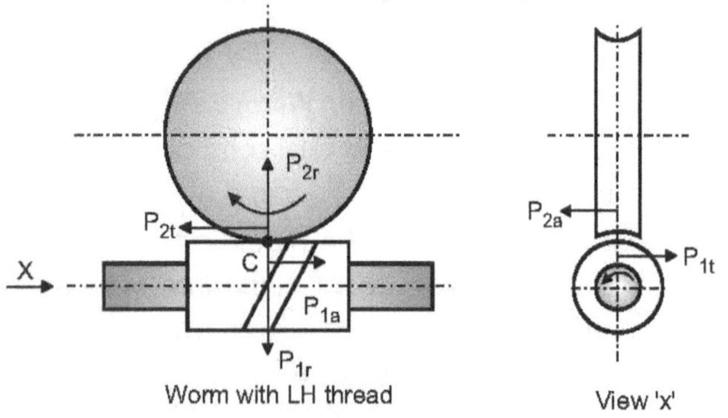

Fig. 6.10: Worm and Worm Gear Force Directions

- As seen from view X, worm (driving) is rotating in anticlockwise direction. Hence at the contact point, linear direction of motion of the worm is to the left. Hence tangential force P_{1t} is towards right. Axial force P_{2a} will be towards left.
- Radial forces P_{1r} and P_{2r} are separating forces. Hence their directions will be from the contact point towards respective centers.
- It can be seen that only P_{1t} and P_{2t} are useful components as they transmit torque. The radial and axial components are present and have to be sustained by the shaft and the bearings. These are considered during design of shaft and bearings.

6.6 COEFFICIENT OF FRICTION AND SLIDING VELOCITY

Thus to find the force components, it is necessary to know the coefficient of friction between the worm and worm wheel. The coefficient of friction depends on gear pair materials, lubricant used, accuracy of mounting and the sliding velocity (rubbing velocity) between the two.

Typical values of coefficient of friction for well lubricated gears are indicated below.

Table 6.4

Rubbing Velocity m/sec	1	2	3	4	5	6
Coefficient of Friction	0.06	0.05	0.04	0.038	0.032	0.03

Sliding Velocity is the relative velocity between worm and worm wheel. Consider top view of a worm with its axis A-A and the gear with axis B-B and that the position of worm is above the gear (Fig. 6.11). The direction of rotations of the worm and the gear are as marked in the figure. Velocities of worm (V_1) and gear (V_2) will then be as shown and sliding velocity will be

$$V_s = V_1 - V_2 \quad \dots (6.17)$$

Pitch line velocity of the worm is given by $V_1 = \dfrac{\pi \cdot d_1 \cdot n_1}{60 \times 1000}$

As seen from the Fig. 6.11, $V_s = V_1/\cos \lambda$

Rubbing velocity is obtained as a function of worm velocity and the lead angle.

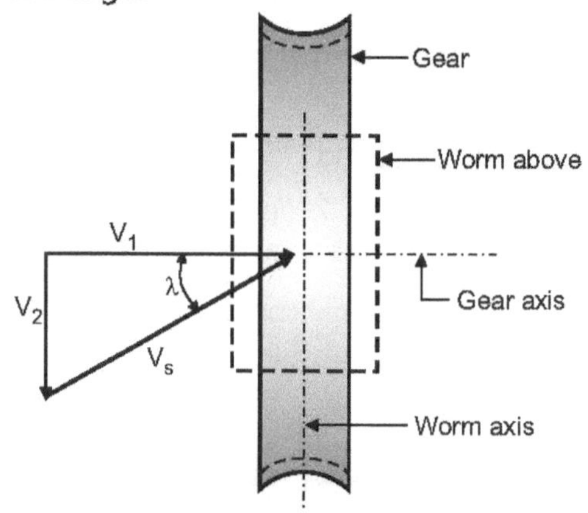

Fig. 6.11: Worm-Gear Sliding Velocity

6.7 MATERIALS FOR WORM AND WORM GEAR

6.7.1 Material for Worm

- **Carbon Steel:** They are inexpensive and strong but susceptible to corrosion.
- **Alloy Steel:** They have superior strength, durability and corrosion resistance. They can also be hardened.

6.7.2 Worm Gear Materials

- **Cast Iron:** It provides durability and ease of manufacture.
- **Brass:** This material is inexpensive, easy to mold and corrosion resistant.
- **Bronze:** This is an alloy consisting primarily of copper and other metals. The addition of other metals (usually tin), produces an alloy much harder than plain copper.
- **Phosphor Bronze** is an alloy of copper with 3.5 to 10% of tin and up to 1% phosphorus. Phosphor bronze has high toughness, strength, low coefficient of friction, and fine grain. The phosphorus also improves the fluidity of the molten metal and thereby improves the castability.

- **Aluminium Bronze:** 5 to 11% aluminium is added to copper as the main alloying element. The alloy has excellent wear resistance and industrial applications.
- Selection of material will depend on the application details. Following guideline can be used.
  - **Light Duty:** For lighter loads and low speed application, worm is made of steel and worm.
  - **Medium Duty:** For such an application, worm is made of case hardened steel and wheel of phosphor bronze.
  - **Heavy Duty:** For heavy duty application, worm is made of alloy steel and wheel of phosphor bronze chilled to increase hardness.

6.8 EFFICIENCY OF WORM/WORM GEAR PAIR

Efficiency of a worm gear can be considered as ratio of output power to input power. (refer Fig. 6.8).

$$\eta = \frac{\text{Output power}}{\text{Input power}} = \frac{(P_2t)\left(\frac{d_2}{2}\right) \cdot N_2}{(P_{1t})\left(\frac{d_1}{2}\right) N_1} \quad \dots (6.18)$$

Now, Speed ratio $\frac{N_1}{N_2} = i$

$$\text{and} \quad \frac{d_2}{d_1} = \frac{Z_2}{q} = \frac{\left(\frac{Z_2}{Z_1}\right)}{\left(\frac{q}{Z_1}\right)} = \frac{i}{\frac{1}{\tan\lambda}} = i \cdot \tan\lambda \quad \dots (6.19)$$

Substituting values from equation (6.19) into (6.18).

$$\eta = \frac{P_2t}{P_1t} \cdot \tan\lambda = \frac{P_1a}{P_1t} \cdot \tan\lambda \quad \dots (6.20)$$

Substituting values from equations (6.14) efficiency is given by

$$\eta = \frac{P_{1a}}{P_{1t}} \cdot \tan\lambda$$

$$= \frac{(\cos\phi_n \cos\lambda - \mu \cdot \sin\lambda)}{(\cos\phi_n \cdot \sin\lambda + \mu \cdot \cos\lambda)} \cdot \tan\lambda$$

This can be written as

$$\eta = \frac{\cos\phi_n - \mu \cdot \tan\lambda}{\cos\phi_n + \mu \cot\lambda} \quad \dots (6.21)$$

Alternately ϕ_v, is defined as the Virtual friction angle, such that

$$\phi_v = \tan^{-1}[\mu/\cos\phi_n] \quad \dots (6.22)$$

On substituting in the above equation equation, η can be reduced to form

$$\eta = \frac{\tan\lambda}{\tan(\phi_v + \lambda)} \quad \dots (6.23)$$

For initial estimation, efficiency of a worm gear pair can taken as

Table 6.5: Worm Gear Efficiency versus Worm Starts

Worm Gear Efficiency	Worm Starts
70 to 75%	For (Z_1 = 1): Single start
75 to 85%	For (Z_1 = 2): Double start
80 to 85%	For (Z_1 = 3): Tripple start
85 to 95%	For (Z_1 = 4): Quadruple start

Efficiency of Worm/Worm Gear Drive (η):

- Overall efficiency of the drive will be total effect of gearing efficiency, together with losses in bearings and losses due to churning of the oil. Thus gear drive efficiency will be

$$\eta = \eta_g \cdot \eta_b \cdot \eta_c$$

$$\eta \approx 0.95 \times \eta_g$$

- Considering other two losses as 5%,

where, η_g = Efficiency of the gear pair considering gearing losses.

η_b = Efficiency considering bearing losses.

η_c = Efficiency considering churning losses.

Combined efficiency at bearings and that due to churning is however quite large ($\approx$ 0.95) compared to gear pair efficiency.

Self Locking Drive:

- Usually for a worm gear drive, worm is the driving member and gear is the driven member. As in case of screw mechanism, self locking will be achieved if the coefficient of friction is greater than tangent of the lead angle. (Friction angle is more than the lead angle).
- Thus if $\mu > \tan\lambda$, self locking is achieved. This property is important and is used with advantage in many material handling equipments where self locking can prevent accidental rotations of the worm wheel.

Effective Load on Worm Gear Tooth:

- Theoretical tangential force acting on the gear, depends on the rated power to be transmitted and the efficiency of worm-gear pair. The dynamic load acting on the gear due to inaccuracies in teeth cutting and tooth deflection is accounted for by the velocity factor. Velocity factor for worm-wheel is usually taken as:

$$C_v = \frac{6}{6 + V_g}$$

- The load value has to be further modified by a load factor (application factor) C_s, depending on nature of load and uncertainty in load. Considering these factors, the tangential load P_t gets modified to effective load which is obtained as

$$P_{eff} = \frac{C_s \cdot P_{2t}}{C_v}$$

- Taking into consideration the factor of safety (FS), the gear strength in bending (S_b) as well as wear strength (S_w) should both be either greater than or equal to ($FS \times P_{eff}$).

  Thus

$$S_b, S_w \geq (FS) \cdot P_{eff}$$

- If the power transmitted is known, tangential tooth load on the worm wheel can be obtained from

$$P_{2t} = \frac{\text{Power}}{V_2}$$

6.9 THERMAL CONSIDERATION IN WORM GEAR DRIVE

- The efficiencies of the worm- worm gear pair can be considerably smaller depending on the geometry and higher friction. Thus there is considerable loss of power and this energy gets converted to heat at the contact area between worm and worm gear. For example, if the efficiency is 70% then 30% of input power is converted into heat. With this the temperature at the contact point can increase.
- Lubrication in worm – worm gear pair therefore has to perform important task of carrying away the heat while reducing the friction. The increasing temperature limits the maximum power that can be transmitted by the pair. Thus heat generated must be equated to heat dissipated while limiting the temperature rise within acceptable limit. Depending on the lubricant selected, usually lubricants will work satisfactorily up to temperatures of 90°C. Considering ambient temperature of 40 to 45°C, temperature rise may be limited to about 45°C.

  Heat Generated,

$$H_g = (1 - \eta) \cdot P_i$$

  Heat Dissipated

$$H_d = C.A. (\Delta T)$$

where, C = Overall heat transfer coefficient of Gear Box casing walls W/ m^2 °C.

= 12 to 18 W/ m^2 °C for natural circulation of air.

20 to 30 W/m^2 °C for forced circulation of air with fins.

A = External surface area of the housing (m^2).

ΔT = Permissible temperature rise of oil.

The heat dissipation capacity can be increased by

- Providing fins on the outer surface of the housing so as to increase surface area.
- Forced circulation by providing a fan on worm shaft.
- Providing cooling system for lubricating oil.

6.10 WORM DRIVE - TYPE OF FAILURES

As in case of spur and helical gears, worm gears are also subjected to damage mainly on two accounts.

1. Lack of proper lubrication and
2. Surface pitting due to long operations at high loads.

Additionally, sliding cannot be totally avoided in a worm drive. The worm wheel usually suffers more damage than the worm. Principal modes of gearing failures are:

- Surface damage caused by initial pitting.
- Tooth wear.
- Seizing due to heavy loads and lack of lubrication.
- Fatigue pitting due to high loads exceeding endurance strength of the material.
- Tooth breakage after severe wear caused by weakening of the section of the gear tooth.

Remedies:

- Adequate wear strength to avoid fatigue pitting. Load should be lesser than the endurance limit for the material.
- Adequate beam strength for gear material.
- Material hardness can be increased so that endurance limit of the material can increase thus avoiding pitting.
- Ensure sufficient and appropriate lubrication to prevent corrosive fatigue, abrasive pitting.

Table 6.6: Recommended Values of $Z_1/Z_2/q/m$ for Worm Gearing

Centre Distance (mm)	Transmission Ratio							
	10	15	20	25	30	40	50	60
80	–	2/30/10/4	2/42/(11)/3	–	1/30/10/4	1/42/(11)/3	–	1/62/(18)/2
100	4/40/10/4	2/30/10/5	2/40/10/4	2/54/(11)/3	1/30/10/5	1/40/10/4	1/55/(11)/3	1/64/16/2.5
125	4/40/10/5	2/31/10/6	2/40/10/5	2/52/10/4	1/31/10/6	1/40/10/5	1/52/10/4	–
160	4/43/10/6	2/30/10/8	–	2/54/10/5	1/30/10/8	–	1/54/10/5	1/62/(18)/4
200	4/40/10/8	2/30/10/10	2/40/10/8	–	1/30/10/10	1/40/10/8	–	–
250	4/40/10/10	–	2/40/10/10	2/52/10/8	–	1/40/10/10	1/52/10/8	–
315	4/43/(9)/12	2/30/(9)/16	–	2/53/10/10	1/30/(9)/16	–	1/53/10/10	1/60/(18)/8

SOLVED EXAMPLES

Example 6.1: *A worn gear box with an effective surface area of 1.5 m² is operating in still air with a heat transfer coefficient of 15 W/m²°C. The temperature rise of lubricant is limited to 60°C. The worm gears are designated as 1/30/10/8. The worm shaft is rotating at 1440 rpm and the normal pressure angle is 20°. Calculate the power rating based on thermal consideration for the drive. Take coefficient of friction as 0.03.*

Given data:

Area = A = 1.5 m^2, C = 15 $W/m^2\,°C$, $\Delta T = 60°C$, $\mu = 0.03$

Gear pair: $Z_1 / Z_2 / q' / m$ 1 / 30 / 10 / 8

Speed = 1440 rpm $\phi_n = 20°$

Solution: Lead angle of worm: $\tan\lambda = \frac{\lambda_1}{q} = \frac{1}{10} = 0.1$

Lead angle, $\lambda = 5.71°$ and $\phi_n = 20°$

Gear pair efficiency

$$\eta = \frac{\cos\phi_n - \mu\tan\lambda}{\cos\phi_n + \mu\cot\lambda}$$

$$= \frac{\cos 20.0° - 0.03\tan 5.71}{\cos 20°.0 + 0.03\cot 5.71}$$

$$= \frac{0.9396 - 0.03 \times 0.09998}{0.9396 + 0.03 \times 10} = 0.755$$

Equating heat balance

Heat generated = Heat dissipated

$$(1 - \eta)\,(kW)_{i/p} = C \cdot A \cdot \Delta T$$

$$(1 - 0.755)\,(kW)_{i/p} = 15 \times 1.5 \times 60$$

$$(kW)_{i/p} = 5510\text{ W}$$

Power rating = 5.51 kW

Based on thermal consideration alone.

Example 6.2: *A worm gear pair 2/40/10/4 is having phosphor bronze gear with ultimate tensile strength 300 MPa. The worm is made of steel with ultimate strength 740 MPa. The coefficient of friction between the worm and the worm gear is 0.03 and normal pressure angle 20°. The worm gear wear factor is 0.9 N/mm². The overall heat transfer coefficient for the gear box is 18 W/m²°C. The permissible temperature rise for lubricating oil is 50°C. The worm rotates as 720 rpm and service factor is 1.5.*

Determine the input power rating based on beam strength, wear strength and thermal consideration.

Assume effective surface area of the gear box as 0.8 m² and factor of safety as 1.5.

Given data: $z_1/z_2/q/m \rightarrow 2/40/10/4$, $i = \frac{z_2}{z_1} = 20$

$$(S_{ult})_g = 300\text{ N/mm}^2$$

$$(S_{ult})_W = 740\text{ N/mm}^2,\ \mu = 0.03$$

$$N_2 = \frac{720}{20} = 36\text{ rpm}$$

$$\phi_n = 20°,\ K_w = 0.9\text{ N/mm}^2,\ C = 18\text{ W/m}^2\,°C$$

$$\Delta T = 50°C,\ N_1 = 720\text{ rpm},\ C_s = 1.5$$

$$FS = 1.5$$

Surface area, $A = 0.8\text{ m}^2$

$$\tan\lambda = \frac{z_1}{q} = \frac{2}{10} = 0.2;\ \lambda = 11.3°$$

Solution:

Beam strength, by Lewis equation is given by (to be applied to worm gear)

$$S_b = \sigma_b \cdot b \cdot Y \cdot M_n$$

where, σ_b = Permissible bending stress for gear material

$$\sigma_b = \frac{1}{3} \times 300 = 100\text{ N/mm}^2$$

$$b = \text{Gear width [larger of]}$$

$$\therefore \quad b = 0.73\, d_1 = 0.73\,(q \times m)$$

$$= 0.73\,(10 \times 4) = 29.2 \text{ mm}$$

OR

$$b = 2m\sqrt{q+1}$$

$$= 2 \times 4\sqrt{10+1} = 26.5 \text{ mm}$$

Hence we select

$$b = 29.2 \text{ mm}$$

$$Y = 0.484 - \frac{2.87}{z_g} = 0.484 - \frac{2.87}{40} = 0.4122$$

$$m_n = m \cos \lambda$$

$$\text{where, } \tan \lambda = \frac{z_1}{q} = \frac{2}{10}$$

$$\therefore \quad \lambda = 11.31°$$

$$m_n = 4 \times 0.98 = 3.922 \text{ mm}$$

∴ Beam strength,

$$S_b = 100 \times 29.2 \times 0.4122 \times 3.922$$

$$S_b = 4720 \text{ N}$$

Wear Strength:

$$S_w = d_g \cdot b \cdot K_w = (m \cdot Z_g) \cdot b \cdot K_w$$

$$= 4 \times 40 \times 29.2 \times 0.9$$

$$S_w = 4204.8 \text{ N}$$

Effective Load:

$$P_{eff} = P_2t \times \frac{C_s}{C_v}$$

$$C_s = 1.5, \quad d_1 = q \times m = 40, \quad n_1 = 720$$

Velocity factor,

$$C_v = \frac{6}{6 + V_2} = \frac{6}{6 + \left(\frac{\pi \cdot d_2 n_2}{60 \times 1000}\right)}$$

$$= \frac{6}{6 + \left(\frac{\pi \times 160 \times 36}{60 \times 1000}\right)}$$

$$= \frac{6}{6 + 0.30} = 0.952$$

$$P_{eff} = P_2t \times \frac{1.5}{0.952} = 1.576\, P_2t$$

Equating beam strength with effective load considering FS = 1.5.

$$S_b = (FS) \cdot P_{eff} = (1.5)\, 1.576 \cdot P_2t$$

$$4720 = 1.5 \times 1.576 \times P_{2t}$$

$$\therefore \quad P_2t = 1997 \text{ N (in bending)}$$

Equating wear strength with effective load considering FS

$$S_w = (FS)\, P_{eff}$$

$$4204.8 = 1.5 \times 1.576 \times P_{2t}$$

$$P_2t = 1779 \text{ N (in wear)}$$

Output power rating based on beam strength

$$= (P_2t) \times V_2 = \frac{1997 \times 0.30}{1000} = 0.599 \text{ kW}$$

Output power rating based on wear strength

$$= (P_2t)_{wear} \times V_2 = \frac{1779 \times 0.3}{1000} = 0.534 \text{ kW}$$

Worm/Gear Pair Efficiency:

$$\eta = \frac{\cos \phi_n - \mu \tan \lambda}{\cos \phi_n + \mu \cot \lambda}$$

$$= \frac{\cos 20° - 0.03 \tan 11.3°}{\cos 20° + 0.03 \cot 11.3°} = \frac{0.9336}{1.09}$$

$$= 0.856$$

Input power rating based on beam strength

$$= \frac{0.599}{0.856} = 0.7 \text{ kW}$$

Input power rating based on wear strength

$$= \frac{0.534}{0.856} = 0.63 \text{ kW}$$

Input power rating based on heat dissipation:

Equating heat generated = Heat dissipated

$$(1 - \eta)\,(\text{Power})_{i/p} = C \cdot A \cdot \Delta T = 18 \times 0.8 \times 50$$

$$(\text{Power})_{i/p} = \frac{18 \times 0.8 \times 50}{(1 - 0.856)} = 5000 \text{ W} = 5.0 \text{ kW}$$

Example 6.3: *A pair of worm and worm gear is designated as 1/30/10/10. The input speed of worm is 1200 rpm. The worm wheel is centrifugally cast and made from phospher bronze having ultimate strength 240 N/mm². Coefficient of friction between worm and worm teeth is 0.04. Application factor and factor of safety are 1.25 and 1.5 respectively. The tooth system is 20° full involute. Calculate power transmitting capacity based on beam strength.*

Given data: Worm/worm gear → 1/30/10/10 → $z_1/z_2/q/m$,

$$i = \frac{z_2}{z_1} = 30$$

Worm: $N_1 = 1200$ rpm, $d_1 = q \times m = 100$, $\tan \lambda = \frac{z_1}{q} = \frac{1}{10}$,

$\lambda = 5.71°$.

Worm wheel: $S_{ult} = 240$ N/mm², $d_2 = m \times z_2 = 300$ mm,

$$N_2 = \frac{1200}{i} = 40$$

$\mu = 0.04$, $C_s = 1.25$, $FS = 1.5$, $\phi_n = 20°$.

Solution: Beam Strength: S_b

$$S_b = \sigma_b \cdot b \cdot Y \cdot m_n = \sigma_b \cdot b \cdot Y \cdot m \cdot \cos \lambda$$

where, $\sigma_b = \frac{1}{3} \times S_{ult} = 80 \text{ N/mm}^2$

$$b = 0.73\ d_1 = 73 \text{ mm}$$

OR

$$b = 2m\sqrt{q+1} = 2 \times 10\sqrt{(10+1)} = 66.3 \text{ mm}$$

Hence we select b = 73 mm (larger of the two)

$$Y = 0.484 - \frac{2.87}{z_2} = 0.484 - \frac{2.87}{30} = 0.388$$

$$m_n = m \cdot \cos\lambda = 10 \times 0.995 = 9.95 \text{ mm}$$

$$\therefore \quad S_b = 80 \times 73 \times 0.388 \times 9.95 = 22546 \text{ N}$$

Effective Load: P_{eff}

$$P_{eff} = \left[\frac{C_s}{C_v}\right] \cdot P_2t$$

$$C_v = \frac{6}{6+V_g} = \frac{6}{6 + \left(\frac{\pi\, d_2 \cdot n_2}{60 \times 1000}\right)}$$

$$= \frac{6}{6 + \left(\frac{\pi \times 300 \times 40}{60 \times 1000}\right)}$$

$$= \frac{6}{6 + 0.6283} = 0.905$$

$$P_{eff} = \left[\frac{1.25}{0.905}\right] P_2t = 1.381\ P_2t$$

Equating beam strength with effective load, considering FS,

$$S_b = (P_{eff}) \cdot FS$$

$$S_b = (1.381)\ P_{2t}\ (1.5)$$

$$S_b = 2.07\ P_2t$$

$$\therefore \quad 22546 = 2.07\ P_2t$$

$$\therefore \quad P_2t = 10892 \text{ N}$$

$\therefore$ Output power rating based on beam strength

$$= (P_{2t})_{bending} \times V_2$$

$$= \frac{10892 \times 0.6283}{1000} = 6.843 \text{ kW}$$

Worm/Worm-Gear Pair Efficiency:

$$\eta = \frac{\cos\phi_n - \mu \cdot \tan\lambda}{\cos\phi_n + \mu\cot\lambda}$$

$$= \frac{\cos 20^\circ - 0.04 \times \tan 5.71}{\cos 20^\circ + 0.04 \cot 5.71}$$

$$= \frac{0.9396 - 0.04 \times 0.0999}{0.9396 + 0.04 \times 10} = 0.6984$$

Input power rating based on beam strength

$$= \frac{\text{Output power}}{\text{Efficiency}} = \frac{6.843}{0.6984} = 9.8 \text{ kW}$$

Example 6.4: *A double start worm made of alloy steel is to mesh with worm gear to be made of phosphor bronze (σ_{ut} = 270 N/mm^2). The gear pair is required to transmit 7 kW power from an electric motor running at 1500 rpm to a machine running at 60 rpm. The service factor is 1.25, while the factor of safety required is 1.5. The face width of the worm gear is 0.73 times the pitch circle diameter of worm. The worm gear wear factor is 0.7 N/mm^2, while the diametrical quotient is 10. The normal pressure angle is 14.5°. If the coefficient of friction between worm and worm gear teeth is 0.03, design the gear and find the power lost. Would you recommend a fan for the gear box? Assume the permissible temperature rise is 50°C.*

Use following data:

- *Lewis form factor:*

$$Y = 0.39 - \frac{2.15}{Z_g}$$

- *Velocity factor:*

$$C_v = \frac{6}{6+V_g}$$

- *Area of housing: $A = 1.14 \times 10^{-4} \times (a)^{1.7}\ m^2$*

  where, a = Centre distance in mm

Given data: 7 kW at worm, i = 25, $\phi_n = 14\frac{1^\circ}{2}$

Worm: $Z_1 = 2$, $N_1 = 1500$ rpm, $q = 10$, $\tan\lambda = \frac{Z_1}{q} = 0.2$.

Gear: Sult = 270 N/mm^2, $N_2 = 60$ rpm, $b = 0.73\ d_1$

$K_w = 0.7$ N/mm^2, $Z_g = i \times Z_1 = 50$, $d_2 = 50 \cdot m$

$C_s = 1.25$, FS = 1.5, $\mu = 0.03$, $\Delta T = 50°C$

Solution:

Beam Strength S_b: (For the gear tooth)

$$S_b = \sigma_b \cdot b \cdot Y \cdot m_n$$

$$\sigma_b = \frac{1}{3} \times 270 = 90 \text{ N/mm}^2$$

$$b = 0.73\ d_1 = 0.73\ q \cdot m = 7.3\ m$$

$$Y = 0.39 - \frac{2.15}{Z_g} = 0.39 - \frac{2.15}{50} = 0.347$$

$$m_n = m \cdot \cos\lambda = m\ (0.98)$$

$$S_b = 90 \times 7.3\ m \times 0.347 \times 0.98\ m = 223.4\ m^2$$

Wear Strength S_w:

$$S_w = d_g\ b \cdot K_w$$

$$= 50 \cdot m \times 7.3\ m \times 0.7 = 255.5\ m^2$$

$\therefore$ Since, $S_b < S_w$, gear is weaker in bending and therefore will be designed against bending failure.

Load on the Gear Tooth:

Efficiency of gear pair ($\phi_n = 14.5°$, $\mu = 0.03$, $\tan\lambda = 0.2$)

$$\eta = \frac{\cos\phi_n - \mu\tan\lambda}{\cos\phi_n + \mu\cdot\cot\lambda} = \frac{0.968 - 0.006}{0.968 + 0.15}$$

$$= \frac{0.962}{1.118} = 0.8605$$

Output power is,

$$= \text{Input power} \times \eta = 7 \times 0.8605$$

$$= 6.02 \text{ kW}$$

Tangential gear tooth load P_t

$$P_t = \frac{\text{Output power}}{V_g} = \frac{6.02 \times 10^3}{\left(\frac{\pi \times d_2 \times n_2}{60 \times 1000}\right)}$$

$$= \frac{6.02 \times 10^3}{\left(\frac{\pi \times 50 \cdot m \times 60}{60 \times 1000}\right)} = \frac{6.02 \times 10^3}{0.157 \text{ m}}$$

$$= \frac{38344}{m} \text{ Newton}$$

Effective gear tooth load

$$P_{eff} = \left(\frac{C_s}{C_v}\right) P_t$$

where, $C_v = \frac{6}{6 + V_g} = \left(\frac{6}{6 + 0.157 \text{ m}}\right)$

$$= \left[1.25 \times \frac{(6 + 0.157 \text{ m})}{6}\right] \cdot P_t$$

$$= 0.208\,(6 + 0.157 \text{ m}) \cdot P_t$$

This should be equated to beam strength considering factor of safety.

$$S_b = (P_{eff})\text{ FS}$$

$$223.4 \text{ m}^2 = 0.208\,(6 + 0.157 \text{ m}) \cdot P_t \times 1.5$$

$$= 0.3125\,(6 + 0.157 \text{ m})\, P_t$$

$$\therefore \quad 223.4 \text{ m}^2 = 0.3125\,(6 + 0.157 \text{ m}) \times \frac{38344}{m}$$

Simplifying,

$$223.4 \text{ m}^3 - 1882 \text{ m} - 71895 = 0$$

$$m^3 - 8.43 \text{ m} - 321.8 = 0$$

By trial and error, m = 7.3 mm = 8.0 mm

Therefore, the gear pair can be designed as,

$Z_1/Z_2/q/m \rightarrow 2/50/10/8$

Power Lost

Input power $(1 - \eta) = 7 \times (1 - 0.8605) = 0.98$ kW

Provision of Forced Circulation (Fan)

Gear box area $= A = 1.14 \times 10^{-4} \times a^{1.7} \text{ m}^2$

Here, $a = \frac{d_1 + d_2}{2} = \frac{q \cdot m + m \cdot Z_2}{2} = \frac{80 + 400}{2}$

$$= 240 \text{ mm}$$

$$\therefore \quad A = 1.27 \text{ m}^2$$

Equating heat balance to find required value of overall heat transfer coefficient

$$\text{Input power } (1 - \eta) = C_3 \cdot A \cdot \Delta T$$

$$7 \times (1 - 0.8605) \times 10^3 = C \times 1.27 \times 50$$

$$\therefore \quad C_{required} = 15.4 \text{ W/m}^2\text{°C}$$

Since this is < 18 W/m²°C.

Natural circulation will be sufficient.

Example 6.5: *A worm 'up' type 2/40/10/5 worm gear pair transmit 5 kW power from worm shaft rotating at 1440 rpm to worm gear shaft. The worm is left hand and rotates in anticlockwise direction, when viewed from right side. Coefficient of friction between worm and worm gear teeth is 0.05. Normal pressure angle is 20°. Determine.*

(i) Components of forces acting on worm and worm gear and show their directions.

(ii) Efficiency of worm gear pair.

(iii) Power lost in friction.

Given data: $Z_1/Z_2/q/m \rightarrow 2/40/10/5$

Worm; 5 kW, 1440 rpm

$\mu = 0.05$, $\phi_n = 20°$, $d_1 = q \times m = 50$ mm

$$\tan\lambda = \frac{Z_1}{q} = \frac{2}{10}, \quad \lambda = 11.3°$$

Solution:

(i) Tangential Force on the Worm

$$P_{1t} = \frac{\text{Power}}{V_1} = \frac{5000}{(\pi \times 50 \times 1440/60 \times 1000)}$$

$$= 1326.4 \text{ N} = P_{2a}$$

$$P_{1a} = \frac{\cos\phi_n \cos\lambda - \mu\sin\lambda}{\cos\phi_n \sin\lambda + \mu\cos\lambda} \cdot (P_{1t})$$

$$= \frac{\cos 20° \cos 11.3° - 0.05 \sin 11.3°}{\cos 20° \sin 11.3° + 0.05 \cos 11.3°} (P_{1t})$$

$$= \frac{0.921 - 0.00979}{0.184 + 0.049} (1326.4)$$

$$= 5188 \text{ N} = P_2t$$

$$P_{1r} = \frac{\sin\phi_n}{\cos\phi_n \sin\lambda + \mu\cos\lambda} (P_{1t}) = \frac{0.342}{0.233} (P_{1t})$$

$$= \frac{0.342}{0.233} (1326.4) = 1947 \text{ N} = P_2r$$

Since worm is above the worm gear.

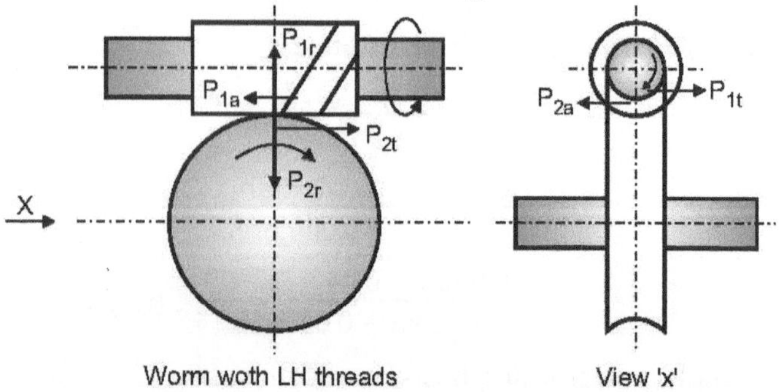

Fig. 6.12

For LH worm and given direction of rotation for worm, worm gear will rotate clockwise.

Other forces and their directions will be as shown.

(ii) Efficiency of Worm-Worm Gear Pair:

$$\eta = \frac{\cos\phi_n - \mu\cdot\tan\lambda}{\cos\phi_n + \mu\cdot\cot\lambda} = \frac{\cos 20° - 0.05\times 0.2}{\cos 20° + 0.05\times 5}$$

$$= \frac{0.8397}{1.1896}$$

$$\eta = 0.706$$

(iii) Power Lost in Friction:

$$= (1-\eta)\text{ Power input} = (1 - 0.706)\ 5\text{ kW}$$

$$= 1.47\text{ kW}$$

Example 6.6: *A worm gear pair 2/52/10/4 transmits 10 kW power from an electric motor rotating at 720 rpm to a machine. The worm is right hand and rotates in clockwise direction when viewed from right side. Sketch the arrangement and show the components of tooth forces. The coefficient of friction between worm and worm wheel is 0.04, while the normal pressure angle is 20°. The overall heat transfer coefficient is 20 W/m²°C. The temperature rise of lubricating oil above the atmospheric temperature is 50°C. Determine:*

(i) Components of tooth forces acting

(ii) Efficiency of worm gear pair

(iii) Minimum surface area of gear box

Is the drive self-locking? Comment. Assume worm is below the worm gear.

Given data:

Gear pair: 2/52/10/4 $\rightarrow z_1/z_2/q/m$, $i = \frac{Z_2}{Z_1} = 26$

Worm: $N_1 = 720$ rpm, $d_1 = q\times m = 40$ mm, $\tan\lambda = \frac{Z_1}{q} = 0.2$, $\lambda = 11.31°$

Wheel: 10 kW, $\mu = 0.04$, $\phi_n = 20°$, $C = 20\ W/m^2°C$, $\Delta T = 50°C$

Solution:

P_{1t} : Tangential force on the worm

$$P_{1t} = \frac{\text{Input power}}{V_1} = \frac{10\times 10^3}{\left(\frac{\pi d_1 N_1}{60\times 1000}\right)}$$

$$= \frac{10\times 10^3}{\left(\frac{\pi\times 40\times 720}{60\times 1000}\right)}$$

$$P_{1t} = \frac{10\times 10^3}{1.507} = 6632\text{ N} = P_{2a}$$

$$P_{1a} = \frac{\cos\phi_n\cos\lambda - \mu\sin\lambda}{\cos\phi_n\sin\lambda + \mu\cos\lambda}\times(P_{1t})$$

$$= \frac{\cos 20°\cdot\cos 11.31° - 0.04\sin 11.31°}{\cos 20°\sin 11.31° + 0.04\cos 11.31°}\times(P_{1t})$$

$$= \frac{0.921 - 0.00784}{0.184 + 0.0392}(P_{1t})$$

$$P_{1a} = \frac{0.913}{0.2232}\times 6632 = 27128\text{ N} = P_{2t}$$

$$P_{1r} = \frac{\sin\phi_n}{\cos\phi_n\sin\lambda + \mu\cos\lambda}(P_{1t})$$

$$= \frac{0.342}{0.2232}\times 6632$$

$$P_{1r} = 10162\text{ N} = P_{2r}$$

$$\eta = \frac{\cos\phi_n - \mu\cdot\tan\lambda}{\cos\phi_n + \mu\cdot\cot\lambda}$$

$$= \frac{\cos 20° - 0.04\tan\lambda}{\cos 20° + 0.04\cot\lambda} = \frac{0.9397 - 0.008}{0.9397 + 0.2}$$

$$= 0.817$$

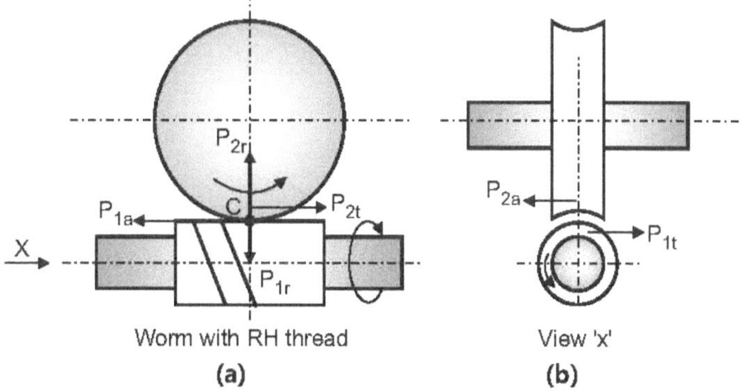

Fig. 6.13

For RH worm and given direction of rotation for worm, worm gear would rotate clockwise.

Other forces and their directions, will be as shown.

Example 6.7: *A worm/worm-gear pair is used to transmit from worm 5 kW power at 1500 rpm. Worm is two start having 60 teeth gear. The axial module is 6 mm and pressure angle is 14.5°. The diametral factor is 10. Worm gear material has ultimate tensile strength of 300 N/mm². Find the factor of safety in bending, wear and heat dissipation. Based on heat dissipation criteria the power transmission capacity is given by,*

$$\text{Input kW} = \frac{a^{1.7}}{34.5\,(i+5)}$$

Take Lewis form factor:

$$Y = 0.39 - \frac{2.15}{Z_g}$$

Assume: $K_w = 0.85\ N/mm^2$, $\mu = 0.05$

Given data: 5 kW, Worm $N_1 = 1500$ rpm, $Z_2 = 60$, $Z_1 = 2$, $m = 6$ mm, $\phi_n = 14\frac{1}{2}°$, $q = 10$

Worm gear: Sult = 300 N/mm², $\mu = 0.05$, $K_w = 0.85$

$$\tan\lambda = \frac{Z_1}{q} = 0.2,\quad d_2 = m\cdot Z_2 = 360\text{ mm},\quad N_2 = 50$$

Solution: Beam Strength: S_b

$$S_b = \sigma_b \cdot b \cdot Y \cdot m_n$$

$$\sigma_b = \frac{300}{3} = 100 \text{ N/mm}^2$$

$$b = 0.73\, d_1 = 0.73\,(q \times m) = 43.8 \text{ mm}$$

OR

$$b = 2\, m\sqrt{q+1} = 2 \times 6\sqrt{11}$$

$$= 39.8 \text{ mm} \simeq 40 \text{ mm}$$

Hence we select

$$b = 43.8 \text{ mm (Larger of the two)}$$

$$Y = 0.39 - \frac{2.15}{60} = 0.354$$

$$m_n = m \cos\lambda = 5.88 \text{ mm}$$

$$S_b = 100 \times 43.8 \times 0.354 \times 5.88$$

$$= 9117 \text{ N (in bending)}$$

Wear Strength:

$$S_w = d_2 \cdot b \cdot K_w$$

$$= (6 \times 60)\,(43.8)\,(0.85)$$

$$= 13403 \text{ N (in wear)}$$

Effective load

$$P_{eff} = \left(\frac{C_s}{C_v}\right) P_{2t}$$

Velocity factor,

$$C_v = \frac{6}{6 + V_g}$$

$$C_v = \frac{6}{\left(6 + \frac{\pi d_2 n_2}{60 \times 1000}\right)} = \frac{6}{\left(6 + \frac{\pi \times 360 \times 50}{60 \times 1000}\right)}$$

$$= \frac{6}{6 + 0.942}$$

$$C_v = 0.864$$

Assuming service factor $C_s = 1$

$$P_{eff} = \left(\frac{1}{0.864}\right) P_{2t} = 1.158\, P_{2t}$$

Based on beam strength

$$P_{eff} = 1.158 \times 9117 = 10557 \text{ N (in bending)}$$

Based on wear strength

$$P_{eff} = 1.158 \times 13403 = 15521 \text{ N (in wear)}$$

Output power rating based on beam strength

$$= P_{eff} \times V_2 = \frac{10557 \times 0.942}{1000} = 9.94 \text{ kW}$$

Output power rating based on wear strength

$$= P_{eff} \times V_2 = \frac{15521 \times 0.942}{1000} = 14.62 \text{ kW}$$

Gear drive efficiency: η ($\phi_n = 14.5°$, $\mu = 0.05$, $\tan\lambda = 0.2$)

$$\eta = \frac{\cos\phi_n - \mu\tan\lambda}{\cos\phi_n + \mu\cot\lambda}$$

$$= \frac{0.968 - 0.01}{0.968 + 0.25} = \frac{0.958}{1.218}$$

$$= 0.786$$

Input power based on beam strength

$$= \frac{9.94}{0.786} = 12.64 \text{ kW}$$

$$FS = \frac{12.64}{5} = 2.528$$

Input power based on wear strength

$$= \frac{14.62}{0.786} = 18.6 \text{ kW}$$

$$FS = \frac{18.6}{5} = 3.72$$

For heat dissipation

$$(\text{kW}) = \frac{a^{1.7}}{34.5\,(i + 5)}$$

where, $a = \frac{d_1 + d_2}{2} = \frac{60 + 360}{2}$

$$= \frac{210^{1.7}}{34.5\,(35)} = 7.35 \text{ kW} = 210 \text{ mm}$$

$\therefore$ FS based on heat dissipation $= \frac{7.35}{5} = 1.47$

Example 6.8: *A worm/worm-gear pair is designated as 1/30/10/10. The worm is rotating at 1000 rpm.*

Find:

(a) Worm and worm gear dimensions.

(b) Centre distance between the worm and the gear.

(c) Estimate the value of coefficient of friction for the pair.

Use following relation.

Sliding Velocity m/sec.	1	2	3	4	5	6
Coefficient of Friction μ	0.042	0.032	0.029	0.027	0.025	0.023

Given data: $Z_1/Z_2/q/m \rightarrow 1/30/10/10$

$$\tan\lambda = \frac{Z_1}{q} = \frac{1}{10} = 0.1, \quad \lambda = 5.71°$$

$$d_1 = q \times m = 10 \times 10 = 100 \text{ mm}$$

$$d_2 = Z_2 \times m = 30 \times 10 = 300 \text{ mm}$$

Solution:

(a) Worm: Addendum $= h_{a1} = m = 10$ mm

Dedendum $= h_{f1} = (2.2\cos\lambda - 1)\, m$

$= 11.89$ mm

Outer diameter $= da_1 = d_1 + 2ha_1 = 120$ mm

Root diameter $= df_1 = d_1 - 2h_{f1} = 76.22$ mm

Clearance $= 0.2\ m \cos\lambda = 1.99$ mm

Length of worm $= m\pi\left[4.5 + \frac{Z_2}{50}\right] = 160.22$ mm

Worm Wheel:

Addendum of wheel $= ha_2 = (2\cos\lambda - 1)\ m$
$= 9.9$ mm

Dedendum of wheel $= h_{f2} = (1 + 0.2\cos\lambda)\ m$
$= 11.99$ mm

Addendum diameter $= da_2 = d_2 + 2ha_2 = 319.8$ mm

Dedendum diameter $= df_2 = d_2 - 2hf_2 = 276.02$ mm

Face width, $b = 0.73\ \ d_1 = 73$ mm

(b) Centre Distance between Worm and Gear

$$a = \frac{d_1 + d_2}{2} = \frac{100 + 300}{2} = 200 \text{ mm}$$

(c) Sliding Velocity

$$v_s = \frac{v_1}{\cos\lambda} = \frac{\frac{\pi d_1 N_1}{60 \times 1000}}{\cos 5.71^\circ} = \frac{5.236}{0.995}$$

$v_s = 5.262$ m/sec.

From chart by interpolation coefficient of friction corresponding to sliding velocity of 5.262 m/sec.

$\mu = 0.0245$

Example 6.9: *The worm of a worm/worm gear pair is required to transmit 6 kW power at 1500 rpm with a speed reduction of 30.*

(a) Select a suitable pair based on heat capacity and

(b) Find the factor of safety for the worm gear based on beam strength.

Assume: Coefficient of friction = 0.04.

Pennissible bending stress for gear material = 80 N/mm^2.

Gear system is 20° full depth involute. Consider service factor as 1.0.

Use following data: Velocity factor $C_v = \frac{\sigma}{\sigma + v}$

Relationship for heat capacity based on centre distance between worm and gear.

$$(P_{input}) = \frac{a^{1.7}}{34.5\ (i + 5)}$$

where, a is centre distance in mm.

Centre Distance a (mm)	Transmission Ratio (i)		
	20	30	40
100	2/40/10/4	1/30/10/5	1/40/10/4
125	2/40/10/5	1/31/10/6	1/40/10/5
160	-	1/30/10/8	-
200	2/40/10/8	1/30/10/10	1/40/10/8

Given data: Power: 6 kW, $N_1 = 1500$ rpm, $i = 30$, $\mu = 0.04$

Solution:

(a) Desired Centre Distance for 6 kW Input Power Based on Heat Capacity.

$$P_{input} = \frac{a^{1.7}}{34.5\ (i + 5)}$$

$$6 = \frac{a^{1.7}}{34.5\ (35)}$$

$\therefore$ $a = 186.5$ mm

$\therefore$ We select a = 200 mm

$\therefore$ Selected gear pair from the table is

1/30/10/10 $\rightarrow Z_1/Z_2/q/m$

(b) $d_1 = q \times m = 100$ mm

$$\tan\lambda = \frac{Z_1}{q} = 0.1$$

$d_2 = m \times Z_2 = 300$ mm

Beam Strength: S_b

$S_b = \sigma_b \cdot b \cdot Y \cdot m_n$

σ_b = Allowable bending stress = 80 N/mm^2

$b = 0.73 \times d_1 = 73$ mm OR

$2m\sqrt{q + 1} = 66.3$ mm

Hence we select

$b = 73$ mm [Larger of the two]

Using standard relationship for Lewis form factor,

$$Y = 0.484 - \frac{2.87}{Z_g} = 0.484 - \frac{2.87}{30} = 0.388$$

$m_n = m\cos\lambda = 10\cos 5.71^\circ = 9.95$ mm

$S_b = 80 \times 73 \times 0.388 \times 9.95$
$= 22546$ N

Gear pair efficiency ($\phi_n = 20°$, $\mu = 0.04$, $\tan\lambda = 0.1$)

$$\eta = \frac{\cos\phi_n - \mu\tan\lambda}{\cos\phi_n + \mu\cot\lambda} = \frac{0.9396 - 0.004}{0.9396 + 0.4}$$
$$= 0.698$$

Output power= Input power $\times\ \eta = 6 \times 0.698 = 4.2$ kW

Gear tangential force ($d_2 = 300$ mm, $N_2 = 1500/30 = 50$ rpm)

$$P_t = \frac{\text{Output power}}{V_g}$$
$$= \frac{4.2 \times 10^3}{\left(\frac{\pi d_2 N_2}{60 \times 1000}\right)} = \frac{4.2 \times 10^3}{\left(\frac{\pi \times 300 \times 50}{60 \times 1000}\right)}$$
$$= \frac{4.2 \times 10^3}{0.785}$$

$P_t = 5348$ N

Effective force $P_{eff} = \left[\frac{C_s}{C_v}\right] P_t$

With $C_s = 1$ and $C_v = \frac{6}{6 + V_g} = \frac{6}{6 + 0.785} = 0.884$

$$= \left[\frac{1}{0.884}\right] 5348 = 6050 \text{ N}$$

Equating beam strength with effective load considering factor of safety.

$$S_b = (P_{eff})\ FS$$
$$22546 = 6060 \times FS$$
$$FS = 3.72$$

Example 6.10: *Design a worm/worm-gear pair based on wear strength and suggest minimum surface area to be provided for the gear box if it has to work with natural circulation. Single start worm is connected to 3.0 kW electric motor rotating at 1500 rpm. Required speed reduction is 30 between the worm and worm wheel. Wear stress factor for the gear material is 0.6 N/mm². Gear tooth system is 20° full depth involute and worm diametral quotient is 10. Consider service factor as 1.2 and required factor of safety is 1.4. Permissible temperature rise for lubricant is 50°C. Use following data. Take coefficient of friction as 0.03.*

Given data: $\mu = 0.03$, $\Delta T = 50°c$

Power: 8kW, $i = 30$, $\phi_n = 20°$, $C_s = 1.2$, $FS = 1.4$.

Worm: $Z_1 = 1$, $N_1 = 1500$, $q = 10$, $d_1 = q \cdot m = 10 \cdot m$.

Gear: $K_w = 0.6$, $N_2 = 50$, $Z_2 = \dfrac{1500}{50} = 30$, $d_2 = 30 \cdot m$

Solution: Wear Strengths S_w:

$$S_w = d_2 \cdot b \cdot K_w$$
$$Z_2 = i \cdot Z_1 = 30 \times 1 = 30$$
$$d_2 = m \cdot Z_2 = 30 \cdot m$$
$$b = 0.73\ d_1 = 7.3\ m \quad \text{OR}$$
$$= 2m\sqrt{q+1} = 2m\sqrt{11} = 6.63\ m$$

Hence we select

$$b = 7.3\ m \text{ (larger of the two)}$$
$$S_w = 30 \cdot m \times 7.3 \times m \times 0.6$$
$$S_w = 131.4\ m^2$$

Gear Pair Efficiency: ($\phi_n = 20°$, $\mu = 0.03$, $\tan\lambda = \dfrac{Z_1}{q} = 0.1$)

$$\eta = \frac{\cos\phi_n - \mu\tan\lambda}{\cos\phi_n + \mu\cot\lambda} = \frac{0.9396 - 0.003}{0.9396 + 0.3} = 0.755$$

∴ Output power is,

$$= \text{Input power} \times \eta = 8 \times 0.755 = 6.04 \text{ kW}$$

Tangential gear tooth load P_t

$$P_t = \frac{\text{Output power}}{V_g} = \frac{6.04 \times 10^3}{\left(\dfrac{\pi d_2 N_2}{60 \times 1000}\right)}$$
$$= \frac{6.04 \times 10^3}{\left(\dfrac{\pi \times 30 \cdot m \times 50}{60 \times 1000}\right)}$$
$$= \frac{6.04 \times 1000}{0.07854 \cdot m} = \frac{76904}{m} \text{ Newton}$$

Effective gear tooth load, P_{eff}:

$$P_{eff} = \left(\frac{C_s}{C_v}\right) \cdot P_t$$

where, $C_s = 1.2$

$$C_v = \frac{6}{6 + V_g} = \left(\frac{6}{6 + 0.0785\ m}\right)$$
$$P_{eff} = \frac{1.2}{1} \times \frac{(6 + 0.0785\ m)}{6} \cdot P_t$$

Equating P_{eff} with P_w = considering factor of safety

$$S_w = (P_{eff}) \cdot FS$$
$$131.4\ m^2 = \left[\frac{1.2}{1} \times \frac{(6 + 0.0785\ m)}{6} \times \frac{76904}{m}\right] 1.4$$
$$131.4\ m^3 = 1690\ m + 129199$$
$$m^3 - 12.85\ m - 983 = 0$$

Solving by trial and error we get m = 10.4 mm

We select next higher standard value as m = 12 mm

Heat Dissipation:

$$\text{Heat generated} = P_{I/P}(1 - \eta)$$
$$= 8(1 - 0.755) = 1.96 \text{ kW}$$
$$\text{Heat generated} = C \cdot A \cdot \Delta T$$

where, $C = 18\ W/m^2{}°C$ (Assumed for natural circulation)

$$= (18 \times A \times 50)\ W$$

Equating Heat generated = Heat dissipated

$$1.96 \times 10^3 = 18 \times A \times 50$$
$$A = 2.18\ m^2$$

EXERCISE

1. What are the types of failures in worm gearing and what remedies you propose to avoid such failures?
2. Why two dissimilar materials are used for worm and worm wheel.
3. Derive an expression for efficiency of worm gear pair.
4. Explain following terms for worm and worm wheels. (a) Diameter quotient, (b) Self locking worm, (c) Rubbing velocity.
5. Why the soft material like Phosphor Bronze is chosen for worm gear and alloy steel for worm.
6. Derive an expression for effective face width of the worm wheel and length of the root of the worm wheel teeth.
7. Explain force analysis for a worm gear drive.
8. Derive equation for efficiency of worm gear in terms of lead angle and virtual friction angle.
9. Explain with neat sketch the difference between single Enveloping and double – Enveloping worm gear pair.
10. With neat sketches explain any two types of construction of worm gear.
11. Discuss the thermal considerations in worm gear drive.

◈◈◈

UNIT IV - BELT AND CHAIN DRIVES

CHAPTER - 7
BELT AND CHAIN DRIVES

BELTS

7.1 INTRODUCTION

- Belt and chain drives are commonly known as *flexible machine elements or flexible drives.* Gear drives have many advantages. However when power is to be transmitted between two shafts at a considerable distance apart, the usage of gear drive results into incompatible solution in terms of space and manufacturing cost. For power transmission over a longer distance, flexible element drives prove to be advantageous.
- Friction between the belt and pulley is responsible for power transmission. However in practical situations, some magnitude of slip is always present between belt and pulley. It is due to this reason that belt drive is not positive drive and hence it is not used in applications that require a constant velocity ratio. To certain extent flexible machine elements can be used as substitutes for gear and other rigid power transmission elements. More or less, these elements possess sufficient amount of elasticity, that helps in absorption/damping of shocks and vibrations. These elements can also tolerate a few degree of misalignment between driving and driven elements. Their applications include but are not limited to lathe machines, mopeds, positive displacement pumps, compressors, radiator cooling fans and many more. Few of the common types in use are flat, round, V and timing or toothed belt cross-sections.

Table 7.1: Details of Typical Fundamental Belt Types

Sr. No.	Belt Cross-Section	Belt Type	Centre Distance	Remarks
1.	Flat 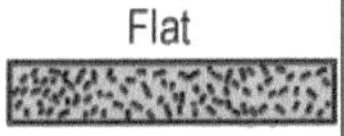		No Upper limit	Thickness up to 5 mm, joints are present.
2.	Round 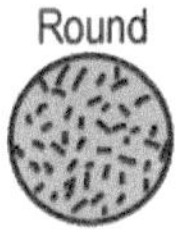		No Upper limit	Diameter up to 20 mm. joints are present.
3.	V 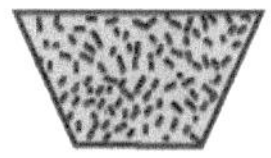		Limited	Manufactured as endless. Width = 5 to 20 mm
4.	Timing/Toothed 		Limited	Pitch: minimum 2 mm, manufactured as endless

7.2 MATERIALS AND CONSTRUCTION OF FLAT AND V-BELTS

- Flat belts were traditionally made of leather or fabric and presently they are mostly made of rubber wherein they are fabric or cotton duck impregnated with rubber. Leather belts are made of excellent quality leather obtained by oak and mineral or chrome tanning process. Leather belts offer moderate and sometimes high coefficient of friction that result into enhanced power transmitting capacity. The number of layers present in a belt are called as plies. These belts are specified according to number of plies or layers, e.g. single-ply, double-ply, triple-ply and four-ply belts. A typical triple-ply belt is shown in Fig. 7.1 below.

Fig. 7.1: A typical Triple-Ply Leather Belt

- Fabric rubber belts are prepared by using a number of layers of canvas. Fabric functions as load transmission element, while the rubber acts as shield against damage and boosts coefficient of friction. Presence of rubber in the belt composition makes the different plies of fabric to act together as one belt. With advantages as high load carrying capacity, long service span and capacity to operate at speeds up to 250 m/sec, the fabric rubber belts are widely used in engineering applications. Different layers are built together by various methods.

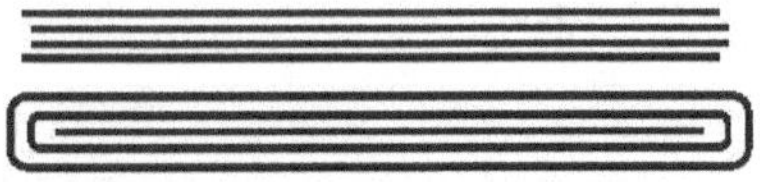

Fig. 7.2: Fabric Rubber Belt

- Ends of flat belts are joined using different processes as cementing, lacing and joining with metal fasteners. Flat belts are marketed in the form of coils and ends can be joined together to make them endless for a specific application as per required length. However care has to be taken to use proper joining method (as per manufacturer's recommendation), otherwise the belts will be weak at the joints and may not give rated capacity. The power rating of flat belts is specified as power transferred per ply of belt.

V-Belts:

- V belts may be completely rubber or polymer throughout the cross-section or there can be fibres of textile materials such as nylon or polyester or even steel embedded inside for achieving better strength. These type of belts mainly consist of three sections as

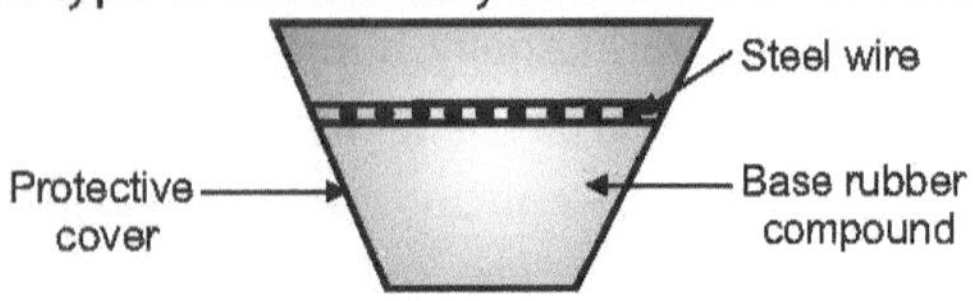

Fig. 7.3: Construction of V-Belt

1. The load carrying layers of polyester/steel, located centrally near C.G of belt cross-section.
2. Layer of rubber that transfers load/force to cords from side walls.
3. Outer cover made of polychloroprene soaked with plastic.

- Analysis of V-belt cross-section shows that the cords which are centrally located acts as load/force transfer members from driving to driven pulleys, thus resulting in power transmission. Furthermore as these steel wires or cords are located near the neutral axis, these are subjected to negligible amount bending stresses. Layers of rubber above the cords are subjected to tension and the layers below cords are subjected to compression and are known as tension and compression layers respectively. Usually, these belts are used as endless with factory manufactured belts to specific lengths.
- Multiple V belts are used for a specific power transmission when one belt is not adequate to transmit desired power. V belts are manufactured to various standards. Trapezoidal section sizes and manufacturer's guidelines need to be followed for their selection. These are very popular for industrial applications.

7.3 GEOMETRIC RELATIONSHIP FOR LENGTH OF BELT

Before proceeding for geometrical relation analysis of belt length, it is necessary to understand the open and crossed belt configurations.

Table 7.2 : Comparison between Open and Cross Belt Drive

Sr. No	Open Belt Drive	Crossed Belt Drive
1.	The belt proceeds from driving to driven pulley without crossing.	The belt proceeds from top of driving to bottom of driven pulley and crosses over itself.
2.	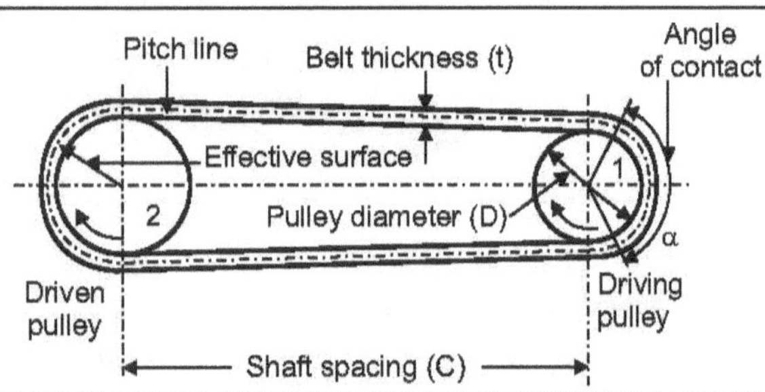	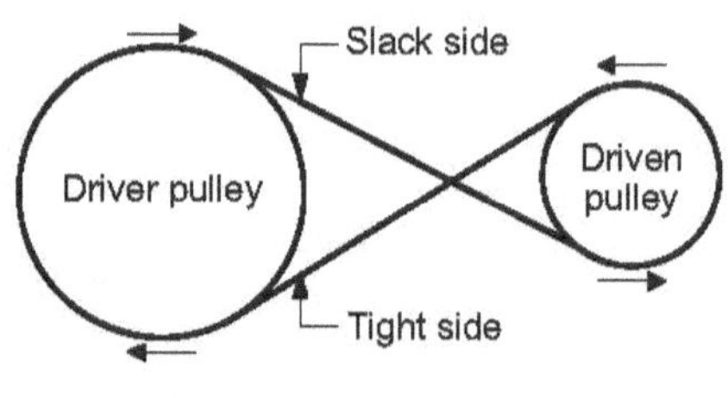
3.	Driving and Driven pulleys rotate in same direction.	Driving and Driven pulleys rotate in opposite direction.
4.	Power transmission capacity is less as compared with the crossed belt drives.	Power transmission capacity is more as compared with the open belt drives.
5.	Magnitude of angle of wrap is less.	Magnitude of angle of wrap is more.
6.	Life of belt is more as compared with crossed belt drives.	Life of belt is less due to bending of belt in two different planes.
7.	Subjected to vibration and slip.	Free from vibration and slip.
8.	More popular than crossed belt drives	Less popular.

7.3.1 Derivation of Equation of Belt Length

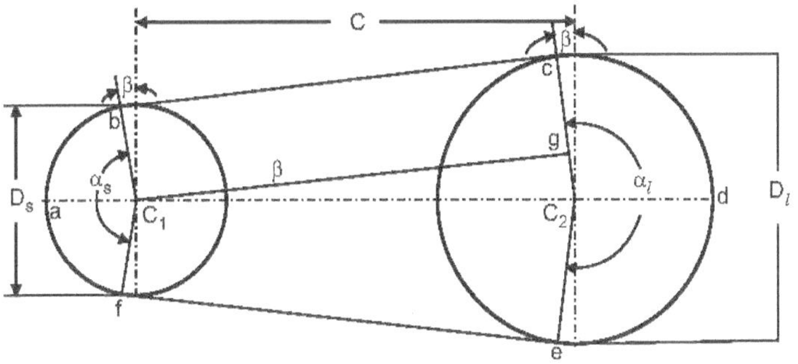

Fig. 7.4: Dimensional Details of Open Belt Drive

Referring the Fig. 7.4 of open belt drive, the belt dimensions are designated as under;

D_s = Diameter of small pulley (mm)

D_l = Diameter of large or big pulley (mm)

α_s = Angle of wrap for small pulley (degrees)

α_l = Angle of wrap for large or big pulley (degrees)

C = Centre distance (mm)

For open belting;

$$\sin\beta = \frac{D_l - D_s}{2C} \qquad \dots (7.1)$$

$$\alpha_s = (180 - 2\beta) \Rightarrow \alpha_s$$

$$= 180 - 2\sin^{-1}\left(\frac{D_l - D_s}{2C}\right) \qquad \dots (7.2)$$

$$\alpha_l = (180 + 2\beta) \Rightarrow \alpha_l$$

$$= 180 + 2\sin^{-1}\left(\frac{D_l - D_s}{2C}\right) \qquad \dots (7.3)$$

Length of belt (L) for open belt drive is given as:

$$L = \text{arc (fab)} + l\text{ (bc)} + \text{arc (cde)} + l\text{ (ef)}$$

$$L = \frac{D_s}{2}(\alpha_s) + l\,(C_1g) + \frac{D_l}{2}(\alpha_l) + l\,(C_1g)$$

$$L = \frac{D_s}{2}(\pi - 2\beta) + C\cos\beta + \frac{D_l}{2}(\pi + 2\beta) + C\cos\beta$$

Rearranging the above equation

$$L = \frac{\pi(D_l + D_s)}{2} + \beta(D_l - D_s) + 2C\cos\beta \quad \dots (7.4)$$

For small values of β; $\beta = \sin\beta = \left(\frac{D_l - D_s}{2C}\right)$

$$\cos\beta = 1 - 2\sin^2\frac{\beta}{2} = 1 - (\beta^2/2)$$

$$= (D_l - D_s)^2/8C^2$$

Using the above values in equation (7.4) and rearranging, we get

$$L = \frac{\pi(D_l + D_s)}{2} + \{(D_l - D_s)^2/2C\} + 2C$$

$$- \{(D_l - D_s)^2/4C\}$$

$$L = 2C + \frac{\pi(D_l + D_s)}{2} + \{(D_l - D_s)^2/4C\} \quad \dots (7.5)$$

Equation (7.5) is the required expression for length of open belt drive.

Equation of Length for Crossed Belts:

Consider the crossed belt configuration. As can be seen from the figure, angle of lap on both pulleys is $\alpha_s = \alpha_l = 180 + 2\beta$.

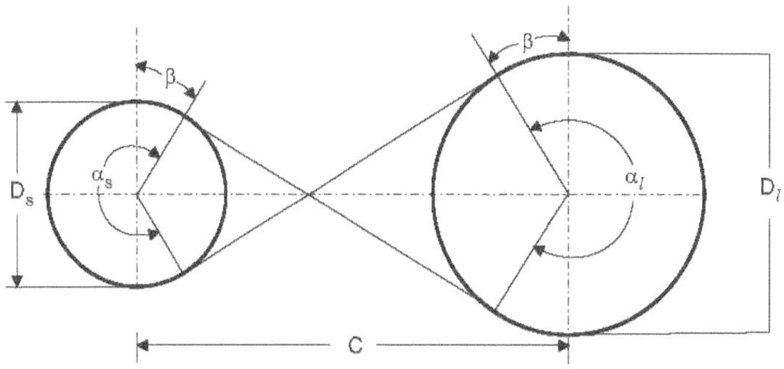

Fig. 7.5: Dimensional Details of Crossed Belt Drive

For evaluation of length of crossed belt drive,

$$\alpha_s = \alpha_l = (180 + 2\beta) \Rightarrow = 180 + 2\sin^{-1}\left(\frac{D_l - D_s}{2C}\right) \dots (7.6)$$

Repeating the procedure followed for evaluation of length of open belt, we get length of crossed belt as

$$L = 2C + \frac{\pi(D_l + D_s)}{2}\{(D_l + D_s)^2/4C\} \quad \dots (7.7)$$

7.4 ANALYSIS OF BELT FORCES AND BELT TENSIONS

7.4.1 Belt Tension Analysis for Flat Belt

The forces acting on an element of a flat belt are shown in Fig. 7.6. Following notations are used in the analysis of belt tensions;

F_1 = Belt tension in the tight side (N)

F_2 = Belt tension in the loose side (N)

m = Mass of the one meter length of belt (kg/m)

v = Belt velocity (m/s)

f, μ = Coefficient of friction

α = Angle of wrap (radians)

Fig. 7.6 shows an element of belt subtending an angle dϕ. This element is in equilibrium under the action of the following forces;

- Tensions F and (F + dF) on loose and tight sides respectively.
- The normal reaction dN between the surface of belt and pulley.
- Frictional force $f \cdot dN$ (or $\mu \cdot dN$) at the belt pulley interface and
- The centrifugal force ($mv^2d\phi$) in radial outward direction

The centrifugal force acting on any element is given as

Centrifugal force = Mass × Acceleration ... (7.8)

Mass of the element is = $mr\, d\phi$ ... (7.9)

Also we know from fundamentals of kinematics of machines, that acceleration is given as

$$\text{Acceleration} = v^2/r \quad \text{... (7.10)}$$

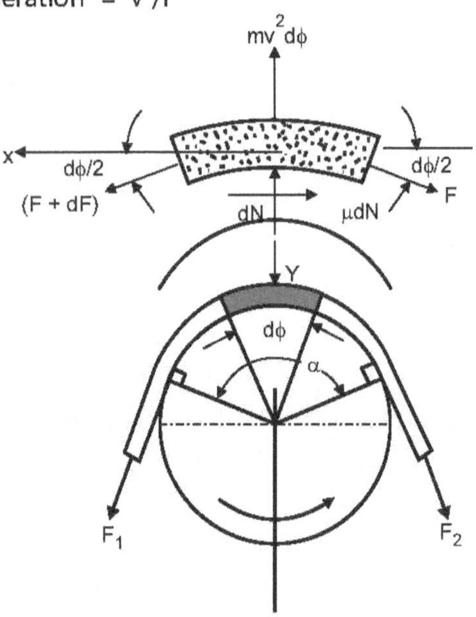

Fig. 7.6: Forces Acting on an Element of a Flat Belt

From equations (7.8), (7.9) (7.10), the magnitude of centrifugal force is given as,

$$\text{Centrifugal force} = mr\, d\phi * v^2/r = mv^2 d\phi$$

Writing the equilibrium by considering the forces in horizontal (X) and vertical (Y) directions; we get

$$(F + dF)\cos\left(\frac{d\phi}{2}\right) - F\cos\left(\frac{d\phi}{2}\right) - \mu dN = 0;$$ Horizontal direction equilibrium

$$(F + dF)\sin\left(\frac{d\phi}{2}\right) + F\sin\left(\frac{d\phi}{2}\right) - mv^2 d\phi - dN = 0;$$ Vertical direction equilibrium

For small values of $\left(\frac{d\phi}{2}\right)$, $\cos\left(\frac{d\phi}{2}\right)$ is nearly = 1 and $\sin\left(\frac{d\phi}{2}\right)$ is nearly = $\left(\frac{d\phi}{2}\right)$

Using these relations in horizontal and vertical equilibrium equations; we get

For horizontal equilibrium;

$$(F + dF) - F - \mu dN = 0$$

$$dF - \mu dN = 0 \Rightarrow dN = \frac{dF}{\mu} \quad \text{... (7.11)}$$

For vertical equilibrium;

$$(F + dF)\left(\frac{d\phi}{2}\right) + F\left(\frac{d\phi}{2}\right) - mv^2 d\phi - dN = 0$$

Neglecting second order differentials, i.e. $dF \times d\phi = 0$; we get

$$F\, d\phi - mv^2 d\phi - dN = 0 \quad \text{... (7.12)}$$

Using the value of dN from equation (7.11) in equation (7.12), and rearranging we get;

$$(F - mv^2)\, d\phi - \frac{dF}{\mu} = 0$$

$$\frac{dF}{F - mv^2} = \mu \cdot d\phi \quad \text{... (7.13)}$$

Equation (7.13) is called variable separable form; integrating the RHS and LHS and setting the limits as

$$\int_{F_2}^{F_1} \frac{dF}{F - mv^2} = \mu \int_0^{\alpha} d\phi$$

Integrating and using the upper and lower limits in the above equation, we get;

$$\frac{(F_1 - mv^2)}{(F_2 - mv^2)} = e^{\mu\alpha} \quad \text{... (7.14)}$$

The above equation is valid for high speed velocity belts. In case if the inertial force is negligible as compared to belt tensions equation (7.14).

$$\frac{F_1}{F_2} = e^{\mu\alpha} \quad \text{... (7.15)}$$

Further the power transmitted is given as

$$kW = \frac{(F_1 - F_2)\, v}{1000}$$

7.4.2 Belt Tension Analysis for V-belt

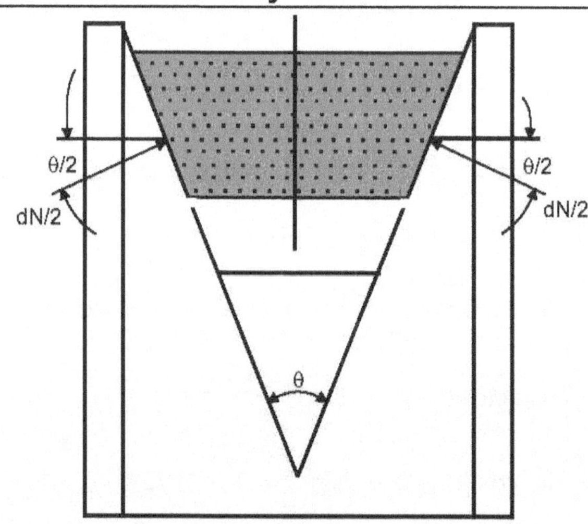

(a)

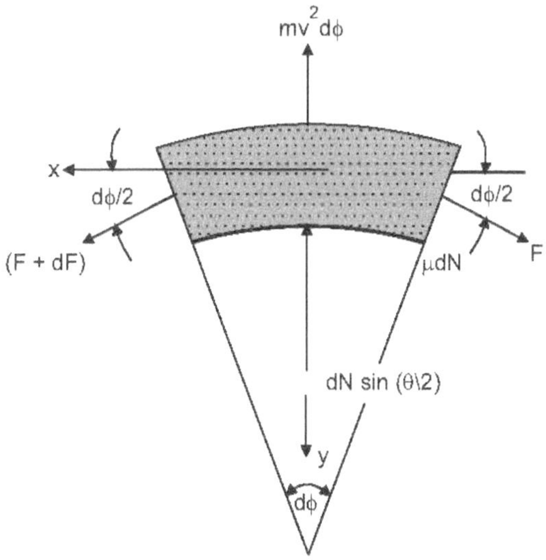

(b)

Fig. 7.7: Force Distribution on V-belt

In order to evaluate the magnitudes of belt tensions in terms of forces acting on the belt, the forces considered are shown in Fig. 7.7 above. Fig. 7.7 represents equilibrium condition of forces acting on V belt. Forces considered in flat belt analysis F, (F + dF) and $mv^2 d\phi$ remain same here also for the analysis of V belts. The sole difference is the magnitude of normal reaction dN, as it acts on both sides of v belt, its magnitude is assumed as (dN/2) on each side. Also the result reaction in X-Y plane is $dN \sin\left(\frac{\theta}{2}\right)$; where θ is called belt angle. As the magnitude of normal reaction is (dN/2); the total frictional force remains unchanged.

Writing the equilibrium by considering the forces in horizontal (X) and vertical (Y) directions; we get

$(F + dF) \cos\left(\frac{d\phi}{2}\right) - F \cos\left(\frac{d\phi}{2}\right) - \mu dN = 0$; Horizontal direction equilibrium

$(F + dF) \sin\left(\frac{d\phi}{2}\right) + F \sin\left(\frac{d\phi}{2}\right) - mv^2 d\phi - dN \sin\left(\frac{\theta}{2}\right); = 0$; Vertical direction equilibrium

For small values of $\left(\frac{d\phi}{2}\right)$, $\cos\left(\frac{d\phi}{2}\right)$ is nearly = 1 and $\sin\left(\frac{d\phi}{2}\right)$ is nearly = $\left(\frac{d\phi}{2}\right)$

Using these relations in horizontal and vertical equilibrium equations; we get

For horizontal equilibrium;

$$(F + dF) - F - \mu dN = 0$$

$$dF - \mu dN = 0 \Rightarrow dN = \frac{dF}{\mu} \quad \dots (7.16)$$

For vertical equilibrium;

$$(F + dF)\left(\frac{d\phi}{2}\right) + F\left(\frac{d\phi}{2}\right) - mv^2 d\phi - dN \sin\left(\frac{\theta}{2}\right) = 0$$

Neglecting second order differentials, i.e. dF x dφ = 0; we get

$$F\, d\phi - mv^2 d\phi - dN \sin\left(\frac{\theta}{2}\right) = 0$$

$$F\, d\phi - mv^2 d\phi - \frac{dF}{\mu} \sin\left(\frac{\theta}{2}\right) = 0 \quad \dots (7.17)$$

Using the value of dN from equation (7.16) in equation (7.17), and rearranging we get;

$$(F - mv^2)\, d\phi - \frac{dF}{\mu} \sin\left(\frac{\theta}{2}\right) = 0$$

$$\frac{dF}{F - mv^2} = \mu\, d\phi / \sin\left(\frac{\theta}{2}\right) \quad \dots (7.18)$$

Equation (7.18) is called variable separable form; integrating the RHS and LHS and setting the limits as

$$\int_{F_2}^{F_1} \frac{dF}{F - mv^2} = \left\{\mu / \sin\left(\frac{\theta}{2}\right)\right\} \int_0^{\alpha} d\phi$$

Integrating and using the upper and lower limits in the above equation, we get;

$$\frac{(F_1 - mv^2)}{(F_2 - mv^2)} = e^{\left(\frac{\mu\alpha}{\sin\frac{\theta}{2}}\right)} \quad \dots (7.19)$$

The above equation is valid for high speed velocity belts. In case if the inertial force is negligible as compared to belt tensions equation (7.19).

$$\frac{F_1}{F_2} = e^{\left(\frac{\mu\alpha}{\sin\frac{\theta}{2}}\right)} \quad \dots (7.20)$$

Furthermore the power transmitted is given as

$$kW = \frac{(F_1 - F_2)\, v}{1000}$$

A relative comparison between flat and V belts is given below.

Table 7.3: Comparison between Flat Belt and 'V' Belt

Sr. No.	Flat Belt	V Belt
1.	Simple and Inexpensive.	Costlier and periodic maintenance is needed.
2.	Low Power transmission capacity.	It can transmit more power than flat belts.
3.	Low velocity ratio (4 : 1).	High velocity ratio than flat belts (7 : 1)
4.	Operation is noisy.	Operation is smooth and noiseless.

...Conti.

5.	It can be used in open environment.	Used in closed environment where inclusion of dirt, dust is not expected.
6.	It can be used for long centre distances up-to 15 m, hence result into robust construction.	V belts are suitable shorter center distance and hence result in compact construction.
7.	Flat belt drives are horizontal.	It can be horizontal and vertical.
8.	Used in belt conveyors, baking machinery, stone crushers, brick manufacturing machinery, textiles, line shafts, flour mills, saw mills etc.	Used to drive compressors from electric motors, pumps, fans, blowers radiators, lathe machinery, mopeds etc.

7.5 POWER RATING AND CONCEPT OF SLIP AND CREEP IN BELTS

7.5.1 Power Rating

Power rating (kW rating) for a flat belt is defined as the power transmission capacity of the belt per mm width per ply at 180° arc of contact. Consider e.g., the power rating (kW rating) of a flat belt is 0.015 kW, this means that a single ply of 1 mm width will transmit 0.015 kW power for 180° arc of contact with the pulley.

7.5.2 Concept of Slip and Creep in Belts

- Creep is a typical phenomenon in belt drives. It is defined as a trivial relative motion of the belt as it passes over the pulley. As the belt moves from tight side to loose side over the pulley, it gets transferred from a zone of higher tension to lower one. It is due to this reduction of tension, belt is shortened and creeps over the surface of the pulley. This results into relative motion between belt and surface of pulley. Due to slight relative motion, the angular velocity of the driven pulley, evaluated from ratio of pulley diameters, is reduced. Creep phenomenon reduces the efficiency of the belt drive by 1% to 2%.
- Slip phenomenon occurs due to overloads. During slipping action, the belt slides over entire arc of contact on the pulley. Slip phenomenon takes place when the difference in tensions of belt is less than force of frictional resistance on the rim of driven pulley, similarly reverse also holds true. If the force which belts deliver, exceeds the sum of limit of friction force between belt and pulley, slip will occur.

7.6 MAXIMUM POWER CONDITION

- The sole objective of belt drives, whether flat or V belt is to transmit power. Many factors are responsible for power losses in belt drives that include belt creep, friction between particles of belt, aerodynamic resistance for pulley motions, bearing friction, variation in belt material properties etc. For any type of belt to transmit maximum power (accounting the above mentioned factors) the initial tension in the belt and the belt velocity are the factors that can be varied to establish maximum power condition.
- For the purpose of power transmission, an initial tension or pretension is provided to the belt. Let this initial tension be denoted by F_i. The magnitude of this pretension depends upon the characteristics of belt such as length of belt, belt material elasticity, centre distance and geometry of belt pulleys. For establishment of relation between the tensions on tight and slack sides and initial tension, it must be assumed that there is no variation in belt length before and after the application of initial tension and elasticity of the belt material is linear.
- As the driving pulley starts rotating, the elongation on the tight side is in proportion with $(F_1 - F_i)$, further on loose side it is $(F_i - F_2)$. With reference to the assumption that the length of the belt is constant, we have;

$$(F_1 - F_i) = (F_i - F_2)$$

$$F_i = \frac{1}{2}(F_1 + F_2) \qquad \ldots (7.21)$$

We know that for a flat belt drive

$$\frac{(F_1 - mv^2)}{(F_2 - mv^2)} = e^{\mu\alpha}$$

Rearranging the above equation as; $\frac{(F_2 - mv^2)}{(F_1 - mv^2)} = e^{-\mu\alpha}$

Simplifying further

$$\frac{[(F_1 + F_2) - 2mv^2]}{(F_2 - F_1)} = \frac{(e^{-\mu\alpha} + 1)}{(e^{-\mu\alpha} - 1)} \qquad \ldots (7.22)$$

But $(F_1 + F_2) = 2\,F_i$

From equation (7.21); using in equation (7.22), we get

$$\frac{[2\,F_i - 2mv^2]}{-(F_1 - F_2)} = \frac{(e^{-\mu\alpha} + 1)}{(e^{-\mu\alpha} - 1)}$$

$$(F_1 - F_2) = [2\,F_i - 2mv^2] \times \frac{(1 - e^{-\mu\alpha})}{(1 + e^{-\mu\alpha})}$$

$$\text{Power} = v\,(F_1 - F_2)$$

$$= [2F_i v - 2mv^3] \times \frac{(1 - e^{-\mu\alpha})}{(1 + e^{-\mu\alpha})} \qquad \ldots (7.23)$$

To obtain the maximum power condition, we differentiate the power with respect to belt velocity v, and equate it to zero. $\frac{d}{dv}$(Power) = 0, $\frac{d}{dv}(F_i v - mv^3) = 0$

$\Rightarrow (F_i - 3mv^2) = 0$ and

Hence, the optimum velocity of the belt at which the maximum power is transmitted is given as $v = \sqrt{\frac{F_i}{3m}}$ required expression.

Alternatively Maximum Power can also be obtained in Terms of Tight Side Tension F_1:

We know that for a flat belt drive

$$\frac{F_1}{F_2} = e^{\mu\alpha} \quad \text{... (7.24)}$$

If F_{max} is the maximum tension to which belt can be subjected

$$F_1 = F_{max} - Fc \quad \text{... (7.25)}$$

where F_c is the centrifugal tension in belt = $m.v^2$

Power that can be transmitted by the belt

$$\text{Power} = (F_1 - F_2) \times v$$

$$= \left(F_1 - \frac{F_1}{e^{\mu\alpha}}\right) \cdot v = F_1\left(1 - \frac{1}{e^{\mu\alpha}}\right) \times v$$

$$= F_1 \cdot v \cdot C$$

where $C = \left(1 - \frac{1}{e^{\mu\alpha}}\right)$

$$\text{Thus, Power} = F_1 \cdot v \cdot C \quad \text{... (7.26)}$$

Substituting values of F_1 from (7.25)

$$\text{Power} = (F_{max} - F_c) \cdot v \cdot C = (F_{max} - m.v^2) \cdot v \cdot C$$

$$= (F_{max} \cdot v - m \cdot v^3) \cdot C$$

For maximum power,

$$\frac{dp}{dv} = 0$$

Hence,

$$(F_{max} - 3\,mv^2) = 0 \quad \text{... (7.27)}$$

Hence $(F_{max} - 3 \cdot F_c) = 0$ OR $F_{max} = 3 \cdot F_c$ ... (a)

From equation (7.27),

$$v = \sqrt{\frac{F_{max}}{3 \cdot m}} \quad \text{... (b)}$$

From equation (7.25)

$$F_1 = F_{max} - F_c$$

$$\therefore \quad F_1 = 3\,F_c - F_c = 2\,F_c \quad \text{... (c)}$$

Equations (a), (b) and (c) in various forms express the conditions for maximum power transmission by a belt.

Initial tension in the belt:

$$F_i = \frac{F_1 + F_2}{2}$$

... (Neglecting centrifugal tension)

$$F_i = \frac{F_1 + F_2 + 2 \cdot F_c}{2}$$

... (Considering centrifugal tension)

7.7 SELECTION OF FLAT AND V BELTS FROM MANUFACTURER'S CATALOGUE

7.7.1 Selection of Flat Belts

Proper belt selection for a given application needs information on (kW) power to be transmitted, input speed, speed ratio, nature and type of load, and the centre distance. The fundamental steps in flat belt selection are given below;

- For optimum performance belt manufacturers recommend belt speed in the range of 18 to 22 m/s, consider e.g., 20 m/s and calculate the diameter of smaller pulley using the following relation

$$d = \frac{60{,}000 \times v}{\pi n};$$

n is the speed (rpm) of smaller pulley or input speed

Further evaluate the diameter of bigger pulley as,

$$D = d \times \frac{\text{Speed of input pulley}}{\text{Speed of output pulley}}$$

After the values of smaller and larger diameters of pulleys are evaluated from the above equations; round-off the diameter values to nearest standard/preferred diameters. The standard diameters for the pulley are found by preferred series R20 and are rounded off. Hence, the preferred diameters are 100, 112, 125, 140, 160, 180, 200, 224, 250, 280, 315, 355, 400, 450, 500,.... These values are available in the manufacturer's catalogue. Having selected the pulley diameters, once again calculate the actual belt velocity. This velocity must be in the prescribed range (18 to 22 m/sec).

Different types of loads act on the belt. In order to account for the nature of load for the application under consideration, a modifying factor known as load correction factor is used. It is denoted by F_a. Using this factor the maximum power is evaluated. The values of load correction factor for different types of load are given in Table 7.4 below

Table 7.4: Typical Values of Load Correction Factor

Sr. No.	Nature of Load	Value (F_a)
1.	Normal Load	1.00
2.	Intermittent Load	1.30
3.	Steady Load	1.20
4.	Impact or Shock load	1.50

The maximum power for the design purpose by considering the given application or load factor is calculated as:

Design Power; $(kW)_{max} = F_a\,(kW)$

- Determine the angle of wrap for smaller pulley using the relationship as

$$\alpha_s = (180 - 2\beta) \Rightarrow = 180 - 2\sin^{-1}\left(\frac{D-d}{2C}\right)$$

- Belt manufacturers consider a value of 180° of arc of contact in developing the power transmitting capacities of the belt. However, the actual arc of contact will vary depending on the arrangement. The result of this is variation in tension in the belt (given by equation 7.15). To account for this, arc of contact factor F_d is considered for evaluating the power. The values for different wrap angles are given in Table 7.5 below

Table 7.5: Values of Contact Arc Factor

α_s (Degrees)	F_d	α_s (Degrees)	F_d	α_s (Degrees)	F_d
90	1.679	160	1.08	210	0.91
120	1.329	170	1.04	220	0.88
130	1.258	180	1.00	230	0.86
140	1.190	190	0.97	240	0.84
150	1.128	200	0.94	250	0.82

Calculate the corrected power by using

$$(kW)_{corrected} = (kW)_{max} \times F_d$$

- Find the power rating from the catalogue for the belt type selected. Belt manufacturer indicates in catalogue the values of belt rating such as 0.023 kW per unit width of belt per ply at some standard velocity (say 10 m/sec).
- The velocity for the application is determined in step 1 above, which could be different from the standard velocity. To account for this determine the corrected kW rating for the flat belt as

  Corrected kW belt Rating

$$= \frac{\text{Power rating per ply} \times \text{Correct belt velocity (v)} \frac{m}{s}}{10}$$

- For the given number of plies in belt, further evaluate the belt width using the relation as

  (Belt width × Number of plies)

$$= \frac{\text{kW corrected power}}{\text{Corrected kW belt rating}}$$

- Standard belt widths are indicated by the manufacturer in the catalogue in the format such as:

3 ply	25 mm	40 mm	50 mm	63 mm
4 ply	40 mm	44 mm	50 mm	63 mm

- The designer can conduct trials with 2/3/4 plies and find required widths. These widths can be compared with standard widths available and proper belt can be selected whose required width matches with standard width available.
- Calculate the length of belt using the formula for length depending upon open or crossed belt drive.

7.7.2 Selection of V belts From Manufacturer's Catalogue

V-belts are prominently used in agitators, hammer mills, crusher, compressor fans, radiator fans and other similar applications. While selecting V belt from manufacturer's catalogue, prior information such as kW power to be transmitted (P), machine to be driven, operating hours per day, input rpm, speed ratio, desired center distance is needed. The steps in selection of V-belt are

- Depending upon the details, such as type of driving unit, driven machine and operating hours per day, the **Correction Factor F_a** according to service is selected from the Table 7.6.

Table 7.6: Values of Load Correction Factor for V-belt

Required Type of Service	Normal Magnitude of Torque for A.C. Motor			High Magnitude of Torque for A.C. Motor		
	Number of Operating shifts			Number of Operating shifts		
	1	2	3	1	2	3
Light duty	1.0 to 1.2			1.1 to 1.3		
Medium duty	1.1 to .3			1.2 to 1.4		
Heavy duty	1.2 to 1.4			1.4 to 1.6		

- **Evaluate the Design Power** by using relationship as under;

  Design Power = $F_a \times P$

  (where, P = Power to be transmitted)

- **V Belt Cross Section** sizes such as A, B, C, D,........ are standardized by manufacturer to take different tensions at various speeds. A selection graph with design power as abscissa and smaller pulley rpm speed as ordinate is published in the catalogue, to select the proper cross-section of the belt. Using this plot (Fig. 7.8) select the belt cross section (B, C, D etc.).

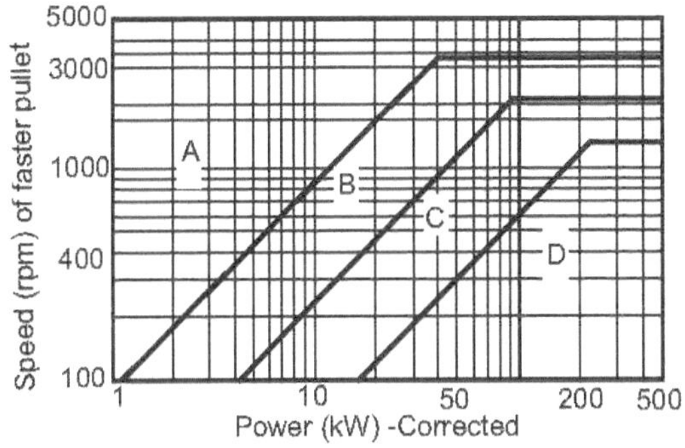

Fig. 7.8: Belt Cross Section Selection

- **Select the Recommended Minimum Pitch Diameter of Smaller Pulley for the Belt Type**

  From the preferred pulley pitch diameters as below.

  B: 125, 132, 140, 150, 160, 170, 180, 200, 224, 250 ... (Recommended Minimum: 200)

  C: 200, 212, 224, 236, 250, 265, 280, 300, 315, 355... (Recommended Minimum: 315)

  D: 355, 375, 400, 425, 450, 475, 500, 530, 560, 600... (Recommended Minimum: 500)

  Knowing the speed ratio determine the pitch diameter of larger pulley. Select the closest standard pulley diameter from the above referred table. Find he velocity ratio obtained.

- **Knowing Final Pulley Diameters, find the Length of Belt by using the Relation**

$$L = 2C + \frac{\pi(D + d)}{2} + \frac{(D - d) \times (D - d)}{4C}$$

Preferred pitch lengths of belts are published in the manufacturer's catalogue such as

B: 930, 1000, 1100, 1210, 1370, 1560, 1690...

C: 1560, 1760, 1950, 2190, 2420, 2720...

D: 2740, 3130, 3330, 3730, 4080, 4620...

Select the length of the belt from preferred lengths that is closest to calculated length.

- Using the value of standard length obtained from step 5, **Find The Correct Centre Distance** C, using relation $L = 2C + \frac{\pi(D + d)}{2} + \frac{(D - d) \times (D - d)}{4C}$ which becomes quadratic equation in terms of C.

- Depending upon the type of cross-section and pitch length of the belt **Select the Correction Factor (F_c)** for the pitch length obtained as under (Manufacturer's catalogue provides this information):

  Lengths of belts with Correction factors

  B : 930 (0.81), 1000 (0.83), 1100 (0.85), 1210 (0.87), 1370 (0.9), 1560 (0.92)...

  C : 1560 (0.82), 1760 (0.84), 1950 (0.87), 2190 (0.9), 2420 (0.92), 2720 (0.94)...

  D : 2740 (0.82), 3130 (0.86), 3330 (0.87), 3730 (0.9), 4080 (0.92), 4620 (0.94)...

- Further **Evaluate the Arc of Contact** for input pulley using the relation

$$\alpha_s = 180 - 2\sin^{-1}\left(\frac{D - d}{2C}\right)$$

Find correction factor F_d for arc of contact as under,

α_s	130	142	160	171	180
F_d	0.86	0.89	0.95	0.98	1

- Find power rating (P_r) from the table in the catalogue for the selected belt depending on the belt cross-section selected. Only typical values corresponding to standard motor speeds are indicated below.

Table 7.7: kW Rating (P_r) for V-belt Sections (B, C, D) at Standard Motor Speeds and Pulley Diameters (Sample Cases)

Smaller Pulley Speed	B Pulley Diameter		C Pulley Diameter		D Pulley Diameter	
	125	160	200	250	355	400
720	1.61	2.48	4.65	6.81	16.26	19.9
960	2.02	3.13	5.76	8.46	19.26	23.45
1440	2.72	4.26	7.49	10.95	21.22	--
2880	3.96	6.11	--	--	--	--

- **Determine Number of Belts by using the Relation**

$$\text{Number of belts} = \frac{P \times F_a}{P_r \times F_c \times F_d}$$

where, F_a = Correction factor for industrial service

P_r = Rating power of single V belt

F_c = Belt length correction factor

F_d = Arc of contact correction factor

7.8 ADJUSTMENT OF BELT TENSIONS

Whenever, a belt is mounted on pulleys it does not transmit any power on its own. This is due to absence of sufficient amount of friction. Hence, the belts are always provided with initial tension which increases the coefficient of friction and further results into power transmission. In course of service, the initial tension gets lost gradually due to many factors and hence there should be a provision to adjust the belt tensions as and when required. Hence, belt tensioning methods are needed. Various methods of belt tensioning are into existence as;

- Cutting a short length of belt by the amount of slack, in case when the belt is laced. However, this method is not applicable to endless belts and is not convenient for regular tensioning. Usually, this method will be employed only when belt tensioning arrangement has reached its limits.
- Using the adjusting screw (Usually on the driver pulley) the center distance between the pulleys is slightly increased. Typically an electric motor with its pulley may be mounted such that the motor position could be shifted using adjusting screw so as to periodically adjust the tension.
- Usage of idler pulley mechanism is also into practice. In this, the idler pulley is located next to the input pulley and is held against the belt by its own and additional weight. Its position can be adjusted to provide a longer belt path as and when belt has slackened.
- Referring the Fig. 7.9; another popular method known as Rockwood Belt drive is shown. From the figure it is observed that the adjustment in belt tension is done by variation in the eccentricity Z. Relationship between the tensions on tight and slack sides of the Rockwood belt drive is given as

$$\therefore \qquad Z = \frac{(T_t) \times y - (T_s) \times x}{W}$$

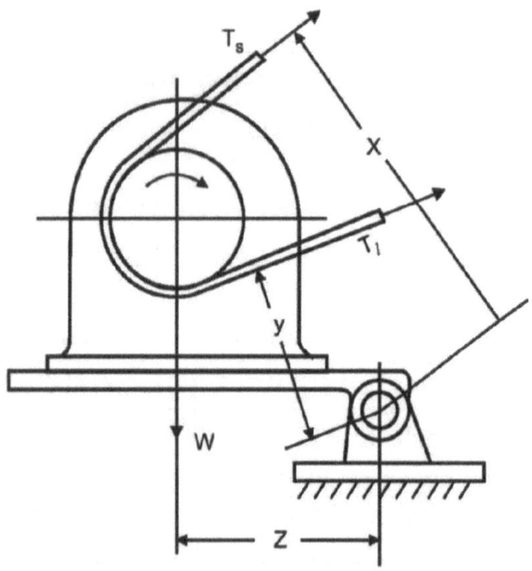

Fig. 7.9: Rockwood Belt Drive

7.9 CONSTRUCTION AND APPLICATIONS OF TIMING BELTS

- These are actually flat belts with a series of evenly spaced circumferential teeth on the inner surface. These teeth consecutively mesh with the teeth on the pulley or sprocket as shown in Fig. 7.10 below. It is also known as synchronous belt, ribbed V-belt or toothed belt.
- The timing belts are positive drives and combine high velocity characteristics of the belt drive with power transmission. The constituents of the timing belts include fibre (steel) cords to transmit load, protecting backing to steel wires especially made of rubber, teeth for engagement with the pulley or sprocket and cover to avoid wear of belts. Advantages of these drives include no slipping due to presence of teeth and high coefficient of friction. Such drives are used in applications that have high pitch line velocities up-to 80 m/s, do not require any arrangement for lubrication, no initial tension is needed, however these are costly as compared with the other belt drives and are sensitive to misalignment.

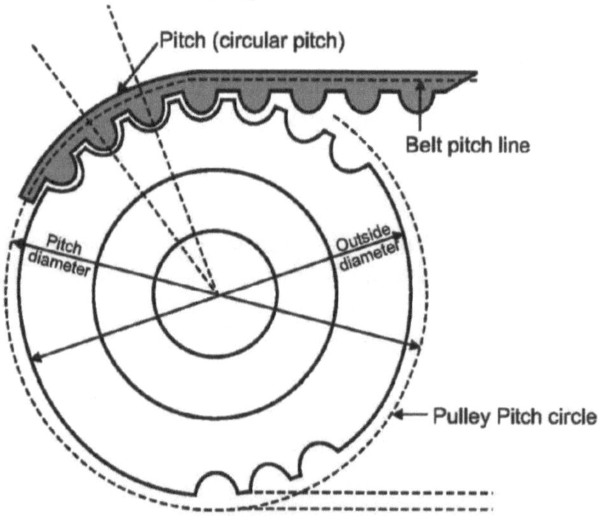

Fig. 7.10: Construction of Timing Belt

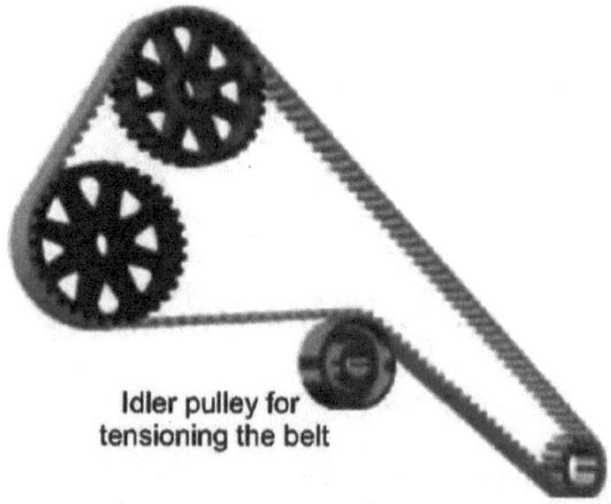

Fig. 7.11: Timing Belt Tensioning Arrangement

- Timing belts are used in automobiles to drive camshaft from crankshaft. Sewing machines also employ V-belts, machine shop, wood working machines and other power transmission devices.

CHAIN DRIVES

7.10 INTRODUCTION

- Chain drives are most reliable machine components that are widely used in industry. These drives are primarily used in power transmission and conveyor systems. Chain drive is a combination of chains, sprockets and a shaft mounting. Chain can be thought as a series of links to form loops; connected and fitted into one another.
- Basically, it consists of two sprockets or toothed wheels on which the loops of the chain fit and work as a drive for transmitting motion or power. Transmission chains have flexibility in one plane.

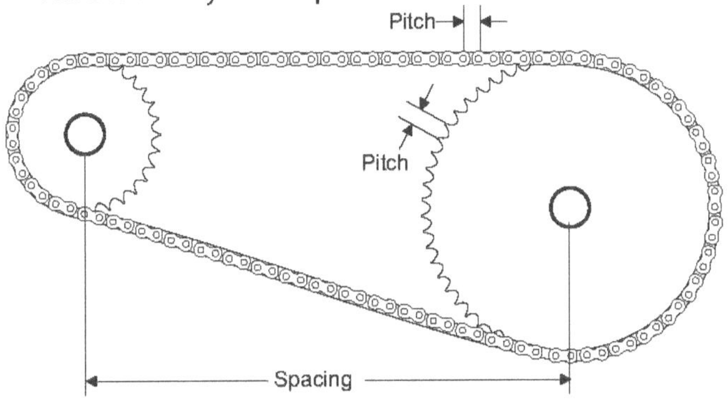

Fig. 7.12: Chain Drive

Chain Drives offer the Following Advantages Over Belt Drives:

- There is no slippage between chain and sprocket teeth.
- As the links are rigid bodies, there is negligible stretch in chain drives and hence chain drives can carry heavy loads.
- No or less sensitivity to external contaminants like dirt, dust, oil, grease etc.
- Can operate /work at elevated temperatures.
- Longer operating life as there is contact between hardened surfaces which are separated by oil film.
- Reliable drive and can take some amount of misalignment.

Disadvantages of Chain Drive are:

- Wearing of links and sprocket teeth leads to elongation of chain.
- These drives need continual or frequent lubrication for certain applications.
- Replacement of sprockets is needed as and when worn out chains are replaced.
- These drives are limited to lower operating speeds than belt drives.

7.11 TYPES OF POWER TRANSMISSION CHAINS

Power transmission chains are applied in transmission of power from one shaft to another.

7.11.1 Roller Chains

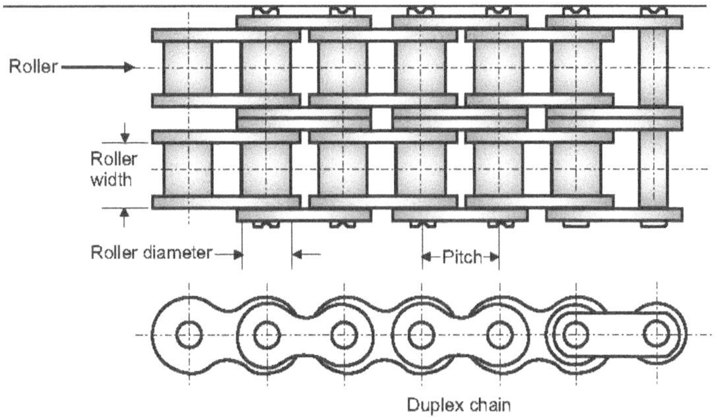

Fig. 7.13: Overview of a Duplex Chain

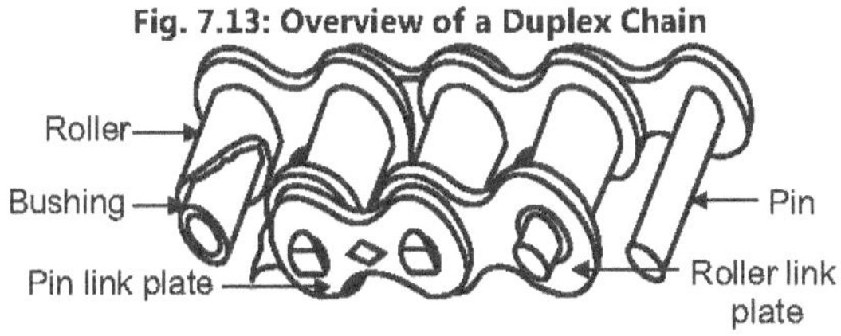

Fig. 7.14: Parts of a Simplex Roller Chain

The roller chain as shown in Figs 7.13 and 7.14 includes four fundamental parts.

1. Pin
2. Roller
3. Bush
4. Inner (bush) and outer (pin) link plates.

Roller chain drives work on rolling friction. The rolling friction is generated due to contact between the roller and the sprocket teeth. Pin is fitted to outer link plates and bush to inner link plates. Both these are fitted using pressing operation. Further, the rollers are fitted on the bushes, which turn freely during the engagement of teeth with the sprocket.

Materials:

- Inner and outer link plates: Medium carbon steel.
- Pin, bush and roller: Alloy steels.

Pitch of the chain is defined as the linear distance between the axes of adjacent rollers. These roller chains are available in single strand or multi strand constructions. (Duplex or triplex constructions)

7.12 GEOMETRC RELATIONSHIP FOR CHAIN DRIVES

A chain engaged on sprocket wheel is shown in Fig. 7.15.

D : PCD of sprocket is defined as the diameter of an imaginary circle which passes through the centers of link pins as the chain is wrapped on the sprocket.

Pitch angle $\alpha = \frac{360}{z}$; where, z is the number of teeth on the sprocket.

From Fig. 7.15 $\sin\left(\frac{a}{2}\right) = \frac{p/2}{D/2}$ or $D = \frac{p}{\sin\frac{\alpha}{2}}$

Velocity ratio of the chain drive is given by $i = \frac{n_1}{n_2} = \frac{Z_2}{Z_1}$.

Where, n_1, n_2 are rotational speeds of driving and driven shafts and corresponding Z_1 and Z_2 are the number of teeth on driving and driven sprockets. Velocity of the chain is one of the significant factor for chain life and operation in any application. The average velocity of the chain is

$$v = \frac{\pi Dn}{60 \times 1000} = \frac{Z.p.n}{60 \times 1000}.$$

Length of chain $L = L_n \times p$;

L_n = Number of links in the chain.

If sometimes the chain length calculated results into odd number of links, an additional link known as offset link is attached. This additional link is generally weaker in terms of strength as compared with other links. Hence, preferably the chain length selected consists of even number of links. Chain tensioning provision is required for longer center distances between sprockets.

7.13 POLYGONAL EFFECT IN CHAINS

In any chain drive, the chain passes or rolls around the sprocket. The length of section of chain, that approaches sprocket, passes over sprocket, and leaves the sprocket, can be presumed as a chord of a circle. Engagement of sprocket and chain can be thought of as non-slipping belt that is wrapped around a rotating polygon. Fig. 7.15 shows this phenomenon with the aid of four teeth on sprocket. The same analysis will hold good for larger number of teeth as well. The link AB in the chain is assumed to be located at a distance of D/2 from sprocket centre and roller B has velocity at this point denoted by V_{max} and is given by

$$V_{max} = \frac{\pi Dn}{60 \times 1000} \text{ m/s}$$

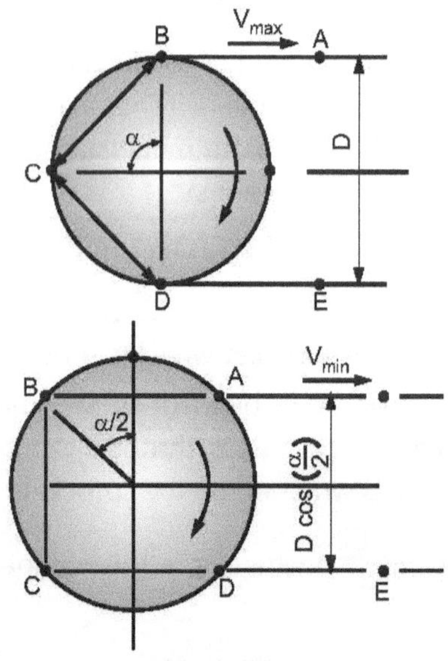

Fig. 7.15

Consider the link with advancement of angle α/2 as described in another figure. For this angular position the centroidal distance of the link AB is $\left(\frac{D}{2}\cos\frac{\alpha}{2}\right)$. The velocity of the link at this position is the minimum velocity V_{min}. The magnitude of this minimum velocity is given as,

$$V_{min} = \frac{\pi Dn \cos\frac{\alpha}{2}}{60 \times 1000}$$

It is thus obvious that velocity of the chain is not constant but varies between these two maximum and minimum limits of velocity. This is the reason for pulsating/fluctuating and jerking or momentary motions of chains. This is called polygonal effect.

The speed variation can be determined as the difference between the maximum and minimum velocities $(V_{max} - V_{min})$.

$$V_{max} - V_{min} = 1 - \cos\frac{\alpha}{2}; \text{ Here } \alpha = \frac{360}{z}$$

To reduce the speed variation it is clear from the above relation that number of teeth on the sprocket is to be increased. Guideline values for speed variation are given in Table 7.8.

Table 7.8

Number of Teeth	Speed Variation %
11	4
17	1.6
24	< 1

7.14 POWER RATING OF ROLLER CHAIN

It is the basic dimension of the chain. The selection of the pitch depends on : power to be transmitted, chain speed, number of revolutions of the smaller chain wheel. Its value is given by the following formulas :

When the centre distance is known, then

$$p = \frac{C}{30 \text{ to } 60}, \text{for slow speed drives} \quad \ldots(7.28)$$

The pitch is also given as,

$$p \leq 10\,(3640/n_1)^{2/3} \quad \ldots(7.29)$$

Pitch can also be determined by the following relations

For bushed and Roller Chains :

$$p = C_1 \cdot \sqrt[3]{\frac{P \times k}{z_1\omega_1 \cdot p_b \cdot x}} \quad \ldots(7.30)$$

For silent chains $p = C_2 \cdot \sqrt[3]{\frac{P \times k}{z_1\omega_1 \cdot p_b \cdot b}}$ ...(7.31)

where P is in kW, k is the load factor, ω_1 is the angular velocity of the driving sprocket, x is the number of chain strands, b is width of chain, p_b is the allowable bearing pressure.

If p and b are in meters, p_b in N/m^2, then

The values of the co-efficients C_1 and C_2 are

$$C_1 = 28, C_2 = 60$$

The allowable bearing pressure, p_b, is given by Table 7.10. But to find its value from this table, the pitch must be known. Therefore, for preliminary calculations, the allowable value of unit pressure, p_b, can be taken is given below :

For Bush-Roller Chains

p_b = 35 N/mm^2 at $n_1 \leq 50$ rev/min. to
= 13.7 N/mm^2 at n_1 = 2800 rev/min

For Silent Chains

p_b = 20 N/mm^2 at $n_1 \leq 50$ rev/min. to
= 7.8 N/mm^2 at n_1 = 2800 rev/min

For silent chains, the width of chain,

$$b = C_3 \times p$$

The value of C_3 ranges from 1.5 to 8.

For preliminary calculations, the recommended value for C_3 = 3 to 4.

The load factor k is given as, $k = k_1 \times k_2$.

k_1, takes into account the nature of load and service conditions, Table 7.9.

Table 7.9 : Factor k_1

Daily Service of Driven Unit	Service Factor k_p for	
	Electric Motor or Steam Turbine	Diesel or Gasoline Engine
Even load, 8-hour day	1.0	1.2
Uneven load, 8 hour day or Even load 16 to 24-hour day	1.2	1.4
Heavy shock load, 8-hour day or Uneven load 16 to 24 hour day	1.4	1.7

Takes into account the method of lubrication

= 1.0 for continuous lubrication
= 1.3 for periodic lubrication

After finding approximate values for 'p' from equation (7.30) or (7.31) the value of p_b is taken from table of equation (7.30) or (7.29) again used to find the exact value for p.

The value obtained by the above formulas or from equations (7.28) to (7.29) is rounded off to the standard value.

The standard pitches used are :

6.25, 8, 9.525, 12.7, 15.875, 19.05, 25.4, 31.75, 38.1, 44.45, 50.8, 63.5, and 76.2 mm

The tentative values or z_1, z_2 and p are adopted as final, if the angular speed n_1 does not exceed the maximum permissible value $n_{1\,max}$ given in Table 7.11.

Table 7.10 : Allowable bearing pressure in chain joints

Speed of Smaller Sprocket rev/min	p_b MN/m^2 with chain pitch, p, mm						
	12 to 15.875		19.05 to 25.4		30 to 38.1		40 to 50.8
	a	b	a	b	a	b	a
50	34.3	19.6	34.3	19.6	34.3	19.6	34.3
200	30.9	17.6	29.4	16.7	28.1	16.1	25.7
400	28.1	16.1	25.7	14.7	23.7	13.7	20.6
800	23.7	13.7	20.6	11.8	18.1	10.3	14.7
1200	20.6	11.8	17.2	9.81	14.7	8.43	-

Note : a – for roller and bush chains (z_1 = 15 to 30)
b – for silent chains (z_1 = 17 to 35)

Table 7.11 : Maximum speed of smaller sprocket, $n_{1\,max}$

Type of Chain	Number of Teeth in Smaller Sprocket, z_1	Chain Pitch p, mm				
		12	15	20	25	30
Roller chain	15	2300	1900	1350	1150	1000
	19	2400	2000	1450	1200	1050
	23	2500	2100	1500	1250	1100
	27	2550	2150	1550	1300	1100
	30	2600	2200	1550	1300	1100
Silent chain	17 to 35	12.7	15.875	19.05	25.4	31.75
		3300	2650	2200	1650	1300

For roller-chains, the maximum chain speed is given as below in Table 7.12.

Table 7.12 : Chain Speed for Roller Chains

Sr. No.	Pitch, mm	Maximum Chain Speed, m/s	Maximum n_1
1	9.525	13.7	3500
2	12.7	13.4	2500
3	15.875	12.0	2300
4	19.05	10.9	1800
5	25.4	9.7	1500
6	31.75	8.1	1000
7	38.1	7.3	800
8	44.45	6.5	600
9	50.8	6.0	400
10	63.5	4.7	300
11	76.2	4.0	200

Design of the Chain Drive

The important parameters in chain drive design are : wear, vibrations and impact. Design calculations are carried out to find the optimal proportions of chain and sprocket at which the drive will be able to operate under the assigned duty without risk of damage. The design is carried out in the following steps :

1. Select the type of chain
2. Find the number of teeth of the smaller sprocket from Table.
3. The chain pitch is selected from formulas (7.28) (7.29) (7.30) or (7.31). With selected values of z_1 and p, the value of $n_{1\,max}$ is to be satisfied as given in Table 7.11. If n_1 is given, then pitch can be selected from Table 7.11, after z_1 has been selected.
4. Total load on the driving side of the chain is determined and factor of safety found out. Total force is

$$F_0 = F + F_c + F_f$$

where F = driving force

= Tension on tight side, tension on slack side is considered zero

$$= \frac{\text{Power in kW} \times 1000}{V}, \text{N}$$

where V = Chain velocity in m/s

and F_c = Chain tension due to centrifugal force of inertia

$$= \frac{wV^2}{g}$$

where w = weight per metre of chain length,

F_f = Tensioning due to sagging

$= k_f \cdot w \cdot C$

where k_f = Co-efficeint to take into account the arrangement of chain drive

k_f = 1.0 for vertical position

= 2.0 for centre line inclined at an angle upto 45°

= 4.0, for horizontal drives

Then
$$FS = \frac{F_u}{F_0}$$

where F_u = Breaking strength of chains

= 106 p^2N, p in mm for roller chains

= 106 pN per mm width of chain for silent chains, p in mm

The values of factor of safety are given in Table 7.13.

Table 7.13 : Factor of Safety for Chain Drives

Type of Chain	Speed of Pinion , rev/min								
	upto 50	200	400	600	800	1000	1200	1600	2000
Bush Roller Chains									
p = 12.7 – 15.875 mm	7	7.8	8.55	9.35	10.2	11.0	11.7	13.2	14.8
= 19.05 – 25.4 mm	7	8.2	9.35	10.3	11.7	12.9	14.0	16.3	-
= 31.75 – 38.1 mm	7	8.55	10.2	13.2	16.3	16.3	19.5	-	-
Silent Chain :									
p = 12.7 = 15.875 mm	20	22.2	24.4	28.7	29.0	31.0	33.0	37.8	42.0
= 19.5 – 25.4 mm	20	23.4	26.7	30.0	33.4	36.8	40.0	46.5	53.5
= 31.75 mm	20	25	32	36.5	41	46	51	-	-

5. Then the chain is checked for wear by unit pressure in the joints given as

$$p_b = \frac{F.k}{A} \text{ N/mm}^2$$

where k = load factor

A = Projection of the joint bearing surface

For roller chains,

$$A = (d_p \times l_b) \times \text{number of chain strands,}$$

For silent chains,

$$A = 0.76\, d_p b$$

where d_p = diameter of joint pivot

b = chain width

l_b = bush length = $(b_i + 2t_p)$

Check the value of p_b with values given in Table 7.10.

6. The force acting on the drive shaft is,

$$F_{sh} = F + 2k_f \cdot w \cdot C$$

7.15 SPROCKET WHEELS

The operation of a chain drive is largely dependent on the quality of the sprockets. Therefore, special attention should be given to : material, manufacture, heat treatment and surface finish of the teeth. The sprockets are made of good grade C.I. and carburized or hardened steel. The shape of the sprocket teeth depends upon the type of chain, the main object is that the teeth may easily disengage from the chain links during operation.

The profile of teeth of the sprockets for bush and Roller Chains is : Convex, Concave, rectilinear and combined rectilinear and convex. For high velocities, the concave profile is preferred due to its greater wear resistance owing to the greater length of the working section of the tooth profile. The form of the wheel tooth gap is shown in Fig. 7.16 (IS 2403-1975).

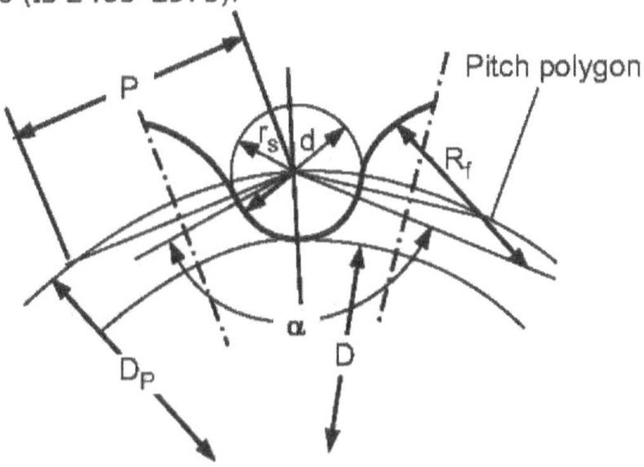

Fig. 7.16 : Form of wheel tooth gap

Pitch diameter of the sprocket is given as,

$$D_p = \frac{p}{\sin \pi/z} \qquad \ldots(7.32)$$

The other dimensions are given as : Roller seating radius , r = 0.505 d.

Roller Seating Angle $\alpha = \left(140° - \frac{90°}{z}\right)$ to $\left(120° - \frac{90°}{z}\right)$

Outer diameter, $D_o = D_p + d\left(0.6 - \frac{2.56}{z}\right)$ to $D_p + d$

Root diameter, $D_r = D_p - 2r$

Tooth flank radius, $R_f = 0.12\ d\ (z + 2)$ to $0.008\ d\ (z^2 + 180)$

The profiles of the wheel rim are shown in Fig. 7.17. (IS 2403-1975).

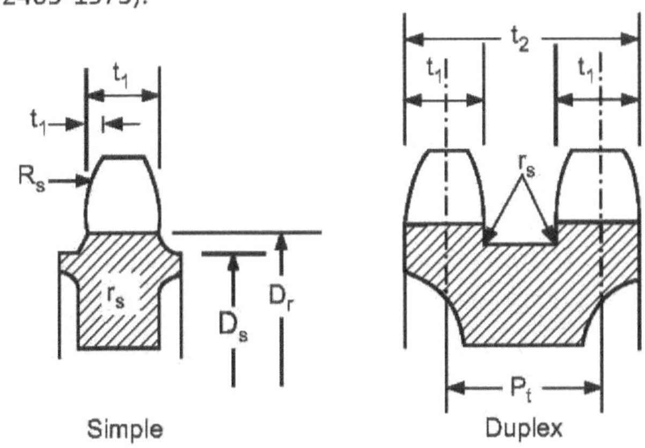

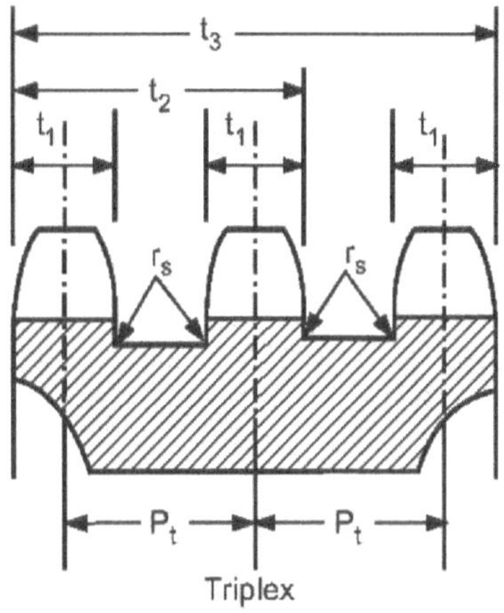

Fig. 7.17 : Wheel rim profiles

The various dimensions are given below :

Shroud diameter $D_s = p \cot \frac{180°}{z} - 13\ h_i$

Minimum tooth side radius R_s = p.

Tooth width t_1

(i) For $p \leq 12.7$ mm

t_1 = 0.93 b_i for simple chain wheels
= 0.91 b_i for duplex and triplex chain wheels
= 0.88 b_i for quadruplex chain wheels and above

(ii) For p > 12.7 mm the corresponding values of t_1 are : 0.95 b_i, 0.93 b_i and 0.93 b_i

Tooth side relief , t_r = 0.1 p to 0.15 p

Shroud radius, r_s = 0.76 mm max

The widths t_2 and t_3 are given as = (Number of chain strands – 1) × p_t + t_1

where p_t = Transverse pitch of strands or strand spacing

Its values are :

p	8.00	9.525	12.70	15.875	19.05	25.4	31.75	38.10	44.45	50.80	63.50	76.20
p_t	5.64	10.24	13.92	16.59	19.46	31.88	36.45	48.36	59.56	58.55	72.29	91.27

The sectioned view of a simple sprocket is shown in Fig. 7.18.

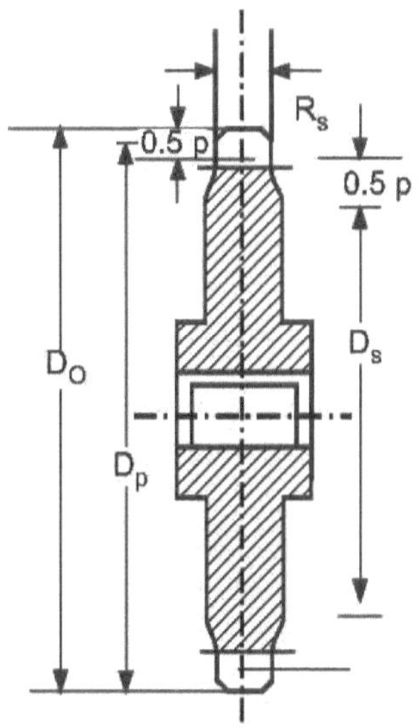

Fig. 7.18 : Sprocket

Table 7.13 : Characteristics of Silent Chains

Pitch, p, mm	Width b, mm	Breaking load, kN	Pitch p, mm	Width b, mm	Breaking Load, kN
12.7	20.5	20.5	19.5	33.0	49.0
	26.5	26.5		39.0	58.0
	33.0	33.0		45.5	68.0
	39.0	39.0		51.5	77.0
	45.6	45.6		63.0	94.5
	51.5	51.5		78.5	117.0
	63.0	63.0		91.0	136
	78.5	78.5		104.5	156.0
	91.0	91.0		123.0	184.0
	104.5	104.5		135.0	203.0
				148.0	222.0
				173.0	259.0
				198.0	297.0
15.87	26.5	31.8	25.4	51.5	87.0
	33.0	39.6		76.5	130.0
	39.0	46.8		93.0	158.0
	45.6	54.6		101.0	171.0
	51.5	61.8		121.0	205.0
	63.0	75.6		154.5	262.0
	78.5	94.2		179.5	305.0
	91.0	109.2		204.5	347.0
	104.5	125.4		253.5	430.0
	123.0	147.6			

7.16 SILENT CHAINS

These chains are not exactly silent but these are much smoother and quieter in action than a roller chain. The construction detail of a silent chain is shown in Fig. 7.19. The chain capacity is proportional to the chain width which is usually taken as $1\frac{1}{2}$ to 12 times the pitch. In design, the silent chains are more complex than bush roller types, more expensive and require more careful maintenance.

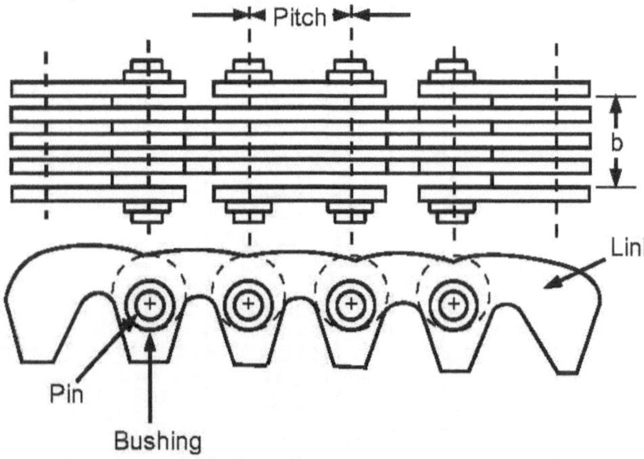

Fig. 7.19 : Construction of a Silent Chain

The dimensions of the various parts of the chain are usually given in terms of the pitch, the approximate values being given below.

(i) Roller diameter, $d = \frac{5}{8} \times \text{pitch}$

(ii) Pin diameter, $dp = \frac{5}{16} \times \text{pitch}$

(iii) Chain width, i.e. the distance between the roller link plates,

$$b_i = \frac{5}{8} \times \text{pitch}$$

(iv) Thickness of link plates, $t_p = \frac{1}{8} \times \text{pitch}$

(v) Width between outer plates, $b_0 = b_i + 2t_p$

(vi) Maximum height of pin link plates of coupling link plates, i.e., the outer plates,

$$h_0 = 0.82 \times \text{pitch}$$

(vii) Maximum height of roller link plates i.e., the inner plates, $h_i = 0.95 \times \text{pitch}$

(viii) Length of Roller, $l = 0.90\, b_i - 0.15$

7.17 MODES OF FAILURE FOR CHAIN

There are various failure criterion for chain. These are described as;

- **Excessive Tensile Load:** If the chain is subjected to excessive tensile load compared to its breaking load capacity, the chain can fail by breakage. The microscopic cracks in the link plate can develop further leading to its failure. Excessive load may be due to shock loads encountered, wrong chain selection, excessive misalignment in mountings etc.
- **Excessive Chain Elongation**: Due to long running and normal wear and tear of the chain parts, the chain starts elongating. If chain extension becomes excessive, the chain starts riding over the sprocket and has a tendency to leave the sprocket. The chain length has to be corrected either through a chain tensioning arrangement or by removing an extra link.
- **Wear:** The pin and bush are held together with the help of pin joint. This articulation is the prime reason for the wear of chain. Due to this wearing, the chain pitch is increased. This results in faulty engagement of teeth. As this wear increases beyond certain limit it becomes mandatory to replace the chain. This limit of elongation is fixed between 1.6 to 2.5%. Periodic lubrication with oil is important remedy for reducing wear as it creates a thin film of oil between the contacting surfaces.

- **Fatigue:** During chain action, the chain passes through two sides i.e. tight side and loose side. At every turn on the sprockets, the chain is subjected to tensile force. This force is maximum on tight side and minimum on the slack side. Hence for one revolution of chain over the sprocket wheel, the chain is subjected to fluctuating stresses. This leads to fatigue failure of side link-plates. To avoid this type of failure, endurance limit of the link plates is kept more than the tensile stress for infinite life.
- **Impact:** During power transmission, the rollers of the chain engage with the teeth. During engagement of rollers and teeth there is slight impact. Improper alignment at two sprockets can result into excessive loads. The magnitude of this impact when exceeds beyond limit, results in to failure of roller or bush. The magnitude of impact can be reduced either by reduction of tension on tight side or by increasing the number of teeth on the sprocket or proper alignment.
- **Galling:** If the tension in the chain is high, microscopic welds are formed between the contacting surfaces/areas. These welds break as a result of relative motion between the contacting surfaces and leads to wear. This wear occurs even in the presence of lubricant. This results into stick-slip phenomenon between the pin and the bush.

7.18 LUBRICATION OF CHAINS

There are three fundamental methods of chain lubrication. These are designated as Type-A, Type-B and Type – C. The type relies on the power rating and velocity of the chain.

1. **Type-A Lubrication:** This is manual or drip lubrication method. In this, the lubricating oil is applied with the aid of brush or oil-can. Frequency of application of oil is so adjusted that the chain is not overheated as well as there is no discoloration. In the drip lubrication process; a pipe with small holes i.e. drip lubricator supplies the oil. The oil emerges out in the form of drops and falls on the link plates. The distance for the travel of oil drop to chain is so adjusted that presence of wind does not affect the lubrication process. Presence of excessive wind may result is spreading of oil on the other surfaces. This method is suitable for low speed applications with velocity upto 2 m/sec.
2. **Type-B Lubrication:** This type of lubrication is called as disk or bath lubrication. In this method of lubrication, the lower side of the chain passes through a sump filled with oil upto the pitch line of the chain. In other method the whole chain is kept above the oil level in the sump and the oil is spilled or deposited on the chain with the help of disk which is attached to one of the shafts. This is known as disk lubrication. This method is suitable for medium speeds upto 5 m/sec.
3. **Type-C Lubrication:** In this method, a stream of lubricating oil is forced on the inside or slack side of the chain loop. A separate oil sump is provided to supply a continuous stream of oil. This method is also known as oil stream lubrication. Forced lubrication of this type is required for higher speeds and larger power transmission capabilities. Chain manufacturer's recommendations need to be followed.

SOLVED EXAMPLES

Example 7.1: *A pulley of 1000 mm diameter is driven by an open type flat belt from 25 kW, 1440 rpm electric motor. The pulley on the motor shaft is 250 mm in diameter and the centre distance between the two shafts is 2.0 m. The allowable tensile stress for the belt material is 2 N/mm² and the coefficient of friction between the belt and pulley is 0.28. The density of the belt material is 900 kg/m³.*

If the width of the belt is 125 mm, determine:

(i) The thickness of belt

(ii) The length of belt and

(iii) The initial tension required in the belt.

Solution: Type of drive: Open type flat belt

D_l = Diameter of driven pulley = 1000 mm

Power = 25 kW

n = 1440 rpm

D_s = 250 mm

C = 2000 mm

σ_{tper} = Permissible tensile stress in belt = 2 N/mm^2

μ = 0.28

ρ = Density of belt material = 900 kg/m^3

b = Width of belt = 125 mm

To find:

(i) t, (ii) L, (iii) F_i

(i) Thickness of Belt (t)

Velocity of belt (v)

$$v = \frac{\pi \times D_s \times n}{60 \times 1000} = \frac{\pi \times 250 \times 1440}{60 \times 1000}$$

$$v = 18.85 \text{ m/s}$$

Power to be transmitted by belt

$$\text{Power} = (F_1 - F_2) \cdot v$$

$$25 \times 10^3 = (F_1 - F_2) \times 18.85$$

$$F_1 - F_2 = 1326.26 \text{ N} \quad \text{... (1)}$$

Angle of contact (α_s)

$$\beta = \sin^{-1}\left(\frac{D_l - D_s}{2C}\right)$$

$$= \sin^{-1}\left(\frac{1000 - 250}{2 \times 2000}\right)$$

$$\beta = 10.807° = 0.1886 \text{ rad}$$

$$\alpha_s = \pi - 2\beta = \pi - 2\,(0.1886)$$

$$\alpha_s = 2.7644 \text{ rad}$$

Limiting tension ratio

$$\frac{F_1}{F_2} = e^{\mu\alpha} = e^{0.28 \times 2.7644}$$

$$F_1 = 2.1685\, F_2 \quad \text{... (2)}$$

From equations (1) and (2),

$$2.1685\, F_2 - F_2 = 1326.26$$

$$F_2 = 1135.04 \text{ N}$$

$$F_1 = 2461.33 \text{ N}$$

Centrifugal tension (F_c)

$$F_c = m \cdot v^2$$

$$= (\rho \times \text{Volume}) \cdot v^2$$

$$= \left(\rho \times \frac{b}{1000} \times \frac{t}{1000} \times 1\right) \times v^2$$

$$= \left(900 \times \frac{125}{1000} \times \frac{t}{1000} \times 1\right) \times (18.85)^2$$

$$F_c = 39.973 \text{ (t) N}$$

Total tension

$$F_t = F_1 + F_c$$

$$T_t = 2461.33 + 39.973 \text{ (t)}$$

$$\sigma_{tper} = \frac{F_t}{b \cdot t}$$

$$2 = \frac{2461.33 + 39.973 \text{ (t)}}{125 \cdot t}$$

$$250\, t = 2461.33 + 39.973 \text{ (t)}$$

$$t = 11.72 \text{ mm}$$

$$t \cong 11.75 \text{ mm}$$

(ii) Length of Belt (L):

$$L = 2C + \frac{\pi\,(D_l + D_s)}{2} + \frac{(D_l - D_s)^2}{4C}$$

$$= 2 \times 2000 + \frac{\pi\,(1000 + 250)}{2} + \frac{(1000 - 250)^2}{4 \times 2000}$$

$$L = 6033.807 \text{ mm}$$

$$L = 6034 \text{ mm}$$

(iii) Initial Tension in Belt (F_i):

$$F_c = 39.973 \times 11.75$$

$$F_c = 469.68 \text{ N}$$

$$F_i = \frac{F_1 + F_2 + 2F_c}{2}$$

$$= \frac{2461.33 + 1135.04 + 2 \times 469.68}{2}$$

$$F_i = 2267.867 \text{ N}$$

Example 7.2: *Two parallel shafts are to be connected by an open flat belt. The diameter of pulleys are 400 mm and 800 mm and they are 1 m apart. The initial tension in belt when it is stationary is 2 kN. The mass of belt is 2 kg/m. The coefficient of friction between the belt and the pulley is 0.3. Calculate the power transmitted if smaller pulley rotates at 1000 rpm.*

Also suggest the speed of the smaller pulley for maximum power transmission by the belt. Determine this maximum power.

Solution: Type of belt drive = Open type flat belt

$$D_s = 400 \text{ mm}$$

$$D_l = 800 \text{ mm}$$

$$C = 1000 \text{ mm}$$

$$F_i = 2000 \text{ N}$$

$$m = \text{Mass of belt} = 2 \text{ kg/m}$$

$$\mu = 0.3$$

$$n = 1000 \text{ rpm}$$

To find:

(i) kW, (ii) n for maximum power, (iii) kW_{max}

Power transmitted by belt:

Belt speed,

$$v = \frac{\pi \cdot D_s \cdot n}{60 \times 1000}$$

$$= \frac{\pi \times 400 \times 1000}{60 \times 1000} = 20.94 \text{ m/s}$$

Centrifugal tension (F_c)

$$F_c = mv^2$$

$$= 2 \times 20.94^2$$

$$F_c = 877.298 \text{ N}$$

Initial tension (F_i)

$$F_i = \frac{F_1 + F_2 + 2F_c}{2}$$

$$2000 = \frac{F_2 + F_1 + (2 \times 877.29)}{2}$$

$$F_1 + F_2 = 2245.4 \text{ N} \quad \text{... (1)}$$

Angle of contact (α_s)

$$\beta = \sin^{-1}\left(\frac{D_l - D_s}{2C}\right) = \sin^{-1}\left(\frac{800 - 400}{2 \times 1000}\right)$$

$$\beta = 11.53° \text{ or } 0.2013 \text{ rad}$$

For an open belt drive

$$\alpha_s = \pi - 2\beta$$

$$= \pi - (2 \times 0.2013) = 2.7388 \text{ rad} = \alpha$$

Limiting tension ratio

$$\frac{F_1}{F_2} = e^{\mu\alpha}$$

$$\frac{F_1}{F_2} = e^{0.3 \times 2.7388}$$

$$F_1 = 2.2742\ F_2 \qquad \ldots (2)$$

Tensions in belt due to friction (F_2) and (F_1) from equations (1) and (2)

$$2.2742\ F_2 + F_2 = 2245.4$$

$$F_2 = 685.78 \text{ N}$$

$$F_1 = 1559.601 \text{ N}$$

(i) Power Transmitted by the Belt (kW)

$$\text{Power} = (F_1 - F_2) \cdot v$$

$$= (1559.601 - 685.78) \cdot 20.94$$

$$\text{Power} = 18297.82 \text{ W} = 18.297 \text{ kW}$$

(ii) Optimum Speed of Driving Pulley

Maximum power transmission capacity of belt is obtained when

$$F_c = \frac{F_i}{3}$$

$$F_c = \frac{2000}{3} = 666.667 \text{ N}$$

$$F_c = m \cdot v^2 \qquad \ldots V = \text{Optimum speed in m/s}$$

$$666.667 = 2\ (v)^2$$

$$v^2 = 333.33$$

$$\therefore \quad v = 18.26 \text{ m/s}$$

$$v = \frac{\pi \cdot D_s \cdot \text{Optimum speed}}{60 \times 1000}$$

$$\therefore \quad 18.26 = \frac{\pi \times 400 \times \text{Optimum speed}}{60000}$$

$\therefore$ Optimum speed = 871.72 rpm

Belt tensions due to belt friction (F_2) and (F_1)

Initial belt tension

$$F_i = \frac{F_2 + F_1 + 2F_c}{2}$$

$$2000 = \frac{F_2 + F_1 + (2 \times 666.667)}{2}$$

$$F_1 + F_2 = 2666.67 \text{ N} \qquad \ldots (3)$$

From equations (2) and (3),

$$2.2742\ F_2 + F_2 = 2666.67$$

$$F_2 = 814.449 \text{ N}$$

$$F_1 = 1852.22 \text{ N}$$

(iii) Maximum Power (kW_{max})

$$\text{Maximum power} = (F_1 - F_2) \cdot v$$

$$= (1852.22 - 814.449) \times 18.26$$

$$= 18949.71 \text{ W}$$

$$\text{Maximum power} = 18.949 \text{ kW}$$

Example 7.3: *A horizontal flat belt is used to transmit 25 kW power from an electric motor running at 1440 rpm to a centrifugal water pump expected to run at 720 rpm. The required centre distance is 4.5 m. Select the flat belt for drive. Use the following data.*

Recommended range of belt speed: 17.8 m/s to 22.9 m/s.

Standard pulley diameter: 90, 100, 112, 125, 140, 160, 180, 200, 224, 250, 280, 315, 355, 400, 450, 500, 560, 630, 710, 800 mm. Load correction factor: F_a = 1.2.

Arc of contact correction factor (F_d)

Arc of Contact	120°	130°	140°	150°	160°	170°	180°
F_d	1.33	1.26	1.19	1.13	1.08	1.04	1.00

Power rating per ply/mm width at 180° arc of contact and v = 10 m/s. For HI-SPEED BELT - 0.023 kW/ply/mm.

Number of plies and standard belt widths.

Number of Ply	Standard Belt Widths 'b' mm
4	40, 44, 50, 63, 76, 90, 100, 125, 152
5	76, 100, 125, 152

Solution:

Type of drive = Open type flat belt drive

$$\text{Power} = 25 \times 10^3 \text{ W}$$

$$n = 1440 \text{ rpm}$$

N = Speed of driven (big) pulley or pump

= 720 rpm

C = Centre distance

= 4500 mm

Assuming, v = 17.8 m/s

$$v = \frac{\pi \cdot D_s \cdot n}{60 \times 1000}$$

$$17.8 = \frac{\pi \times D_s \times 1440}{60000}$$

$$D_s = 236.08 \text{ mm}$$

Selecting nearest higher standard diameter

$\therefore \quad D_s = 250$ mm

Now, $\quad \frac{N}{n} = \frac{D_s}{D_l}$

$\frac{720}{1440} = \frac{250}{D_l}$

$\therefore \quad D_l = 500$ mm

Actual belt speed

$$v = \frac{\pi \cdot D_s \cdot n}{60000}$$

$$v = \frac{\pi \cdot 250 \times 1440}{60000} = 18.85 \text{ m/s}$$

Load correction factor (F_a) = 1.2

$$\text{Design power} = F_a \times kW = 1.2 \times 25 \times 10^3$$

$$\text{Design power} = 30 \times 10^3 \text{ W}$$

Angle of Contact (α_s):

$$\beta = \sin^{-1}\left(\frac{D_l - D_s}{2C}\right) = \sin^{-1}\left(\frac{500 - 250}{2 \times 4500}\right)$$

$$= 1.5918°$$

$$\beta = 0.02778 \text{ rad}, \ \alpha = 180° = 2 \times 1.5918$$

$$= 176.814$$

Arc of contact factor,

$$F_d = 1.04 - \frac{(1.04 - 1.00)(170° - 176.8144°)}{(170° - 180°)}$$

$$F_d = 1.0127$$

Power rating (kW rating)

$$\text{kW rating} = 0.023 \times 10^3 \text{ W/Ply/mm}$$

$$\text{Modified kW rating} = \frac{\text{kW rating}}{F_d} \times \frac{v}{10}$$

$$= \frac{0.023 \times 10^3}{1.0127} \times \frac{18.85}{10}$$

$$= 0.0428 \times 10^3 \text{ W/Ply/mm}$$

Belt width × Number of plies

$$= \frac{\text{Design power}}{\text{Modified kW rating}}$$

$$= \frac{30 \times 10^3}{0.0428 \times 10^3}$$

$$= 700.93 \text{ ply-mm}$$

For 4 plies, belt width = $\frac{700.93}{4}$ = 175.23 mm

For 5 plies, belt width = $\frac{700.93}{5}$ = 140.187 mm

Selecting number of plies = 5

Width of belt (b) = 152 mm

Length of Belt (L)

$$L = 2C + \pi\left(\frac{D_l + D_s}{2}\right) + \frac{(D_l - D_s)^2}{4C}$$

$$= 2 \times 4500 + \frac{\pi (500 + 250)}{2} + \frac{(500 - 250)^2}{4 \times 4500}$$

$$L = 10181.57 \text{ mm}$$

$$L = 10182 \text{ mm}$$

Example 7.4: *A V-belt drive is used to transmit 38 kW power from a three phase induction motor to a centrifugal pump. Speed of the motor is 1440 rpm and the centrifugal pump is required to be operated at 360 rpm. For the motor pulley pitch diameter is 225 mm and the groove angle is 38°. Centre distance between the pulley is 1 m. Coefficient of friction for the belt pulley combination is 0.2 and the density of the belt material is 0.97 gm/cc. Allowable tension in the belt is 800 N.*

Determine:

(i) Number of belts required.

(ii) Pitch length of the belt.

Assume C/S for the belt as shown in the Fig. 7.20

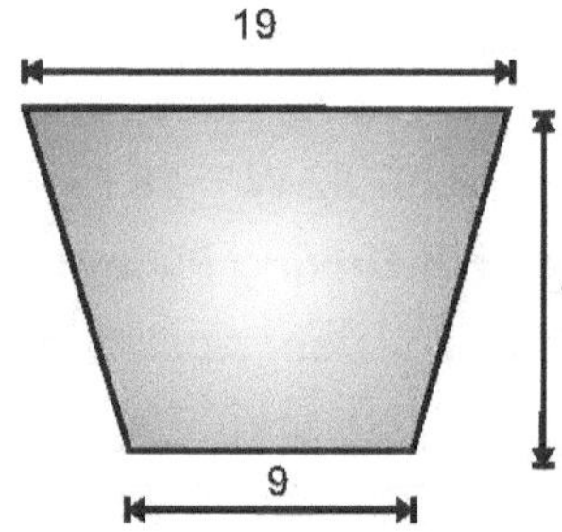

Fig. 7.20

Solution: Type of drive = Open type V-belt

Power = 38 kW

n = 1440 rpm

N = Speed of driven (big) pulley = 360 rpm

D_s = 225 mm

$\theta = 38°$ = Groove angle of pulley

C = 1000 mm

μ = Coefficient of friction = 0.2

ρ = 970 kg/m^3

F_{max} = Maximum allowable tension = 800 N

To find:

(i) Number of belts required.

(ii) Pitch length of belt.

(i) Number of Belts Required:

Cross-sectional Area of V belt

$$A = \frac{1}{2} \times (9 + 19) \times 15$$

$$= 210 \text{ mm}^2$$

Velocity of belt (v)

$$v = \frac{\pi \cdot D_s \cdot n}{60 \times 1000} = \frac{\pi \times 225 \times 1440}{60000}$$

$$v = 16.96 \text{ m/s}$$

Centrifugal tension (F_c)

$$F_c = m \cdot v^2$$

$$= (\rho \cdot A) \cdot v^2$$

$$= (970 \times 10^{-6} \times 210) \times (16.96)^2$$

$$F_c = 58.62 \text{ N}$$

Speed ratio $\frac{N}{n} = \frac{D_s}{D_l}$

$$\frac{360}{1440} = \frac{225}{D_l}$$

$$D_l = 900 \text{ mm}$$

Angle of contact (α_s)

$$\beta = \sin^{-1}\left(\frac{D_l - D_s}{2C}\right) = \sin^{-1}\left(\frac{900 - 225}{2 \times 1000}\right)$$

$$\beta = 19.7246° = 0.3442 \text{ rad}$$

$$\alpha_s = \pi - 2\beta = 2.453 \text{ rad}$$

Limiting tension ratio

$$\frac{F_1}{F_2} = e^{\left(\frac{\mu \cdot \alpha}{\sin \theta/2}\right)}$$

$$\frac{F_1}{F_2} = e^{\left(\frac{0.2 \times 2.453}{\sin 19}\right)}$$

$$F_1 = 4.5127 \, F_2$$

$$F_{max} = F_1 + F_c$$

$$800 = F_1 = 58.62$$

$$F_1 = 741.4074 \text{ N}$$

$$\therefore \quad F_2 = 164.2935 \text{ N}$$

Power transmitted

$$\text{Power/belt} = (F_1 - F_2) \times v$$

$$= (741.4074 - 164.2935) \times 16.96$$

$$= 9787.85 \text{ W} = 9.787 \text{ kW}$$

$$\text{Number of belts required} = \frac{\text{kW}}{\text{kW/belt}}$$

$$= \frac{38}{9.787}$$

$$= 3.88$$

Number of belts required ≅ 4

(ii) Pitch Length of Belt

$$L = 2C + \frac{\pi (D_l + D_s)}{2} + \frac{(D_l - D_s)^2}{4C}$$

$$= (2 \times 1000) + \frac{\pi (900 + 225)}{2} + \frac{(900 - 225)^2}{4 \times 1000}$$

$$L = 3881.052 \text{ mm}$$

$$L = 3882 \text{ mm}$$

Example 7.5: *A single V belt is used to transmit power from a grooved pulley of pitch diameter 200 mm running at 1500 rpm to a flat pulley of diameter 600 mm. The centre distance between the pulleys is 1000 mm. The mass of the belt is 0.3 kg per metre. The coefficient of friction between the belt and pulley is 0.25. The V-belt pulley groove angle is 38°. If the allowable tension in the belt is 800 N. Determine.*

(i) Power transmitting capacity of the belt, and

(ii) Initial tension required in the belt.

Solution: Type of drive = Open type V-belt

Number of belts = 1

$$D_s = 200 \text{ mm}$$

$$D_l = 600 \text{ mm}$$

$$n = 1500 \text{ rpm}$$

$$C = 1000 \text{ mm}$$

$$m = 0.3 \text{ kg/m}$$

$$\mu = 0.25$$

$$2\theta = 38°$$

$$F_{max} = 800 \text{ N}$$

To find: (i) kW, (ii) F_i

Angle of contact (α_s):

$$\beta = \sin^{-1}\left(\frac{D_l - D_s}{2C}\right) = \sin^{-1}\left(\frac{600 - 200}{2 \times 1000}\right)$$

$$\beta = 11.53° = 0.2014 \text{ rad}$$

$$\alpha_s = \pi - 2\beta = \pi - 2 \times 0.2014$$

$$= 2.7388 \text{ rad}$$

For driven flat belt

$$e^{\frac{\mu \cdot \alpha_s}{\sin \theta}} = e^{\frac{0.25 \times 2.7388}{\sin 19}} = 8.1915$$

$$\alpha_l = \pi + 2\beta$$

$$= \pi + 2\,(0.2014) = 3.5444 \text{ rad}$$

$$e^{\mu \cdot \alpha l} = e^{0.25 \times 3.5444} = 2.4256$$

Limiting tension ratio:

$$\frac{F_1}{F_2} = e^{0.25 \times 3.544} \quad \ldots (1)$$

$$F_1 = 2.4256 \, F_2$$

Velocity of belt

$$v = \frac{\pi \cdot D_s \cdot n}{60000} = \frac{\pi \times 200 \times 1500}{60000}$$

$$v = 15.71 \text{ m/s}$$

Centrifugal tension (F_c)

$$F_c = m \cdot v^2$$

$$= 0.3 \times (15.71)^2$$

$$F_c = 74.022 \text{ N}$$

$$F_{max} = F_1 + F_c$$
$$800 = F_1 + 74.022$$
$$F_1 = 725.9588 \text{ N}$$

From equation (1)

$$F_2 = 299.29 \text{ N}$$

(i) Power Transmitting Capacity of Belt (kW):

$$\text{Power} = (F_1 - F_2) \times v$$
$$= (725.9588 - 299.29) \times 15.71$$
$$\text{Power} = 6702.96 \text{ W}$$
$$= 6.703 \text{ kW}$$

(ii) Initial Tension (F_i):

$$F_i = \frac{F_1 + F_2 + 2F_c}{2}$$
$$= \frac{725.9588 + 299.29 + 2\,(74.022)}{2}$$
$$F_i = 586.6464 \text{ N}$$

Example 7.6: *A V belt drive is used to transmit 30 kW power from an electric motor running at 1440 rpm to a machine running at 480 rpm. The central distance between the shaft is 1 m. Groove angle for pulley is 38° and coefficient of friction between the belt and the pulley is 0.2. The density of the belt material is 1000 kg/m³ and allowable tensile stress for the belt is 1.53 N/mm². The c/s dimensions of the belt are as follows:*

Width of the belt at the top: 37 mm.

Width of the belt at the bottom: 19 mm.

Depth of the belt: 25 mm.

Find: (i) Diameter of pulleys, (ii) Minimum number of belts required.

Assume maximum power transmission capacity conditions for belts.

Solution: Type of drive: Open type V-belt

$$\text{Total Power} = 30 \text{ kW}$$
$$n = 1440 \text{ rpm}$$
$$N = 480 \text{ rpm}$$
$$C = 1000 \text{ mm}$$
$$2\theta = 38°$$
$$\mu = 0.2$$
$$\rho = 1000 \text{ kg/m}^3$$
$$\sigma_t = 1.53 \text{ N/mm}^2$$
$$b_t = 37 \text{ mm (Width of belt at top)}$$
$$b_b = 19 \text{ mm (Width of belt at bottom)}$$
$$t = \text{Thickness of belt} = 25 \text{ mm}$$

To find: (i) D_l and D_s, (ii) Number of belts required.

$$A = \frac{1}{2} \times (37 + 19) \times 25$$
$$A = 700 \text{ mm}^2$$

Maximum permissible tension (F_{max})

$$F_{max} = \sigma_t \times A$$
$$= 1.53 \times 700$$
$$F_{max} = 1071 \text{ N}$$

Centrifugal tension (F_c)

$$F_c = \frac{F_{max}}{3}$$
$$F_c = \frac{1071}{3} = 357 \text{ N}$$

Velocity of belt

$$F_c = m \cdot v^2$$
$$F_c = (\rho \cdot A) \cdot v^2$$
$$357 = \left(1000 \times \frac{700}{1000 \times 1000} \times 1\right)^{v^2}$$
$$v = 22.58 \text{ m/s}$$
$$v = \frac{\pi \cdot D_s \cdot n}{60000}$$
$$22.58 = \frac{\pi \times D_s \times 1440}{60000}$$
$$D_s = 299.518 \text{ mm}$$
$$\therefore \quad D_s = 300 \text{ mm}$$

From speed ratio

$$\frac{N}{n} = \frac{D_s}{D_l}$$
$$D_l = \frac{1440 \times 300}{480}$$
$$D_l = 900 \text{ mm}$$

Angle of contact (α_s)

$$\beta = \sin^{-1}\left(\frac{D_l - D_s}{2C}\right) = \sin^{-1}\left(\frac{900 - 300}{2 \times 1000}\right)$$
$$\beta = 17.45° \text{ or } 0.3046 \text{ rad}$$
$$\alpha_s = \pi - 2\beta = 2.5322 \text{ rad}$$

We know

$$\frac{F_1}{F_2} = e^{\frac{\mu \cdot \alpha_s}{\sin \theta}}$$
$$= e^{\frac{0.2 \times 2.5332}{\sin 19}}$$
$$F_1 = 4.7377\, F_2 \qquad \ldots (1)$$
$$F_{max} = F_1 + F_c$$
$$F_1 = F_{max} - F_c$$
$$F_1 = 1071 - 357$$

$$F_1 = 714 \text{ N}$$

$$\therefore \quad F_2 = 150.706 \text{ N}$$

Power transmission capacity of each belt

$$\text{Power/belt} = (F_1 - F_2) \cdot v$$

$$= (714 - 150.706) \cdot 22.58$$

$$= 12719.177 \text{ W} = 12.72 \text{ kW/belt}$$

Number of belts required

$$= \frac{\text{Total power to be transmitted (kW)}}{\text{Power transmission capacity of each belt (kW/belt)}}$$

$$\therefore \quad \text{Number of belts} = \frac{30}{12.72} = 2.35$$

$$\therefore \quad \text{Number of belts} = 3$$

Example 7.7: *The following data is given for an open type V-belt drive.*

Diameter of driving pulley = 120 mm

Diameter of driven pulley = 240 mm

Centre distance = 0.8 m

Groove angle = 40°

Mass of belt = 0.25 kg/m

Maximum possible tension = 800 N

Coefficient of friction = 0.18

Plot the graph of maximum tension and power transmitted against the belt velocity. Neglect power losses.

Solution:

Type of drive = Open type V-belt drive

$$D_s = 120 \text{ mm}$$

$$D_l = 240 \text{ mm}$$

$$C = 800 \text{ mm}$$

$$2\theta = 40°$$

$$m = 0.25 \text{ kg/m}$$

$$F_{max} = 800 \text{ m}$$

$$\mu = 0.18$$

Angle of contact (α_s)

$$\beta = \sin^{-1}\left(\frac{D_l - D_s}{2C}\right) = \sin^{-1}\left(\frac{240 - 120}{2 \times 800}\right)$$

$$\beta = 4.3012° \text{ or } 0.07507$$

$$\alpha_s = \pi - 2\beta = 2.9914 \text{ rad}$$

Limiting tension ratio

$$\frac{F_1}{F_2} = e^{\frac{\mu \cdot \alpha_s}{\sin \theta}} = e^{\frac{0.18 \times 2.9914}{\sin 20}}$$

$$F_1 = 4.8274 \, F_2$$

Initial tension in belt (F_i).

Initial tension in the belt occurs at v = 0.

$$F_{max} = F_1 + F_c$$

$$800 = F_1 + mv^2$$

$$800 = F_1 + 0.25 \times (0)$$

$$F_1 = 800 \text{ N}$$

$$F_2 = 165.717 \text{ N}$$

Initial tension in belt,

$$F_i = \frac{F_1 + F_2 + 2F_c}{2}$$

$$= \frac{800 + 165.717}{2}$$

$$F_i = 482.8585 \text{ N}$$

Belt tension on tight side

$$F_{max} = F_1 + F_c$$

$$F_1 = F_{max} - F_c$$

$$F_1 = F_{max} - mv^2$$

$$2F_1 = F_1 + F_2 = 2F_c$$

$$F_2 = 2(F_i - F_c) - F_1$$

$$= 2[F_i - m(v^2)] - F_1$$

$$F_2 = 2[482.8586 - mv^2] - [F_{max} - mv^2]$$

$$F_2 = 965.7172 - mv^2 - F_{max}$$

$$\frac{F_1}{F_2} = 4.8275$$

$$\frac{F_{max} - mv^2}{965.712 - mv^2 - F_{max}} = 4.8275$$

$$F_{max} - mv^2 = 4661.9998 - 3.8275 \times mv^2$$

$$F_{max} = 800 - 0.1642 \, v^2$$

Belt tension on slack side

$$\text{Tension on slack slide} = F_2 + F_c$$

$$= 165.7172 - 0.0858 \, v^2 + 0.25 \, v^2$$

$$= 165.7172 + 0.1642 \, v^2$$

$$\text{Power} = (F_1 - F_2) \times v$$

$$= \{[800 - 0.1642 (v^2)] - [165.7172 + 0.1642 \, v^2]\} \, v$$

$$\text{Power} = 634.2828 \, v - 0.3284 \, v^3$$

$$F_{max} = 800 - 0.1642 \, (v)^2$$

At v = 0, F_{max} = 800 N

$\therefore$ The equation for power transmitted by belt is

$$\text{Power} = 634.2828 \, (v) - 0.3284 \times v^3 \quad \ldots (1)$$

At v = 0, kW = 0

$$v = \sqrt{\frac{F_i}{3m}}$$

For maximum power

$$v = \sqrt{\frac{482.8586}{3 \times 0.25}} = 25.3734 \text{ m/s}$$

$$\text{Maximum power} = 634.2828\ v - 0.3284\ v^2$$
$$= 15882.48\ \text{W}$$
$$\text{Maximum power} = 15.882\ \text{kW}$$

Speed versus power chart is drawn below using equation (1).

Belt Speed v (m/sec)	0	10	20	25.37	30	40	43.48
F_{max} (N)	800	783.6	734.6	694.3	652.2	537.3	482.4
kW	0	6.014	10.058	10.729	10.162	4.334	0

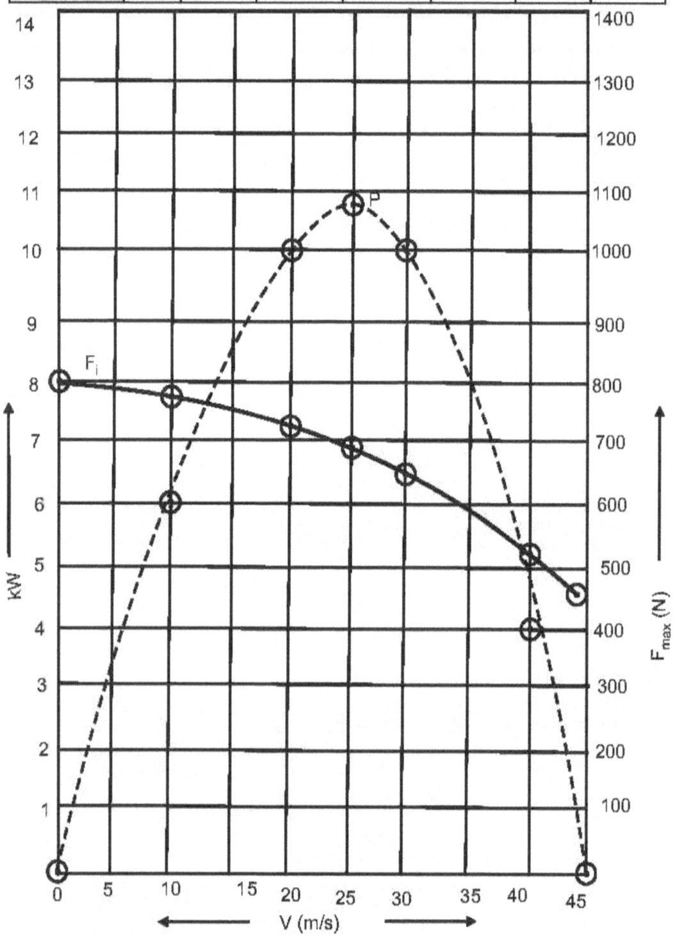

Fig. 7.21

EXERCISE

1. Derive the condition for maximum power transmitting capacity of the belt.
2. Derive a relation for optimum velocity of a belt for maximum power transmission capacity in terms of total tension in tight side of the belt and mass per unit length of the belt.
3. Discuss different belt-tightening (tensioning) methods.
4. Explain any one belt tensioning method with suitable sketch.
5. Derive the condition for maximum power capacity of belt.
6. Prove that, the maximum power transmitted by the belt at a given initial tension Ti.
$$P(Kw) = \frac{(T1 - T2)}{1000}\sqrt{\frac{Ti}{3m}}$$
where, T1 and T2 = Tension in tight and slack side (N), m - Mass per unit length of belt (kg/m).
7. Explain the procedure of selection of flat belt from manufacturer's catalogue.
8. Explain the reasons for speed variations in chain drives.
9. Compare belt, chain and gear drives with respect to the power transmitting capacity, performance, losses and noise and areas of application. Give two applications of each drive.
10. Write a note on 'silent chains'.
11. Discuss relative advantages and limitations of chain drives.
12. State and explain four basic-modes of failure in roller chain.
13. Explain polygonal effect in chain with sketch.
14. What are inverted tooth chain? Sketch and explain their advantages over roller chains.

◈◈◈

CHAPTER - 8

FLYWHEEL

8.1 INTRODUCTION

- A flywheel is used to control the variations in speed during each cycle of an operation. A flywheel acts as a reservoir of energy which stores energy during the period when the supply of energy is more than the requirement and releases the energy during the period when the supply energy is less than the requirement.
- Flywheel which minimizes fluctuations of speed within the cycle but it cannot minimize fluctuations due to load variation. This means flywheel does not exercise any control over mean speed of the engine.
- To minimize fluctuations in the mean speed which may occur due to load variation, governor is used. The governor has no influence over cyclic speed fluctuations but it controls the mean speed over a long period during which load on the engine may vary.
- The function of governor is to increase the supply of working fluid going to the primemover when the load on the prime-mover increases and to decrease the supply when the load decreases so as to keep the speed of the prime-mover almost constant at different loads.

Table 8.1 : Difference between Governor and Flywheel

Sr. No.	Governor	Flywheel
1.	Maintains the variation of mean speed within prescribed limit.	Controls the speed variations in an engine caused due to fluctuations of turning moment.
2.	It regulates the speed over a period of time.	It regulates the speed during a cycle only.
3.	It regulates speed by charge of the engine or prime mover.	It stores energy and give it up whenever required in a cycle.

Example : When the load on an engine increases, its speed decreases, therefore it becomes necessary to increase the supply of working fluid. On the other hand, when the load on the engine decreases, its speed increases and hence less working fluid is required.

When there is change in load, variation in speed also takes place then governor operates a regulatory control and adjusts the fuel supply to maintain the mean speed nearly constant. Therefore, the governor automatically regulates through linkages, the energy supply to the engine as demanded by variation of load so that the engine speed is maintained nearly constant.

8.2 TURNING MOMENT DIAGRAM

The turning moment diagram (also known as **crank effort diagram**) is the graphical representation of the turning moment or crank-effort for various positions of the crank.

Turning Moment Diagrams

- Turning moment diagram for single cylinder double acting steam engine
- A turning moment diagram for a single acting cylinder is as shown in Fig. 8.1.
- The vertical ordinate represents the turning moment and the horizontal ordinate represents the crank angle.
- The turning moment can be as,

$$T = F_p \times r\left(\sin\theta + \frac{\sin 2\theta}{2\sqrt{n^2 - \sin\theta^2}}\right)$$

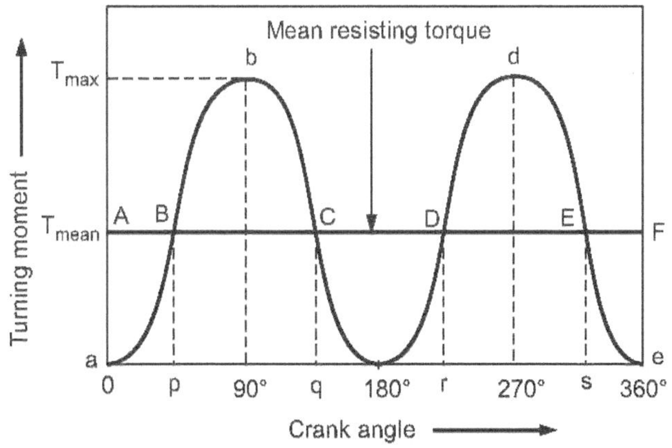

Fig. 8.1 : Turning moment diagram for a single cylinder, double acting steam engine

where, F_p = Piston effort,

r = Radius of crank,

n = Ratio of the connecting rod length and radius of crank, and

θ = Angle turned by the crank from inner dead centre.

From the above expression, it can be seen that the turning moment (T) is zero when the crank angle (θ) is zero.

- It is maximum when the crank angle is 90° and it is again zero when crank angle is 180°.

- This is shown by the curve abc in Fig. 8.1 and it represents the turning moment diagram for outstroke. The curve cde is the turning moment diagram for in stroke and is somewhat similar to the curve abc.
- Since the work done is the product of the turning moment and the angle turned, therefore the area of the turning moment diagram represents the work done per revolution.
- In actual practice, the engine is assumed to work against the mean resisting torque, as shown by a horizontal line A-f. The height of the ordinate a-A represents the mean height of the turning moment diagram.
- Since it is assumed that the work done by the turning moment per revolution is equal to the work done against the mean resisting torque, therefore the area of the rectangle a-A-F-e is proportional to the work done against the mean resisting torque.
- Turning Moment Diagram for a Four Stroke Cycle Internal Combustion Engine
- A turning moment diagram for a four stroke cycle internal combustion engine is shown in Fig. 8.2.
- As we know that in a four stroke cycle internal combustion engine, there is one working stroke after the crank has turned through two revolutions, i.e. 720° (or 4π radians).
- Since the pressure inside the engine cylinder is less than the atmospheric pressure during the suction stroke, therefore a negative loop is formed as shown in Fig. 8.2.

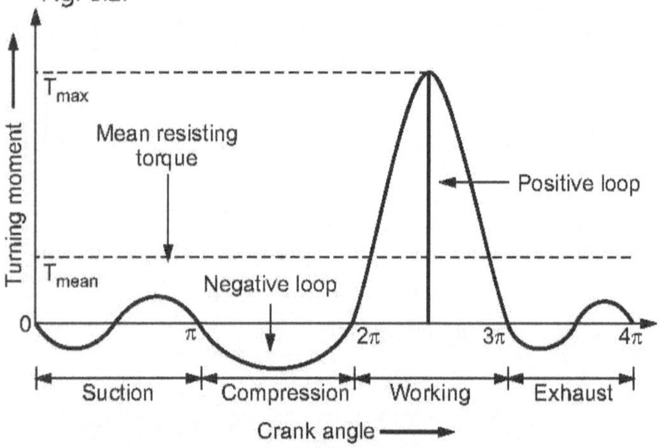

Fig. 8.2: Turning moment diagram for a four stroke cycle internal combustion engine

- During the compression stroke, the work is done on the gases, therefore a higher negative loop is obtained.
- During the expansion or working stroke, the fuel burns and the gases expand, therefore a large positive loop is obtained. In this stroke, the work is done by the gases.
- During exhaust stroke, the work is done on the gases, therefore a negative loop is formed. It may be noted that the effect of the inertia forces on the piston is taken into account in Fig. 8.2.

8.3 FLUCTUATION OF ENERGY AND SPEED

- The fluctuation of energy may be determined by the turning moment diagram for one complete cycle of operation. Consider the turning moment diagram for a single cylinder double acting steam engine as shown in Fig. 8.3.

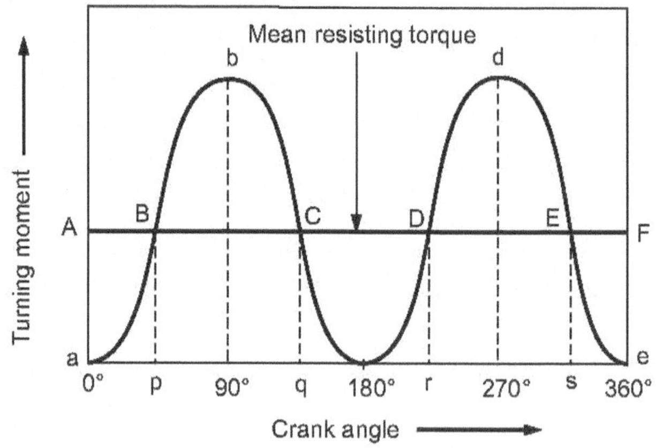

Fig. 8.3

- We see that the mean resisting torque line A-F cuts the turning moment diagram at points B, C, D and E.
- When the crank moves from a to p, the work done by the engine is equal to the area a-B-p, whereas the energy required is represented by the area a-A-B-p.
- In other words, the engine has done less work (equal to the area a-A-B) than the requirement. This amount of energy is taken from the flywheel and hence the speed of the flywheel decreases.
- Now the crank moves from p to q, the work done by the engine is equal to the area p-B-b-C-q, whereas the requirement of energy is represented by the area p-B-C-q.
- Therefore, the engine has done more work than the requirement. This excess work (equal to the area B-b-C) is stored in the flywheel and hence the speed of the flywheel increases while the crank moves from p to q.
- Similarly, when the crank moves from q to r, more work is taken from the engine than is developed. This loss of work is represented by the area C-c-D. To supply this loss, the flywheel gives up some of its energy and thus the speed decreases while the crank moves from q to r.

- As the crank moves from r to s, excess energy is again developed given by the area D-d-E and the speed again increases.
- As the piston moves from s to e, again there is a loss of work and the speed decreases.
- The variations of energy above and below the mean resisting torque line are called fluctuations of energy. The areas B-b-C, C-c-D, D-d-E, etc. represent fluctuations of energy.
- A little consideration will show that the engine has a maximum speed either at q or at s. This is due to the fact that the flywheel absorbs energy while the crank moves from p to q and from r to s.
- On the other hand, the engine has a minimum speed either at p or at r. The reason is that the flywheel gives out some of its energy when the crank moves from a to p and q to r.
- The difference between the maximum and the minimum energies is known as maximum fluctuation of energy.

  Determination of Maximum Fluctuation of Energy

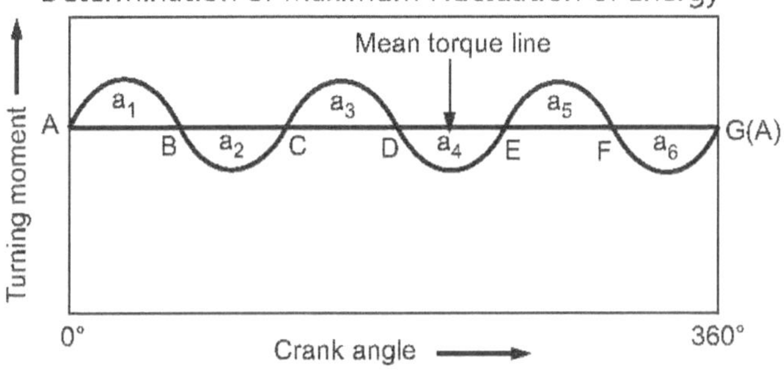

Fig. 8.4

- Turning moment diagram for a multi-cylinder engine.
- A turning moment diagram for a multi-cylinder engine is shown by a wavy curve in Fig. 8.4.
- The horizontal line AG represents the mean torque line.

  Let a_1, a_3, a_5 be the areas above the mean torque line and a_2, a_4 and a_6 be the areas below the mean torque line.
- These areas represent some quantity of energy which is either added or subtracted from the energy of the moving parts of the engine.

  Let the energy in the flywheel at A = E, then from Fig. 8.4, we have.

$$\text{Energy at B} = E + a_1$$
$$\text{Energy at C} = E + a_1 - a_2$$
$$\text{Energy at D} = E + a_1 - a_2 + a_3$$
$$\text{Energy at E} = E + a_1 - a_2 + a_3 - a_4$$
$$\text{Energy at F} = E + a_1 - a_2 + a_3 - a_4 + a_5$$
$$\text{Energy at G} = E + a_1 - a_2 + a_3 - a_4 + a_5 - a_6$$
= Energy at A (i.e. cycle repeats after G)

Let us now suppose that the greatest of these energies is at B and least at E. Therefore,

Maximum energy in flywheel = $E + a_1$

Minimum energy in the flywheel = $E + a_1 - a_2 + a_3 - a_4$

Therefore, Maximum fluctuation of energy,

$$\Delta E = \text{Maximum energy} - \text{Minimum energy}$$
$$= (E + a_1) - (E + a_1 - a_2 + a_3 - a_4)$$
$$= a_2 - a_3 + a_4$$

8.3.1 Coefficient of Fluctuation of Energy

It may be defined as the **ratio of the maximum fluctuation of energy to the work done per cycle.**

Mathematically, coefficient of fluctuation of energy,

The work done per cycle (in N-m or joules) may be obtained by using the following two relations :

Work done per cycle = $T_{mean} \times \theta$

Where,

T_{mean} = Mean torque, and

θ = Angle turned (in radians), in one revolution.

= 2π, in case of steam engine and two stroke i. c. engines,

= 4π, in case of four stroke internal combustion engines.

The mean torque (T_{mean}) in N-m may be obtained by using the following relation :

Where,

P = Power transmitted in watts,

N = Speed in r.p.m., and

ω = Angular speed in rad/s = $2\pi N/60$

The work done per cycle may also be obtained by using the following relation:

Where,

n = Number of working strokes per minute,

= N, in case of steam engines and two stroke i. c. engines,

= N/2, in case of four stroke internal combustion engines.

Note:

The following table shows the values of coefficient of fluctuation of energy for steam engines and internal combustion engines.

Table 8.2 : Coefficient of Fluctuation of Energy (C_E) for Steam and Internal Combustion Engines

Sr. No.	Type of Engine	Coefficient of Fluctuation of Energy (C_E)
1.	Single cylinder, double acting steam engine.	0.21
2.	Cross-compound steam engine.	0.096
3.	Single cylinder, single acting, four stroke gas engine.	1.93
4.	Four cylinder, single acting, four stroke gas engine.	0.066
5.	Six cylinder, single acting, four stroke gas engine.	0.031

8.3.2 Coefficient of Fluctuation of Speed

The difference between the maximum and minimum speeds during a cycle is called the **maximum fluctuation of speed.**

The ratio of the maximum fluctuation of speed to the mean speed is called the coefficient of fluctuation of speed.

Let,

N_1 and N_2 = Maximum and minimum speeds in r.p.m. during the cycle, and

Therefore, coefficient of fluctuation of speed

$$C_S = \frac{N_1 - N_2}{N} = \frac{2(N_1 - N_2)}{N_1 + N_2}$$

$$= \frac{\omega_1 - \omega_2}{\omega} = \frac{2(\omega_1 - \omega_2)}{\omega_1 + \omega_2}$$

...In terms of angular speeds

$$= \frac{v_1 - v_2}{v} = \frac{2(v_1 - v_2)}{v_1 + v_2}$$

...In terms of linear speeds

The coefficient of fluctuation of speed is a limiting factor in the design of flywheel. It varies depending upon the nature of service to which the flywheel is employed.

Note:

The reciprocal of the coefficient of fluctuation of speed is known as **coefficient of steadiness** and is denoted by m.

$$\therefore \quad m = \frac{1}{C_s} = \frac{N}{N_1 - N_2} = \frac{\omega}{\omega_1 - \omega_2} = \frac{v}{v_1 - v_2}$$

Determination of flywheel size for different types of prime movers and machines.

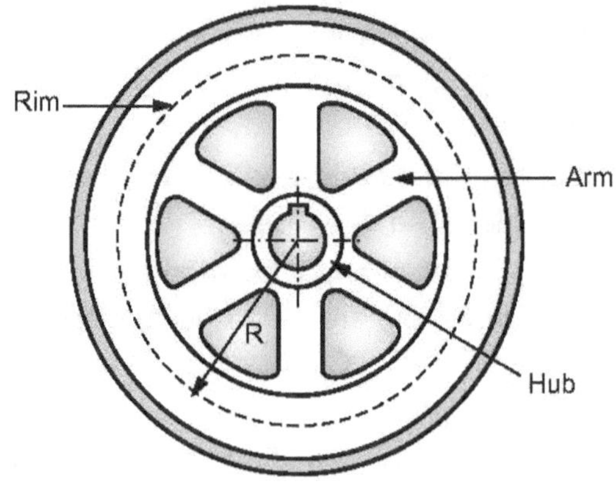

Fig. 8.5 : Flywheel

A flywheel is shown in Fig. 8.5. We know that in fluctuation of energy that when a flywheel absorbs energy, its speed increases and when it gives up energy, its speed decreases.

Let,

m = Mass of the flywheel in kg,

k = Radius of gyration of the flywheel in metres,

I = Mass moment of inertia of the flywheel about its axis of rotation in

kg-m^2 = m.k^2,

N_1 and N_2 = Maximum and minimum speeds during the cycle in r.p.m.,

ω_1 and ω_2 = Maximum and minimum angular speeds during the cycle in rad/s,

N = Mean speed during the cycle in r.p.m.

$$= \frac{N_1 + N_2}{2}$$

ω = Mean angular speed during the cycle in rad/s

$$= \frac{\omega_1 + \omega_2}{2}$$

C_s = Coefficient of fluctuation of speed

$$= \frac{N_1 + N_2}{2} \text{ or } \frac{\omega_1 + \omega_2}{2}$$

We know that the mean kinetic energy of the flywheel,

$$E = \frac{1}{2} \times I \cdot \omega^2 = \frac{1}{2} \times m \cdot k^2 \cdot \omega^2$$

(in N-m or joules)

As the speed of the flywheel changes from ω_1 to ω_2, the maximum fluctuation of energy,

$$\Delta E = \text{Maximum K.E.} - \text{Minimum K.E.}$$

$$= \frac{1}{2} \times I(\omega_1)^2 - \frac{1}{2} \times I(\omega_2)^2$$

$$= \frac{1}{2} \times I[(\omega_1)^2 - (\omega_2)^2]$$

$$= \frac{1}{2} \times I\,(\omega_1 + \omega_2)\,(\omega_1 - \omega_2)$$

$$= I \cdot \omega\,(\omega_1 - \omega_2)$$

$$\ldots\left(\because \frac{\omega_1 + \omega_2}{2}\right) \quad \ldots(8.1)$$

$$= I \cdot \omega^2 \left(\frac{\omega_1 - \omega_2}{\omega}\right)$$

...(Multiplying and dividing by ω)

$$= I \cdot \omega^2 \cdot C_s = m \cdot k^2 \cdot \omega^2 \cdot C_s$$

$$\ldots(\because I = m \cdot k^2) \quad \ldots(8.2)$$

$$= 2 \cdot E \cdot C_s \text{ (in N-m or joules)}$$

$$\ldots\left(\because E = \frac{1}{2} \times I \cdot \omega^2\right) \quad \ldots(8.3)$$

The radius of gyration (k) may be taken equal to the mean radius of the rim (R), because the thickness of rim is very small as compared to the diameter of rim. Therefore, substituting k = R, in equation (ii), we have

$$\Delta E = m.R^2.\omega^2.C_S = m.v^2.C_S$$

Where, v = Mean linear velocity (i.e. at the mean radius) in m/s = ω.R

Since $\omega = 2\,\pi\,N/60$, therefore may be written as

$$\Delta E = I \times \frac{2\pi N}{60}\left(\frac{2\pi N_1}{60} - \frac{2\pi N_2}{60}\right)$$

$$= \frac{4\pi^2}{3600} \times I \times N\,(N_1 - N_2)$$

$$= \frac{\pi^2}{900} \times m \cdot k^2 \cdot N\,(N_1 - N_2)$$

$$= \frac{\pi^2}{900} \times m \cdot k^2 \cdot N^2 \cdot C_s$$

$$\ldots\left(\because C_s = \frac{N_1 - N_2}{N}\right)$$

8.4 STRESSES IN FLYWHEEL

The stresses in the flywheel rims are due to the following elements of stresses as

(i) Centrifugal stresses

(ii) Stresses due to change of speed and load

(iii) Shrinkage stresses while casting

Centrifugal stresses can be evaluated due to change of speed and load can be calculated with fair accuracy. The shrinkage stresses, might be quite large cannot be computed with any degree of accuracy. It is taken into account by carefully selecting dimension arrangement of flywheel parts and with a large factor of safety.

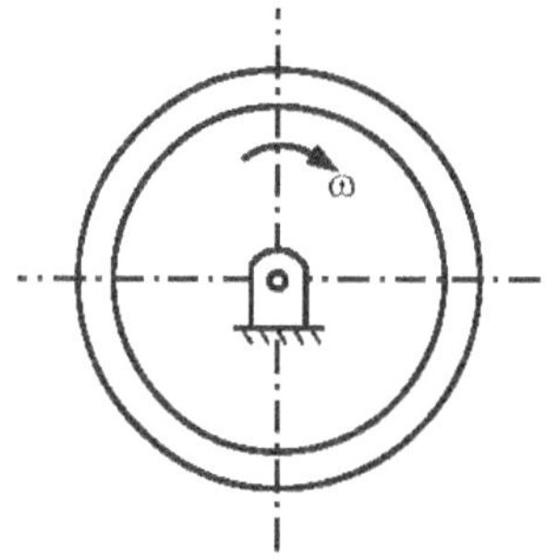

Fig. 8.6

The stresses in the rim due to centrifugal stresses depend upon two options.

1. The arms are flexible and without a restraining effect on the rim.
2. The arms are rigid and these completely restrain the rim at the places of attachment.

(i) The wheel rim will act as a free rotating ring subjected to internal pressure. The stress induced in the rim will be tangential stress and it may be evaluated in the following manner.

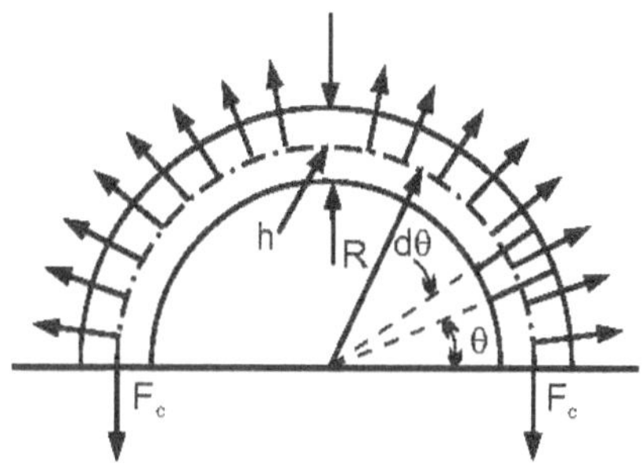

Fig. 8.7

Consider a small section of half as in Fig. 8.7 rim, so that $\Delta\theta$ (radians) is very small.

Now mass of section $dm = w \cdot h \cdot R\,\Delta\theta \cdot \rho$ kg

$\therefore$ Centrifugal force or inertia load on the small section

$$= \text{mass} \times \text{acceleration} = (w \cdot h \cdot R\,\Delta\theta \cdot \rho)\,(R\omega^2)$$

Its horizontal components will balance. Its vertical component is

$$= (W \cdot h \cdot R\Delta\theta \cdot \rho)\,(R\omega^2) \sin\theta$$

$\therefore$ Vertical component of the inertia load of half the rim

$$= \int_0^{\pi} (w \cdot h \cdot R\,\Delta\theta \cdot \rho)\,(\omega^2 R) \sin\theta$$

$$= 2R^2 wh\rho\omega^2$$

This is balanced by the tensile forces at the cut section

$$\therefore \quad 2F_c = 2R^2\,wh\rho\omega^2$$

Assuming that the tensile stress is uniformly distributed across a section

$$2F_c = 2(whf_1)$$

$$\therefore \quad \sigma_t = \rho\omega^2 R^2 = \rho V^2 \text{ N/m}^2$$

(ii) When the arms are rigid, the portion of the rim between two arms will act as a fixed beam and uniformly loaded, as shown in Fig. 8.8.

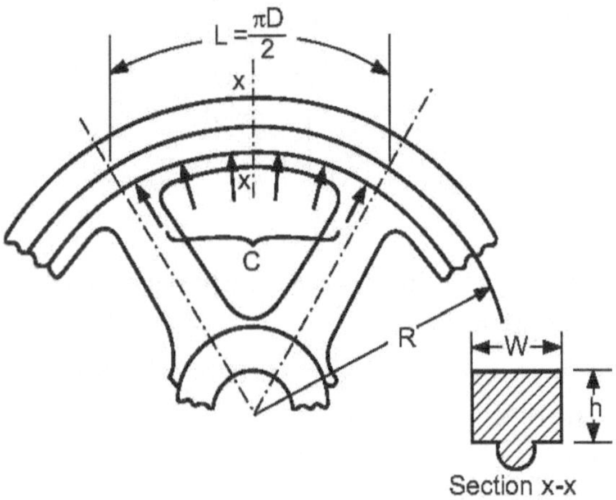

Fig. 8.8

The uniformly distributed load,

$$C = \frac{\rho \cdot w \cdot h \cdot v^2}{R} \text{ N/m of arc of rim.}$$

Now the maximum bending moment due to this will act at the arms and is approximately equal to

$$= \frac{Cl^2}{12}$$

where l = distance between the arms along the rim arc

$$= \frac{2\pi R}{z} \quad \text{where } z = \text{number of arms}$$

$$\therefore \quad \text{Max. B. M.} = \frac{\rho \cdot w \cdot h \cdot v^2}{R} \cdot \left(\frac{2\pi R}{z}\right)^2 \cdot \frac{1}{12}$$

$$= \sigma_b \times Z,$$

where σ_b = bending stress, $Z = \frac{1}{6} wh^2$

$$\sigma_b = \frac{2\pi^2 \rho v^2 R^2}{z^2 Rh} = \frac{2\pi^2 \rho v^2 R}{z^2 h} \text{ N/m}^2$$

(iii) It has been seen experimentally that the arms stretch approximately three fourth the amount required if the rim expanded freely under the action of the centrifugal force. Therefore, the total stress in the rim at junctions with the arms will be,

$$\sigma = \frac{3}{4}\sigma_t + \frac{1}{4}\sigma_b$$

Now $\sigma_t \not> 7$ MN/m^2 for C.I. wheels

$\not> 28$ MN/m^2 for C.S. wheels

(iii) Rim Velocity

It is clear from the above analysis that the stress in the rim depends upon the speed. It has been experimentally determined that the "bursting speed" for two piece cast iron wheels with thick, heavy rims was about 88 m/s and about 69.5 m/s for two piece wheels with wide thin rims divided along a diameter through diametrically opposite arms. Using a factor 3 with respect to the bursting speed, which is equivalent to a factor of safety of 9 with respect to the stress, the safe speed will be about 29 m/s for the first type of wheel and 22.5 m/s for the second type of wheel. The safe rim speeds are given below :

25 m/s	for C.I. wheels of engines under 75 kW
30 m/s	for C.I. wheels of larger machines
35 m/s	with special design of rim which eliminates the danger of blowholes
50 m/s	with a special arm design, this and the above values are for C.I.
60 m/s	with case steel wheels
75 m/s	with C.S. arms are laminated steel arms and flywheels of all welded construction with forged steel rims.
50 m/s	with C.I. flywheels machined all over in automotive engines
75 m/s	with semisteel flywheels of automotive engines
60 m/s	with steel flywheels of automotive engines

Stresses in Arms

The arms of the flywheel are subjected to types of stresses given :

1. Direct pull due to the expansion or enlargement of the rim under centrifugal action.
2. Bending action due to the transmission of torsional moment from the rim to the shaft or from the shaft to to the rim.
3. Initial stresses due to unequal cooling and shrinkage.
4. Acceleration forces transmitted to the rim.

Initial stress due to shrinkage and unequal cooling and the accelerating forces are difficult to be ascertained correctly. For the later case, the worst happens when the piston or bearings sieze and the engine comes to a halt. The flywheel seldom fails due to this and it is assumed that if the flywheel is strong enough to withstand the direct pull and bending action due to respectively, it will be able to resist the acceleration forces. The shrinkage and cooling stresses will be higher for one-piece wheel than for two-piece wheel. In proportioning flywheel arms low allowable stresses are used to cover the uncertainties regarding the initial casting stress and bending action on the arms. Moreover, the arms are subjected to complete reversal of stresses, therefore, the minimum factor of safety for C.I.

wheels should be 10 and for steel wheels 8. It is quite common to take the allowable stress of 1/12 the ultimate strength of the material. When the flywheel is used on punches and other machines subjected to severe shock, the factor of safety should be as high as 15 for C.I. and 13 for W.I. and steel.

The direct stress due to the expansion or enlargement of the rim under centrifugal action is determined on the assumption that the stretch of the arms is equal to the corresponding to free, unrestrained radial expansion of the rim. As already indicated this will be equal to three-fourths full radial expansion.

$$\text{Direct stress, } \sigma_d = \frac{3}{4}\sigma_t = \frac{3}{4}\rho v^2 \text{ N/m}^2$$

The arms will be under bending action due to transmission of torsional moment from the rim to the shaft or from the shaft to the rim, therefore, they will support maximum full torsional moment capacity of the machine. This bending stress will depend upon two distinct possibilities :

(i) For wheels with thick rims, the arms act as catilevers fixed at the hub but free and guided at the rim joint.

(ii) For wheels with wide, thin rims, the arms act more like cantilevers fixed at the hub, but free and unguided at the outer ends.

The arm will bend as shown in Fig. 8.9 for the second case and for the first case, as shown in Fig. 8.10

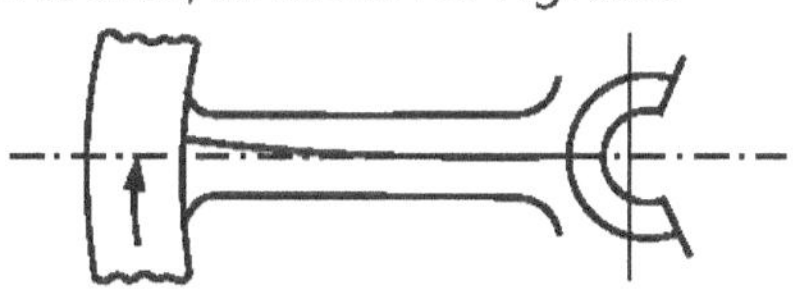

Fig. 8.9 : Arm as a free cantilever

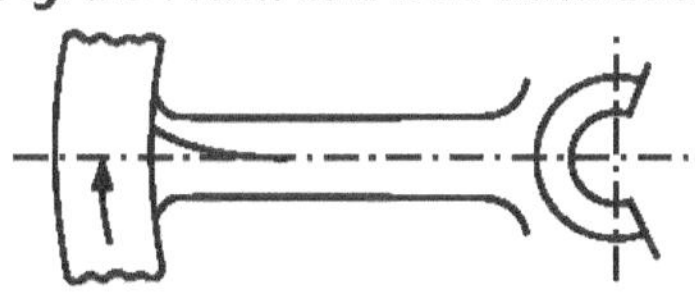

Fig. 8.10 : Arm as a guided cantilever

To calculate the bending moment on the arm, the usual assumption is that load applied at the rim is sufficient to produce a torsional moment at the shaft equal to the maximum torsional moment. Therefore, force at the mean radius of the rim, is $F_r = \frac{M_t}{R}$, where M_t = maximum driving torsional moment on the shaft.

$$\therefore \quad \text{Load per arm} = \frac{F_r}{z} = \frac{M_t}{R \cdot z}$$

∴ Bending moment on the arm at the hub, for free cantilever,

$$M_l = \frac{F_r}{z} \times \frac{(D - d_h)}{2} = \frac{M_t}{R \cdot z} \cdot \frac{(D - d_h)}{2} \quad \ldots(8.4)$$

$$= \frac{M_t}{z} \cdot \frac{D - d_h}{D}$$

where d_h = hub diameter

If the arm is taken as guided cantilever, the maximum bending moment will be one-half that for the free cantilever, and is given as

$$M_2 = \frac{M_t}{2z} \cdot \frac{D - d_h}{D} \quad \ldots(8.5)$$

The sum of the direct stress and bending stress should not exceed the safe allowable stress. The bending moment M_2 will occur at both the ends of the arm.

8.5 FLYWHEEL IN PUNCHING PRESS

- We have discussed that the function of a flywheel in an engine is to reduce the fluctuations of speed, when the load on the crankshaft is constant and the input torque varies during the cycle.
- The flywheel can also be used to perform the same function when the torque is constant and the load varies during the cycle. Such an application is found in punching press or in a riveting machine.
- A punching press is shown diagrammatically in Fig. 8.11.
- The crank is driven by a motor which supplies constant torque and the punch is at the position of the slider in a slider-crank mechanism.
- From Fig. 8.11 we see that the load acts only during the rotation of the crank from $\theta = \theta_1$ to $\theta = \theta_2$, when the actual punching takes place and the load is zero for the rest of the cycle.

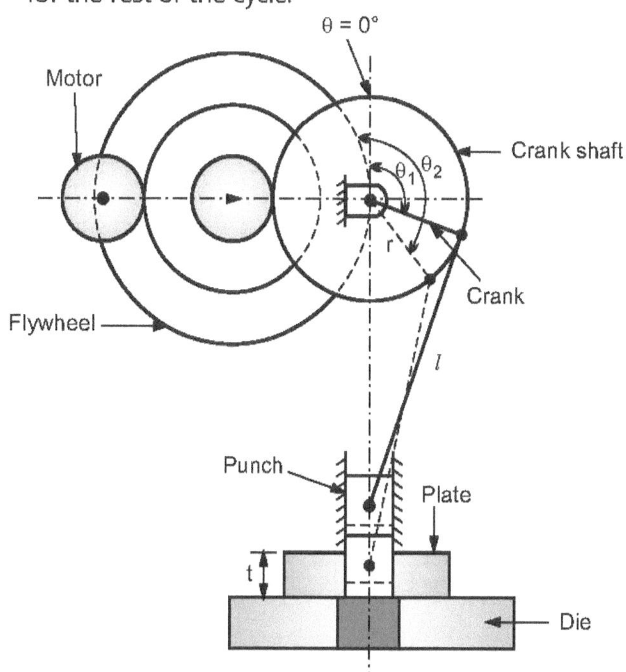

Fig. 8.11 : Operation of flywheel in a punching press

- Unless a flywheel is used, the speed of the crankshaft will increase too much during the rotation of crank from $\theta = \theta_2$ to $\theta = 2\pi$ or $\theta = 0$ and again from $\theta = 0$ to $\theta = \theta_1$, because there is no load while input energy continues to be supplied.
- On the other hand, the drop in speed of the crankshaft is very large during the rotation of crank from $\theta = \theta_1$ to $\theta = \theta_2$ due to much more load than the energy supplied.
- Thus the flywheel has to absorb excess energy available at one stage and has to make up the deficient energy at the other stage to keep to fluctuations of speed within permissible limits.
- This is done by choosing the suitable moment of inertia of the flywheel.
- Let E_1 be the energy required for punching a hole. This energy is determined by the size of the hole punched, the thickness of the material and the physical properties of the material.

  Let, d_1 = Diameter of the hole punched,

  t_1 = Thickness of the plate, and

  τ_u = Ultimate shear stress for the plate material.

- Maximum shear force required for punching,

$$F_S = \text{Area sheared} \times \text{Ultimate shear stress} = \pi \cdot d_1 \cdot t_1 \cdot \tau_u$$

- It is assumed that as the hole is punched, the shear force decreases uniformly from maximum value to zero.
- Work done or energy required for punching a hole,

$$E_1 = \frac{1}{2} \times F_s \times t$$

- Assuming one punching operation per revolution, the energy supplied to the shaft per revolution should also be equal to E_1.
- The energy supplied by the motor to the crankshaft during actual punching operation,

$$E_2 = E_1\left(\frac{\theta_2 - \theta_1}{2\pi}\right)$$

- Balance energy required for punching,

$$= E_1 - E_2$$

$$= E_1 - E_1\left(\frac{\theta_2 - \theta_1}{2\pi}\right)$$

$$= E_1\left(1 - \frac{\theta_2 - \theta_1}{2\pi}\right)$$

- This energy is to be supplied by the flywheel by the decrease in its kinetic energy when its speed falls from maximum to minimum.
- Thus maximum fluctuation of energy,

$$\Delta E = E_1 - E_2 = E_1\left(1 - \frac{\theta_2 - \theta_1}{2\pi}\right)$$

- The values of θ_1 and θ_2 may be determined only if the crank radius (r), length of connecting rod (l) and the relative position of the job with respect to the crankshaft axis are known.
- In the absence of relevant data, we assume that

$$\frac{\theta_2 - \theta_1}{2\pi} = \frac{t}{2s} = \frac{t}{4r}$$

  Where,

  t = Thickness of the material to be punched,

  s = Stroke of the punch

  = 2 × Crank radius

  = 2r.

- By using the suitable relation for the maximum fluctuation of energy (ΔE), we can find the mass and size of the flywheel.

SOLVED EXAMPLES

Example 8.1 : *A flywheel of a steam engine weights 2000 N and has a radius of gyration of 76 cm. The starting torque of steam engine is 130 Kg and m is assumed to be constant. Determine the angular acceleration of flywheel along with speed and kinetic energy after 10 g.*

Solution : Given data : K = 0.76 m

Starting torque T= 130 kgm

$$\omega = 130 \times 9.81 = 1274 \text{ N.m}$$

mass of flywheel,

$$m = \frac{\omega}{g} = \frac{2000}{9.81} = 204.08 \text{ Kg}$$

moment of inertia of flywheel,

$$I = mK^2 - 204.08 \times (0.76)^2 = 117.88 \text{ kg m}^2$$

Angular acceleration of flywheel (α),

We have $T = I \cdot \alpha$

$$\alpha_1 = \frac{T}{L} = \frac{1274}{117.88} = 10.8 \text{ rad/s}^2$$

The flywheel started from rest, so $\omega_1 = 0$

and $\omega_2 = \omega_1 + \alpha t$

$$= 0 + 10.8 \times 10 = 108 \text{ rad/s}$$

$$\text{Kinetic energy (K.E.)} = \frac{1}{2} I\omega_2^2$$

$$= \frac{1}{2} \times 117.88 \times 108^2$$

$$= 687476.16 \text{ N-m}$$

***Example 8.2 :** The crank effort diagram for a 4-stroke cycle gas engine may be assumed to for simplicity of four rectangles, areas of which from line of zero pressure are power stroke = 6000 mm², exhaust stroke = 500 mm², Suction stroke = 300 mm², compression stroke = 1500 mm². Each Sq mm represents 10 Nm. Assuming the resisting torque to be uniform, find*

(a) Power of the engine

(b) Energy to be stored by the flywheel

(c) Mass of a flywheel rim of 1m radius to limit the total fluctuation of speed to ±2% of the mean speed of 150 rpm.

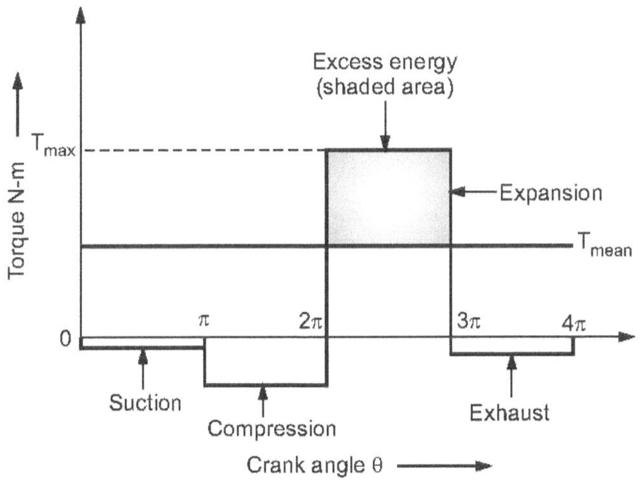

Fig. 8.12

Data : 4 stroke cycle → crank angle per cycle = 4π radians

Radius of gyration k = 1 meter, Mean speed n = 150 rpm,

$$C_s = \pm 2\,\% = 4\% = 0.04$$

(∵ Total fluctuation = 2 × Fluctuation on either side)

Solution : Angular velocity of flywheel $\omega = \dfrac{2\pi n}{60}$

i.e. $$\omega = \frac{2\pi \times 150}{60} = 15.71 \text{ rad/sec}$$

WD/cycle = W.D. during expansion (W. D. during other strokes)

$$\Rightarrow \quad \text{W.D./ cycle} = [6000 - (300 + 1500 + 500)]$$

$$= 3700 \text{ mm}^2$$

$$\text{W.D./cycle} = 3700 \times \text{scale of diagram}$$

$$= 3700 \times 10 = 37000 \text{ N-m}$$

∴ Mean Torque

$$T_m = \frac{\text{W.D./cycle}}{\text{Crank angle/cycle}} = \frac{37000}{4\pi}$$

$$= 2944.4 \text{ N-m}$$

(a) Power developed by engine :

$$P = T_m \times \omega = 2944.4 \times 15.71 = 46.256 \text{ KW}$$

(b) Energy stored by flywheel :

$$e = \text{shaded area} = \pi\,(T_{max} - T_m)$$

But W.D. during expansion = $T_{max} \times \pi$

$$\Rightarrow \quad 6000 \times 10 = T_{max} \times \pi$$

$$\therefore \quad T_{max} = 19098.6 \text{ N-m}$$

Substituting for T_{max},

$$e = \pi\,(T_{max} - T_m) = \pi\,(19098.6 - 2944.6)$$

$$\therefore \quad e = 50749.27 \text{ N-m}$$

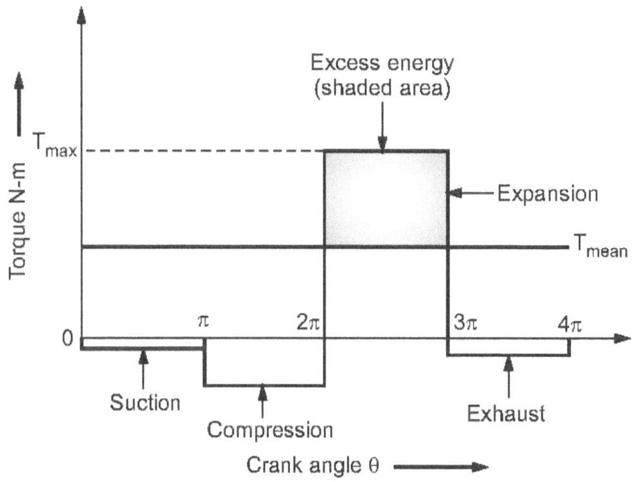

Fig. 8.13

(c) Mass of flywheel,

We know that energy stored by flywheel

$$e = mk^2\,\omega^2\,C_s$$

$$50749.27 = m \times (1)^2\,(15.71)^2 \times 0.04$$

∴ Mass of flywheel m = 5140.64 Kg

***Example 8.3 :** A multi cylinder engine is to run at a speed of 600 rpm. On drawing the TMD to scale of 1mm = 250 Nm and 1 mm = 3°, the areas above and below the mean torque line are +160, –172, +168, –191, +197, –162 mm² respectively.*

The speed is to be kept within ± 1% of the mean speed. Density of cast iron flywheel = 7250 Kg/mm³ and hoop stress is 6 MPa. Assuming that the rim contributes to 92% of the flywheel effect, determine the dimensions of the rectangular cross section of the rim assuming width to be twice the thickness.

Solution :

Let the Energy at 1 = E

Energy at 2 = (E + 160)

Energy at 3 = (E + 160) – 172 = (E – 12)

Energy at 4 = (E – 12) + 168 = (E + 156)

Energy at 5 = (E + 156) – 191 = (E – 35)

Energy at 6 = (E – 35) + 197 = (E + 162)

Energy at 7 = (E + 162) – 162 = E = Energy at 1

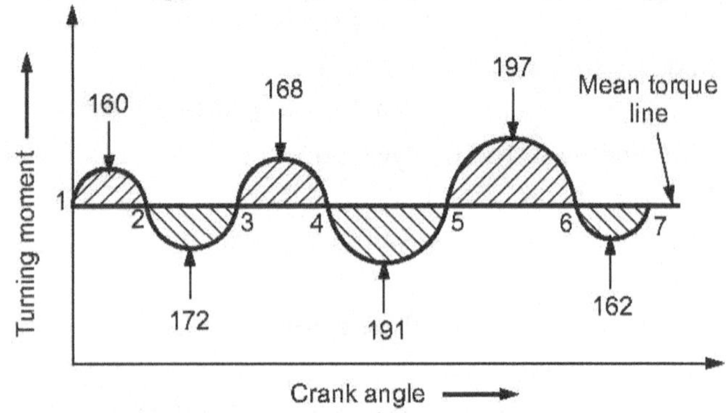

Fig. 8.14

Hence, Maximum fluctuation of energy

(in terms of area) = (E + 162) – (E – 35)

= 197 Sq. mm

Energy stored by the flywheel :

$$\text{Angular velocity } \omega = \frac{2\pi N}{60} = \frac{2\pi \times 600}{60} = 62.84 \text{ rad/sec,}$$

$$C_s = \pm 1\,\% = 2\% = 0.02$$

Scale of the diagram is

$$1\text{mm}^2 = 250 \times \left(\frac{3 \times \pi}{180}\right) = 13.1 \text{ Nm}$$

Max Fluctuation of energy e = (Max. K.E – Min K.E)

$$\Rightarrow \quad e = (E + 162) - (E - 35) = 197 \text{ mm}^2$$

$$\therefore \quad e = 197 \times 13.1 = 2581 \text{ Nm}$$

i.e. $2581 = I\omega^2 C_s = I \times (62.84)^2 \times 0.02$

$\therefore$ Mass moment of inertia I = 32.7 Kg – m^2

Diameter of the flywheel :

Using $\sigma = gv^2$; $6 \times 10^6 = 7250 \times v^2$

$\Rightarrow$ Velocity V= 28.8 m/sec

$$\text{Also} \quad v = \frac{\pi DN}{60} \Rightarrow 28.8 = \frac{\pi \times D \times 600}{60}$$

$\therefore$ Mean dia. of flywheel D = 0.92 m

Given 92% of the flywheel effect is provided by the rim,

$$e_{rim} = 0.92 \times 2581 = 2375 \text{ Nm}$$

$$e_{rim} = mk^2\,\omega^2\,C_s = m(k\omega)^2\,C_s = mv^2\,C_s$$

$$\Rightarrow \quad 2375 = m \times (28.8)^2 \times 0.02$$

$\therefore$ Mass of rim m = 143 Kg.

Dimensions of the cross section of the rim :

We know that mass of the flywheel rim

$$m = \text{Volume of rim} \times \text{density} = (\pi DA) \times g$$

$$\Rightarrow \quad 143 = (\pi \times 0.92 \times A) \times 7250$$

$$\therefore \quad A = 0.006824 \text{ m}^2$$

As cross section of rim is rectangular width b = 2t,

$$A = (b \times t) = 2t^2 \Rightarrow 0.006824 = 2t^2$$

Hence t = 58.4 mm, b = 2t = 116.8 mm.

Example 8.4 : *The torque required for a machine is shown in Fig. 8.15. The motor driving the machine has a mean speed of 1500 rpm and develop constant torque. The flywheel on the motor shaft is of rim type with mean diameter of 40 cm and mass 25 Kg. Determine :*

(a) Power of motor

(b) % variation in motor speed per cycle.

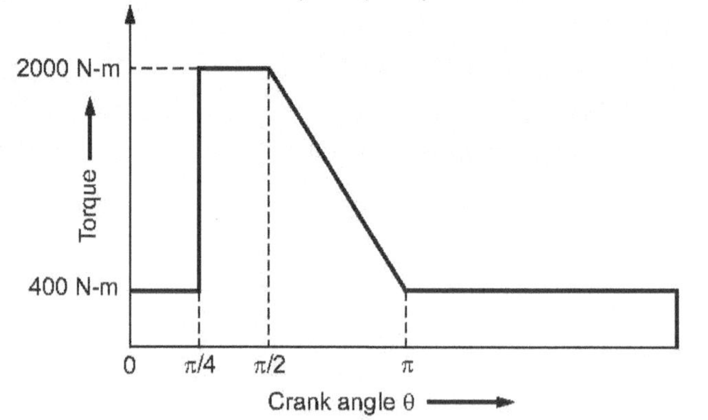

Fig. 8.15

Solution : Data : Crank angle per cycle = 2π radians

Mean speed n = 1500 rpm,

$$\text{Radius of gyration } K = \frac{40}{2} = 20\text{cm} = 0.2 \text{ m} \quad \text{(for rim type)}$$

$$m = 25 \text{ Kg}$$

$$\text{Angular velocity of flywheel } \omega = \frac{2\pi n}{60}$$

$$\text{i.e.} \quad \omega = \frac{2\pi \times 1500}{60} = 157.08 \text{ rad/sec}$$

(a) Power developed by the engine :

W.D./cycle = area 1 + area 2 + area 3

$$= 400 \times 2\pi + (2000 - 400)\left(\frac{\pi}{4}\right) + \left(\frac{1}{2}\right) \times \left(\frac{\pi}{2}\right) \times (2000 - 400)$$

W.D./cycle = 1600 π N – m

$$\text{Mean torque } T_m = \frac{\text{W. D per cycle}}{\text{Crank angle per cycle}} = \frac{1600\pi}{2\pi}$$

$$\Rightarrow \quad T_m = 800 \text{ N} - \text{m}$$

$\therefore$ Power developed by the engine

$$P = T_m \times \omega = 800 \times 157.08$$

$$= 125.664 \text{ KW}$$

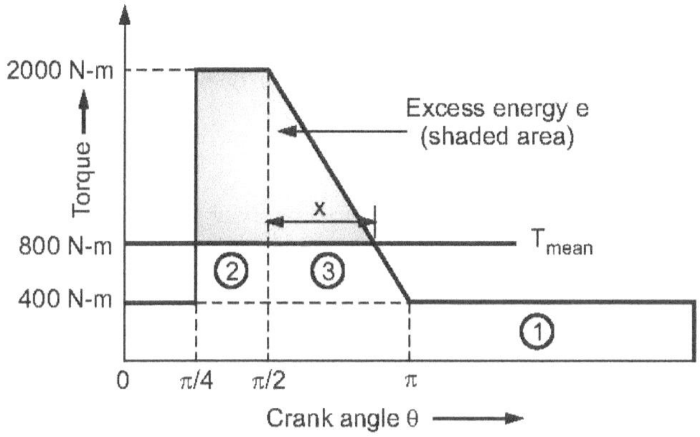

Fig. 8.16

(b) Coefficient of fluctuation of speed C_S

From the similar triangles,

$$\frac{x}{\left(\frac{\pi}{2}\right)} = \frac{1200}{1600} \Rightarrow x = 1.178 \text{ rad}$$

Energy stored by flywheel e = Shaded area

$$e = (2000 - 800) \times \frac{\pi}{4} + \frac{1}{2} \times 1.178 \times (2000 - 800)$$

$\therefore$ $e = 1649.28$ Nm

We know that energy stored by the flywheel

$$e = mk^2 \omega^2 C_s$$

$$1649.28 = 25 \times (0.2)^2 (157.08)^2 \times C_s$$

$\therefore$ Coefficient of fluctuation of speed = 0.0668 = 6.68 %

Example 8.5 : *A torque delivered by a two stroke is represented by T = (1000 + 300 sin 2θ – 500 cos 2θ) N-m where θ is the angle turned by crank from IDC. The engine speed is 250 rpm. The mass of the flywheel is 400 Kg and the radius of gyration is 400 mm. Determine*

(a) The power developed

(b) Total percentage fluctuation of speed

(c) The angular acceleration and retardation of flywheel when the crank has rotated through an angle of 60° from the IDC

(d) Maximum and Minimum angular acceleration and retardation of flywheel.

Solution : Data : As the torque is a function of 2θ,

equate $2\theta = 360° \Rightarrow \theta = 180°$

$\therefore$ The crank angle per cycle = 180° = π radians

Mean speed n = 250 rpm,

Radius of gyration k = 400 mm = 0.4 m, m = 400 Kg

Solution : Angular velocity of flywheel $\omega = \frac{2\pi n}{60}$

i.e. $$\omega = \frac{2\pi \times 250}{60} = 26.18 \text{ rad/sec}$$

(a) Power developed by engine :

$$\text{Mean torque } T_m = \frac{\text{W.D per cycle}}{\text{Crank angle per cycle}}$$

$$= \frac{1}{\pi} \int_0^{\pi} T d\theta$$

$$= \frac{1}{\pi} \int_0^{\pi} (1000 + 300 \sin 2\theta - 500 \cos 2\theta)\, d\theta$$

$\Rightarrow$ Mean torque,

$$T_m = 1000 \text{ N-m}$$

$\therefore$ Power developed by engine $P = T_m \times \omega$

$$P = 1000 \times 26.18 = 26.18 \text{ KW}$$

Sr. No.	Angle θ	Torque T N-m
1	0	500
2	30	1010
3	60	1510
4	90	1500
5	120	990
6	150	490
7	180	500

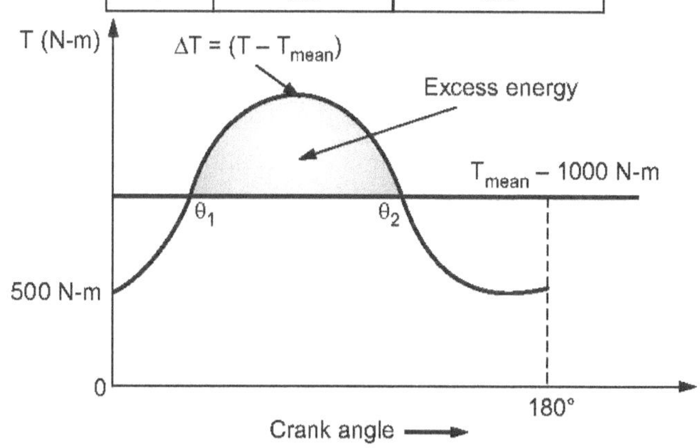

Fig. 8.17

(b) Coefficient of fluctuation of speed :

The excess energy stored by the flywheel is given by integrating ΔT between the limits θ_1 and θ_2 where θ_1 and θ_2 correspond to points where

$$T = T_{mean} \text{ or } (T - T_{mean}) = \Delta T = 0$$

(As the torque curve intercepts the mean torque line at these points)

i.e. (300 sin 2θ – 500 cos 2θ) = 0

$$\text{Hence } \tan 2\theta = \left(\frac{500}{300}\right) = 1.667$$

$\Rightarrow$ $2\theta_1 = 59°$ and $2\theta_2 = (180° + 59°) = 239°$

Hence, $\theta_1 = 29.5°$ and $\theta_2 = 119.5°$

$$\text{Excess energy } e = \int_{\theta_1}^{\theta_2} \Delta T \cdot d\theta$$

$$\Rightarrow \quad e = \int_{29.5}^{119.5} (300 \sin 2\theta - 500 \cos 2\theta)\, d\theta$$

(The above integration may be performed using calculator by keeping in radian mode and substituting the limits of integration in radians)

$\therefore \quad e = 583.1$ N–m

Also $e = mk^2 \omega^2 C_s \Rightarrow 583.1 = 400 \times (0.4)^2 \times (26.18)^2 \times C_s$

$\therefore$ Coefficient of fluctuation of speed $C_s = 0.0133 = 1.33$ %

(c) Angular acceleration at 60° crank position

Acceleration (or retardation) is caused by excess (or deficit) torque measured from mean torque at any instant. i.e. ΔT.

At $= \theta = 60°$, $\Delta T = 300 \sin (2 \times 60) - 500 \cos (2 \times 60)$

$\Rightarrow \quad \Delta T = 509.8$ N-m

Now $\quad \Delta T = I\alpha$

$\Rightarrow \quad 509.8 = \{400 \times (0.4)^2\} \times \alpha$

Hence Angular acceleration at 60° crank position

$$\alpha_{60°} = 7.965 \text{ rad/sec}^2$$

(d) Maximum angular acceleration :

Maximum acceleration (or retardation) is caused by maximum fluctuation of torque from mean, i.e. ΔT_{max} (To find ΔT_{max} first find the crank positions at which ΔT is maximum and then substitute those values in the equation of ΔT.)

For max value of ΔT, $\frac{d}{d\theta} (\Delta T) = 0$

$$\Rightarrow \frac{d}{d\theta} (300 \sin 2\theta - 500 \cos 2\theta) = 0$$

i.e. $600 \cos 2\theta + 1000 \sin 2\theta = 0$

Hence $\tan 2\theta = -0.6 \Rightarrow 2\theta = -31°$ and

$2\theta = (180° + (-31°)) = 149°$

At $2\theta = -31°$, $\Delta T = -583.1$ N-m (causes retardation)

At $2\theta = 149°$, $\Delta T = +583.1$ (Causes acceleration)

$\therefore$ Max angular acceleration of flywheel

$$\alpha_{max} = \frac{\Delta T_{max}}{I} = \frac{583.1}{400(0.4)^2} = 9.11 \text{ rad/sec}^2$$

$\therefore$ Max angular retardation of flywheel

$$\alpha_{min} = \frac{\Delta T_{min}}{I} = \frac{-583.1}{400(0.4)^2} = -9.11 \text{ rad/sec}^2$$

(– ve sign indicates retardation)

EXERCISE

1. What is the function of a flywheel ? How does it differ from that of a governer.
2. Discuss the turning moment diagram.
3. Prove that the maximum fluctuation of energy

   $$\Delta E = E \times 2\, Cs$$

   where E = Mean kinetic energy of the flywheel,

   cs = coefficient of fluctuation of speed.
4. Derive equations for
   (i) Stresses in disc flywheel
   (ii) Stresses in armed flywheel
5. The flywheel of a steam engine has a radius of gyration of 1 m and mass 2500 kg. The starting torque of the steam engine is 1500 N-m and may be assumed constant. Determine: the angular acceleration of the flywheel and the kinetic energy of the flywheel after 10 seconds from the start.
6. A single cylinder internal combustion engine working on the four stroke cycle develops 100 kW at 400 r.p.m. The fluctuation of energy can be assumed to be 0.9 times the energy developed per cycle. If the fluctuation of speed is not to exceed 1 per cent and the maximum centrifugal stress in the flywheel is to be 7 MPa, estimate the mean diameter and the cross-sectional area of the rim. The material of the rim has a density of 8.0 Mg/m^3.

◈◈◈

UNIT V - BRAKES AND CLUTCHES

CHAPTER - 9
BRAKES AND CLUTCHES

PART A : CLUTCHES

9.1 FRICTION CLUTCHES

- A friction clutch has its principal application in the transmission of power of shafts and machines which must be started and stopped frequently.
- In friction clutches, one should take care, so that the friction surfaces engage easily and gradually brings the driven shaft upto proper speed.
- The proper alignment of the bearing must be maintained and it should be located as close to the clutch as possible.

Requirements from Clutch Surfaces :

- Contact surfaces should develop a frictional force that can pick up and hold the load with reasonably low pressure between the contact surfaces.
- Heat of friction should be rapidly dissipated and tendency to grab should be at a minimum.
- The surfaces should be backed by a material stiff enough to ensure a reasonably uniform distribution of pressure.

"A clutch is a mechanical device used to engage or disengage the driven shaft from the driving shaft at the will of the operator while power is transmitted from driving to driven shaft".

- In friction clutches, the connection of the engine shaft to the gear box shaft is affected by friction between two or more concentric surfaces.
- The surfaces can be pressed firmly against one another when engaged and the clutch tends to rotate as a single unit.
- When the vehicle is to be stopped for a while or to change the gear, it is required that the driven shaft should stop but the engine should run naturally under the no load condition.
- This is achieved by using clutch mounted between engine shaft and gear box shaft and which is operated by a lever.

9.2 TYPES OF CLUTCHES

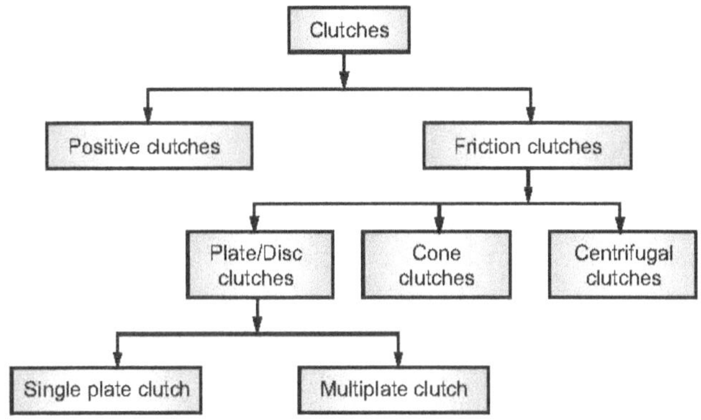

[A] Positive Clutches : Refer Fig. 9.1

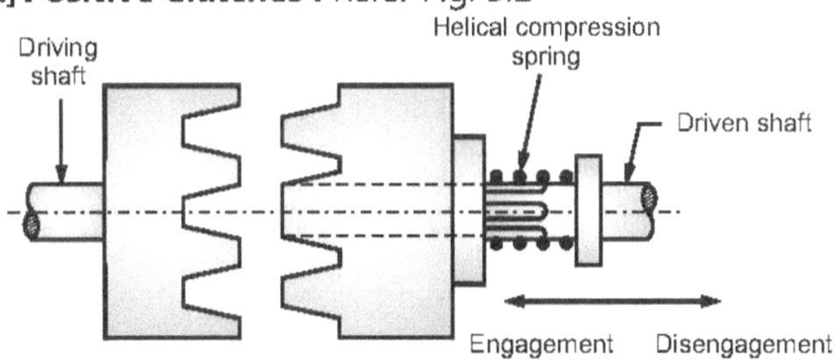

Fig. 9.1 : Gear tooth shaped jaw clutch

- The name in itself suggests that it is a positive type of drive and there is no loss of power due to slipping of friction surfaces as observed in the other type viz. friction clutches. Jaw clutch is the simplest type of positive clutches which transmits the torque from driving shaft to the driven shaft through interlocking jaws.
- Jaw clutch consists of two interlocking jaws, one jaw is permanently fastened to the driving shaft and the other jaw is free to slide axially on the splined driven shaft. The jaw mounted on the driven shaft side can slide axially on driven shaft enabling it to engage or disengage.

Jaw Shapes :

- **Square Shaped Jaws :** They move in either direction and are used when engagement and disengagement of clutch in motion and under load is not necessary.
- **Spiral or Gear Toothed Shaped Jaws :** They transmit motion in only one direction. Their use is restricted when engagement and disengagement of clutch in motion is required. Because of their profile like a gear tooth they offer better engagement and disengagement as compared to square shaped jaws.

Merits of Positive Clutches :

- They do not slip.
- No heat generation at clutch surface.

Demerits of Positive Clutches :

- Shock and noise is observed during engagement in motion.
- They cannot be aligned at high speeds and they can't be engaged at rest when the jaws are not properly aligned.

Application Areas of Positive (Jaw) Clutches :

Power presses, punches and some machine tools.

[B] Friction Clutches

The principle of friction clutch can be understood with the help of Fig. 9.2 (a) and (b).

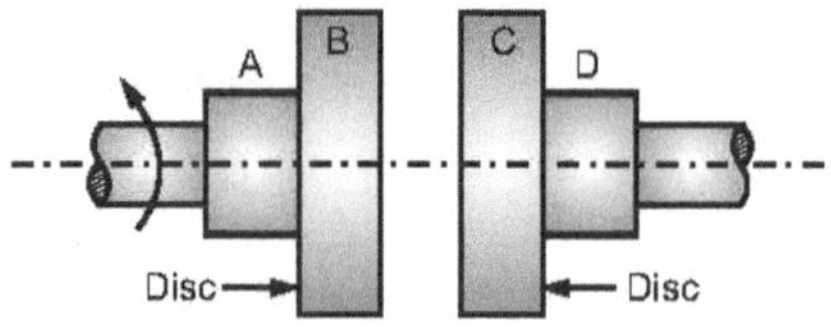

Fig. 9.2 (a) : Before engagement

In Fig. 9.2 (a), shaft A is rotating with some speed and disc B which is a part of it rotates with the same speed. Shaft D is stationary having a disc 'C' attached to it rigidly. When some axial force W is applied to the shaft D and it is brought in contact with the rotating shaft A as shown in Fig. 9.2 (b), the shaft D also starts rotating because of the force of friction between discs B and C. The force of friction depends upon the magnitude of axial force W. The speed of shaft D depends upon the value of W. For a particular value of W, the speed of shaft D is equal to the speed of shaft A, then the clutch is engaged.

Let W Axial load

μ = Coefficient of friction between discs

T = Torque transmitted

r = Effective mean radius of friction surface

So, $P = T \times \omega$

∴ P = Power lost in friction

ω = Angular speed of shaft

and $T = \mu \omega r$

So in this type of clutch, friction plays an important role and hence, the name friction clutch. These clutches are used primarily to engage or disengage the drive shaft from the driver as per the requirement.

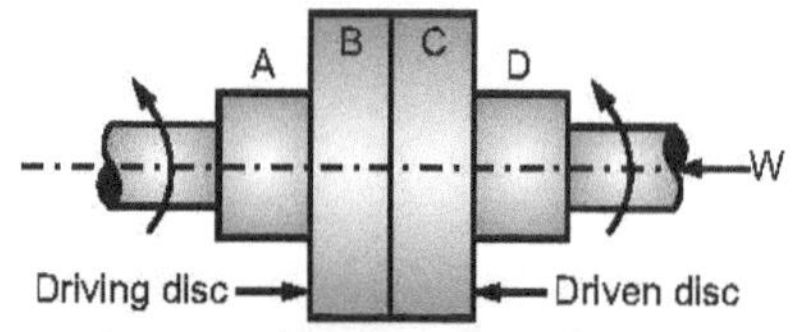

Fig. 9.2 (b) : After engagement

9.3 DISC CLUTCH (SINGLE-PLATE CLUTCH) / DRY CLUTCH

Refer Fig. 9.3.

Parts of a Single Plate Clutch :

- Driving or input shaft.
- Clutch Plate (or Friction Plate) - Driven plates with friction lining usually Ferrodo.
- Pressure Plate - Driving Plates.
- Driven or Output Shaft with Splines.
- Cover/Body.
- Spring.
- Flywheel.
- Clutch Pedal/Clutch Lever.
- Thrust Bearing.
- Disengagement Sleeve.

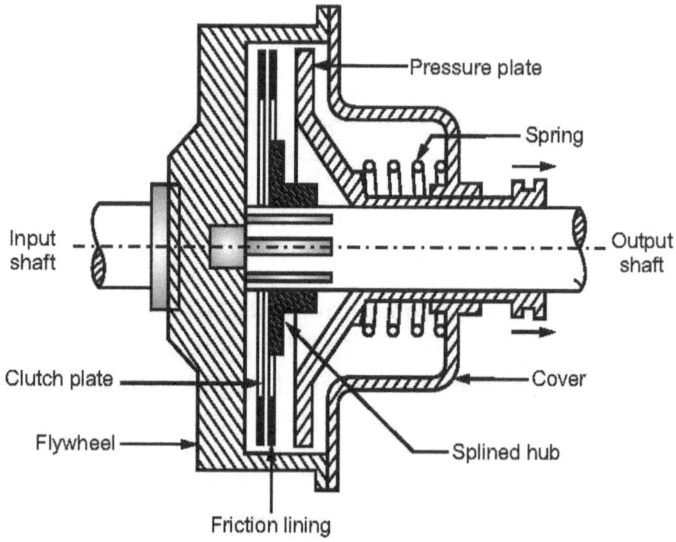

Fig. 9.3 : Disc clutch [Single-plate clutch]

Construction and Working :

- A disc clutch consists of a clutch plate [or friction plate] attached to a splined hub which is free to slide axially on splines cut on the driven shaft.
- The clutch plate is made of steel and has a ring at friction lining usually of ferrodo on each side. The engine shaft supports a rigidly fixed flywheel.
- When the clutch is engaged, the spring loaded pressure plate presses the clutch plate firmly against the flywheel. When disengaged, the springs press against a cover attached to the flywheel.

- Both flywheel and pressure plate rotate with the input shaft. When the clutch pedal is pressed/released the movement of clutch pedal, is transferred to the pressure plate through a thrust bearing.
- When the clutch pedal is pressed, the disengagement sleeve moves to the right and clutch is disengaged. Fig. 9.3 shows the pressure plates pulled back by release levers and the friction lining on the clutch plate are not in contact with the pressure plate or flywheel. The flywheel rotates without driving the clutch plate and thus, the driven shaft.
- When the clutch pedal is released, the pressure on the thrust bearing is released. The springs are now free to move the pressure plate to bring it in contact with the clutch plate.
- The clutch plate (or friction plate lined with ferrodo material) slides on the splined hub and is tightly gripped between the pressure plate and the flywheel. The friction between the lining on the clutch plate, and the flywheel on one side and the pressure plate on the other cause the clutch plate and hence the driven shaft to rotate.
- Clutch slip condition occurs when the resisting torque on the driven shaft exceeds the torque at the clutch.

  Application Area : They are used where large radial space is available such as trucks, buses etc. Here sufficient surface area is available for heat dissipation which is produced at the surfaces due to friction.

  As such, no cooling oil is required for cooling. Hence, single plate clutches (or disc clutches) are also called as **'Dry Clutches'.**

9.4 TORQUE TRANSMITTING CAPACITY OF SINGLE PLATE CLUTCH

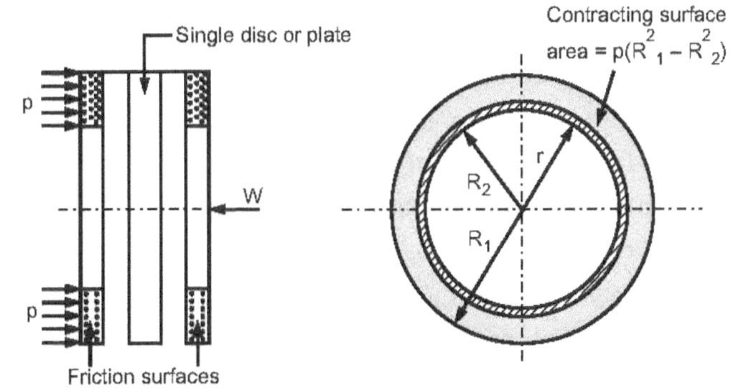

Fig. 9.4 : Forces on single plate/disc clutch

Note : Expression for torque transmitting capacity of single plate clutch is similar to multiple flat collar bearing.

Let,

T = Torque transmitted by clutch (N-m)

W = Axial load (or thrust) on clutch plate (N)

R_1 = External radius of clutch plate (m)

R_2 = Internal radius of clutch plate (m)

p = Intensity of pressure between contacting surfaces (N/m^2)

μ = Coefficient of friction between contacting surfaces.

Consider a small elemental ring of radius 'r' and thickness δr on the contacting surface.

Area of contacting ring is,

$$\delta A = 2\pi r \times \delta r \quad \ldots (9.1)$$

Normal or axial load on the ring is

$$\delta W = \text{Pressure} \times \text{Area of ring} = p \times 2\pi r\delta r \quad \ldots (9.2)$$

Frictional resistance offered by ring is,

$$F_r = \mu \cdot \delta W$$

$$\therefore \quad F_r = \mu \cdot p \cdot 2\pi r\delta r = 2\pi\mu p r\delta r$$

Frictional torque acting on the ring is,

$$T_r = F_r \cdot r$$

$$T_r = 2\pi\mu p r\delta r \cdot r$$

$$T_r = 2\pi\mu p r^2\delta r \quad \ldots (9.3)$$

Torque transmitting capacity of a single plate clutch can be obtained by using two assumptions :

1. Uniform pressure theory.
2. Uniform wear theory.

1. Uniform Pressure Theory : (Pressure, p = constant)

Assume that the intensity of pressure over the entire contacting surface area is constant, then p = constant.

i.e. pressure is uniformly distributed over the contacting surface area.

$$\therefore \quad p = \frac{\text{Load}}{\text{Contacting surface area}} = \frac{W}{\pi(R_1^2 - R_2^2)}$$

Total torque transmitted by single plate clutch,

$$\int_0^T T_r = \int_{R_2}^{R_1} 2\pi\mu p r^2\delta r$$

$$\therefore \quad T = 2\pi\mu p \int_{R_2}^{R_1} r^2\delta r \quad [\because p = \text{constant}]$$

$$\therefore \quad T = 2\pi\mu p \left[\frac{r^3}{3}\right]_{R_2}^{R_1}$$

$$\therefore \quad T = \frac{2\pi\mu p}{3} [R_1^3 - R_2^3]$$

$$\therefore \quad T = \frac{2}{3}\pi\mu \times \frac{W}{\pi(R_1^2 - R_2^2)} (R_1^3 - R_2^3)$$

$$\therefore \quad T = \frac{2}{3}\mu W \left[\frac{R_1^3 - R_2^3}{R_1^2 - R_2^2}\right] \quad \ldots (9.4)$$

Equation (9.30) gives the torque transmitted by single plate (Disc) clutch due to friction assuming uniform pressure theory.

2. **Uniform Wear Theory :** Wear rate is uniform over the entire contacting surface i.e.

$$p \cdot r = \text{constant}$$

$$\therefore \quad p \cdot r = C$$

$$\therefore \quad p = \frac{C}{r} \quad \ldots (9.5)$$

Axial load acting on the ring is,

$$\delta W = p \times 2\pi r \delta r = \frac{C}{r} \times 2\pi r \delta r$$

$$\therefore \quad \delta W = 2\pi C \delta r \quad \ldots (9.6)$$

Integrating equation (9.6) within limits, we get

$$\int_0^W \delta W = \int_{R_2}^{R_1} 2\pi C \delta r$$

$$\therefore \quad W = 2\pi C [r]_{R_2}^{R_1} = 2\pi C [R_1 - R_2]$$

$$\therefore \quad C = \frac{W}{2\pi(R_1 - R_2)} \quad \ldots (9.7)$$

From equation (9.3), frictional torque acting on the ring is,

$$T_r = 2\pi\mu p r^2 \delta r$$

$$T_r = 2\pi\mu \times \frac{C}{r} \times r^2 \delta r$$

$$T_r = 2\pi\mu C r \delta r \quad \ldots (9.8)$$

Torque transmitted by single plate clutch can be obtained by integrating equation (9.8), within its limits,

$$\int_0^T T_r = \int_{R_2}^{R_1} 2\pi\mu C r \delta r$$

$$\therefore \quad T = 2\pi\mu C \left[\frac{r^2}{2}\right]_{R_2}^{R_1}$$

$$\therefore \quad T = \frac{2\pi\mu C}{2} [R_1^2 - R_2^2]$$

$$\therefore \quad T = \pi\mu C [R_1^2 - R_2^2]$$

$$\therefore \quad T = \pi\mu \times \frac{W}{2\pi(R_1 - R_2)} (R_1^2 - R_2^2)$$

[Using equation (9.8)]

$$\therefore \quad T = \frac{1}{2}\mu W (R_1 + R_2) \quad \ldots (9.9)$$

Equation (9.9) gives the torque transmitted by single plate (disc) clutch due to friction assuming uniform wear theory.

Power transmitted by single plate clutch is

$$P = \frac{2\pi NT}{60} = \omega T \text{ watt} \quad \ldots (9.10)$$

where, N = Speed of driving shaft

9.5 MULTIPLATE CLUTCH

- A multiplate disc clutch may be used when a large torque is to be transmitted.
- The inside discs (clutch plates usually of steel) are fastened to the driven shaft to permit axial motion (except for the last disc).
- The outside discs (pressure plate usually of bronze) are held by bolts and are fastened to the housing which is keyed to the driving shaft.
- As number of frictional linings and the metal plates is increased which increases the capacity of the clutch to transmit torque.
- The friction rings are splined on their outer circumference and engage with corresponding splines on the flywheel. They are free to slide axially.
- The friction material rotates with the flywheel and the engine shaft, the number of friction rings depends upon the torque to be transmitted.
- The driven shaft also supports discs on the splines which rotate with the driven shaft and can slide axially.
- When the clutch pedal is released, the spring presses the discs into contact with the friction rings and the torque is transmitted between the engine shaft and the driven shaft.

Important : If 'n' is the total number of plates both on the driving and the driven members, the number of active surfaces will be (n – 1). First friction plate has friction lining on one side only while all others have both side friction lining.

- For given torque capacity radial size of multiplate clutch is smaller than that of single plate clutch.

Therefore, it is compact in construction and hence heat dissipation Example is severe. So, in order to overcome this Example cooling oil is used to overcome heat dissipation Example.

So multiple clutch is also sometimes called as '**Wet Clutch'.**

- In case of multiplate clutches, two sets of plates of the driving plates and driven plates are arranged alternately. It consist of more than one driving as well as driven plates. As such, number of pair of contacting surfaces is more than two.
- Both friction plates and pressure plates are splined but the former is splined on outer circumference and engage with corresponding splines on the casing and the letter is splined from inside and engage with corresponding splines on the driven shaft. Both friction plates and pressure plates are free to slide axially.
- Working of multiplate clutch is similar to single plate clutch :
  - ➢ Press the clutch pedal, clutch disengages as the sleeve moves to right. Driving shaft alone rotates. No axial force between pressure plates and friction plates.
  - ➢ Release the clutch pedal, clutch engages as the sleeve moves towards left. Pressure plate and clutch plates come close due to spring force. Power transmission takes place from driving to driven shaft.

Application Area : Extensively used in motor cars, machine tools etc.

9.5.1 Torque Transmitting Capacity of Multiple Clutch

$$T = n \times \frac{2}{3} \mu W \left[\frac{R_1^3 - R_2^3}{R_1^2 - R_2^2} \right] \quad \dots (9.11)$$

Considering uniform pressure theory, and

$$T = n \times \frac{1}{2} \mu W (R_1 + R_2) \quad \dots (9.12)$$

considering **uniform wear theory.**

Here, n = Number of pairs of contact surfaces

$$P = \frac{2\pi NT}{60}$$

N = Speed of shaft (rpm)

∵ P = Power (W)

T = Torque (N-m)

$$n = n_1 + n_2 - 1$$

∵ n_1 = Number of plates on driving shaft

n_2 = Number of plates on driven shaft

Note : (1) Power is transmitted by clutches. So for safer design, it is assumed that less amount of power is transmitted and hence, from above equation (9.11) and (9.12) equations, uniform wear theory is used as it gives less torque and in turn less power.

(2) In case of a new clutch, the intensity of pressure is approximately uniform.

Hence, for a new clutch uniform pressure theory is used but for an old and worn out clutch uniform wear theory is applicable.

(3) Both the sides of the plate are effective in case of a single plate clutch normally. Hence, single plate clutch has two pairs of surfaces in contact i.e. n = 2.

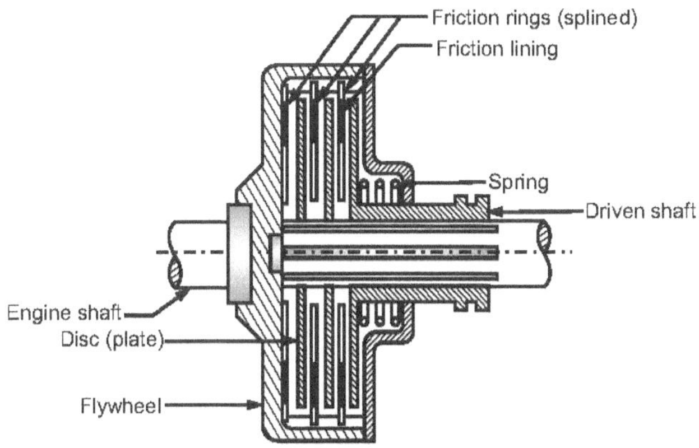

Fig. 9.5 : Multiplate clutch

9.6 CONE CLUTCH

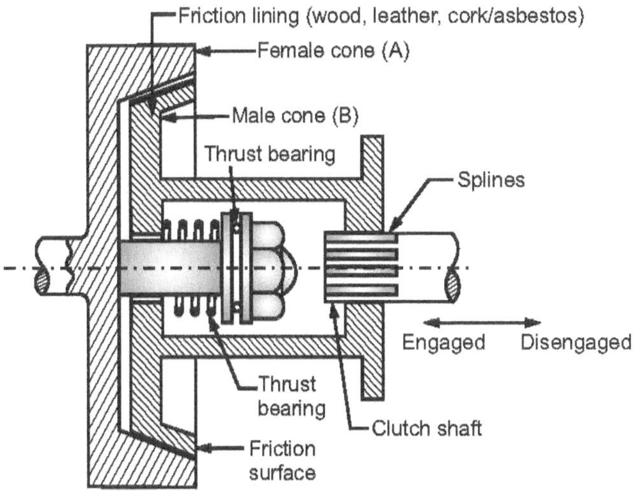

Fig. 9.6 : Cone clutch

- In a cone clutch, contact surfaces are in the shape of cones. The male and female cones are in contact when the clutch is engaged as shown in Fig. 9.6.

- The contact is complete and tight with the help of springs.
- Cone clutch was extensively used in automobiles but now a days has been replaced completely by the disc clutch.
- Reason of cone clutches becoming obsolete is that they have very small cone angles (α varying from 8° to 15°) and exposure to dust and dirt tend to bind the two cones and it becomes difficult to disengage them.
- In a cone clutch, the driver is keyed to the driving shaft by a sunk key and has an inside conical surface or face which exactly fits into the outside conical surface of the driven.
- The driven member resting on the feather key in the driven shaft, may be shifted along the shaft by a forked lever provided, in order to engage the clutch by bringing the two conical surfaces in contact.
- Due to the frictional resistance set up at the contact surface, torque transmission takes place from one shaft to another. Contact surfaces of the clutch may be metal to metal contact but most of the times, the driven members is lined with material like wood, leather, cork or asbestos etc. material of the clutch faces depends upon the allowable normal pressure and the coefficient of friction.
- Clutch is disengaged by means of fork fitted in shifting collar or groove.
- When the clutch pedal is pressed down the male cone 'B' is pulled back through a lever system against the force of the spring.
- When the clutch pedal is released [clutch pedal position upwards], the friction surfaces of the two female (cone A) cone and male cone (cone B) are in complete contact due to spring pressure that keeps one cone pressed against the other all time. During this engaged position, the torque transmission takes place from the driving shaft to the driven shaft through the flywheel and the friction cones.

Merits :

- Cone clutches are more compact than a disc clutch.
- They have better torque transmitting capacity than disc clutches.
- Wedging action due to conical working surfaces results into considerable normal pressure and friction force with small engaging force.

Demerits :

- Tendency of sticking of male and female cone.
- Small amount of wear on conical surface, results in considerable axial play between clutch plate and flywheel.

Applications : Earlier they were used in larger vehicles, such as trucks and buses. They are obsolete these days because of demerits stated above.

9.6.1 Torque Transmitting Capacity of Cone Clutch

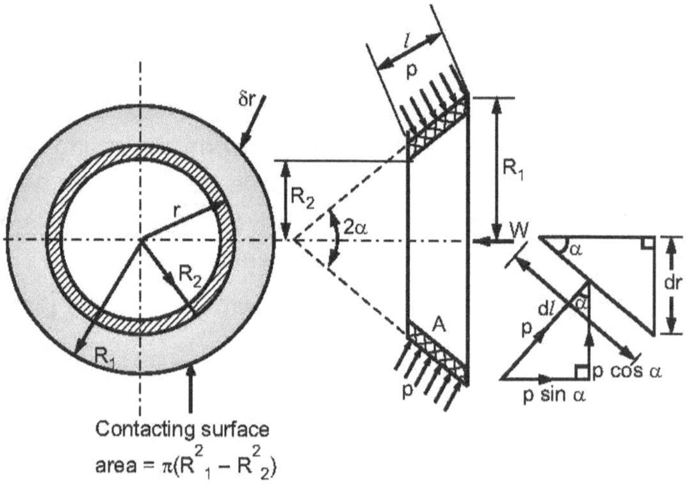

(a) Cone clutch analysis **(b) Enlarged view of 'A'**

Fig. 9.7

Fig. 9.7 shows a cone having friction lining on the outer surface.

Let,

W = Axial load on cone clutch (N)

p = Intensity of pressure between contacting surfaces in N/mm²

R_1 and R_2 = Outer and inner radius of cone respectively (mm)

l = Width of contact surfaces or face width (mm)

2α = Cone angle

α = Semi-cone angle or face angle of cone

μ = Coefficient of friction

Consider on elemental ring of radius 'r' and thickness 'δr'. Let 'δr' be the length of ring of friction surface such that,

$$\sin\alpha = \frac{\delta r}{\delta l} \text{ or } \delta l = \frac{\delta r}{\sin\alpha}$$

$$\text{Area of ring is,} dA = 2\pi r \cdot \delta r = 2\pi r \times \frac{\delta r}{\sin\alpha}$$

Normal load acting on the ring is,

$$\delta W_n = \text{Normal pressure} \times \text{Area}$$

$$= p \times 2\pi r \frac{\delta r}{\sin \alpha} \qquad \ldots (9.13)$$

Horizontal load acting on the ring is,

$$dW = \text{Horizontal component of pressure} \times \text{Area}$$

$$= p \sin \alpha \times 2\pi r \frac{\delta r}{\sin \alpha} = 2\pi pr\, \delta r \qquad \ldots (9.14)$$

Frictional force or frictional resistance offered by the ring is,

$$dF_r = \mu \times \delta W_n = \mu \times p \times 2\pi r \cdot \frac{\delta r}{\sin \alpha}$$

...... Using equation (9.13)

Corresponding frictional torque on the ring is,

$$dT_r = dF_r \times r = \mu_p \times 2\pi r \cdot \frac{\delta r}{\sin \alpha} \times r$$

$$\therefore \quad dT_r = 2\pi\mu pr^2 \cdot \frac{\delta r}{\sin \alpha} \qquad \ldots (9.15)$$

Equation of torque transmitted can be obtained by using the uniform pressure and uniform wear assumptions.

[A] Uniform Pressure Theory (Pressure p = Constant)

Assumption for this theory, pressure is uniformly distributed over entire contacting surface area is constant. i.e. p = constant.

$$\text{But,} \quad p = \frac{\text{Load}}{\text{Contacting surface area}}$$

$$p = \frac{W}{\pi(R_1^2 - R_2^2)} \qquad \ldots (9.16)$$

To obtain the torque transmitting capacity, we integrate equation (9.15) within limits R_2 to R_1 of the variable radius.

$$\int_0^T dT_r = \int_{R_2}^{R_1} \frac{2\pi\mu pr^2}{\sin \alpha} \delta r$$

$$\therefore \quad [Tr]_0^T = \frac{2\pi\mu p}{\sin \alpha}\left[\frac{r^3}{3}\right]_{R_2}^{R_1}$$

$$= \frac{2}{3}\frac{\pi\mu p}{\sin \alpha}[R_1^3 - R_2^3]$$

$$= \frac{2}{3}\frac{\pi\mu}{\sin \alpha} \times \frac{W}{\pi(R_1^2 - R_2^2)}(R_1^3 - R_2^3)$$

$$T = \frac{2}{3}\frac{\mu W}{\sin \alpha}\left[\frac{R_1^3 - R_2^3}{R_1^2 - R_2^2}\right] \qquad \ldots (9.17)$$

Equation (9.17) represents the torque transmitting capacity of cone clutch by using uniform pressure theory.

[B] Uniform Wear Theory (Pressure p = Constant)

Underlying principle of this theory is wear over the entire contacting surface is constant i.e. $p \cdot r$ = constant = C or $p = \frac{C}{r}$.

We know horizontal load acting on the ring is,

$$dW = 2\pi pr\, \delta r$$

$$\therefore \quad dW = 2\pi \frac{C}{r} \times r\, \delta r$$

$$\therefore \quad dW = 2\pi C \delta r \qquad \ldots (9.18)$$

Integrating equation (9.18) within limits of variable R_2 to R_1.

$$\int_0^W \delta W = \int_{R_2}^{R_1} 2\pi C\, dr$$

$$\therefore \quad W = 2\pi C\, [r]_{R_2}^{R_1}$$

$$\therefore \quad W = 2\pi C\, [R_1 - R_2]$$

$$\therefore \quad C = \frac{W}{2\pi(R_1 - R_2)} \qquad \ldots (9.19)$$

Frictional torque acting on the ring is

$$dT_r = 2\pi\mu pr^2 \cdot \frac{\delta r}{\sin \alpha}$$

$$\therefore \quad dT_r = 2\pi\mu \frac{C}{r} \times r^2 \times \frac{\delta r}{\sin \alpha}$$

$$\therefore \quad dT_r = 2\pi\mu Cr \cdot \frac{\delta r}{\sin \alpha}$$

To get torque transmitting capacity of cone clutch, we have to integrate the above equation within the limits R_2 to R_1.

$$\int_0^T dT_r = \int_{R_2}^{R_1} 2\pi\mu Cr \cdot \frac{dr}{\sin \alpha}$$

$$\therefore \quad \int_0^T dT_r = \frac{2\pi\mu C}{\sin \alpha}\int_{R_2}^{R_1} r\, \delta r$$

$$\therefore \quad (T_r)_0^T = \frac{2\pi\mu C}{\sin \alpha}\left[\frac{r^2}{2}\right]_{R_2}^{R_1}$$

$$\therefore \quad T = \frac{2\pi\mu C}{\sin \alpha}\frac{(R_1^2 - R_2^2)}{2}$$

$$\therefore \qquad T = \frac{\pi\mu C}{\sin\alpha}(R_1^2 - R_2^2)$$

$$\therefore \qquad T = \frac{\pi\mu}{\sin\alpha} \times \frac{W}{2\pi(R_1 - R_2)} \times (R_1^2 - R_2^2) \quad \ldots (9.20)$$

Using value of C from equation (9.20)

$$\therefore \qquad T = \frac{\mu}{\sin\alpha}\frac{W}{2(R_1 - R_2)}(R_1 + R_2)(R_1 - R_2)$$

$$\therefore \qquad \boxed{T = \frac{1}{2}\frac{\mu W}{\sin\alpha}(R_1 + R_2)} \quad \ldots (9.21)$$

Equation (9.21) gives us the torque transmitting capacity of a cone clutch using uniform wear theory.

9.7 CENTRIFUGAL CLUTCH

Centrifugal Clutch:

It is required to engage a load after the driving member attained a particular speed. In that condition, the centrifugal clutch is used. This clutch is helpful with an engine that has trouble starting under load. They are usually used in motor pulleys.

The centrifugal clutch consists of a number of shoes or friction pads arranged radially symmetrical position inside the rim. It can slide along the guides integral with the boss on the driving shaft. The shoes are held against boss by using a spring that exerts a radially inward force. As the inner hub rotates, the weight of the shoe causes a radially outward force known as centrifugal force. This force depends on the weight of the shoe and the speed at which it rotates.

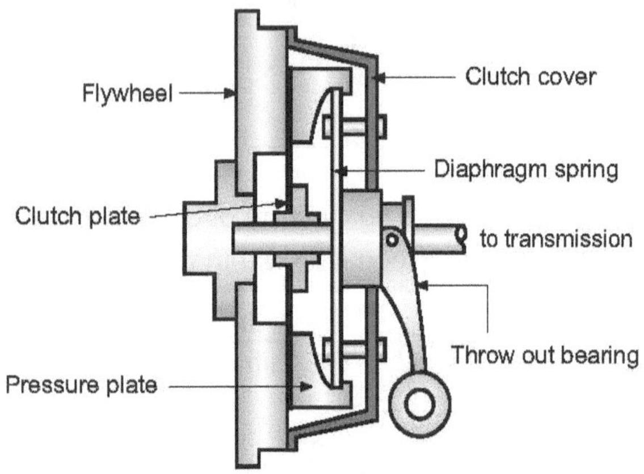

Fig. 9.8

At low speed, the centrifugal force also low, the shoes remain in the same position. As speed increases, the centrifugal force also increases, when centrifugal force becomes equal to spring force the shoes start floating. When the driver rotates fast enough the centrifugal force exceeds the spring force the shoes moves outward. At a certain speed, it gets contact with the inner surface of the drum and torque is transmitted. As the load increases, speed decreases; the shoes return to their original position and clutch gets disengaged

Application of Centrifugal Clutch

Centrifugal clutch found application in the chain saw, lawn mower. The centrifugal clutch engages after a particular speed limit, so they used to disengage the engine when starting and idling.

Advantages and Disadvantages of Centrifugal Clutch

Advantage :

- Simple and inexpensive and need little maintenance.
- The centrifugal clutch is automatic any kind of control mechanism is not necessary.
- They help to prevent the engine from stalling.
- The engagement speed can precisely control by selecting spring.

Disadvantage :

- Loss of power due to friction and slipping.
- This type of clutch not appropriate for the high amount of torque, the shoes will slip at the heavy loaded condition.
- They engage at full or near-full power, shoes get heated very quickly may cause overheating.

9.7.1 Torque Transmitting Capacity of Centrifugal Clutch

(1) Mass of the Shoes :

Consider one shoe of a centrifugal clutch as shown in Fig. 9.9.

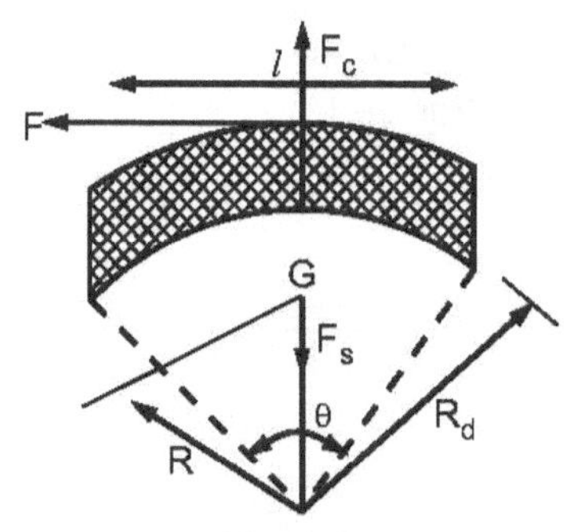

Fig. 9.9

Let,

m = Mass of each shoe (kg)

n = Number of shoes

R_g = Distance of centre of gravity of the shoe from the centre of spider (mm)

R_D = Inside radius of the pulley rim (mm)

N = Running speed of pulley (rpm)

ω = Angular speed of pulley (rad/sec)

ω_1 = Angular speed at which engagement begins to take place

and μ = Coefficient of friction between shoe and rim

We know that centrifugal force acting on each shoe at the running speed is,

$$F_C = mr\omega^2 = mR_g\,\omega^2 \quad \dots (9.22)$$

Inward force on each shoe exerted by the spring at the speed at which engagement begins to take place,

$$F_C = m \cdot \omega_1^2 \cdot R_g \quad \dots (9.23)$$

The net radial force with which the shoe presses against the rim is,

$$\text{Net force} = F_C - F_S \quad \dots (9.24)$$

Frictional force acting tangentially on each shoe is,

$$F = \mu \times \text{Net force} = \mu \times (F_C - F_S) \quad \dots (9.25)$$

Frictional torque on each shoe is

$$T = F \times R_D = \mu \times (F_C - F_S) \times R_D$$

Similarly, total frictional torque transmitted by clutch is,

$$T = n \times \mu\,(F_C - F_S) \times R_D \quad \dots (9.26)$$

By use of equation (9.26), mass of the shoe may be evaluated.

(2) Size of the Shoes :

Let, l = Contact length of the shoes (mm)

b = Width of the shoes (mm)

R = Contact radius of shoes. It is same as inside radius of rim of pulley (mm)

θ = Angle subtended by shoes at the centre of spider (Rad.)

p = Intensity of pressure exerted on shoe

In order to ensure reasonable life, intensity of pressure may be taken as 0.1 N/mm^2.

We know, $\theta = \dfrac{l}{R}$ rad or $l = \theta \cdot R$

$\therefore$ Area of contact of shoe

$$A = l \cdot b$$

and the force with which the shoe presses against rim

$$= A \times p = l \cdot b \cdot p.$$

Since the force with which shoe presses against the rim at the running speed is $(F_C - F_S)$

$$\therefore \quad l \cdot b \cdot p = F_C - F_S \quad \dots (9.27)$$

From this expression, the width of the shoe (b) may be obtained.

Table 9.1 : Showing Equations for all Types of Clutches

Sr. No.	Type of Clutch	Uniform Pressure Theory (p = Constant)	Uniform Wear Theory (p · r = constant)
1.	Single plate clutch	$p = \dfrac{W}{\pi(R_1^2 - R_2^2)}$	$T = \dfrac{1}{2}\mu W\,(R_1 + R_2)$
2.	Multiplate clutch	$T = n \times \dfrac{2}{3}\mu W \left[\dfrac{R_1^3 - R_2^3}{R_1^2 - R_2^2}\right]$	$T = n \times \dfrac{1}{2}\mu W\,(R_1 + R_2)$
3.	Cone clutch	$p = \dfrac{W}{\pi(R_1^2 - R_2^2)}$ $T = \dfrac{2}{3}\dfrac{\mu W}{\sin\alpha}\left[\dfrac{R_1^3 - R_2^3}{R_1^2 - R_2^2}\right]$	$T = \dfrac{1}{2}\dfrac{\mu W}{\sin\alpha}(R_1 + R_2)$
4.	Centrifugal clutch	$T = n \times \mu\,(F_C - F_S) \times R_D$ $F_C - F_S = p \times l \times b$	

SOLVED EXAMPLES

Example 9.1 : *An automobile single plate clutch consists of a pair of contacting surfaces. The inner and outer radii of friction plate are 120 mm and 250 mm respectively. Coefficient of friction is 0.25 and the total axial force is 15 kN. Calculate the power transmitting capacity of the clutch at 500 rpm by using (1) uniform wear theory, (2) uniform pressure theory.*

Solution : Data : Number of contacting surfaces, n = 2

Inner radius of friction plate R_2 = 120 mm = 0.12 m

Outer radius of friction plate R_1 = 250 mm = 0.25 m

Coefficient of friction μ = 0.25

Axial force, W = 15 kN = 15 × 10^3 N

Speed of engine N = 500 rpm

To Find : Power transmitting capacity

Power Transmitting Capacity :

1. Uniform Wear Theory :

$$T = n \times \frac{1}{2}\mu W\,(R_1 + R_2)$$

$$= 2 \times \frac{1}{2} \times 0.25 \times 15 \times 10^3\,(0.25 + 0.12)$$

$$T = 1387.5 \text{ N-m}$$

$$\text{Power, } P = \frac{2\pi NT}{60} = \frac{2\pi \times 500 \times 1387.5}{60} \text{ W}$$

$$P = 72.6493 \text{ kW}$$

2. Uniform Pressure Theory : As per uniform pressure theory,

$$T = n \times \frac{2}{3}\mu W \left[\frac{R_1^3 - R_2^3}{R_1^2 - R_2^2}\right]$$

$$\therefore \quad T = \frac{2 \times 2}{3} \times 0.25 \times 15 \times 10^3 \left[\frac{(0.25)^3 - (0.12)^3}{(0.25)^2 - (0.12)^2}\right]$$

$$\therefore \quad T = 5000 \times \left[\frac{0.013897}{0.0481}\right]$$

$$\therefore \quad T = 1444.594 \text{ N-m}$$

$$\text{Power, } P = \frac{2\pi NT}{60} = \frac{2\pi \times 500 \times 1444.594}{60}$$

$$P = 75.6387 \text{ kW}$$

Example 9.2 : *A single plate clutch transmits 25 kW at 900 rpm. The maximum pressure intensity is 85 kN/m². The outer diameter of plate is 360 mm with both sides effective and coefficient of friction 0.25, find the inner diameter of plate and the axial force to engage the clutch.*

Solution : Power, $P = 25 \text{ kW} = 25 \times 10^3$ W

Speed of clutch,

$$N = 900 \text{ rpm}$$

Maximum intensity of pressure $p_{max} = 85 \text{ kN/m}^2$

$$= 85 \times 10^3 \text{ N/m}^2$$

Outer diameter,

$$D_1 = 360 \text{ mm} = 0.36 \text{ m}$$

$$\text{Outer radius, } R_1 = \frac{0.36}{2} = 0.18 \text{ m}$$

Number of effective sides = n = 2

Coefficient of friction = μ = 0.25

To Find : Inner diameter D_2

(1) Torque Transmitting Capacity of Clutch :

$$P = \frac{2\pi NT}{60}$$

$$\therefore \quad 25 \times 10^3 = \frac{2\pi \times 900 \times T}{60}$$

$$\therefore \quad T = 265.2582 \text{ N-m}$$

As per uniform wear theory, pressure is maximum at inner radius i.e.

$$p_{max} = \frac{C}{R_2} \qquad \therefore C = 85 \times 10^3 \times R_2$$

$$\text{Axial load, } W = 2\pi C (R_1 - R_2)$$

$$= 2\pi \times 85 \times 10^3 R_2 (0.18 - R_2)$$

$$W = 534.07 \times 10^3 R_2 (0.18 - R_2)$$

(2) Inner Diameter of Plate :

$$\text{Torque T (by uniform wear theory)} = n \times \frac{1}{2}\mu W (R_1 + R_2)$$

$$\therefore \quad 265.2582 = 2 \times \frac{1}{2} \times 0.25 \times 534.07 \times 10^3 R_2 (0.18 - R_2) \times (0.18 + R_2)$$

$$\therefore \quad 1.98669 \times 10^{-3} = R_2 (0.18^2 - R_2^2)$$

$$\therefore \quad 1.98669 \times 10^{-3} = 0.0324 R_2 - R_2^3$$

$$\therefore \quad R_2^3 - 0.0324 R_2 + 1.98669 \times 10^{-3} = 0$$

$$R_2 = 0.1315, -0.2051 \text{ m and } 0.0736 \text{ m}$$

Selecting larger value of R_2,

$$R_2 = 0.1315 \text{ m} = 131.5 \text{ mm}$$

$$D_2 = 2 \times R_2 = 2 \times 131.5 \text{ mm}$$

$$D_2 = 263 \text{ mm}$$

So,

$$W = 534.07 \times 10^3 R_2 (0.18 - R_2)$$

$$= 534.07 \times 10^3 \times 0.1315 [0.18 - 0.1315]$$

$$W = 3406.1649 \text{ N}$$

Example 9.3 : *A conical clutch has a cone angle of 30°, if the maximum intensity of pressure between the contact surface is limited to 70 kPa and the width of conical surface is not to exceed one third of the mean radius, find the dimensions of the contact surfaces to transmit 22 kW at 2000 rpm assuming uniform wear and take coefficient of friction as 0.15.*

Solution : Data : Cone clutch : Cone angle = 2α = 30°

$\therefore$ Semi-cone angle = α = 15°,

p = Maximum intensity of pressure = 70×10^3 Pa,

Width of conical surface, $l = \frac{1}{3}$ (mean radius) $= \frac{R_m}{3}$.

$$\therefore \quad l = \frac{1}{3}\left[\frac{R_1 + R_2}{2}\right] = \left(\frac{R_1 + R_2}{6}\right)$$

Power $P = 22 \times 10^3$ W, Speed of clutch = N = 2000 rpm, μ = coefficient of friction = 0.15

To Find : Dimensions of the conical surface.

Power, $P = \frac{2\pi NT}{60}$ $\therefore\ 22 \times 10^3 = \frac{2\pi \times 2000 \times T}{60}$

$\therefore$ $T = 105.0422$ N-m

As per uniform wear theory,

$$T = \frac{1}{2}\frac{\mu W}{\sin\alpha}(R_1 + R_2) = \frac{\mu W}{\sin\alpha} \times R_m \quad \ldots (1)$$

$$\left(\because \frac{R_1 + R_2}{2} = R_m\right)$$

and $$P = \frac{W}{\pi(R_1^2 - R_2^2)} = \frac{W}{\pi(R_1 - R_2)(R_1 + R_2)}$$

$$P = \frac{W}{\pi(R_1 - R_2) \times 2R_m} = \frac{W}{\pi \times l \sin\alpha \times 2R_m}$$

(as $(R_1 - R_2) = l \sin\alpha$)

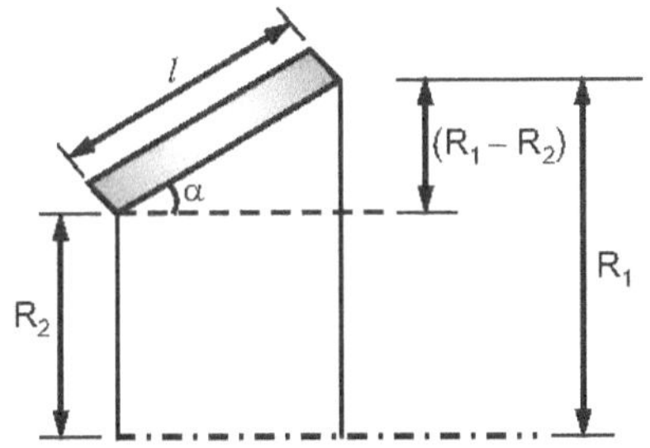

Fig. 9.10

From equation (1), $\sin\alpha = \frac{R_1 - R_2}{l}$

$\therefore$ $(R_1 - R_2) = l \sin\alpha$

$\therefore$ $W = 2\pi \times \frac{R_m}{3} \sin\alpha \times R_m \times p$

Substituting this value in equation (1),

$$T = \frac{\mu}{\sin\alpha} \times R_m \times \frac{2\pi}{3} R_m^2 \times p \sin\alpha$$

$\therefore$ $T = \frac{2}{3}\pi\mu p\, R_m^2$

$\therefore$ $105.042 = \frac{2}{3} \times \pi \times 0.15 \times 70 \times 10^3 \times R_m^3$

$\therefore$ $R_m = 0.16841$

$$l = \frac{R_m}{3} = \frac{0.16841}{3} = 0.056137 \text{ m}$$

$= 56.1371$ mm

From Fig. 9.10, we can write,

$$\sin\alpha = \frac{R_1 - R_2}{l} \qquad \therefore l = \frac{R_1 - R_2}{\sin 15°}$$

Also, $$l = \frac{R_1 + R_2}{6}$$

$\therefore$ $$\frac{R_1 + R_2}{6} = \frac{R_1 - R_2}{\sin 15°}$$

$\therefore$ $$\frac{R_1 + R_2}{6} = \frac{R_1 - R_2}{0.2588}$$

$\therefore$ 0.2588 R_1 to 2588 $R_2 = 6R_1 - 6R_2$

$\therefore$ $6.2588\, R_2 = 5.7412\, R_1$

$\therefore$ $R_2 = 0.9173\, R_1$ ... (2)

Also, $T = \frac{1}{2}\mu W (R_1 + R_2) \operatorname{cosec}\alpha$

$\therefore$ $105.04 = \frac{1}{2} \times 0.15 \times 439.82 \times 10^3\, R_2 (R_1 - R_2) \operatorname{cosec} 15°$

$\therefore$ $8.2416 \times 10^{-4} = R_2 (R_1^2 - R_2^2)$ ... (3)

Putting value of R_2 from equation (2) in equation (3),

$8.2416 \times 10^{-4} = 0.9173\, R_1 [R_1^2 - (0.9173)^2 R_1]$

$\therefore$ $8.2416 \times 10^{-4} = 0.9173\, R_1^3 - 0.7718\, R_1^3$

$\therefore$ $8.24616 \times 10^{-4} = 0.1455\, R_1^3$

$\therefore$ $R_1 = 0.17829$ m

$R_2 = 0.16354$ m

Length of contact surface is,

$$l = \frac{R_1 + R_2}{6}$$

$$= \frac{0.17829 + 0.16354}{6} = 0.05697 \text{ m}$$

$l = 56.97$ mm

Example 9.4 : *Determine the axial force required to engage a cone clutch transmitting 25 kW power at 750 rpm. Average friction diameter of the cone is 400 rpm, semicone angle 10° coefficient of friction 0.25.*

Solution : Power transmitted = P = 25000 W

Speed of clutch = N = 750 rpm

Average diameter,

$D_m = 400 \text{ mm} = 400 \times 10^{-3}$ m

Semicone angle,

$\alpha = 10°$ and

μ = Coefficient of friction = 0.25

$$P = \frac{2\pi NT}{60}$$

$\therefore$ $$25000 = \frac{2\pi \times 750 \times T}{60}$$

$T = 318.309$ N-m

Torque transmitted by cone clutch considering uniform wear theory is,

$$T = \frac{1}{2} \mu W (R_1 + R_2) \operatorname{cosec} \alpha$$

$$\therefore \quad 318.309 = \frac{1}{2} \times 0.25 \times W \times 0.4 \times \operatorname{cosec} 10°$$

$$W = 1105.475 \text{ N}$$

Example 9.5 : *A cone clutch transmits 20 kW at 1600 rpm. The following data apply. Cone angle = 30°, maximum intensity of pressure = 0.8 × 10^5 N/m^2, the mean radius is twice the width of the friction surface. Coefficient of friction is 0.3. Determine :*

(1) The dimensions of the contact surfaces.

(2) The axial load or force to keep the clutch engaged when transmitting power.

(3) The width of the friction surface.

Assume uniform wear.

Solution : Data : Cone Clutch : Cone angle 2α = 30°

∴ Semicone angle = α = 15°,

maximum intensity of pressure, $p_{max} = 0.8 \times 10^5$ N/m^2,

Width of conical surface,

$$l = \frac{1}{2} R_m = \frac{1}{2}\left(\frac{R_1 + R_2}{2}\right) = \frac{R_1 + R_2}{4}$$

Power transmitted,

$$P = 20 \text{ kW} = 20 \times 10^3 \text{ watt}$$

Speed of clutch,

$$N = 1600 \text{ rpm}$$

Coefficient of friction,

$$\mu = 0.3$$

Power transmitted by cone clutch is

$$P = \frac{2\pi NT}{60}$$

$$\therefore \quad 20 \times 10^3 = \frac{2\pi \times 1600 \times T}{60}$$

$$\therefore \quad T = 119.366 \text{ N-m}$$

Maximum intensity of pressure is,

$$p_{max} = \frac{C}{r_{min}} = \frac{C}{R_2} \qquad \therefore 0.8 \times 10^5 = \frac{C}{R_2}$$

$$\therefore \quad C = 0.8 \times 10^5 R_2$$

Axial load on contacting surface is,

$$W = 2\pi C (R_1 - R_2)$$

$$= 2\pi \times 0.8 \times 10^5 R_2 (R_1 - R_2)$$

$$= 502.65 \times 10^3 R_2 (R_1 - R_2) \qquad \ldots(1)$$

Torque transmitted by clutch using uniform wear theory is,

$$T = \frac{1}{2} \mu W (R_1 + R_2) \operatorname{cosec} \alpha$$

$$119.366 = \frac{1}{2} \times 0.3 \times 502.65 \times 10^3 R_2 (R_1 - R_2) (R_1 + R_2) \operatorname{cosec} 15°$$

$$\therefore \quad 4.0975 \times 10^{-4} = R_2 (R_1^2 - R_2^2) \qquad \ldots (2)$$

Width l of contacting surface is $l = \frac{R_1 + R_2}{4}$ ... (3)

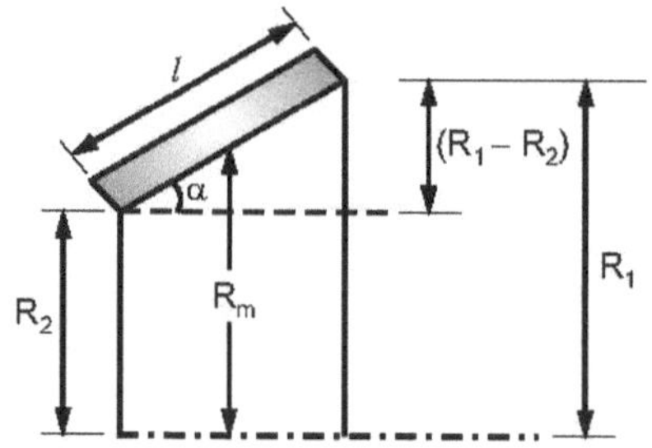

Fig. 9.11

From Fig. 9.11,

$$\sin \alpha = \frac{R_1 - R_2}{l}$$

$$\therefore \quad l = \frac{R_1 - R_2}{\sin 15°} \qquad \ldots (4)$$

Equating equations (2) and (3),

$$\frac{R_1 + R_2}{4} = \frac{R_1 - R_2}{\sin 15°}$$

$$\therefore \quad \frac{R_1 + R_2}{4} = \frac{R_1 - R_2}{0.2588}$$

$$\therefore 0.2588 R_1 + 0.2588 R_2 = 4R_1 - 4R_2$$

$$\therefore \quad 4.2588 R_2 = 3.7412 R_1$$

$$\therefore \quad R_2 = 0.878463 R_1 \qquad \ldots (5)$$

Putting value of R_2 from equation (4) in equation (2),

$$4.0975 \times 10^{-4} = 0.878463 R_2^3 [R_1^2 - (0.87846)^2 R^2]$$

$$\therefore 4.0975 \times 10^{-4} = 0.878463 R_1^3 - 0.67790 R_1^3$$

$$\therefore \quad R_1 = 0.2736 \text{ m} = 273.629 \text{ mm}$$

$$R_2 = 0.8784 \times 273.629$$

$$R_2 = 240.356 \text{ mm}$$

l = Length of contact surface

$$= \frac{R_1 + R_2}{4} = \frac{273.623 + 240.356}{4}$$

$$l = 128.49 \text{ mm}$$

From Equation (1)

$$w = 502.65 \times 10^3 R_2 (R_1 - R_2)$$
$$= 502.65 \times 10^3 \times 240.356 (273.629 - 240.356)$$
$$= 4.0198 \text{ kN}$$

Example 9.6 : *A centrifugal clutch transmits 15 kW power at 900 rpm. The shoes are four in number and engages with the drum at 75% of running speed of input shaft. The inner radius of drum is 15 cm and the radial distance of centre of gravity of shoe from centre of rotation of spider in engaged position is 12 cm. Take coefficient of friction as 0.25 and maximum permissible pressure intensity as 0.1 N/mm². Determine : (1) Mass of each shoe. (2) Size of shoes if angle subtended by the shoes at the centre of spider is 60°.*

Solution : Data : Centrifugal Clutch :

Power, P = 15 kW; Speed of clutch = N = 900 rpm, Number of shoes = n = 4.

Angular speed of clutch $= \omega = \frac{2\pi N}{60} = \frac{2\pi \times 900}{60}$

$= 94.2477$ rad/sec

Angular speed of engagement $= 0.75 \times \omega$

$= 70.68577$ rad/s

Inner radius of drum $= R_D = 0.15$ m

Radius of C.G. of shoe in engaged position,

$R_g = 0.12$ m.

Coefficient of friction $= \mu = 0.25$

Intensity of pressure $= p = 0.1 \text{ N/mm}^2 = 0.1 \times 10^6 \text{ N/m}^2$

Angle subtended by the shoes,

$$\theta = 60° = \frac{60 \times \pi}{180} = 1.04719 \text{ rad}$$

To Find : (1) Mass of each shoe = m. (2) Size of shoe.

$$\text{power, } P = \frac{2\pi NT}{60} = T \cdot \omega$$

$\therefore \quad 15 \times 10^3 = T \times 94.2477$

$\therefore \quad T = 159.155$ N-m

Centrifugal force acting on each shoe is,

$$F_c = m R_g \omega^2 = m \times 0.12 \times (94.2477)^2$$

$\therefore \quad F_c = m\,[1065.91547]$ N

And centrifugal force at the engagement speed is,

$$F_s = m R_g \omega_1^2 = m \times 0.12 \times (70.6857)^2$$

$\therefore \quad F_s = 599.5761\, m$ N

Torque transmitted by centrifugal clutch is

$$T = n \times \mu \times (F_c - F_s) \times R_d$$

$\therefore \quad 159.155 = 4 \times 0.25 \times m\,[1065.19547 - 599.5761] \times 0.15$

$\therefore \quad 159.155 = 69.9509$ m

$\therefore \quad m$ = Mass of each shoe

$= 2.275238$ kg

Length of arc is, $l = \theta \times R_D$

$l = 1.0471 \times 0.15$

$= 0.157$ m

Force with which shoe is pressed against the rim is

$$F_c - F_s = p \times l \times b$$

$\therefore \quad (1065.9154 - 599.5761) \times 2.275238 = 0.1 \times 10^6 \times 0.157 \times b$

$\therefore \quad b = 0.06758$ m

Example 9.7 : *A centrifugal clutch transmits 25 kW power at 900 rpm. The speed at which engagement begins is 70% of the running speed. The inner radius of drum is 15 cm and the radial distance of the centre of gravity of shoe beam the axis of rotation of the spider in the engaged position is 12 cm. Determine mass of each shoe. Take number of shoes as four and μ = 0.25 and intensity of pressure as 0.1 N/mm².*

Solution : Data : Centrifugal Clutch :

Power, $P = 25 \text{ kW} = 25 \times 10^3$ watts

Speed of clutch,

N = 900 rpm

Angular speed of clutch = ω

$$= \frac{2\pi N}{60} = \frac{2\pi \times 900}{60} = 94.2477 \text{ rad/sec}$$

Speed of engagement,

$$\omega = 0.7 \times \omega = 0.7 \times 94.2477$$

$\therefore \quad \omega = 65.97344$ rad/s

Inner radius of drum,

$$R_d = 15 \times 10^{-2} \text{ m} = 0.15 \text{ m}$$

Radius of C.G. of shoe in engaged position,

$R_g = 12$ cm

$\therefore \quad R_g = 0.12$ m

Coefficient of friction = μ

= 0.25

Intensity of pressure,

$$p = 0.1 \text{ N/mm}^2 = 0.1 \times 10^6 \text{ N/m}^2$$

To Find : Mass of each shoe, m.

$$\text{Power, } P = \frac{2\pi NT}{60} = T \cdot \omega$$

$$\therefore \quad 25 \times 10^3 = T \times 94.2477$$

$$\therefore \quad T = 265.258 \text{ N-m}$$

Centrifugal force acting on each shoe,

$$F_c = m\, R_g\, \omega^2$$

$$= m \times 0.12 \times (94.2477)^2$$

$$F_c = 1065.915 \text{ m, N}$$

Centrifugal force at engagement speed is,

$$F_s = m \cdot R_g \cdot \omega_1^2 = m \times 0.12 \times (65.9734)^2$$

$$= 522.2987 \text{ m N}$$

Torque transmitted by centrifugal clutch is,

$$T = n \times \mu \times (F_c - F_s) \times R_d$$

$$\therefore \quad 265.2584 = 4 \times 0.25 \times (1065.9154\text{ m} - 522.2987\text{ m}) \times 0.15$$

$$\therefore \quad 265.2584 = 81.542505 \text{ m}$$

$$\therefore \quad m = 3.253 \text{ kg} = \text{mass of each shoe}$$

Example 9.8 : *A centrifugal clutch has four shoes which slide radially in a spider keyed to the driving shaft and make contact with the internal surface of a rim keyed to the driven shaft. When the clutch is at rest, each shoe is pulled against a stop by using a spring and leave a clearance of 9 mm between the shoe and rim. The pull force exerted by the spring is 600 N. The C.G. of the shoe is 180 mm from the axis of the clutch. The internal diameter of the rim is 450 mm, stiffness of each spring is 50 kN/m. If the mass of each shoe is 10 kg and coefficient of friction between the rim and shoe is 0.4. Find the power transmitted by the clutch. Take speed of shaft as 600 rpm.*

Solution : Data :

Number of shoes, $n = 4$,

Clearance between shoe and rim, $C = 9 \text{ mm} = 0.009 \text{ m}$,

Pull force of spring, $F_s = 600$ N,

Radius up to C.G. of the shoe, $R_g = 180 \text{ mm} = 0.18 \text{ m}$,

Internal diameter of the rim, $D_d = 450 \text{ mm} = 0.45 \text{ m}$,

Stiffness of spring, $k = 50 \text{ kN/m} = 50 \times 0^3 \text{ N/m}$,

Mass of each shoe, $m = 10$ kg,

Coefficient of friction, $\mu = 0.4$,

Speed of shaft, $N = 600$ rpm.

$\therefore$ Angular speed of shaft,

$$\omega = \frac{2\pi N}{60} = \frac{2\pi \times 600}{60} = 62.8318 \text{ rad/s}$$

To Find : Power P.

Radius of C.G. of shoe during engagement is,

$$R_{g_1} = R_g + c = 0.18 + 0.009 = 0.189 \text{ m}$$

Centrifugal force on each shoe is,

$$F_c = m\, R_{g_1}\, \omega^2 = 10 \times 0.189 \times (62.8318)^2$$

$$\therefore \quad F_c = 7461.408 \text{ N}$$

Total spring force during engagement is,

$$F_s = \text{Pull force of spring} + \text{Force due to stiffness}$$

$$F_s = 600 + (\text{Stiffness} \times \text{Radial clearance})$$

$$F_s = 600 + 50 \times 10^3 \times 0.009 = 1050 \text{ N}$$

Frictional force acting tangentially on each shoe,

$$\therefore \quad F = \mu\,(F_c - F_s)$$

$$\therefore \quad F = 0.4 \times (7461.408 - 1050)$$

$$\therefore \quad F = 2564.563 \text{ N}$$

Torque transmitted by the clutch is,

$$T = n \times F \times R_d = n \times F \times \frac{D_d}{2}$$

$$\therefore \quad T = 4 \times 2564.563 \times \frac{0.45}{2}$$

$$\therefore \quad T = 2308.1067 \text{ N-m}$$

Power transmitted by the clutch is,

$$P = T \times \omega = 2308.1067 \times 62.8318$$

$$\therefore \quad P = 145.02249 \times 10^3 \text{ W}$$

$$P = 145.022 \text{ kW}$$

Example 9.9 : *Determine the minimum value for the radius ratio $\left(\frac{R_2}{R_1}\right)$ of a single plate clutch at which the capacity of clutch will decrease by not more than 10% during the initial wear period.*

Solution : We know that, toque transmitted by the clutch is,

$$T = \frac{2}{3} \times n\mu W \left[\frac{R_1^3 - R_2^3}{R_1^2 - R_2^2}\right] \quad \text{... Uniform pressure theory}$$

$$\text{and} \quad T = \frac{1}{2} n\mu W\,(R_1 + R_2) \quad \text{... Uniform wear theory}$$

Assumption : A new clutch before initial wear period will have uniform pressure distribution.

Given : Capacity of clutch during initial wear = 0.9 × Uniform pressure.

$$(\text{Torque})_{\text{Uniform wear}} = 0.9 \times (\text{Torque})_{\text{Uniform pressure}}$$

$$n \times \frac{1}{2} \mu W (R_1 + R_2) = 0.9 \times n \times \frac{2}{3} \times \mu W \left[\frac{R_1^3 - R_2^3}{R_1^2 - R_2^2}\right]$$

$$\therefore \quad (R_1 + R_2) = 1.2 \times \left[\frac{R_1^3 - R_2^3}{R_1^2 - R_2^2}\right]$$

$$\therefore \quad R_1\left[1 + \frac{R_2}{R_1}\right] = \frac{1.2 \times R_1^3}{R_1^2}\left[\frac{1 - \left(\frac{R_2}{R_1}\right)^3}{1 - \left(\frac{R_2}{R_1}\right)^2}\right]$$

$$\therefore \quad \left(1 + \frac{R_2}{R_1}\right) = 1.2 \times \left[\frac{1 - \left(\frac{R_2}{R_1}\right)^3}{1 - \left(\frac{R_2}{R_1}\right)^2}\right]$$

Let $\frac{R_2}{R_1} = x$

$$\therefore \quad (1 + x) = 1.2 \times \left[\frac{1 - x^3}{1 - x^2}\right]$$

$$(1 + x)(1 - x^2) = 1.2 \times (1 - x^3)$$

$$\therefore (1 + x)(1 + x)(1 - x) = 1.2 \times (1 - x^3)$$

$$\therefore (1 + x)(1 - x)(1 + x) = 1.2 \times (1 - x)(1 + x + x^2)$$

$$\therefore \quad (1 + 2x + x^2) = 1.2 \times (1 + x + x^2)$$

$$\therefore \quad 0.2x^2 - 0.8x + 0.2 = 0$$

Solving above quadratic equation,

If $ax^2 + bx + c = 0$ then $x = \frac{-b \pm \sqrt{b^2 - 4ac}}{2a}$

We get, $x = 3.732$ and $x = 0.2679$

As ratio of $\frac{R_2}{R_1}$ should never be greater than one,

$$x = 0.2679 = \frac{R_2}{R_1}$$

Example 9.10 : *Determine the time required to accelerate a counter shaft of rotating mass 500 kg and radius of gyration of 200 mm to the full speed of 250 rpm from rest through a single plate clutch of internal and external radii 125 mm and 200 mm, taking coefficient of friction as 0.3 and axial spring force of 600 N. Assume that only one side of clutch is effective.*

Solution : Data : Single Plate Clutch :

Mass, m = 500 kg, N = 250 rpm,

k = Radius gyration = 200 mm = 0.2 m

R_1 = External radius of plate = 200 mm = 0.2 m

R_2 = Internal radius of plate = 125 mm = 0.125 m

μ = Coefficient of friction = 0.3

W = Axial spring force = 600 N

n = Number of effective sides of plate = 1

To Find : t = time.

Torque transmitted by clutch,

$$T = n \times \frac{1}{2} \mu W (R_1 + R_2)$$

$$= 1 \times \frac{1}{2} \times 0.3 \times 600 \times (0.2 + 0.125)$$

$$= 29.25 \text{ N-m}$$

Mass moment of inertia = $I = mk^2$

$$\therefore \quad I = 500 \times (0.2)^2 = 20 \text{ kg-m}^2$$

Angular acceleration,

$$\alpha = \frac{T}{I} = \frac{29.25}{20} = 1.46 \text{ rad/s}^2 \quad (\because T = I\alpha)$$

Assume, that this acceleration uniform and let 't' be the time taken by clutch to reach to the speed of 250 rpm,

$\omega_{final} = \omega_{initial} + \alpha \cdot t$ (Similar to v = u + at in linear quantities)

$$\frac{2\pi \times N_{final}}{60} = 0 + 1.46 \times t$$

$$\therefore \quad \frac{2\pi \times 250}{60} = 0 + 1.46 \times t$$

$$\therefore \quad t = 17.93 \text{ sec}$$

Example 9.11 : *A single plate clutch, effective on both sides, is required to transmit 25 kW at 3000 rpm. Determine the outer and inner radii of frictional surface if the coefficient of friction is 0.225, the ratio of radii is 1.25 and the maximum pressure is not to exceed 0.1×10^6 N/m². Also determine the axial thrust to be provided by springs. Assume uniform wear theory.*

Solution : Data : A Singe Plate Clutch :

P = Power transmitted = 25 kW = 25000 W

n = Number of effective substances = 2

N = Speed = 3000 rpm

μ = Coefficient of friction = 0.225

$\frac{R_1}{R_2}$ = Ratio of radii = 1.25

p_{max} = Maximum pressure intensity

$= 0.1 \times 10^6 \text{ N/m}^2$

To Find : W = Axial thrust

Assumption : Uniform wear theory is to be assumed.

$$\omega = \frac{2\pi N}{60} = \frac{2\pi \times 3000}{60} = 314.159 \text{ rad/s}$$

$$\text{Power, } P = \omega T = 314.159 \times T$$

$$\therefore \quad 25000 = 314.159 \times T$$

$$\therefore \quad T = 79.577 \text{ N-m}$$

Torque transmitted by clutch considering uniform wear theory is,

$$T = n \times \frac{1}{2} \mu W (R_1 + R_2)$$

$$79.57 = 2 \times \frac{1}{2} \times 0.225 \times W (1.25 R_2 + R_2)$$

$$\therefore \quad 353.644 = W (2.25 R_2)$$

$$\therefore \quad 353.644 = 2.25 WR_2 \quad \ldots (1)$$

Maximum axial thrust,

$$W = 2\pi \cdot c (R_1 - R_2) = 2\pi \times c (1.25 R_2 - R_2)$$

$$\frac{353.644}{2.25 R_2} = 2\pi c R_2$$

$$\therefore \quad 353.644 = 14.137 R_2^2 c \quad \ldots (2)$$

Maximum pressure intensity,

$$p_{max} = \frac{c}{R_{min}} = \frac{c}{R_2}$$

$$\therefore \quad c = p_{max} \times R_2 = 0.1 \times 10^6 \times R_2 \quad \ldots (3)$$

Use value of c from equation (3) into equation (2),

$$353.644 = 14.137 R_2^2 \times 0.1 \times 10^6 \times R_2$$

$$\therefore \quad R_2^3 = 2.50154 \times 10^{-4}$$

$$\therefore \quad R_2 = 0.063 \text{ m}$$

$$R_1 = 1.25 \times R_2 = 1.25 \times 0.063 = 0.07876 \text{ m}$$

Maximum axial thrust is,

$$W = 2\pi \times 0.1 \times 10^6 \times (0.063) \times 0.25 (0.063)$$

$$= 623.445 \text{ N}$$

Example 9.12 : *A friction clutch is used to rotate a machine from a shaft rotating at a uniform speed of 240 rpm. The disc type clutch has both of its sides effective, the coefficient of friction being 0.3. The outer and the inner diameters of the friction plate are 200 mm and 120 mm respectively. Assuming uniform wear of the clutch, the intensity of pressure is not to be more than 100 kN/m². If the moment of inertia of the rotating parts of the machine is 5.5 kg.m², determine the time to attain the full speed by the machine and the energy lost in slipping of the clutch.*

What will be the intensity of pressure if the condition of uniform pressure of the clutch is considered ? Also, determine the ratio of power transmitted with uniform wear to that with uniform pressure.

Solution : Data : Disc Clutch :

N = 240 rpm, n = Number of effective sides = 2

μ = Coefficient of friction = 0.3

R_1 = External radius of plate

$= 100 \times 10^{-3}$ m

R_2 = Internal radius of plate = 60×10^{-3} m

p_{max} = Maximum pressure intensity

$= 100 \times 10^3 \text{ N/m}^2$

I = Moment of inertia of rotating parts

$= 5.5 \text{ kg-m}^2$

To Find : t = Time required to attain full speed

$= \dfrac{\text{Power with uniform wear}}{\text{Power with uniform perssure}}$

(1) With uniform wear conditions,

W = Maximum axial thrust

$$\therefore \quad W = 2\pi C (R_1 - R_2)$$

$$\text{But} \quad C = p_{max} \times R_2$$

$$\therefore \quad W = 2\pi \times p_{max} \times R_2 (R_1 - R_2)$$

$$= 2\pi \times 100 \times 10^3 \times 60 \times 10^{-3} \times [100 \times 10^{-3} - 60 \times 10^{-3}]$$

$$= 1507.96 \text{ N}$$

$$T = n \times \frac{1}{2} \mu W (R_1 + R_2)$$

$$= 2 \times \frac{1}{2} \times 0.3 \times 1507.96 \times (100 + 60) \times 10^{-3}$$

$$= 72.3820 \text{ N-m}$$

$$P = T \times \omega$$

$$= 72.38 \times \frac{2\pi \times 240}{60} = 1819.1078 \text{ watts}$$

$$\text{Also,} \quad T = I \times \alpha \quad (\alpha \text{ is angular acceleration})$$

$$72.38 = 5.5 \times \alpha$$

$$\therefore \quad \alpha = 13.16 \text{ rad/s}^2$$

$$\frac{\omega}{t} = 13.16 \text{ rad/s}^2$$

$$\frac{2\pi \times 240}{60 \times t} = 13.16 \quad \left(\because \omega = \frac{2\pi N}{60}\right)$$

$\therefore \quad t = 1.909$ sec

So, the full speed is attained by the machine is 1.909 sec.

(2) During the slipping period.

Angle turned by driving shaft,

$$\theta_1 = \omega t = \frac{2\pi N}{60} \times t = \frac{2\pi \times 240}{60} \times 1.909$$

$$= 47.978 \text{ rad}$$

Angle turned by driven shaft,

$$\theta_2 = \omega_0 t + \frac{1}{2}\alpha t^2$$

(Similar to $s = ut + \frac{1}{2}at^2$ in linear units)

$$\therefore \quad \theta_2 = 0 + \frac{1}{2} \times 13.16 \times (1.909)^2$$

$$\therefore \quad \theta_2 = 23.979 \text{ rad}$$

$$\text{Energy lost in friction} = T(\theta_1 - \theta_2)$$

$$= 72.3820\ [47.978 - 23.979]$$

$$= 1737.095 \text{ N-m}$$

$$= 1.737095 \text{ kN-m}$$

(3) With uniform pressure theory,

$$p = \frac{W}{\pi (R_1^2 - R_2^2)} = \frac{1507.96}{\pi [(0.1)^2 - (0.06)^2]}$$

$$\therefore \quad p = 74999.77 \text{ N/m}^2$$

$$= 74.999 \text{ kN/m}^2 \approx 75 \text{ kN/m}^2$$

and

$$T = n \times \frac{2}{3}\mu W \left[\frac{R_1^3 - R_2^3}{R_1^2 - R_2^2}\right]$$

$$= 2 \times \frac{2}{3} \times 0.3 \times 1507.96 \left[\frac{(0.1)^3 - (0.06)^3}{(0.1)^2 - (0.06)^2}\right]$$

$$= 603.184\ [0.1225]$$

$$= 73.890 \text{ N-m}$$

$$P = T \times \omega$$

$$= 73.89 \times \frac{2\pi \times 240}{60} = 1857.058 \text{ W}$$

$$\frac{\text{Power with uniform wear}}{\text{Power with uniform pressure}} = \frac{1815.1078}{1857.058} = 0.979$$

***Example 9.13 :** A friction clutch is used to rotate a machine from a shaft rotating at a uniform speed of 250 rpm. The disc-type clutch has both of its sides effective, the coefficient of friction being 0.3. The outer and the inner diameter of the friction plate are 200 mm and 120 mm respectively. Assuming uniform wear of the clutch, the intensity of pressure not to be more than 100 kN/m². If the moment of inertia of the rotating parts of the machine is 6.5 kg.m², determine the time to attain the full speed by the machine and the energy lost in slipping of the clutch.*

What will be the intensity of pressure if the condition of uniform pressure of the clutch is considered ? Also, determine the ratio of power transmitted with uniform wear to that with uniform pressure.

Solution :

$$p_{max} = 100 \times 10^3 \text{ N/m}^2$$

$$R_1 = 0.1 \text{ m}$$

$$I = 6.5 \text{ kg.m}^2$$

$$R_2 = 0.06 \text{ m}$$

$$\mu = 0.3$$

$$N = 250 \text{ rpm}$$

$$n = 2 \text{ (both sides effective)}$$

(1) With Uniform Wear :

Force surface,

$$W = 2\pi P_{max} \times R_2 (R_1 - R_2)$$

$$= 2\pi \times 100 \times 10^3 \times 0.06\ (0.1 - 0.06)$$

$$= 1508 \text{ N}$$

$$T = \frac{\mu W}{2}(R_1 + R_2) \times n$$

$$= \frac{0.3 \times 1508}{2}(0.1 + 0.06) \times 2$$

$$= 72.38 \text{ N.m}$$

$$p = T \times \omega = 72.38 \times \frac{2\pi \times 250}{60} = 1895 \text{ W}$$

Also, $T = I\alpha$ (α is angular acceleration)

$$72.38 = 6.5 \times \alpha$$

or $\alpha = 11.135 \text{ rad/s}^2$

or $\frac{\omega}{t} = 11.135$

or $\frac{2\pi \times 250}{60 \times t} = 11.135$

or $t = 2.35$ s

Thus, the full speed is attained by the machine in 2.35 seconds.

(2) During the Slipping Period :

Angle turned by the driving shaft,

$$\theta_1 = \omega t = \frac{2\pi N}{60} t = \frac{2\pi \times 250}{60} \times 2.35$$

$$= 61.5 \text{ rad}$$

Angle turned by the driven shaft,

$$\theta_2 = \omega_o t + \frac{1}{2}\alpha t^2 \qquad (\omega_o = 0)$$

$$= 0 + \frac{1}{2} \times 11.135 \times (2.35)^2$$

$$= 30.75 \text{ rad}$$

Energy lost in friction $= T(\theta_1 - \theta_2)$

$$= 72.38 \times (61.5 - 30.75)$$

$$= 2226 \text{ N.m or } 2.226 \text{ kN.m}$$

(3) With Uniform Pressure :

$$P = \frac{F}{\pi\left(R_1^2 - R_2^2\right)} = \frac{1508}{\pi\,[(0.1)^2 - (0.06)^2]}$$

$$= 75000 \text{ N/m}^2 \text{ or } 75 \text{ kN/m}^2$$

$$T = \frac{2}{3}\mu W\left[\frac{R_1^3 - R_2^3}{R_1^2 - R_2^2}\right] \times n$$

$$= \frac{2}{3} \times 0.3 \times 1508\left[\frac{(0.1)^3 - (0.06)^3}{(0.1)^2 - (0.06)^2}\right] \times 2$$

$$= 73.89 \times \frac{2\pi \times 250}{60} = 1934 \text{ W}$$

$$\frac{\text{Power with uniform wear}}{\text{Power with unifor pressure}} = \frac{1895}{1934} = 0.98$$

Example 9.14 : *A single-plate clutch used to drive a rotor from the motor shaft has the following data :*

Internal diameter of the plate = 200 mm

External diameter of the plate = 240 mm

Spring force pressing the plates = 600 N

Mass of the rotor = 1200 kg

Radius of gyration of the rotor = 200 mm

Mass of motor armature and shaft = 750 kg

Radius of gyration of motor and shaft = 220 mm

Coefficient of friction = 0.32

Both sides of the plate are effective. The driving motor is brought to a speed of 1260 rpm and then suddenly the current is switched off and the clutch is engaged. Determine the :

(i) final speed of the rotor and the motor and the time to attain this speed

(ii) kinetic energy lost during slipping

(iii) slipping time, if a constant resisting torque of 64 N-m exists on the armature shaft

(iv) slipping time if instead of a resisting torque, a constant driving torque of 64 N.m exists on the armature shaft.

Solution : $\omega = \frac{2\pi N}{60} = \frac{2\pi \times 1260}{60} = 132$ rad/s

Moment of inertia of motor,

$$I_m = 750 \times (0.22)^2 = 36.3 \text{ kg.m}^2$$

Moment of inertia of rotor,

$$I_r = 1200 \times (0.2)^2 = 48 \text{ kg.m}^2$$

Friction torque,

$$T = \frac{\mu\omega}{2}(R_1 + R_2) \times n$$

$$= \frac{0.32 \times 600}{2}(0.12 + 0.1) \times 2$$

$$= 42.24 \text{ N.m}$$

(1) Final speed of motor = Final speed of rotor

or $\omega_m + \alpha_m t = \omega_r + \alpha_r t$

or $\omega_m - \frac{T}{I_m}t = \omega_r + \frac{T}{I_r}t$

or $132 - \frac{4224}{36.3}t = 0 + \frac{42.24}{48}t$

or $2.0436\,t = 132$

$$t = 64.6 \text{ s}$$

$$\text{Final speed} = \frac{42.24}{48}t = \frac{42.24}{48} \times 64.6$$

$$= 56.85 \text{ rad/s}$$

$$= \frac{60}{2\pi} \times 56.85 = 452.9 \text{ rpm}$$

(2) Angle turned by the driving shaft,

$$\theta_1 = \omega t - \frac{1}{2}\alpha t^2$$

$$= 132 \times 6.46 - \frac{1}{2} \cdot \frac{42.24}{36.3} \times 64.6^2$$

$$= 6099 \text{ rad}$$

Angle turned by the rotor,

$$\theta_2 = \omega t + \frac{1}{2}\alpha r^2$$

$$= 0 + \frac{1}{2} \cdot \frac{42.24}{48} \times 64.6^2 = 1836 \text{ rad}$$

Energy lost in friction

$$= T(\theta_1 - \theta_2) = 42.24 \times (6099 - 1836)$$

$$= 180060 \text{ N or } 180.06 \text{ kN}$$

(3) Total resisting torque on the armature

$$= -42.24 - 64 = -106.24 \text{ N.m}$$

$$\therefore \quad 132 - \frac{106.24}{36.3}t = 0 + \frac{42.24}{48}t$$

$$3.8067\, t = 132$$

$$\therefore \quad t = 34.7 \text{ sec}$$

(4) Net resisting torque on the armature

$$= -42.24 + 64 = 21.76 \text{ N.m}$$

i.e. it is accelerating torque.

$$\therefore \quad 132 + \frac{21.76}{36.3}t = 0 + \frac{42.24}{48}t$$

$$0.2806\, t = 132$$

$$t = 470.5 \text{ s}$$

Example 9.15 : *An engine is coupled to a rotating drum by a single disc friction clutch having both of its sides lined with friction material. Axial pressure on the disc is 1 kN. Inner and outer diameters of the disc are 280 mm and 360 mm respectively. The engine develops a constant torque of 36 N.m and the inertia of its rotating parts is equivalent to that of a flywheel of 30 kg mass and a radius of gyration of 280 mm. The mass and radius of gyration of the drum are 50 kg and 420 mm respectively and the torque to overcome the friction is 6 N.m. The clutch is engaged when the engine speed is 480 rpm and the drum is stationary. Assuming the coefficient of friction to be 0.3, determine the*

(i) duration of slipping

(ii) speed when the clutch slip ceases

(iii) total time taken for the drum to reach a speed of 480 rpm.

Solution : $W = 1000 \text{ N/m}^2$, $N = 480$ rpm, $R_1 = 0.18$ m, $R_2 = 0.14$ m, $\mu = 0.3$.

$$n = 2 \text{ (both sides effective)}$$

$$\omega = \frac{2\pi N}{60} = \frac{2\pi \times 480}{60} = 16\,\pi \text{ rad/s}$$

Moment of inertia of engine,

$$I_e = 30 \times (0.28)^2 = 2.35 \text{ kg.m}^2$$

Moment of inertia of drum,

$$I_d = 50 \times (0.42)^2 = 8.82 \text{ kg.m}^2$$

Friction torque,

$$T = \frac{\mu W}{2}(R_1 + R_2) \times n$$

$$= \frac{0.3 \times 1000}{2}(0.18 + 0.14) \times 2$$

$$= 96 \text{ N.m}$$

(i) Net resisting torque on the armature $= -96 + 36$

$$= -60 \text{ N.m}$$

Net accelerating torque on the drum = 96 – 6 = 90 N.m

First speed of engine = Final speed of rotor

$$\text{or} \quad \omega_e + \alpha_e t = \omega_d + \alpha_d t$$

$$\text{or} \quad \omega_e - \frac{T}{I_e}t = \omega_d + \frac{T}{I_d}t$$

$$\therefore \quad 16\pi - \frac{60}{2.35}t = 0 + \frac{90}{8.82}t$$

$$\therefore \quad 35.714\, t = 16\pi$$

$$t = 1.407 \text{ s}$$

(ii) The Speed when the Slip Ceases

$$= \frac{90}{8.82} \times 1.407 = 14.357 \text{ rad/s}$$

$$\text{or} \quad \frac{60}{2\pi} \times 14.357 = 137.1 \text{ rpm}$$

(iii) After the Slipping Ceases,

Net accelerating torque on the engine and the drum = 36 – 6

$$= 30 \text{ N.m}$$

Net moment of inertia of the engine and the drum

$$= 2.35 + 8.82$$

$$= 11.17 \text{ kg.m}^2$$

Now, Final speed = Initial speed + Acceleration × Time

$$16\pi = 14.357 + \frac{30}{11.17}t$$

$$2.685\, t = 35.908$$

$$t = 13.37 \text{ s}$$

∴ Total time taken = 1.41 + 13.37 = 14.78 s

Example 9.16 : *A multi-plate disc clutch transmits 30 kW of power at 1800 rpm. It has four discs on the driving shaft and three discs on the driven shaft providing six pairs of contact surfaces. The external and internal diameters of the contact surfaces are 200 mm and 100 mm respectively. Assuming the clutch to be new, find the total spring load pressing the plates together. Coefficient of friction 0.3.*

Also, determine the maximum power transmitted when the contact surfaces have worn away by 0.4 mm. There are 8 springs and the stiffness of each spring is 15 kN/m.

Solution : $$\omega = \frac{2\pi \times 1800}{60} = 60\pi \text{ rad/s}$$

A new clutch has a uniform pressure distribution, but after the initial wear the clutch exhibits the characteristics of uniform wear.

$$P = T\omega \text{ or } 30000$$

$$= T \times 60\pi \text{ or } T = 159.15 \text{ N.m}$$

Thus, torque transmitted by new clutch

$$= \frac{2}{3}\mu W\left[\frac{R_1^3 - R_2^3}{R_1^2 - R_2^2}\right] \times n$$

or $$159.15 = \frac{2}{3} \times 0.3W\left[\frac{100^3 - 50^3}{100^2 - 50^2}\right] \times 6$$

or $$W = 1137 \text{ N}$$

When the surfaces are worn out.

Contact surfaces = Number of pairs of contact

$$2 = 6 \times 2 = 12$$

Total wear = Number of surfaces × Wear of each surface

$$= 12 \times 0.4 = 4.8 \text{ mm}$$

Stiffness of each spring = 15 kN/m = 15 N/mm

Total stiffness of springs = Stiffness × Number of springs

$$= 15 \times 8 = 120 \text{ N/mm}$$

∴ Reduction in spring force = 120 × 4.8 = 576 N

New axial load = 1137 – 576 = 561 N

$$T = \frac{1}{2}\mu W (R_1 + R_2) \cdot n$$

$$= \frac{1}{2} \times 0.3 \times 561 (100 + 50) \times 6$$

$$= 75.735 \text{ N.mm}$$

$$= 75.735 \text{ N.m}$$

Maximum power transmitted

$$= 75.735 \times 60\pi$$

$$= 14276 \text{ W} = 14.276 \text{ kW}$$

Example 9.17 : *The semi-cone angle of a cone clutch is 12.5° and the contact surfaces have a mean diameter of 80 mm. Coefficient of friction is 0.32. What is the minimum torque required to produce slipping of the clutch for an axial force of 200 N ?*

If the clutch is used to connect an electric motor with a stationary flywheel, determine the time needed to attain the full speed and the energy lost during slipping. Motor speed is 900 rpm and the moment of inertia of the flywheel is 0.4 kg.m².

Solution : $R_m = 0.04$ m, $\mu = 0.32$, $W = 200$ N, $\alpha = 12.5°$, $N = 900$ rpm, $I = 0.4$ kg.m²

$$T = I\alpha_a \qquad (\alpha_a = \text{Angular acceleration})$$

or $$\mu \frac{W}{\sin \alpha} R_m = I \cdot \alpha_a$$

or $$0.32 \times \frac{200}{\sin 12.5°} \times 0.04 = 0.4\alpha_a$$

$$\alpha_a = 29.57 \text{ rad/s}^2$$

and $$T = 0.4 \times 29.57$$

$$= 11.828 \text{ N.m}$$

or $$\frac{\omega}{t} = 29.57$$

or $$\frac{2\pi \times 900}{60t} = 29.57$$

$$t = 3.187 \text{ s}$$

During slipping period.

Angle turned by driving shaft,

$$\theta_1 = \omega t$$

$$= \frac{2\pi \times 900}{60} \times 3.187 = 300.4 \text{ rad}$$

Angle turned by driven shaft,

$$\theta_2 = \omega_0 t + \frac{1}{2}\alpha_a t^2$$

$$= 0 + \frac{1}{2} \times 29.57 \times (3.187)^2$$

$$= 150.2 \text{ rad}$$

Energy lost in friction

$$= T(\theta_1 - \theta_2)$$

$$= 11.828 (300.4 - 150.2)$$

$$= 1776.5 \text{ N.m}$$

Alternatively,

Energy supplied = $T\omega \times$ Time

$$= 11.828 \times \frac{2\pi \times 900}{60} \times 3.187$$

$$= 3553 \text{ N.m}$$

Energy of flywheel

$$= \frac{1}{2}I\omega^2$$

$$= \frac{1}{2} \times 0.4 \times \left(\frac{2\pi \times 900}{60}\right)^2$$

$$= 1776.5 \text{ N.m}$$

Energy lost = 3553 – 1776.5

$$= 1776.5 \text{ N.m}$$

PART B : BRAKES

9.9 INTRODUCTION TO BRAKES

A brake is an appliance used to apply frictional resistance to a moving body to stop or retard it by absorbing its kinetic energy or potential energy or both.

A dynamometer is a brake incorporating device to measure the frictional resistance applied. It is used to determine the power developed by the machine, while maintaining its speed at the rated value.

The functional difference between a clutch and a brake is that a clutch connects two moving members of a machine whereas a brake connects a moving member to a stationary member.

In case of clutches, power is transmitted from one shaft to another shaft by a clutch but in case of brakes the power is absorbed by the brake. Hence, brake is a friction device for converting kinetic energy of moving vehicle into heat by means of friction.

Capacity of brake depends on unit pressure between braking surfaces, coefficient of friction, peripheral velocity of brake drum, projected area of friction surfaces and heat dissipating capacity of brake.

9.10 CLASSIFICATION OF BRAKES

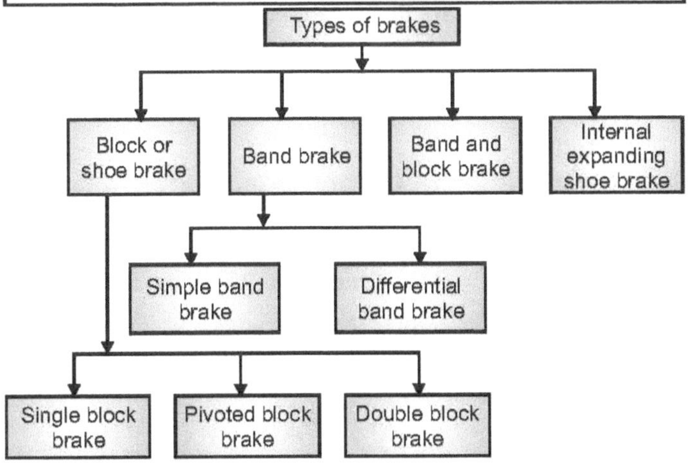

9.10.1 Classification of Brakes Based on Mode of Actuation

Mechanical brakes, hydraulic brakes, disc brakes, pneumatic brakes, vacuum brakes, electrical brakes.

9.11 REQUIREMENTS OF BRAKING SYSTEM

- Quick stoppage of vehicle.
- Good working on all kinds of roads.
- Less pedal effort by the driver.
- Less wearing parts.
- Less maintenance.
- Proper working of brakes in all weather conditions.
- Minimum sound should be produced after applying brakes.
- Steering geometry should be maintained after braking.

9.12 REQUIREMENTS OF BRAKE LINER

- It should have high coefficient of friction at all temperatures.
- Low wear rate.
- High heat resistance.
- High heat dissipation capacity.
- Adequate mechanical strength.
- It should not be affected by moisture and oil.

9.13 MATERIAL FOR BRAKE LINER

Bronze on cast iron, steel on cast iron, wood on cast iron, fiber on metal, cork on metal, asbestos on metal, leather on metal.

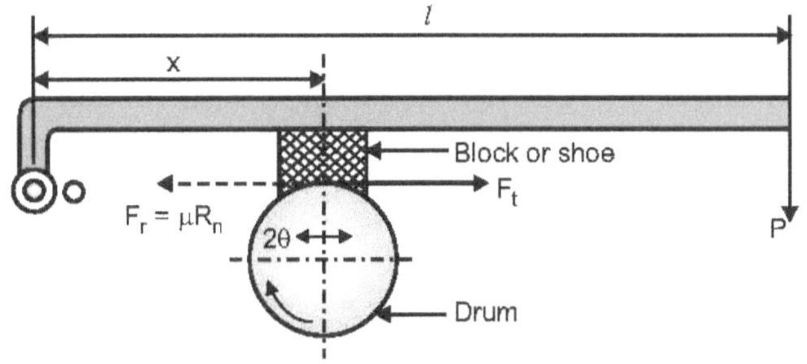

Fig. 9.12 : Block or shoe brake

Block or Shoe Brake :

It consists of a block or shoe which is pressed against a rotating drum. The force on the drum is increased by using a lever. The block which is rigidly attached or pivoted to the lever is lined with friction material. Friction between friction lining on block and drum retards the rotation of the drum.

A material softer than that of the drum or the rim of the wheel is used to make the blocks so that these can be replaced easily on wearing. Wood and rubber are used for light and slow vehicles and cast steel for heavy and fast ones.

Application Area : In bicycles and railway trains.

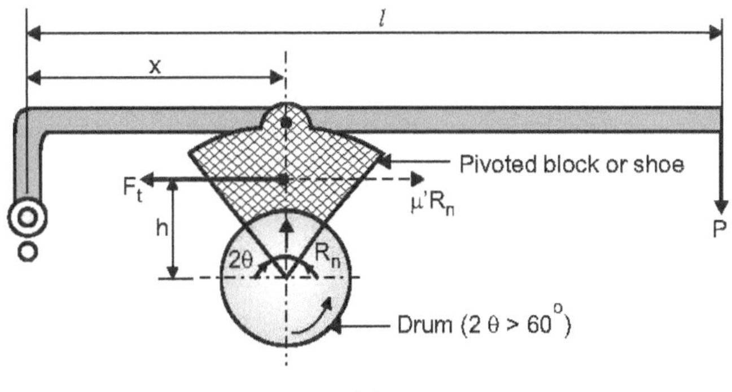

(a)

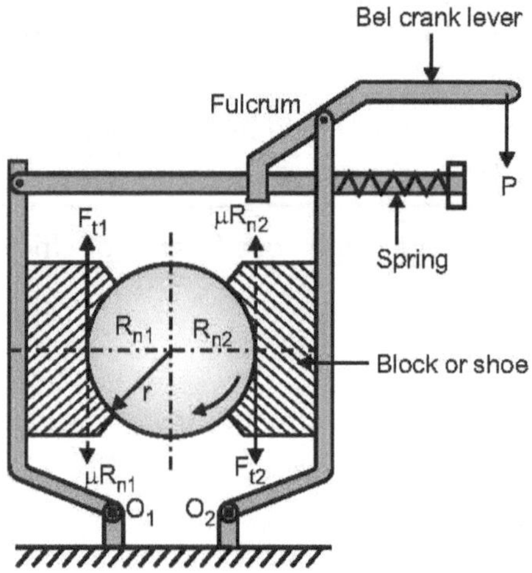

(b)

Fig. 9.13

9.14 SINGLE BLOCK OR SHOE BRAKE

Shoe Brake :

In a shoe brake the rotating drum is brought in contact with the shoe by suitable force. The contacting surface of the shoe is coated with friction material. Different types of shoe brakes are used, viz., single shoe brake, double shoe brake, internal expanding brake, external expanding brake. These are sketched in Fig.

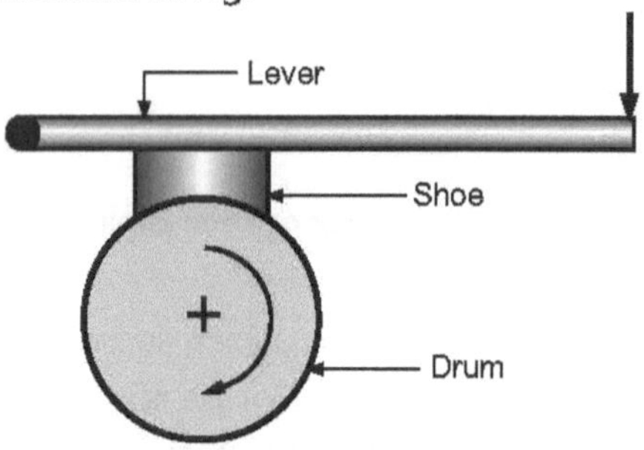

Fig. 9.14 : Single shoe brake

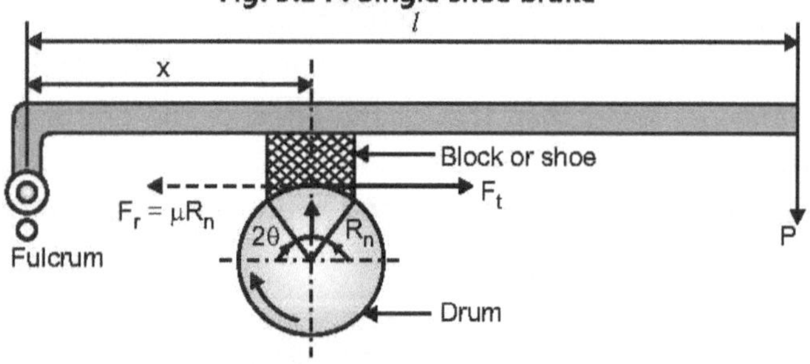

(a) Clockwise rotation of drum

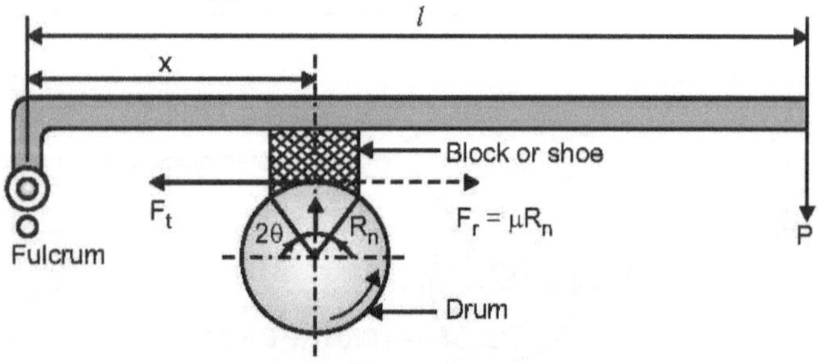

(b) Anticlockwise rotation of drum

Fig. 9.15

P = Force applied at lever

R_n = Normal reaction on block

μ = Coefficient of friction

r = Radius of drum or wheel

2θ = Angle of contact surface of block

$F_r = \mu R_n$ = Frictional force at the contact surface of block and wheel

F_t = Tangential braking force

= F_r but opposite in direction at contact surface between block and wheel

$$F_t = F_r$$

If angle of contact is less than 60° (i.e. $2\theta < 60°$), it can be assumed that normal pressure between block and drum or wheel is uniform.

$$\therefore \quad F_t = \mu R_n$$

Braking torque on wheel is,

$$T_B = F_t \times r$$

$$T_B = \mu R_n \times r \quad \ldots (9.28)$$

Depending on position of fulcrum of lever w.r.t. line of action of braking or frictional force three cases arise :

(1) Line of action passes through fulcrum of lever.

(2) Line of action of F_t is at a distance 'a' below the fulcrum 'o'.

(3) Line of action of F_t is at a distance 'a' above the fulcrum 'o'.

Case I : (a) $\Sigma m_o = 0$ for Clockwise Rotation of Drum.

$$\therefore \quad (R_n) \times (x) = (P) \times (l)$$

$$\therefore \quad R_n = \frac{Pl}{x} \quad \ldots (9.29)$$

$$\therefore \quad T_B = \text{Braking torque} = \mu R_n \times r$$

$$T_B = \frac{\mu P l r}{x} \quad \ldots (9.30)$$

(b) $\Sigma m_o = 0$ for Anticlockwise Rotation of Drum.

$$R_n \times x = (P) \times l$$

$$\therefore \quad R_n = \frac{Pl}{x}$$

and $$T_B = \mu R_n \times r$$

$$\therefore \quad T_B = \frac{\mu P l r}{x} \quad \ldots (9.31)$$

Equation (9.30) = equation (9.31)

Case II : (a) Clockwise Rotation of Drum :

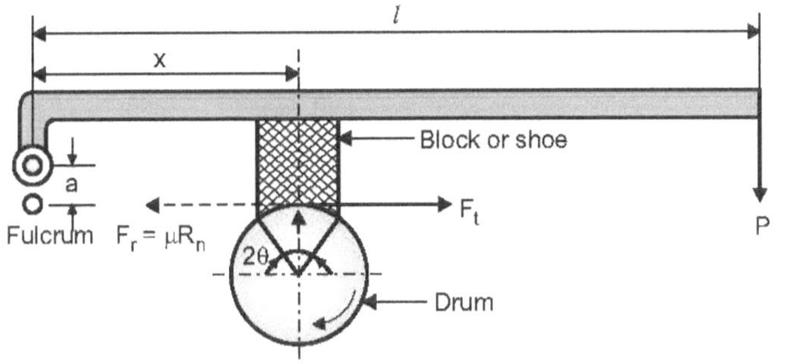

Fig. 9.16

$$\therefore \quad \sum m_o = 0$$

$$\therefore \quad (P \times l) - R_n \times x - (F_t \times a) = 0$$

$$\therefore \quad (R_n \times x) + (F_t \times a) = P \times l$$

$$\therefore \quad (R_n \times x) + (\mu R_n \times a) = P \times l$$

$$\therefore \quad R_n (x + \mu_a) = P \times l$$

$$\therefore \quad R_n = \frac{P \times l}{(x + \mu_a)} \qquad \ldots (9.32)$$

$$\therefore \quad T_B = \text{Braking torque} = \mu R_n \times r$$

$$T_B = \mu \times r \times \frac{P \times l}{(x + \mu_a)} = \frac{\mu P l r}{(x + \mu_a)} \qquad \ldots (9.33)$$

(b) Anticlockwise Rotation of Drum :

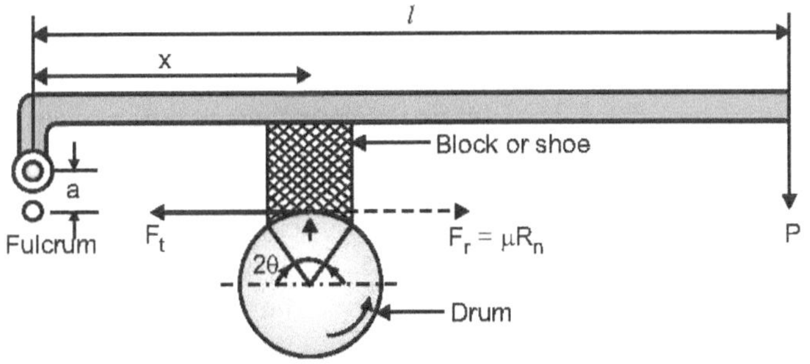

Fig. 9.17

$$\sum m_o = 0$$

$$\therefore P \times l - (R_n \times x) + F_t \times a = 0$$

$$\therefore \quad (R_n \times x) = (P \times l) + (F_t \times a)$$

$$\therefore (R_n \times x) - (F_t \times a) = P \times l$$

$$\therefore (R_n \times x) - (\mu R_n \times a) = P \times l$$

$$\therefore \quad R_n (x - \mu_a) = P \times l$$

$$\therefore \quad R_n = \frac{P \times l}{(x - \mu_a)} \qquad \ldots (9.34)$$

$\therefore$ Braking torque,

$$T_B = \mu R_n \times r$$

$$T_B = \frac{\mu \times r \times P \times l}{(x - \mu_a)} \qquad \ldots (9.35)$$

Case III : (a) Clockwise Rotation of Drum :

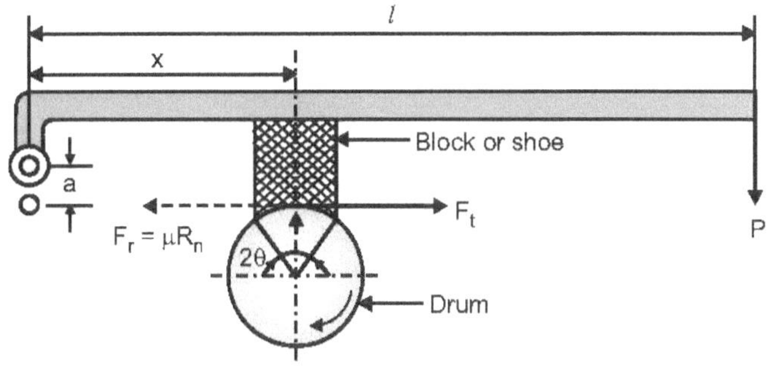

Fig. 9.18

$$\sum m_o = 0$$

$$\therefore \quad P \times l + F_t \times a = R_n \times x$$

$$\therefore \quad (P \times l) + (\mu R_n \times a) = (R_n \times x)$$

$$\therefore (R_n \times x) - (\mu R_n \times a) = P \times l$$

$$\therefore \quad R_n (x - \mu_a) = P \times l$$

$$\therefore \quad R_n = \frac{P \times l}{(x - \mu_a)} \qquad \ldots (9.36)$$

$$\therefore \quad T_B = \text{Braking torque} = \mu R_n \times r$$

$$T_B = \frac{\mu \times r \times P \times l}{x - \mu_a} \qquad \ldots (9.37)$$

(b) Anticlockwise Rotation of Drum :

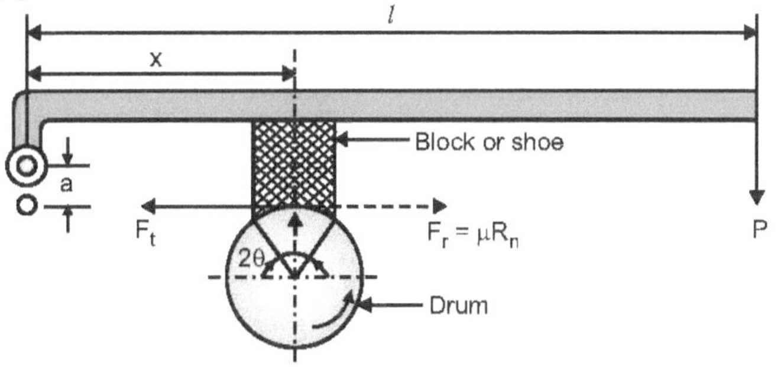

Fig. 9.19

$$\therefore \quad \sum m_o = 0$$

$$\therefore P \times l - (R_n \times x) - (F_t \times a) = 0$$

But $\quad F_t = \mu R_n$

$$\therefore (P \times l) - (R_n \times x) - (\mu R_n \times a) = 0$$

$$\therefore \quad (P \times l) = R_n (x + \mu_a)$$

$$\therefore \quad R_n = \frac{P \times l}{(x + \mu_a)} \qquad \ldots (9.38)$$

$$\therefore \quad \text{Braking torque} = T_B = \mu R_n \times r$$

$$T_B = \frac{\mu \times r \times P \times l}{(x + \mu_a)} \qquad \ldots (9.39)$$

Self Locking and Self Energizing of Brakes

(a) Self Locking of Brakes :

When drum rotates in anticlockwise direction and fulcrum is below line of action of tangential braking force referring equation (9.34) and when drum rotates in clockwise direction and fulcrum is above line of action of tangential braking force referring equation (9.36),

$$R_n = \frac{P \times l}{(x - \mu_a)}$$

$$\therefore \quad P = \frac{R_n (x - \mu_a)}{l} \qquad \dots (9.40)$$

In equation (9.40), if $x < \mu_a$, the effort 'P' becomes negative or zero. Hence, no effort is required to apply brakes and hence brake is self locking.

Hence, condition for self locking is

$$x \leq \mu \cdot a$$

(b) Self Energizing Brakes or Self Actuating Brakes :

$$P = \frac{R_n (x - \mu_a)}{l}$$

$$\therefore \quad P \times l = (R_n \times x) - (R_n \times \mu_a)$$

$$\therefore \quad (R_n \times x) = Pl + \mu R_n \times a \qquad \dots (9.41)$$

From equation (9.41), we can say that the frictional force ($\mu R_n \times a$) adds to the moment force ($P \times l$). Hence, frictional force helps to apply the brake. Such types of brakes are called self energizing brakes.

For this $P > 0$.

9.15 PIVOTED BLOCK BRAKE (2θ > 60°)

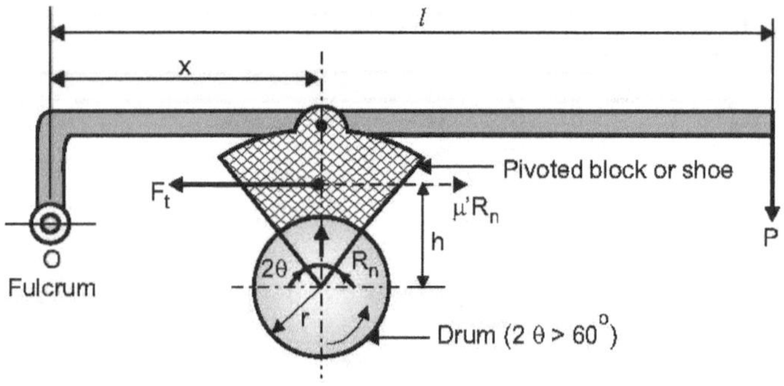

Fig. 9.20 : Pivoted block or shoe brake

Here, $T_B = F_t \times r$

$= \mu' R_n \times r$

μ' = Equivalent coefficient of friction

$$\mu' = \frac{4\mu \sin \theta}{2\theta + \sin 2\theta} \qquad \dots (9.42)$$

μ = Actual coefficient of friction

The tangential frictional force acts at a distance 'h' or equivalent brake drum radius h.

$$h = \frac{4r \sin \theta}{2\theta + \sin 2\theta} \qquad \dots (9.43)$$

and resultant vertical reaction,

$$R_n = \frac{1}{2} P_{max}\, b \times r\,(2\theta + \sin 2\theta)$$

$\because$ b = Shoe width

and P_{max} = Maximum intensity of pressure

9.16 DOUBLE BLOCK OR SHOE BRAKE

If the angle of contact between block or shoe and brake drum $2\theta < 60°$, then pressure normal to the surface of contact is less at the ends than at the centre i.e. intensity of pressure is not uniform. In this case, it is assumed that the wear is uniform. To get uniform wear in the direction of applied force, block or shoe is pivoted to the lever instead of being rigidly attached to lever.

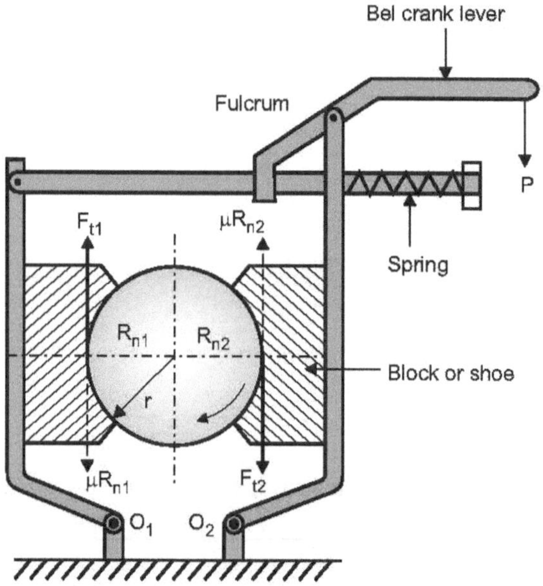

Fig. 9.21 : Double block or shoe brake

$$T_B = (F_{t_1} + F_{t_2})\, r = (\mu R_{n_1} + \mu R_{n_2})\, r$$

$\therefore$ F_{t_1} and F_{t_2} = Braking force on two blocks

When a single block brake is pressed against rotating drum, an unbalanced force on the bearing of shaft supporting the drum will act due to normal reaction R_n.

This unbalanced load produces bending of shaft. It can be prevented by a double block or shoe brake having two blocks on 2 sides of drum. Upper ends of brake arms are held together by using a spring.

When force 'P' is applied on bell crank lever, spring gets compressed and brake is released. When there is no force P on bell crank lever, the brake is engaged automatically. Here, T_B = Braking torque is doubled due to two shoes.

Application Area : They are used in electric cranes.

Example 9.18 : *A bicycle and rider, travelling at 12 km/hr on a level road, have a mass of 105 kg. A brake is applied to the rear wheel, which is 800 mm in diameter. The pressure force on brake is 80 N and coefficient of friction is 0.06. Find the distance covered by the bicycle and the number of turns of its wheel before coming to rest.*

Solution : Data : Speed of bicycle = v = 12 km/hr

$$= \frac{12000 \text{ m}}{3600 \text{ sec}} = 3.33 \text{ m/sec.}$$

Mass of bicycle and rider = m = 105 kg

Diameter of drum = D = 800 mm = 0.8 m

Pressure applied to brake = R_n = 80 N

Coefficient of friction = μ = 0.06

When bicycle comes to rest, K.E. of bicycle = Frictional work generated in brakes

Distance travelled = Displacement of bicycle = S

$$\therefore \quad \mu R_n \times S = \frac{1}{2} m \times v^2$$

(Here, μR_n = Frictional force, S = Displacement)

$$\therefore \quad 0.06 \times 80 \times S = \frac{1}{2} \times 105 \times (3.33)^2$$

$$= \frac{582.16725}{(0.06 \times 80)}$$

$$S = 121.28 \text{ m}$$

Distance travelled to move by one turn

$$= \pi \times d$$

= Circumference of wheel

$$= \pi \times 0.8$$

$$\therefore \text{ No. of turns of wheels} = \frac{\text{Total distance travelled}}{\text{Distance travelled in one turn}}$$

$$= \frac{121.28}{\pi \times 0.8} = 48.25 \text{ rotations}$$

Number of turns of wheels = 48.25 rotations

Example 9.19 : *A bicycle and rider of mass 125 kg are travelling at a speed of 10 km/hr. The rider applies a brake to the rear wheel of 1 m diameter. How far the bicycle will travel before it comes to rest ? The pressure applied to the brake is 120 N and μ = 0.05. Also find the number of revolutions made by bicycle before coming to rest.*

Solution : Mass of bicycle and rider = 125 kg

Velocity or speed of bicycle = 10 km/hr = $\frac{10000}{3600}$ = 2.77 m/s

Diameter of wheel or drum = 1 m,

Pressure applied on brake shoe = R_N = 120 N

Coefficient of friction = μ = 0.05.

To Find : Displacement of bicycle and number of turns made by wheel.

When bicycle comes to rest, total K.E. of bicycle and rider will be obtained in the form of heat generated.

K.E. of bicycle= Frictional work generated in brakes

$$\frac{1}{2} \times m \times v^2 = \mu R_n \times S$$

$$\therefore \frac{1}{2} \times 125 \times (2.77)^2 = 0.05 \times 120 \times S$$

$$\therefore \quad S = \frac{0.5 \times 125 \times (2.77)^2}{0.05 \times 120}$$

$$S = 79.926 \text{ m}$$

Distance travelled in one turn = Circumference of wheel

$$= \pi \times d$$

$$= \pi \times 1 \text{ m}$$

Number of rotation of wheel

$$= \frac{S}{\pi} = \frac{79.926}{3.14}$$

Number of rotations = 25.45

Example 9.20 : *A hand operated block brake is shown in Fig. 9.22. Face width of brake liner is 45 mm and maximum permissible intensity of pressure is 0.55 N/m^2 with coefficient of friction = 0.24. Calculate :*

(1) Actuating force.

(2) Braking torque.

(3) If the speed of the drum is 100 rpm and brake is applied for 5 sec of full capacity to bring the shaft to rest. How much heat is generated ?

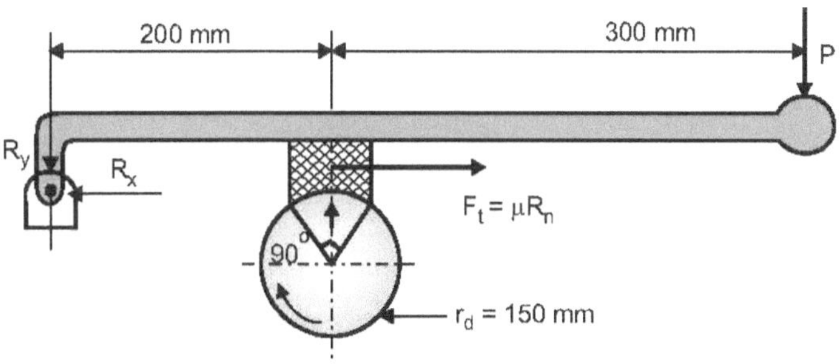

Fig. 9.22

Solution : Data : b = face width of brake liner = b = 45 mm = 0.045 m

$$P_{max} = 0.55 \text{ N/m}^2$$

$$\mu = 0.24, \; r = 150 \text{ mm} = 0.15 \text{ m}$$

$2\theta > 60°$ $\therefore$ Frictional force cuts at a distance 'h' from centre of drum. It is a pivoted block or shoe brake.

$$h = \frac{4r \sin\theta}{2\theta + \sin 2\theta}$$

$$= \frac{4 \times 0.15 \times \sin 45}{2 \times \left(45° \times \frac{\pi}{180}\right) + \sin 90°}$$

$$= 0.1650 \text{ m}$$

Normal reaction

$$R_n = \frac{1}{2} P_{max}\, br\, (2\theta + \sin 2\theta)$$

$$= \frac{1}{2} \times 0.55 \times 0.045 \times 0.15 \left[90 \times \frac{\pi}{180} + \sin 90°\right]$$

$$R_n = 4772 \text{ N}$$

$$\Sigma m_0 = 0$$

$$-R_n \times 200 + F_t (h - r) + P \times 500 = 0$$

$$\therefore \quad R_n \times 200 = \mu R_n (h - r) + P \times 500$$

$$\therefore \quad 4772 \times 200 = 0.24 \times 4772 \times (0.1650 - 0.15) + P \times 500$$

$$\therefore \quad 954400 = 17.1792 + 500\, P$$

$$\therefore \quad P = 1908.76 \text{ N}$$

Braking torque, $T_B = \mu R_n \times h$

$$\therefore \quad T_B = 0.24 \times 4772 \times 0.1650$$

$$= 188.9712 \text{ N-m}$$

Assume all frictional at work i.e. heat generated = K.E.

$$H_g = T_B \times \text{Angular displacement}$$

$$= T_B \times \omega_{avg} \times t$$

$$\omega_{avg} = \frac{2\pi (N_1 + N_2)}{2}$$

$$H_g = T_B \left[\frac{2\pi \times (100 + 0)}{60 \times 2} \times t\right]$$

$$= 189 \left[\frac{2\pi}{60} \times \frac{100}{2} \times 5\right]$$

$$= 4948.00 \text{ J}$$

$$H_g = 4.948 \text{ kJ}$$

9.17 BAND BRAKE

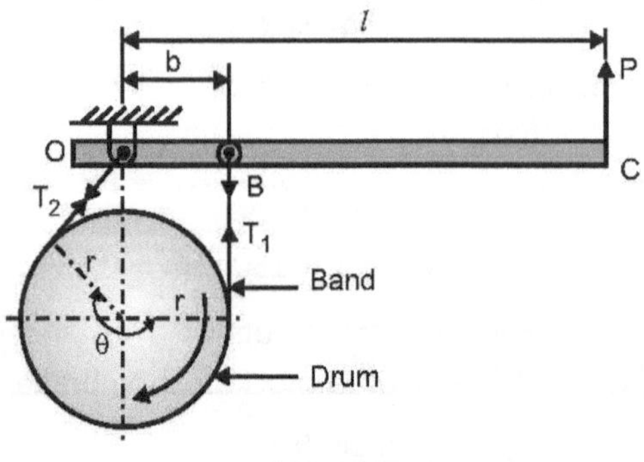

Fig. 9.23

Elements of a Band Brake : A rope or a belt or a flexible steel band lined with frictional material wrapped partly round the drum. Drum or wheel, lever.

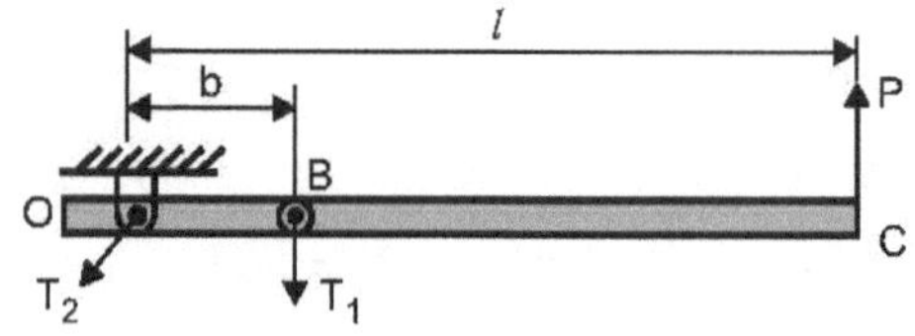

(a)

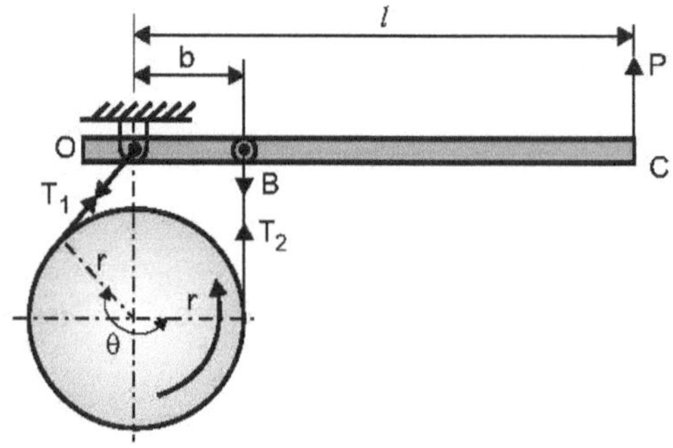

Fig. 9.24 (b)

When band is pressed against the external surface of drum, the frictional force between drum and band will generate braking torque on drum.

Band brakes are of two types :

1. Simple band brake.
2. Differential band brake.

Application of Band Brakes : In winch drum and chain saws and also for some bicycle brakes. They are prone to chatter and loss of brake force when hot.

9.18 SIMPLE BAND BRAKE

Here, one end of band is attached to fulcrum of lever and other end is at a distance b from fulcrum. As force 'P' is applied at free end of lever, band gets tightened on drum and brakes are applied. Braking force is obtained by friction between band and drum.

Notations used :

θ = Angle of lap or angle of wrap or angle of embrace of band on drum

T_1 = Tension on tight side of band

T_2 = Tension on slack side of band

R = Radius of drum

t = Thickness of band

r_e = Effective radius of drum = $R + \frac{t}{2}$

μ = Coefficient of friction

P = Force at end of lever

We know, $\frac{T_1}{T_2} = e^{\mu\theta}$

Limiting ratio of tension for belt

$\therefore$ Braking torque,

$T_B = (T_1 - T_2)\, R$ (if thickness of belt is neglected)

$T_B = (T_1 - T_2)\, r_e \qquad (\because r_e = R + t)$

(If thickness of belt is considered)

Case (A) : Clockwise Rotation of Drum :

For clockwise rotation of drum, the tight side of band is near point 'B' and slack side of band will be the side where band is connected with fulcrum 'o'.

$$\sum m_0 = 0$$

$$\therefore \quad (-P \times l) + (T_1 \times b) = 0$$

$$\therefore \quad Pl = T_1 b$$

$$\therefore \quad P = \frac{T_1 \times b}{l} \qquad \ldots (9.44)$$

Case (B) : Anticlockwise Rotation of Drum :

Here, tight side of band is on the fulcrum side while the slack side of band is the side where band is attached with the lever at point 'B'.

$$\sum m_0 = 0$$

$$\therefore (T_2 \times b) - (P \times l) = 0$$

$$\therefore \quad P \times l = T_2 b$$

$$\therefore \quad P = \frac{T_2 b}{l} \qquad \ldots (9.45)$$

9.19 DIFFERENTIAL BAND BRAKE

Here, the two ends of bands are attached at two different points on opposite side of fulcrum as shown in Fig. 9.25.

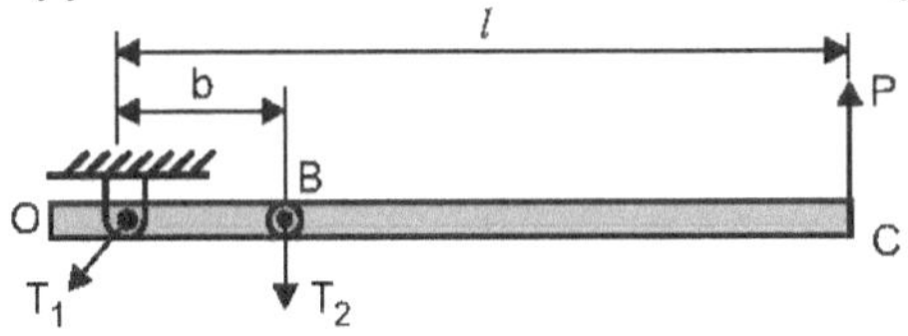

(a)

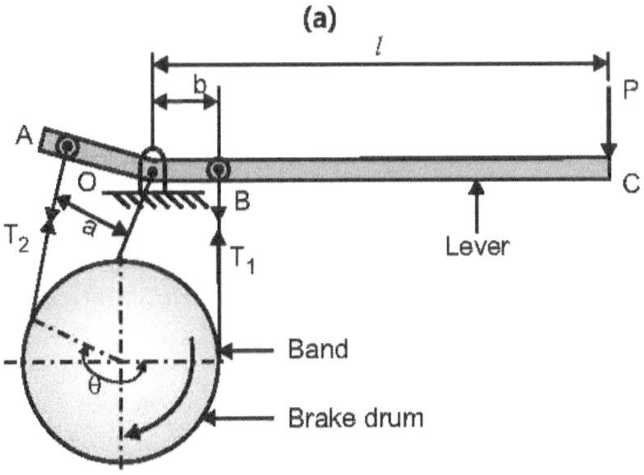

(b)

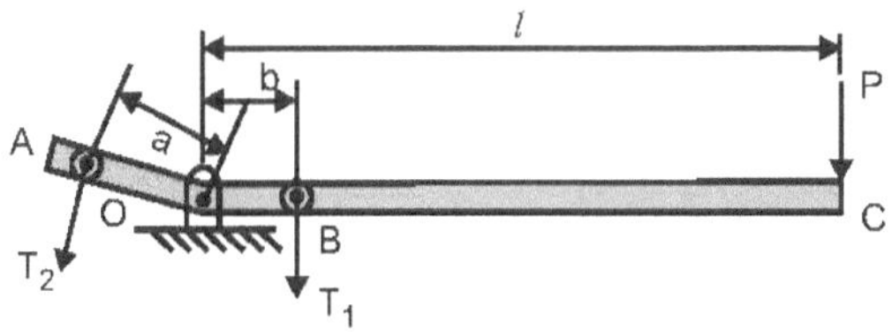

(c)

Fig. 9.25

The lever of AOC is pivoted at fulcrum 'o' and two ends of band are attached at point 'A' and 'B'.

Referring Fig. 9.25 for clockwise rotation of drum,

$$\sum m_0 = 0$$

$$\therefore \quad P \times l + T_1 \times b - T_2 \times a = 0$$

$$\therefore \quad P \times l = T_2 a - T_1 b$$

$$\therefore \quad P = \frac{T_2 a - T_1 b}{l} \qquad \ldots (9.46)$$

Referring Fig. 9.26 for anticlockwise rotation of drum,

$$\sum m_0 = 0$$

$$\therefore \quad P \times l + T_2 \times b - T_1 \times a = 0$$

$$P \times l = T_1 \times a - T_2 b$$

$$\therefore \quad P = \frac{T_1 a - T_2 b}{l} \qquad \ldots (9.47)$$

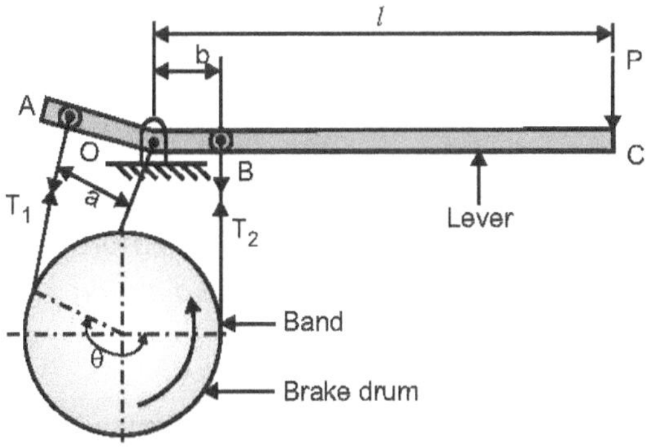

Fig. 9.26

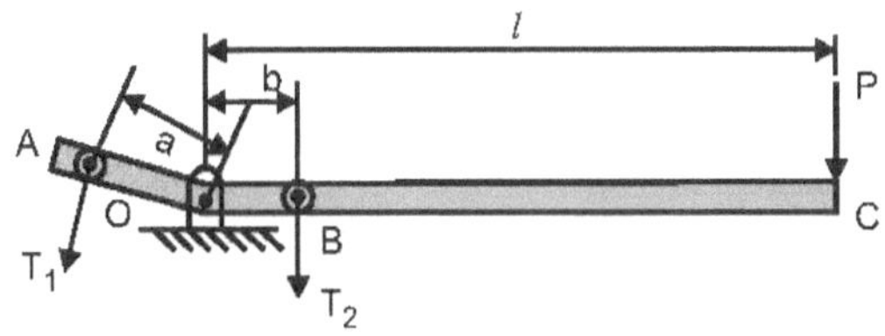

Fig. 9.27

Self Locking or Self Energizing of Brakes :

Here, $P \leq 0$

For clockwise rotation at drum,

$$P = \frac{T_2 a - T_1 b}{l}$$

As $P \leq 0$

$\therefore \quad T_2a - T_1b \leq 0$

$\therefore \quad T_2a \leq T_1b$

$\therefore \quad \frac{T_2}{T_1} \leq \frac{b}{a}$

For anticlockwise rotation of drum,

$$P = \frac{T_1a - T_2b}{l}$$

Also $\quad P \leq 0$

$T_1a - T_2b \geq 0$

$\therefore \quad T_1a \leq T_2b$

$$\frac{T_1}{T_2} \leq \frac{b}{a}$$

(b) Self Energizing Brake :

From FBD of lever for clockwise and anticlockwise rotation of drum,

T_1b and T_2b are the moments helping in applying brakes.

From FBD of lever :

If $a > b \rightarrow$ P acts vertically downwards

and If $a < b \rightarrow$ P acts vertically upwards

***Example 9.21 :** A band brake acts on 3/4th circumference of drum of 450 mm diameter, which keyed to the shaft. The band brake provides a braking torque of 225 N-m. One end of band is attached to a fulcrum pin of the lever and the other end to a pin, 100 mm from the fulcrum. If the operating force is applied at 500 mm from the fulcrum and μ = 0.25. Find the operating force when drum rotates in (i) anticlockwise direction, (ii) clockwise direction.*

Solution :

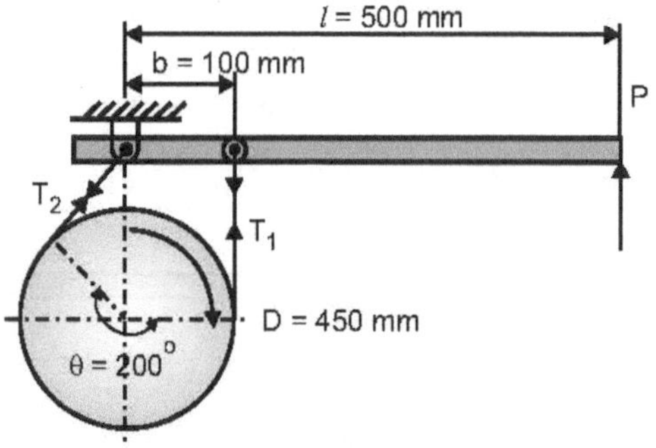

Fig. 9.28

T_B = 22.5 N-m

b = 100 mm, l = 500 mm

μ = 0.25, θ = 270° = 4.712 rad

D = 450 mm

r = 225 mm = 0.225 m

(a) Clockwise Rotation of Drum :

$$\Sigma m_0 = 0$$

$$-P \times 500 + T_1 \times 100 = 0$$

$\therefore \quad P \times 500 = T_1 \times 100$

$\therefore \quad P = \frac{T_1 \times 100}{500} = \frac{1}{5}T_1$

Also $\quad \frac{T_1}{T_2} = e^{\mu\theta} = e^{0.25 \times 0.7123} = 3.2481$

$\therefore \quad T_1 = T_2 \times 3.2481$

$T_B = 225 = (T_1 - T_2)\, r$

$225 = (3.2481\, T_2 - T_2) \times 0.225$

$\therefore \quad 225 = 2.2481 \times T_2 \times 0.225$

$\therefore \quad T_2 = \frac{225}{2.248 \times 0.225} = 444.80 \text{ N}$

$T_1 = 1444.75$ N

$P = 288.55$ N

(b) Anticlockwise Rotation of Drum :

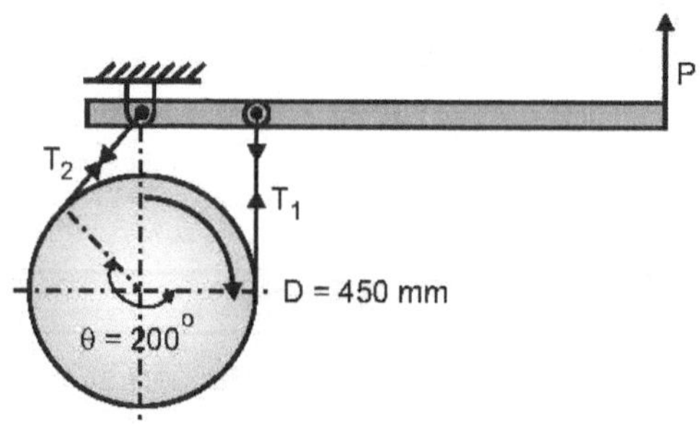

Fig. 9.29

$-P \times l + T_2 \times b = 0$

$\therefore \quad P \times l = T_2 \times b$

$\therefore \quad P = \frac{T_2 \times b}{l} = \frac{444.80 \times 0.1}{5}$

$P = 88.96$ N

***Example 9.22 :** A double shoe brake as shown in Fig. 9.38 is capable of absorbing a torque of 1400 N-m. The diameter of brake drum is 350 mm and the angle of contact for each shoe is 100°, if coefficient of friction between break drum and lining is 0.4. Find :*

(1) Spring force necessary to set the brake.

(2) The width of brake shoes if braking pressure on lining material is not to exceed 0.3 N/mm².

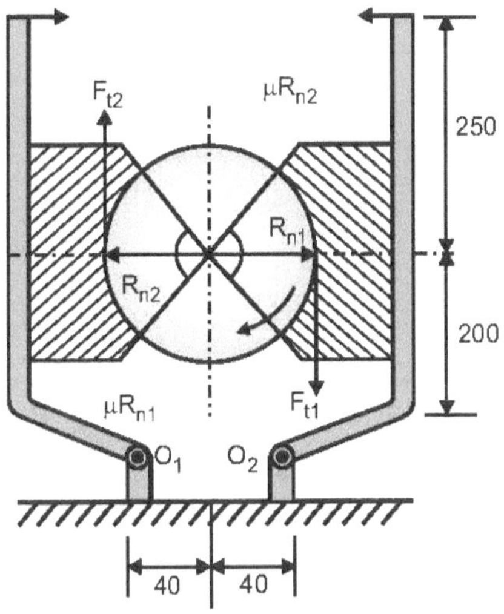

Fig. 9.30

Solution : Given : T_B = 1400 N-m = 1400 × 10^3 N-mm, d = 350 mm, 2θ = 100° = $100 \times \frac{\pi}{180}$ = 1.75 rad, μ = 0.4, P_{max} = 0.3 N/mm².

$$\mu' = \frac{4\mu \sin\theta}{2\theta + \sin 2\theta} = \frac{4 \times 0.4 \times \sin 50}{1.75 + \sin 100} = 0.45$$

$\Sigma m_{O_1} = 0.$

$$-S \times 450 + R_{N_1} \times 200 + F_{t_1} \times (175 - 40) = 0$$

$$\therefore \quad S \times 450 = \frac{F_{t_1}}{0.45} \times 200 + F_{t_1} \times 135 \quad \left(\because R_{N_1} = \frac{F_{t_1}}{\mu'}\right)$$

$$\therefore \quad F_{t_1} = 0.7765\ S$$

$\Sigma m_{O_2} = 0$

$$S \times 450 + F_{t_2} \times (175 - 40) - R_{n_2} \times 200 = 0$$

$$\therefore \quad S \times 450 + F_{t_2}(135) - 200 \times \frac{F_{t_2}}{\mu'(= 0.45)} = 0$$

$$\therefore \quad F_{t_2} = 1.454\ S$$

But $\quad T_B = (F_{t_1} + F_{t_2}) \times r$

$$= (0.7765 + 1.4545) \times 175$$

$$1400 \times 10^3 = 390.25\ S$$

$$\therefore \quad S = 3587\ N$$

Projected braking area for one shoe is

$$A_b = (2bl \sin\theta)$$

$$= 2 \times b \times 175 \sin 50$$

$$= 268\ b\ mm^2$$

$$R_{N_1} = \frac{F_{t_1}}{\mu'} = \frac{0.776 \times S}{0.45} = 6186\ N$$

$$R_{N_2} = \frac{F_{t_2}}{\mu'}$$

$$= \frac{1.454\ S}{0.45} = 11590\ N$$

$$P_b = \frac{R_{N_2}}{A \cdot b} = \frac{11590}{268\ b} = 0.3$$

$$\therefore \quad b = 144.2\ mm$$

Example 9.23 : *A simple band brake is operated by a lever of length 500 mm. The brake drum has a diameter of 500 mm and the brake band embraces 5/8 of the circumference. One end of the band is attached to the fulcrum of the lever, while the other end is attached to a pin on the lever 100 mm from the fulcrum. If the effort applied to the end of the lever is 2 kN and the coefficient of friction is 0.25, find the maximum braking torque on the drum.*

Solution : Data : l = 500 mm = Length of lever, D = 500 mm = Diameter of brake drum,

$$\theta = \frac{5}{8} \times 360° = 225° = 225 \times \frac{\pi}{180} = 3.9269 \text{ radians}$$

r = Radius of brake drum = 250 mm, b = 100 mm

P = 2 kN = Effort applied on lever = 2000 N

μ = Coefficient of friction = 0.25

To find : T_B (Maximum braking torque on drum).

$$T_B = (T_1 - T_2) \times r$$

Assumption : Drum rotates clockwise.

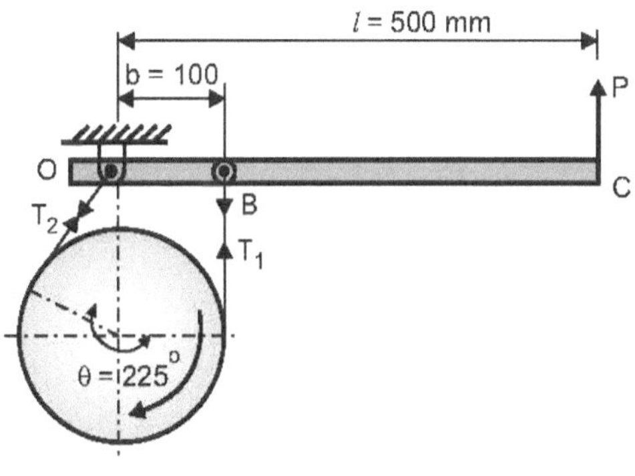

(a)

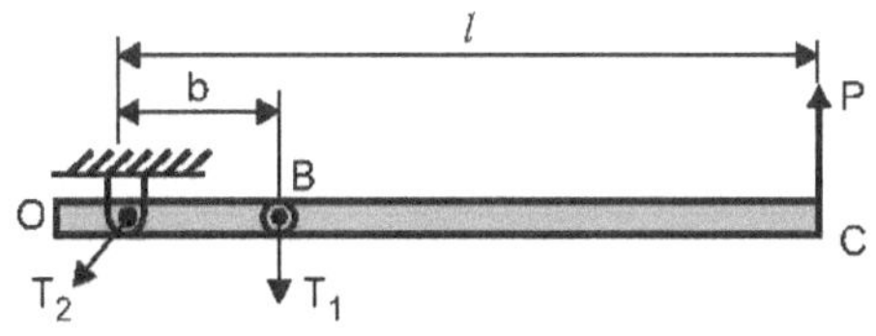

(b)

Fig. 9.31

$$\frac{T_1}{T_2} = e^{\mu\theta}$$

$$\therefore \quad \frac{T_1}{T_2} = e^{0.25 \times 3.9269}$$

$$\frac{T_1}{T_2} = 2.668$$

$$\therefore \quad T = 2.668\ T_2$$

$$T_1 \times 100 - 2000 \times 500 = 0$$

$$100\ T_1 = 1000000$$

$$\therefore \quad T_1 = 10000\ \text{N}$$

$$\therefore \quad T_2 = \frac{T_1}{2.668} = 3748.125\ \text{N}$$

and
$$T_B = (T_1 - T_2) \times r$$
$$= [10000 - 3748.125] \times 0.25\ \text{m}$$
$$= 1562.96\ \text{N-m}$$

Example 9.24 : *A simple band brake is applied to a shaft carrying a flywheel of mass 250 kg and radius of gyration 300 mm. The shaft speed is 200 rpm. The drum diameter is 200 mm and coefficient of friction is 0.25. The free end of band is attached at 100 mm from fulcrum and effort of 120 N is applied on lever at 280 mm from fulcrum. The angle embraced by belt is 225°. Determine for counterclockwise rotation of drum : (i) Braking torque, (ii) The number of turns of flywheel before it comes to rest, (iii) The time taken by flywheel to come to rest.*

Solution :

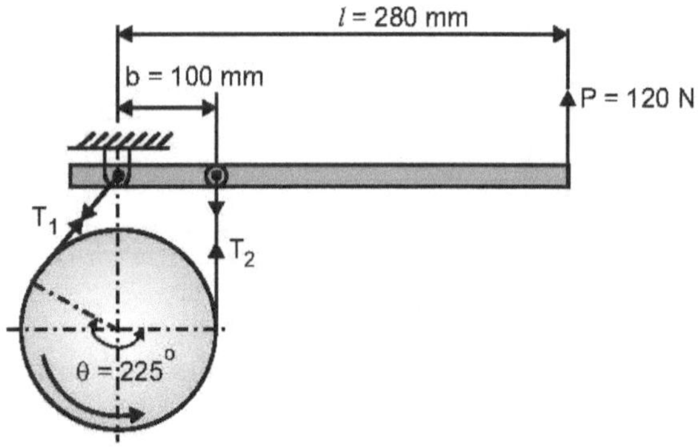

Fig. 9.32

m = Mass of flywheel = 250 kg

k = Radius of gyration = 300 mm

N = 200 rpm = shaft speed

D = 200 mm = diameter of drum

μ = 0.25, b = 100 mm, P = 120 N,

l = 280 mm

θ = 225°

$$= 225 \times \frac{\pi}{180} = 3.9269\ \text{rad}$$

To Find : T_B, n, t :

$$T_B = (T_1 - T_2) \times r$$

$$\therefore \quad T_B = (T_1 - T_2) \times 100$$

Also, $-120 \times 280 + T_2 \times 100 = 0$

$$\therefore \quad T_2 = \frac{120 \times 280}{100} = 336\ \text{N}$$

$$\frac{T_1}{T_2} = e^{\mu\theta} = e^{0.25 \times 3.926} = 2.69729$$

$$\therefore \quad T_1 = T_2 \times 2.69729$$
$$= 336 \times 2.69729$$
$$T_1 = 906.29\ \text{N}$$
$$T_B = (T_1 - T_2) \times r$$
$$= (906.29 - 336) \times 100$$
$$= 57.029\ \text{N-m}$$

K.E. of revolution of drum = Work done against friction

$$\frac{1}{2} \times I \times \omega^2 = T_B \times \theta_B$$

$$\therefore \quad \frac{1}{2} \times mk^2 \times \omega^2 = T_B \times 2\pi n$$

$$\therefore \quad \frac{1}{2} \times 250 \times (0.30)^2 \times (20.9439)^2 = 57.029 \times 2\pi n$$

$$\left[\because \omega = \frac{2\pi N}{60} = \frac{2\pi \times 200}{60} = 20.9439\ \text{rad/s}\right]$$

$$\therefore \quad n = 13.77184\ \text{revolutions}$$

200 revolutions need 1 min. So, 13.77184 revolutions will need 0.068 min i.e. 4 sec.

Example 9.25 : *In a simple band brake applied to a shaft carrying a flywheel of mass 250 kg and radius of gyration 300 mm. One end of the band is attached to the fulcrum and the other at a distance 80 mm from the fulcrum. The force is applied to the brake lever at a distance 280 mm. The angle of embrace by band is 225° and drum diameter is 200 mm. Coefficient of friction is 0.25. The shaft speed is 200 rpm.*

Determine : (i) The brake torque when a force of 120 N is applied.

(ii) The number of turns of flywheel before it comes to rest.

(iii) Time taken by the flywheel to come to rest.

Solution : Data : Simple Band Brake :

m = Mass of flywheel = 250 kg

k = Radius of gyration = 300 mm

Assume that drum rotates in anticlockwise direction.

$$\mu = \text{Coefficient of friction} = 0.25$$

$$P = 120 \text{ N}, \ N = 200 \text{ rpm}$$

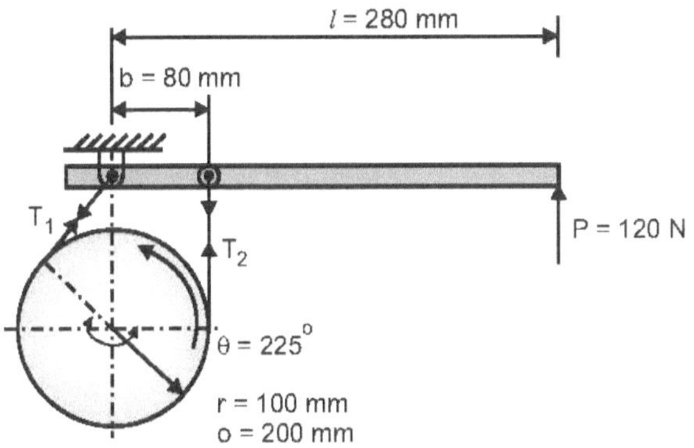

Fig. 9.33

To find : T_B, n, t :

$$\theta = 225°$$

$$= 225 \times \frac{\pi}{180} \text{ rad}$$

$$\frac{T_1}{T_2} = e^{\mu\theta}$$

$$\therefore \quad \frac{T_1}{T_2} = e^{0.25 \times 3.9269} = \mathbf{2.669}$$

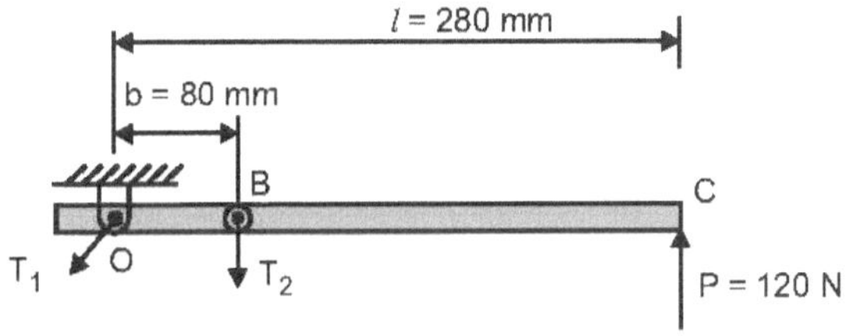

Fig. 9.34

Also, $\Sigma m_O = 0$

$$\therefore \quad T_2 \times 80 - 120 \times 280 = 0$$

$$\therefore \quad T_2 \times 80 = 120 \times 280$$

$$\therefore \quad T_2 = \frac{120 \times 280}{80} = 420 \text{ N}$$

$$\therefore \quad \frac{T_1}{T_2} = 2.669$$

$$\therefore \quad T_1 = 1120.98$$

$$\therefore \quad T_B = \text{Braking torque} = (T_1 - T_2) \times r$$

$$= (1120.98 - 420) \times 100$$

$$= 70098 \text{ N-mm}$$

$$T_B = 70.098 \text{ N-m}$$

K.E. of rotation of drum = Work done due to braking torque

$$\frac{1}{2} I\omega^2 = T_B \times \theta_B$$

(Angular displacement in revolutions)

$$\therefore \frac{1}{2} mk^2\omega^2 = T_B \times 2\pi n$$

$$\therefore \quad \frac{1}{2} \times 250 \times (300 \times 10^{-3})^2 \times \left[\frac{2\pi \times 200}{60}\right]^2$$

$$= 70.098 \times 2\pi \times n$$

$$\therefore \quad n = 11.204 \text{ revolutions}$$

Time taken by flywheel to come to rest : For 200 revolutions 1 min. = 60 sec are required. So for 11.204 revolutions, time required is $\frac{11.204 \times 60}{200}$ = 3.3612 sec.

So, t = time taken by flywheel to come to rest = 3.3612 sec.

Example on Differential Band Brake

***Example 9.26 :** The hand operated brake shown in Fig. 9.35 is fitted to a shaft carrying a flywheel of mass 400 kg with a radius of gyration 45 cm running at 360 rpm. Determine :*

(i) Torque applied due to pull of 100 N.

(ii) Number of revolutions of the flywheel before it is brought to rest.

(iii) Time required to bring the wheel to rest.

Take coefficient of friction 0.2.

Solution :

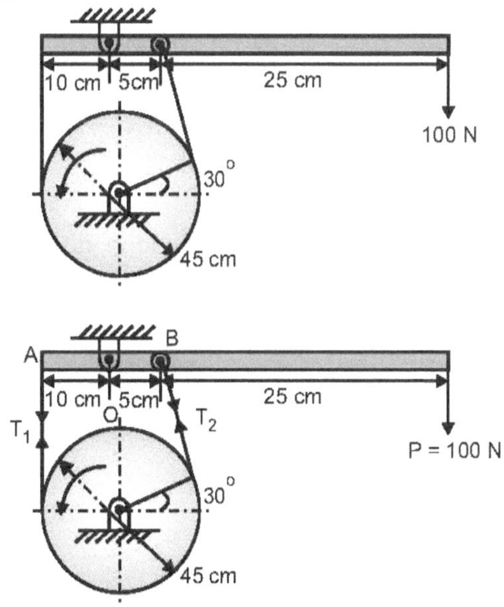

Fig. 9.35

Data : Differential Band Brake :

$$m = \text{Mass of flywheel} = 400 \text{ kg},$$

$$k = \text{Radius of gyration} = 45 \text{ cm} = 0.45 \text{ m}$$

$$N = 360 \text{ rpm}, \ P = 100 \text{ N}, \ \mu = 0.2$$

To find : T_B, n, t :

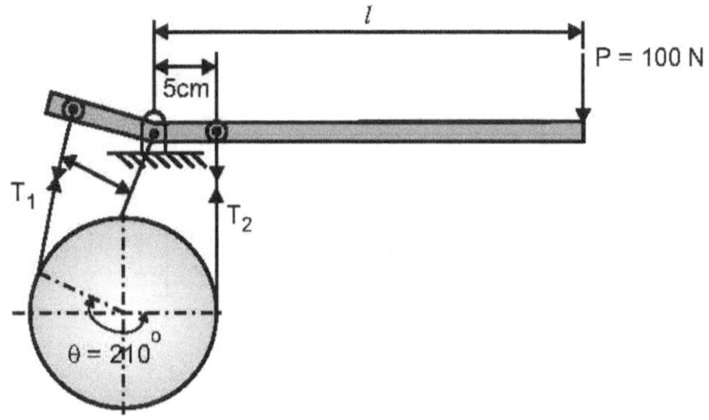

Fig. 9.36

Here, $\theta = (180 + 30)° \times \frac{\pi}{180} = 3.665$ rad

$$\therefore \quad \frac{T_1}{T_2} = e^{\mu\theta} = e^{(0.2 \times 3.665)} = 2.0813$$

$$\therefore \quad \frac{T_1}{T_2} = 2.0813 \quad \ldots (A)$$

$$\therefore \quad (100 \times 0.30) + (T_2 \times 0.05) - (T_1 \times 0.1) = 0$$

$$\therefore \quad 30 + 0.05\, T_2 = 0.1\, T_1 \quad \ldots (B)$$

From equations (A) and (B),

$$30 + 0.05\, T_2 = 0.1 \times 2.0813\, T_2$$

$$\therefore \quad T_2 = 189.717 \text{ N}$$

$$\therefore \quad T_1 = 394.858 \text{ N}$$

$$\therefore \quad T_B = (T_1 - T_2) \times r$$

$$= (394.858 - 189.717) \times 45 \times 10^{-2}$$

$$= 92.31 \text{ N-m}$$

$$T_B = 92.31 \text{ N-m}$$

K.E. of rotation of drum = Work done due to braking torque

$$\therefore \quad \frac{1}{2} I\omega^2 = T_B \times \theta$$

(Angle displacement in 'n' revolutions)

$$\frac{1}{2} \times mk^2 \times \omega^2 = T_B \times 2\pi n \quad (\because I = mk^2)$$

$$\therefore \quad \frac{1}{2} \times 400 \times (0.45)^2 \times \left(\frac{2\pi \times 360}{60}\right)^2 = 92.31 \times 2\pi \times n$$

$$\therefore \quad n = 99.24 \text{ revolutions}$$

$$x = \frac{99.24 \times 60}{360} = 16.54$$

$\therefore$ Time taken to complete 99.24 revolutions by the flywheel = 16.54 sec

Example 9.27 : *A differential band brake is shown in Fig. 9.45. The width and the thickness of the steel band are 120 mm and 4 mm respectively and the permissible tensile stress in the band is limited to 60 N/mm². The coefficient of friction between the friction lining and the brake drum is 0.25. Find the maximum braking torque that can be applied and the corresponding actuating force P. Is the brake self locking ? What actuating force is needed if the direction of rotation is reversed ?*

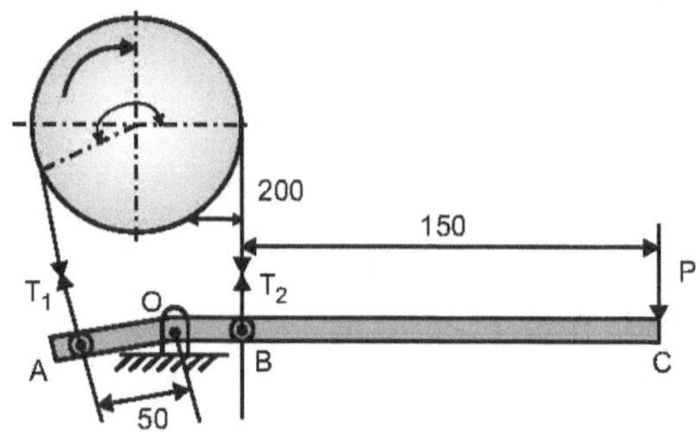

Fig. 9.37

Here $a = 50$ mm, $b = 200$ mm, $l = 350$ mm

As b > a 'P' must act downward.

Data : b = Width of steel band = 120 mm

t = Thickness of steel band = 4 mm

$\sigma_t = 60$ N/mm²

$= 60 \times 10^6$ N/m²,

$\mu = 0.25$, $\theta = 240°$

To Find : T_B, P : P when drum rotates counterclockwise.

$$\frac{T_1}{T_2} = e^{\mu\theta} = e^{0.25 \times 4.188} = 2.84909$$

$$\Sigma m_0 = 0$$

$$T_1 \times 0.05 - T_2 \times (200 \times 10^{-3}) + P \times 0.35 = 0 \quad \ldots (A)$$

But $\sigma_{av} = \dfrac{\text{Maximum tension in band}}{\text{Area of cross-section}}$

$$60 \times 10^6 = \frac{T_1}{b \times t}$$

$$60 \times 10^6 = \frac{T_1}{120 \times 4 \times 10^{-6}}$$

$$\therefore \quad T_1 = 28800 \text{ N}$$

$$\therefore \quad T_2 = \frac{T_1}{2.84909}$$

$$\therefore \quad T_2 = \frac{28800}{2.84905}$$

$$= 10108.49 \text{ N}$$

$$\therefore \quad T_B = (T_1 - T_2) \times \left(r + \frac{t}{2}\right)$$

$$= (28800 - 10108.49) \left[\left(300 + \frac{4}{2}\right) \times 10^{-3}\right]$$

$$T_B = 5644.836 \text{ N-m}$$

From equation (A),

$$28800 \times 0.05 - 10108.49 \times (200 \times 10^{-3}) + P \times 0.35 = 0$$

$$\therefore \quad P = 1661.99 \text{ N}$$

For Self Locking of Brake :

$$\therefore \quad [O \times OC + T_1 \times OA - T_2 \times OB] = 0$$

$$\therefore \quad 0 + 28800 \times 0.05 - 10108.49 \times OB = 0$$

$$\therefore \quad OB = 0.14245 \text{ m} = 142.45 \text{ mm}$$

As OB is 200. Brake is not self locking.

For Anticlockwise Rotation of Drum :

$\sum m_0 = 0.$

$$\therefore \quad -T_1 \times 200 + P \times 350 + T_2 \times 50 = 0$$

$$\therefore \quad -28800 \times 200 + P \times 350 + 10108.49 \times 50 = 0$$

$$\therefore \quad P = 15013.0728 \text{ N}$$

Testing Question : Identify the direction of applied force 'P' on lever, whether it is upward or downward on the simple band brake is shown in Fig. 9.38 and Fig. 9.39.

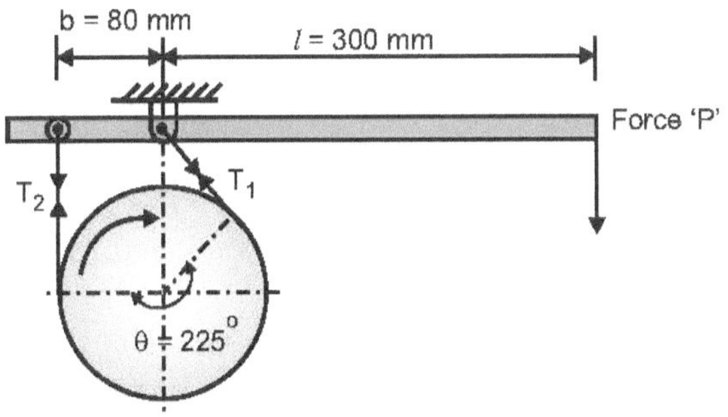

Fig. 9.38

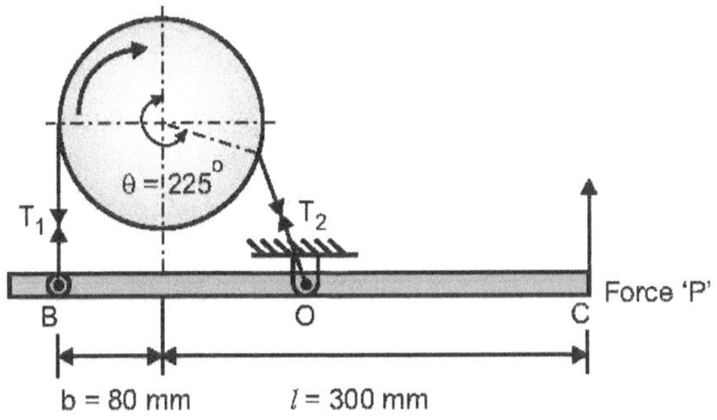

Fig. 9.39

9.20 BAND AND BLOCK BRAKE

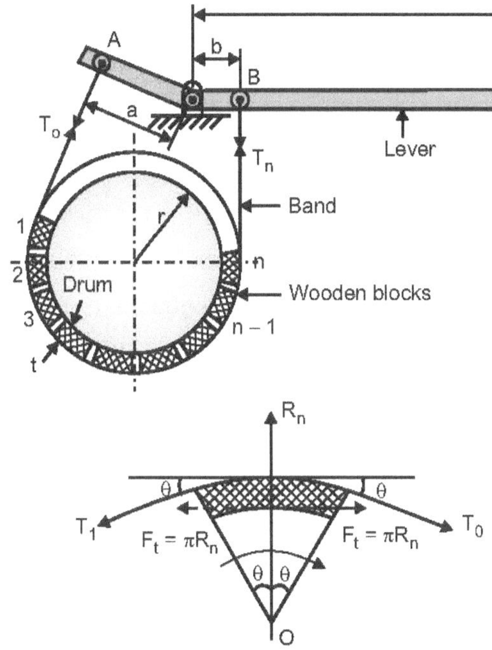

Fig. 9.40

It consists of number of wooden blocks fixed inside a flexible band. When the brake is applied, blocks are pressed against the drum. The friction between block and drum provides the braking action.

Wooden blocks provide higher coefficient of friction and they can be easily replaced after being worn out. Because of higher coefficient of friction of wooden blocks, effectiveness of this brake increases.

Let drum rotates clockwise. End of band attached to B will be tight side with tension T_n and end attached to A will be slack side tension T_0.

Notations :

- n = Number of blocks
- T_0 = Tension on slack side
- T_1 = Tension on tight side after one block
- T_2, T_3, T_4 etc. = Tension after second, third, fourth block etc.
- T_n = Tension on tight side after n blocks
- μ = coefficient of friction
- 2θ = Angle subtended by one block at centre
- R_n = Normal reaction of drum on block
- r = Radius of drum
- t = Thickness of block
- r_e = Effective radius of drum = r + t

Consider FBD of one block. Refer Fig. 9.40.

Resolving forces radially,

$$T_1 \sin\theta + T_0 \sin\theta = R_n$$

$$(T_1 + T_0)\sin\theta = R_n \quad \dots (9.48)$$

Resolving forces tangentially,

$$-T_1\cos\theta + F_t + T_0\cos\theta = 0$$

$$\therefore \quad \mu R_n = (T_1 - T_0)\cos\theta \quad (\because F_t = \mu R_n) \quad \dots (9.49)$$

$\therefore$ Dividing equation (9.75) by equation (9.74),

$$\frac{(T_1 - T_0)\cos\theta}{(T_1 + T_0)\sin\theta} = \mu$$

$$\therefore \quad \frac{T_1 - T_0}{(T_1 + T_0)\tan\theta} = \mu$$

$$\therefore \quad T_1 - T_0 = \mu\tan\theta\,(T_1 + T_0)$$

$$\therefore \quad T_1 - T_0 = \mu\tan\theta \times T_1 + T_0 \times \mu\tan\theta$$

$$\therefore \quad T_1 - T_1\mu\tan\theta = T_0 + T_0\mu\tan\theta$$

$$\therefore \quad T_1(1 - \mu\tan\theta)) = T_0(1 + \mu\tan\theta)$$

$$\therefore \quad \frac{T_1}{T_0} = \frac{(1 + \mu\tan\theta)}{(1 - \mu\tan\theta)} \quad \dots (9.50)$$

Likewise for second block ratio of tensions will be

$$\frac{T_2}{T_1} = \left(\frac{1 + \mu\tan\theta}{1 - \mu\tan\theta}\right) \cdots \frac{T_n}{T_{n-1}} = \left(\frac{1 + \mu\tan\theta}{1 - \mu\tan\theta}\right)$$

But $$\frac{T_n}{T_0} = \frac{T_1}{T_0} \times \frac{T_2}{T_1} \times \frac{T_3}{T_2} \times \frac{T_4}{T_3} \times \dots \times \frac{T_n}{T_{n-1}}$$

$$\therefore \quad \frac{T_n}{T_0} = \left(\frac{1 + \mu\tan\theta}{1 - \mu\tan\theta}\right)^n$$

Braking torque,

$$T_B = (T_n - T_0) \times r$$

($\because$ If thickness of block is neglected)

or $$T_B = (T_n - T_0) \times r_e \; (= r + t)$$

(If thickness is considered)

Example 9.28 : *A band and block brake having 12 blocks each of which subtends an angle of 18° at the centre, is applied to a rotating drum 600 mm diameter. The blocks are 80 mm thick. The two ends of the band are attached to pins on opposite sides of the fulcrum at distances of 50 mm and 150 mm. Determine maximum braking torque if a force of 300 N is applied to the lever at distance of 800 mm from the fulcrum. Assume μ = 0.25.*

Solution :

Data : n = Number of blocks = 12, $2\theta = 18°$,

$\theta = 9°$

D = 600 mm, r = 300 mm = 0.3 m,

a = 150 mm = 0.15 m

b = 50 mm = 0.05 m, P = 300 N,

l = 800 mm = 0.8 m

t = Thickness of block = 80 mm

= 0.08 m,

μ = 0.25

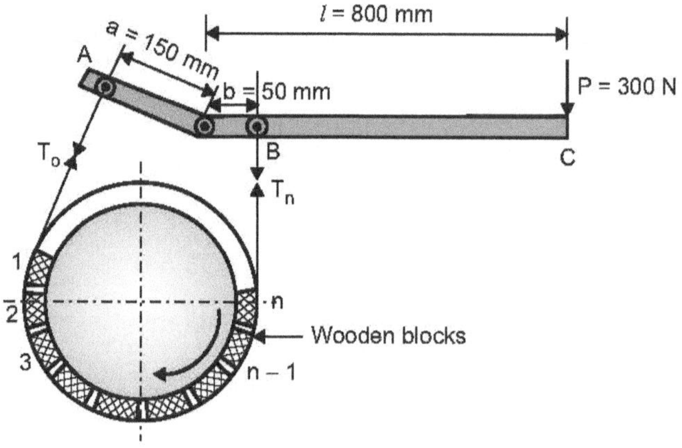

Fig. 9.41

To Find : T_B.

$$\frac{T_n}{T_0} = \left(\frac{1 + \mu\tan\theta}{1 - \mu\tan\theta}\right)^n$$

$$\frac{T_n}{T_0} = \left(\frac{1 + 0.25 \times \tan 9°}{1 - 0.25 \times \tan 9°}\right)^{12}$$

$$= 2.5877$$

$$\therefore \quad T_n = 2.5877\,T_0 \quad \dots (1)$$

Resolving forces,

$$\Sigma m_0 = 0.$$

$$300 \times 0.8 + (T_n \times 0.05) = T_0 \times 0.15$$

$$\therefore \quad 300 \times 0.8 + (2.5877\,T_0 \times 0.05) = T_0 \times 0.15$$

$$\therefore \quad T_0 = 11.622 \text{ kN}$$

$$\therefore \quad T_n = 4142.552 \text{ N}$$

$$T_B = (T_n - T_0) \times (r \times t)$$

$$= (4142.552 - 1600.86259) \times (0.3 + 0.08)$$

$$= 965.842 \text{ N-m}$$

Example 9.29 : *A band and block brake having 14 blocks each of which subtends an angle of 16° at the centre is applied to a drum of 1 m effective diameter. The drum and flywheel mounted on the shaft has a mass of 2000 kg and combined radius of gyration of 600 mm. The two ends of band are attached to pins on opposite sides of the brake lever at distances of 40 mm and 120 mm from the fulcrum.*

The coefficient friction between blocks and drum may be taken as 0.25. If a force of 250 N is applied at a distance of 800 mm from the fulcrum.

Find :

(i) Maximum braking torque.

(ii) Angular retardation of the drum.

(iii) Time taken by the system to come to rest from the rated speed of 350 rpm.

Solution : Band and Block Brake :

n = Number of blocks = 14, $2\theta = 16°$, $\theta = 8°$, $D_e = 1$ m, $r_e = 0.5$ m, m = 2000 kg, k = 600 mm = Radius of gyration, a = 120 mm, b = 40 mm, $\mu = 0.25$, P = 250 N, $l = 800$ mm.

To Find : T_B, α, t :

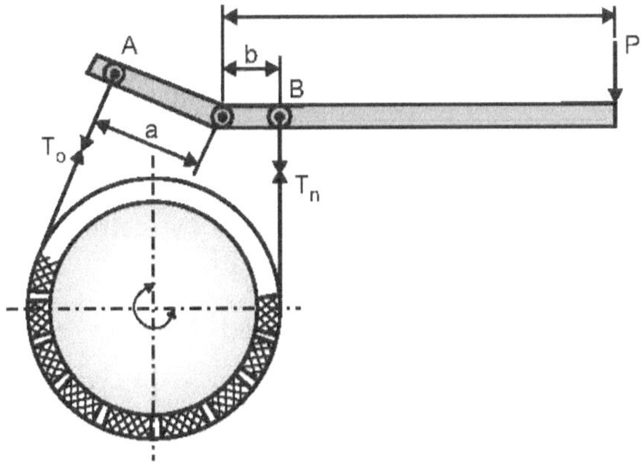

Fig. 9.42

$$\frac{T_n}{T_0} = \left(\frac{1 + \mu \tan \theta}{1 - \mu \tan \theta}\right)^n$$

$$\therefore \quad \frac{T_n}{T_0} = \left(\frac{1 + 0.25 \tan 8}{1 - 0.25 \tan 8}\right) = 2.67564$$

$\sum m_0 = 0$,

$$250 \times 0.8 + T_n \times 0.04 = T_0 \times 0.12$$

$$\therefore \quad (250 \times 0.8) + (2.67564\, T_0 \times 0.04) = 0.12\, T_0$$

$$200 + 0.1070258\, T_0 = 0.12\, T_0$$

$$\therefore \quad T_0 = 15420.2 \text{ N}$$

$$T_n = 41258.905 \text{ N}$$

$$\therefore \quad T_B = (T_n - T_0) \times r_e$$

$$= (41258.905 - 15420.2) \times 0.5$$

$$T_B = 12919.3525 \text{ N-m}$$

$$T_B = I\alpha = mk^2\alpha$$

$$= 2000 \times (0.6)^2 \times \alpha = 12919.3525 \text{ N-m}$$

$$\therefore \quad \alpha = 17.9435$$

$$\omega_1 = \text{Rated initial angular velocity}$$

$$= \frac{2\pi \times 350}{60} = 36.6519$$

$$\omega_0 = 0$$

$$v = u + at \quad \text{or} \quad (-\alpha) = \frac{\omega_0 - \omega_1}{t}$$

$$\omega_0 = \omega_1 - \alpha t$$

$$\therefore \quad t = \frac{\omega_0 - \omega_1}{-\alpha} = \frac{\omega_1 - \omega_0}{\alpha} = \frac{36.6519 - 0}{17.9435}$$

$$t = 2.04262 \text{ sec}$$

Example 9.30 : *A band and block brake is lined with 14 blocks, each subtending an angle of 14° at the drum centre. The drum is 500 mm in diameter and the radial thickness of the block is 50 mm. The coefficient between the block and drum is 0.35. The two ends of the band are attached to pins on opposite sides of the fulcrum of brake lever at distance 200 mm and 40 mm respectively. Find the least force required to be applied at the end of the brake lever at a distance of 500 mm from fulcrum to absorb 75 kW at 500 rpm.*

Solution :

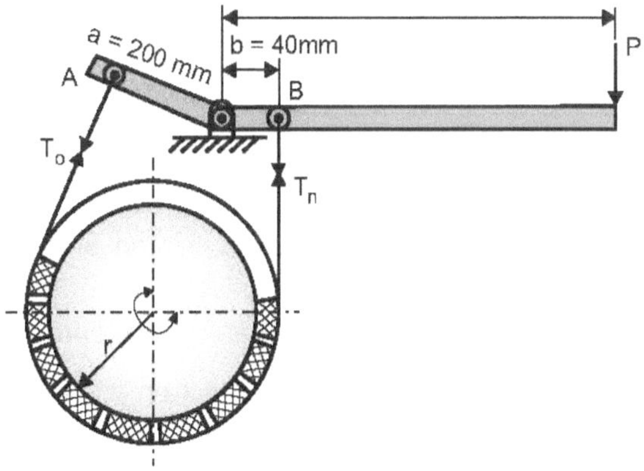

Fig. 9.43

Data : n = 14, $2\theta = 14°$, $\theta = 70$, D = 0.5 m, r = 0.25 m, t = 0.05 m, $r_e = r + t = 0.250 + 0.05 = 0.3$ m, $\mu = 0.35$, P = 75×10^3 W, N = 500 rpm.

Assume drum rotates clockwise,

$$\frac{T_n}{T_0} = \left(\frac{1 + \mu \tan \theta}{1 - \mu \tan \theta}\right)^n = \left(\frac{1 + 0.35 \tan 7}{1 - 0.35 \tan 7}\right)^{14}$$

$$T_n = 3.3335\, T_0$$

$$P = \frac{2\pi NT}{60}$$

$$\therefore \quad 75 \times 10^3 = \frac{2\pi \times 500}{60} \times T_B$$

$$T_B = 1432.39 \text{ N-m}$$

$$T_B = (T_n - T_0)\, r_e$$

$$1432.39 = (3.3335\, T_0 - T_0) \times 0.3$$

$$T_0 = 2046.12 \text{ N}$$

$$T_n = 3.335 \times 2046.12$$

$$= 6820.80 \text{ N}$$

$$\Sigma m_0 = 0$$

$$P \times OC + T_n \times 0\,b = T_0 \times OA$$

$$P \times 0.5 + 6820.880 \times 0.04 = 2046.12 \times 0.2$$

$$P = 272.78 \text{ N}$$

9.21 INTERNAL EXPENDING SHOE BRAKE

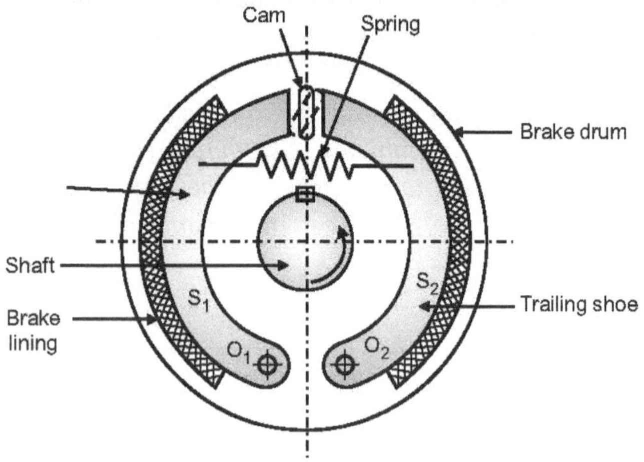

Fig. 9.44

It consists of two semicircular shoes S_1 and S_2 pivoted at the fixed fulcrum O_1 and O_2 respectively. The outer surface of the shoes is lined with friction material like Ferrodo.

One end of the shoe is pivoted at fulcrum while the other end is subjected to the actuating force.

The actuating force on both shoes is applied by cam mechanism or hydraulic cylinder. When cam rotates the shoes are pushed outward against the rim of the drum. Friction between shoes and drum produces braking torque causing the drum to retard or stop completely.

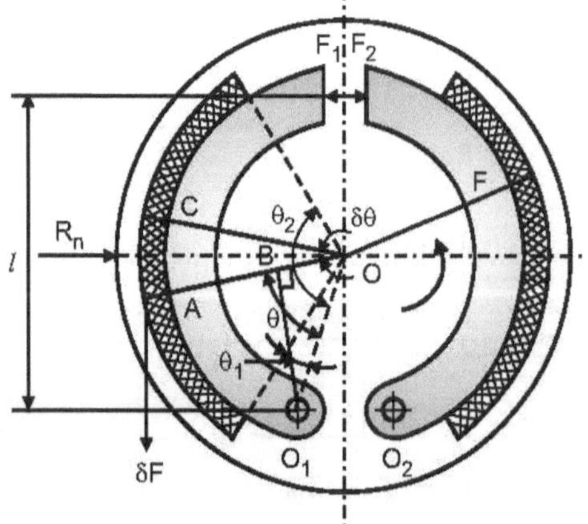

Fig. 9.45

When this actuating force through the rotation of cam is released, the helical tension spring retracts the shoes to the original position. The drum encloses the entire system to keep out of dust and moisture.

For anticlockwise rotation of drum, left shoe is known as leading or forward or primary shoe and right shoe as trailing or rear or secondary shoe.

They are used in motor cars and light trucks.

Braking Torque of an Internal Expending Shoe Brake :

Let the drum rotates in anticlockwise direction, the left hand shoe is leading or forward or primary shoe and right hand shoe is trailing or rear or secondary shoe.

Let r = Internal radius of break drum

b = Width of brake lining

P_{max} = Maximum intensity of normal pressure

P_n = Normal pressure

F_1 = Force exerted by cam on leading shoe

F_2 = Force exerted by cam on trailing shoe

μ = Coefficient of friction

Consider an elementary portion AC of leading shoe, at an angle θ with OO_1. It subtends on angle $\delta\theta$ at centre 'O' of brake drum.

Due to application of force F_1 at top of the shoe as shown, let δR_n be the normal reaction and $\mu \delta R_n$ as corresponding frictional force.

It is assumed that the pressure distribution on the shoe is nearly uniform however, the rate of wear at top or free end of shoe is more.

The rate of wear of shoe lining at A will be directly proportional to perpendicular distance from O_1 to OA i.e. distance O_1B.

From Fig. 9.45, $O_1B = OO_1 \sin\theta$

Rate of wear at A $\propto O_1B$ or $OO_1 \sin\theta$ or $\sin\theta$

$\therefore$ Normal pressure at A, $P_n \propto \sin\theta$

$$\therefore \quad P_n = P_{max} \sin\theta$$

Normal force acting on element,

$$\delta R_n = \text{Normal pressure} \times \text{Area of element}$$

$$\delta R_n = P_{max} \sin\theta \times (b \times l \times \delta\theta)$$

Braking or friction force on element,

$$\delta F = \mu \delta R_n$$

$$= \mu P_{max} \sin\theta \, br \, \delta\theta$$

Braking torque due to element about 'O',

$$\delta T_B = \delta F \times r$$

$$= \mu P_{max} \times \sin\theta \times b \times r\delta\theta \times r$$

$$\delta T_B = \mu P_{max} \times b \times r^2 \sin\theta \, \delta\theta$$

Total braking torque about O for whole of one shoe (i.e. leading shoe).

$$T_B = \int_{\theta_1}^{\theta_2} \mu P_{max} br^2 \sin\theta \, \delta\theta$$

$$= \mu P_{max} br^2 \int_{\theta_1}^{\theta_2} \sin\theta \, \delta\theta$$

$$= \mu P_{max} br^2 [-\cos\theta]_{\theta_1}^{\theta_2}$$

$$\therefore \quad T_B = \mu P_{max} b \times l^2 [\cos\theta_1 - \cos\theta_2] \quad \ldots (9.51)$$

Moment of Normal Force δR_n about Fulcrum θ

$$\delta m_n = \delta R_n \times O_1B$$

$$= \delta R_n \times (OO_1 \sin\theta)$$

$$\therefore \quad \delta m_n = P_{max} \sin^2\theta \, br \, \delta\theta \times OO_1$$

Total Moment of Normal Forces about Fulcrum is

$$m_n = \int_{\theta_1}^{\theta_2} P_{max} br \, OO_1 \sin^2\theta \, \delta\theta$$

$$\therefore \quad m_n = P_{max} br (OO_1) \int_{\theta_1}^{\theta_2} \sin^2\theta \, \delta\theta$$

$$= P_{max} br (OO_1) \int_{\theta_1}^{\theta_2} \left(\frac{1-\cos 2\theta}{2}\right) \delta\theta$$

$$= \frac{1}{2} P_{max} br (OO_1) \left[\theta - \frac{\sin 2\theta}{2}\right]_{\theta_1}^{\theta_2}$$

$$m_n = \frac{1}{2} P_{max} b \times r \times OO_1 \left[\theta_2 - \frac{\sin 2\theta_2}{2} - \theta_1 + \frac{\sin 2\theta_1}{2}\right]$$

$$\therefore m_n = \frac{1}{2} P_{max} br (OO_1) \left[(\theta_2 - \theta_1) + \frac{1}{2}(\sin 2\theta_1 - \sin 2\theta_2)\right] \quad \ldots (9.52)$$

Moment of Frictional Force δF about Fulcrum O_1,

$$\delta m_F = \delta F \times AB$$

$$AB = r - OO_1 \cos\theta$$

$$\therefore \quad \delta m_F = \mu P_{max} \sin\theta \times br \times \delta\theta \, (r - OO_1 \cos\theta)$$

$$\delta m_F = \mu P_{max} br (r \sin\theta - OO_1 \sin\theta \cos\theta) \, \delta\theta$$

$$\therefore \quad \delta m_F = \mu P_{max} bl \left(r \sin\theta - \frac{OO_1}{2} \sin 2\theta\right) d\theta$$

$$\therefore \quad m_F = \mu P_{max} br \int_{\theta_1}^{\theta_2} \left(r \sin\theta - \frac{OO_1 \sin 2\theta}{2}\right) \delta\theta$$

$$= \mu P_{max} br \left[-r\cos\theta + \frac{OO_1 \cos 2\theta}{4}\right]_{\theta_1}^{\theta_2}$$

$$= \mu P_{max} br \left[-r\cos\theta_2 + \frac{OO_1 \cos 2\theta}{4} + r\cos\theta_1 - \frac{OO_1}{4}\cos 2\theta_1\right]$$

$$= \mu P_{max} br \left[r(\cos\theta_1 - \cos\theta_2) + \frac{OO_1}{4}(\cos 2\theta_2 - \cos 2\theta_1)\right] \quad \ldots (9.53)$$

for leading shoe taking moment about O_1

$$F_1 \times l = m_n - m_F \quad \ldots (9.54)$$

For trailing shoe taking moment about O_2,

$$F_2 \times l = m_n + m_F \quad \ldots (9.55)$$

If $m_F > m_N \rightarrow$ break becomes self locking.

i.e. F_1 is negative or zero.

The maximum normal pressure over the trailing shoe $(P_{max})_2$ is different from maximum normal pressure for the leading shoe $(P_{max})_1$.

So, torque applied by trailing shoe is

$$T_B = \mu (P_{max})_2 br^2 (\cos\theta_1 - \cos\theta_2)$$

$$\therefore \quad T_{B_2} = T_{B_1} + T_{B_2}$$

$$T_B = [(P_{max})_1 + (P_{max})_2] \mu \cdot b \cdot r^2 (\cos\theta_1 - \cos\theta_2)$$

Example 9.31 : *Fig. 9.46 shows the arrangement of an internal expanding shoe brake in which the brake shoe is pivoted at a fixed point C. The distance of this point from the centre O of the rotating drum is 0.075 m. The internal radius of drum is 0.1 m. The friction lining extends over an arc AB such that angle AOC is 135° and the angle BOC is 45°. The brake is applied by means of a force F at Q as shown in Fig. 9.54. Assuming that brake lining obeys Hook's law in compression. Calculate braking torque on trailing shoe of brake when braking force of 500 N is applied at Q. The drum rotates in clockwise direction.*

Assume μ = 0.35.

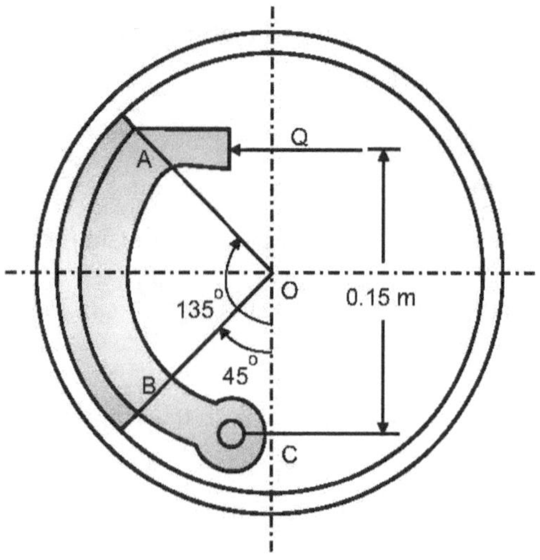

Fig. 9.46

Solution : l = 0.15 m, OO_2 = 0.075 m, r = 0.1 m, θ_2 = 135°, θ_1 = 45°, F = 500 N.

Assume : μ = 0.35.

To Find : T_B.

$$\theta_1 = 45° \times \frac{\pi}{180} = \frac{\pi}{4} \text{ rad}, \quad \theta_2 = 135°$$

$$= 130 \times \frac{\pi}{180} = \frac{3\pi}{4} \text{ rad}$$

Consider the trailing shoe.

The braking torque on trailing shoe is given by :

$$T_{B_2} = \mu (P_{max})_2 \, b \cdot r^2 (\cos\theta_1 - \cos\theta_2)$$

Here, $\Sigma m_e = 0$ or $\Sigma m_{O_2} = 0$.

$$-F_2 \times l + m_{n_2} + m_{F_2} = 0$$

$$\therefore \quad m_{n_2} + m_{F_2} = F_2 \times l$$

But let us find $(P_{max})_2$ b.

$$\therefore \quad F_2 \times l = m_{n_2} + m_{F_2}$$

$$m_{n_2} = \frac{1}{2}(P_{max})_2 \, br \, (OO_2) \left[(\theta_2 - \theta_1) + \frac{1}{2}(\sin 2\theta_1 - \sin 2\theta_2)\right]$$

$$= \frac{1}{2}(P_{max})_2 \times b \times 0.1 \times 0.075 \left[\left(\frac{3\pi}{4} - \frac{\pi}{4}\right) + \frac{1}{2}(\sin 90 - \sin 270)\right]$$

$$\therefore \quad m_{n_2} = (P_{max})_2 \times b \times 9.640486 \times 10^{-3}$$

$$m_{F_2} = \mu (P_{max})_2 \, br \left[r(\cos\theta_1 - \cos\theta_2) + \frac{OO_2}{4}(\cos 2\theta_2 - \cos 2\theta_1)\right]$$

$$\therefore \quad m_{F_2} = 0.35 \times (P_{max})_2 \times b \times 0.1 \left[0.1 \times (0.45 - \cos 135) + \frac{0.075}{4}(\cos 90 - \cos 270)\right]$$

$$\therefore \quad m_{F_2} = P_{max_2} \times b \times 0.0529974$$

Let $\quad x = P_{max_2} \times b$

$\therefore$ Putting value in $F_2 \times l = m_{n_2} + m_{F_2}$

$$\therefore \quad 500 \times 0.15 = x \times [(9.640486 \times 10^{-3}) + (4.9497 \times 10^{-3}]$$

$$\therefore \quad x = 5140.44 = P_{max_2} \times b$$

$$\therefore \quad T_{B_2} = \mu (P_{max})_2 \times br^2 [\cos\theta_1 - \cos\theta_2]$$

$$= 0.35 \times 5140.44 \times (0.1)^2 [\cos 45 - \cos 135]$$

$$= 25.44 \text{ N-m}$$

Similarly, $F_1 \times l + m_F - m_n = 0$

$$\therefore \quad F_1 \times l = m_{n_1} - m_{F_1}$$

$$\therefore \quad m_{n_1} = \frac{1}{2}(P_{max})_1 \times b \times r \times OO_1 \left[(\theta_2 - \theta_1) + \frac{1}{2}(\sin 2\theta_1 - \sin 2\theta_2)\right]$$

$$= \frac{1}{2} \times x \times 0.1 \times 0.075 \left[\left(\frac{3\pi}{4} - \frac{\pi}{4}\right) + \frac{1}{2}(\sin 90 - \sin 270)\right]$$

$$= x \times 9.640 \times 10^{-3}$$

$$m_{F_1} = \mu (P_{max})_1 \, br \left[r(\cos\theta_1 - \cos\theta_2) + \frac{OO_1}{4}(\cos\theta_2 - \cos 2\theta_1)\right]$$

$$= 0.35 \times x \times 0.1 \left[0.1\left(\cos\frac{\pi}{4} - \cos\frac{3\pi}{4}\right) + \frac{0.075}{4}\left(\cos\frac{3\pi}{4} - \cos\frac{2\pi}{4}\right)\right]$$

$$\therefore \quad 500 \times 0.15 = x [9.64 \times 10^{-3} - 4.48570 \times 10^{-3}]$$

$$x = 14550.95$$

$$T_{B_1} = 0.35 \times 14550.95745 \times (0.1)^2 \times [\cos 45 - \cos 135]$$

$$= 26.074$$

$$T_B = T_{B_1} + T_{B_2} = 2 \times 0.74 + 5.9265$$

$$T_B = 32 \text{ N-m}$$

Example 9.32 : *Fig. 9.47 shows a brake drum 330 mm diameter, acted on by 2 brake shoes mounted on pin A and pushed apart by 2 hydraulically operated piston at B, each exerting a force of 1200 N on the shoe on which it makes contact.*

The brake lining on each shoe extends 60° above and 60° below the horizontal centre line. The coefficient of friction is 0.2. The material of brake lining obeys Hook's law in compression.

Find the braking torque exerted on brake drum.

Solution : Data : D = Diameter of drum = 330 mm = 0.33 m.

$$r = \text{Radius of drum} = \frac{0.33}{2} = 0.165 \text{ m}$$

Force exerted on both shoes,

$$F_1 = F_2 = 1200 \text{ N}$$

$$\mu = 0.2 = \text{Coefficient of friction}$$

$$\theta_1 = 30°$$

$$\theta_1 = 30 \times \frac{\pi}{180} = \frac{\pi}{6} \text{ rad}$$

$$\theta_2 = 90° + 60° = 150°$$

$$= 150 \times \frac{\pi}{180} = \frac{5\pi}{6} \text{ rad}$$

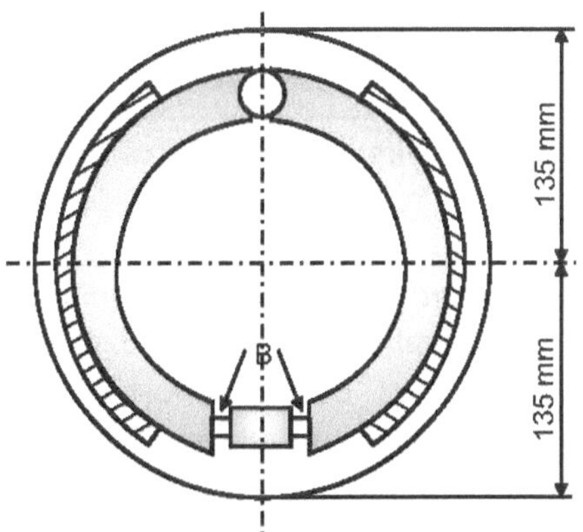

Fig. 9.47

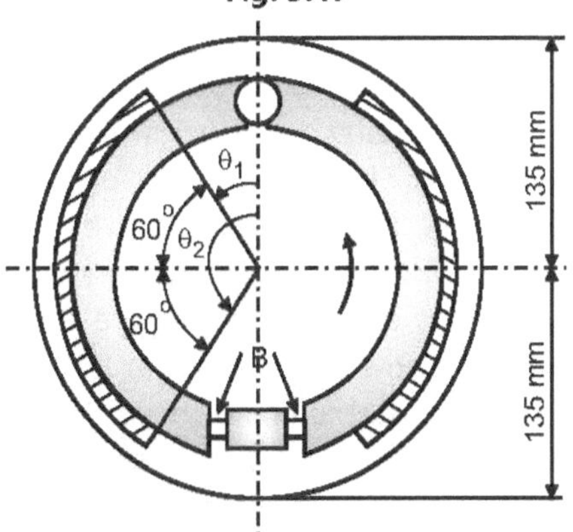

Fig. 9.48

For Leading Shoe :

$$T_{B_1} = \mu (P_{max}) br^2 (\cos\theta_1 - \cos\theta_2)$$

To find $(P_{max})_1$ b : $\Sigma m_{o_1} = 0$.

$$-F_1 \times l + m_{n_1} - m_{F_1} = 0$$

$$\therefore \quad F_1 \times l = m_{n_1} - m_{F_1}$$

$$m_{n_1} = \frac{1}{2} (P_{max})_1 \, br \, (OO_1) \left[(\theta_2 - \theta_1) + \frac{1}{2}(\sin 2\theta_1 - \sin 2\theta_2)\right]$$

$$= \frac{1}{2} \times (P_{max_1}) \, b \times 0.165 \times 0.135 \left[\left(\frac{5\pi}{6} - \frac{\pi}{6}\right) + \frac{1}{2}(\sin(2 \times 30) - \sin 2 \times 160)\right]$$

$$\therefore m_{n_1} = \frac{1}{2} (P_{max})_1 \, b \times 0.165 \times 0.135 \left[\left(\frac{5\pi}{6} - \frac{\pi}{6}\right) + \frac{1}{2}(\sin 60 - \sin 300)\right]$$

$$= 0.03297168 \times x \qquad (\because x = P_{max})_1 \times b)$$

$$m_{F_1} = \mu (P_{max})_1 \, br \left[r(\cos\theta_1 - \cos\theta_2) + \frac{OO_1}{4}(\cos\theta_2 - \cos 2\theta_1)\right]$$

$$\therefore \quad m_{F_1} = 0.2 (P_{max})_1 \, b \times 0.165 \left[0.165(\cos 30 - \cos 150) + \frac{0.135}{4}(\cos 150 - \cos 30)\right]$$

$$\therefore \quad m_{F_1} = 0.0094310166 (P_{max})_1 \times b$$

$$\therefore \quad F_1 \times l = m_{n_1} - m_{F_1}$$

$$\therefore \quad 1200 \times 0.27 = 0.03297168 (P_{max})_1 \times b - 0.0094310166 (P_{max})_1 \, b$$

$$\therefore \quad (P_{max})_1 \times b = 13763.4184$$

$$\therefore \quad T_{B_1} = \mu (P_{max})_1 \, br^2 (\cos\theta_1 - \cos\theta_2)$$

$$= 0.2 \times 13763.4184 \times (0.165)^2 [\cos 30 - \cos 150]$$

$$= 129.803 \text{ N-m}$$

For Trailing Shoe :

$$T_{B_2} = \mu (P_{max})_2 \, br^2 (\cos\theta_1 - \cos\theta_2)$$

To find $(P_{max})_2$ b :

$$F_2 \times l - m_{F_2} - m_{n_2} = 0$$

$$\therefore \quad F_2 \times l = m_{n_2} + m_{F_2}$$

$$m_{n_2} = \frac{1}{2} (P_{max})_2 \, br \, (OO_1) \left[(\theta_2 - \theta_1) + \frac{1}{2}(\sin 2\theta_1 - \sin 2\theta_2)\right]$$

$$= \frac{1}{2} \times (P_{max})_2 \times b \times 0.165 \times 0.135$$

$$\left[\left(\frac{5\pi}{6} - \frac{\pi}{6}\right) + \frac{1}{2}(\sin(2 \times 30) - \sin(2 \times 150))\right]$$

$$\therefore \quad m_{F_2} = \mu (P_{max})_2 \, br$$

$$\left[l(\cos\theta_1 - \cos\theta_2) + \frac{OO_1}{4}(\cos 2\theta_2 - \cos 2\theta_1)\right]$$

$$\therefore \quad m_{F_2} = 0.2 \times (P_{max})_2 \, b \times 0.165$$

$$\left[0.165(\cos 30 - \cos 150) + \frac{0.135}{4}(\cos 60 - \cos 300)\right]$$

$$\therefore \quad m_{F_2} = 0.0094310166 \, (P_{max})_2 \times b$$

$$F_2 \times l - m_{n_2} - m_{F_2} = 0$$

$$\therefore \quad F_2 \times l = m_{n_2} + m_{F_2}$$

$$\therefore \quad 1200 \times 0.27 = 0.03297168 \, (P_{max})_2 + (0.00943\ 10/66)\, (P_{max})_2 \times 3$$

$$(P_{max})_2 \, b = 1741.02347$$

$$T_{B_2} = 72.062619 \text{ N-m}$$

$$T_B = T_{B_1} + T_{B_2}$$

$$= 129.80 + 72.06$$

$$T_B = 201.86 \text{ N-m}$$

9.22 FRICTION MATERIALS

(i) Brake Materials

Following table gives various operating limits for Brakes

Table 9.2 : Operating limits for brakes

Sr. No.	Operating Condition	Upper Limit of pV
1.	Continuous application of load and poor dissipation of energy	10.5×10^5
2.	Intermittent application of load, comparatively longer periods of rest, poor dissipation	12×10^5
3.	Continuous application of load good dissipation of heat, as in an oil bath.	29.4×10^5

Some typical values of the maximum temperature are given below :

(i) Leather, fibre and wood facing = 65 – 70°C

(ii) Asbestos = 90 – 105°C

(iii) Asbestos block lining for automotives = 180 – 225°C

Some values for co-efficients of friction and allowable pressure are given below for friction materials of brackes.

Table 9.3 : Co-efficients of friction and allowable pressures

Material in Contact	Co-efficient of Frictions			Allowable Pressure N/mm²
	Dry	Greasy	Oily	
Cast Iron on Cast Iron	0.2-0.15	0.10-0.06	0.10-0.05	1.05-1.75
Bronze on Cast Iron	-	0.10-0.05	0.10-0.05	0.56-0.84
Steel on Cast Iron	0.35-0.25	0.12-0.07	0.10-0.06	0.84-1.40
Wood on Cast Iron	0.35-0.20	0.12-0.08	-	0.42-0.63
Fibre on metal	-	0.20-0.10	-	0.07-0.28
Cork on metal	0.35	0.30-0.25	0.25-0.22	0.055-0.105
Leather on metal	0.5-0.3	0.20-0.15	0.15-0.12	0.07-0.28
Wired asbestos on metal	0.5-0.35	0.30-0.25	0.25-0.20	0.21-0.56
Asbestos in rubber compound on metal	0.3-0.4	-	-	0.525-0.700
Asbestos in resin binder, on metal	0.3-0.4	-	0.10	0.525-0.700 4.2 of oily
Powered metal on metal	0.2-0.4	-	0.05-0.08	2.8

(ii) Materials for Brake Drums and for Brake Linings

The most common material for brake durms is cast iron with some alloying elements added. When good heat conduction is needed stainless steel, aluminium and model can also be used. Brake linings are usually made of some from of asbestos and binders. Most of the time, the asbestos based linings are moulded. These lining are used upto temperatures of about 200°C. For temperature range of 200°C to 400°C, sintered metal lining are usually employed. Sintered metal linings with ceramic particles can be used for temperature range of 400°C to 540°C. By increasing the ceramic content, temperatures as high as 1000°C can be tolerated. Ceramics linings have high thermal conductivity, long life and very stable friction properties. However, these and expensive, have lower co-efficient of friction and must be made to shape because of their rigidity. Linings are attached to the drum usually by riveting or bonding. Riveting is cheaper, and simple but its drawback is that the usable lining thickness depends upon the depth to which the rivet head can be counter sunk. Bonded linings have greater effective thickness and more friction area but more expensive.

(iii) Clutch Materials

Friction Materials : The three factors for selecting a material for facing of a friction clutch disc are : Co-efficient of friction, design pressure of contact, and maximum operating temperature. A good friction material should have the following qualities :

1. A high co-efficient of friction capable of uniform over of wide range of surface velocities temperatures and loads.
2. Adequate mechanical and thermal strength.
3. Little wear and no scoring.
4. High heat conductivity for rapid transfer of heat from the friction surfaces.

Some of the most widely used friction materials are given in Table 9.4 .

Table 9.4 : Friction Materials

Material of Friction Surfaces		Operating Conditions	Co-efficient of Friction	Unit Pressure N/mm²	Maximum Operating Temperature °C
Hardened Steel	Hardened steel	In oil	0.08	0.6-0.8	250
Cast Iron	Cast / Iron or Steel	In oil	0.06	0.6-0.8	250-300
Cast Iron	-do-	Dry	0.15	0·25-0.4	250-300
Bronze	-do-	In oil	0.05	0.4	150
Pressed Asbestos	-do-	Dry	0.3	0.2-0.3	150-250
Powder-Metal	-do-	Dry	0.4	0.3	550
Power Metal	-do-	In oil	0.1	0.8	550

9.23 THERMAL CONSIDERATIONS

For the useful service by a brake, it should not be allowed to overheat, Heat dissipating capacity of the brake depends upon the braking surface and the unit pressure. Increased braking surface and reduced unit pressure results in longer life for the brake.

Thus, heat generated is,

$$H_g = \mu \cdot F_n \cdot V, \text{ watts}$$

$$= \mu \cdot p \cdot A_c \cdot V \text{ watts}$$

where μ = co-efficient of friction,

p = average contact pressure, N/m^2

A_c = contact area, m^2,

V = rubbing velocity, m/s

The heat generated can also be found out by considering the amount of kinetic energy and/or potential energy that is being absorbed. Thus,

$$H_g = (E_k + E_p) \text{ watts}$$

Now, the heat energy dissipated may be estimated by

$$H_d = C \, \Delta T \cdot Ar \text{ watts}$$

where C = Co-efficient of heat transfer, W/m^2K

ΔT = temperature difference between the exposed radiating surface and surrounding air

A_r = area of radiating surface, m^2

The value of C may be about 29.5 for a ΔT of 40°C and 44 for a ΔT of 200°C.

Another simple criterion of brake capacity is,

$$= \frac{kW}{wD}, \text{ which is limited to } 350 \text{ W/m}^2$$

where w = width of band or shoe, m

D = diameter of drum, m

Or the temperature rise of the brake assembly can be found out by equation

All the above relations are approximate due to a large number of factors involved.

Now in a unit time, the heat generated will be equal to the frictional force times the rubbing velocity equation. Now if the co-efficient of friction is fixed, then

Heat generated in a unit time = Normal force × rubbing velocity. For a given braking surface.

Heat generated in a unit time = Unit pressure × rubbing velocity = $p \times V$

Thus the product pV is the common criterion for the design of brackes. The upper limits of pV that will give a reasonable wear life for the brake are given in table , p is in N/m^2 and V in m/s.

EXERCISE

1. Derive an expression for frictional torque of a collar thrust bearing assuming uniform pressure and uniform wear theory.
2. Explain multiple clutch with the help of neat sketch.

3. Justify the statement, "Though cone clutches provide high frictional torque, yet they have become obsolete".
4. Derive an expression for frictional torque of a collar thrust bearing assuming (i) uniform pressure, (ii) uniform wear.
5. Derive an expression for frictional torque of conical pivot bearing, assuming uniform wear theory.
6. Draw a neat sketch of centrifugal clutch, give one application.
7. Deduce an expression for frictional torque for truncated conical pivot collar bearing under uniform pressure condition.
8. Derive the expression for frictional torque for uniform pressure and wear theory for flat pivot bearing.
10. Explain 'self energizing' and 'self-locking' effect on block brakes.
11. Discuss energy equation for the break.
12. Explain the materials used for
    (i) Brakes
    (ii) Brake drum
    (iii) Clutches
13. Explain 'self energizing' and 'self-locking' effect on block brakes.
14. Explain internal expanding shoe brake.

◈◈◈

UNIT VI - PRESSURE VESSEL

CHAPTER - 10
PRESSURE VESSEL

10.1 PRESSURE VESSELS AND CYLINDERS

- It is used to store fluids under pressure. e.g. low pressure boilers, reaction vessels, nuclear vessels, vessels in chemical plants.
- Pressure vessels are designed with great care because rupture of a pressure vessel means an explosion, which may cause loss of life and property.

Applications of Storage Vessels:

- Petrochemical Plants
- Pharmaceutical plants
- Oil & gas industry
- Nuclear Power plants
- Biotech industry
- Food Industry
- Domestic applications

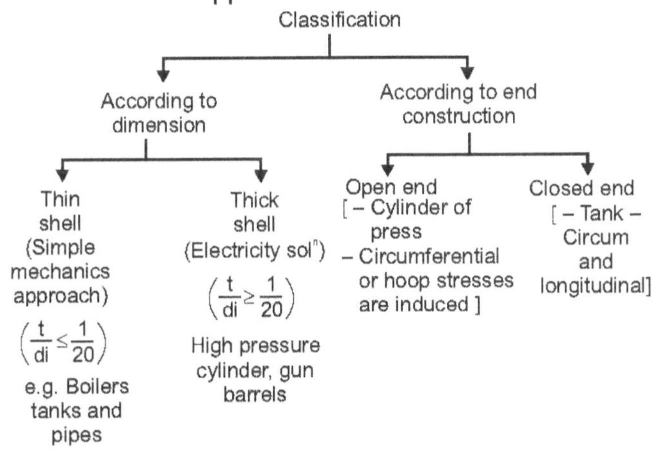

Fig. 10.1

PART A : DESIGN OF CYLINDERS

10.2 DESIGN OF THIN-CYLINDRICAL VESSEL

- It is a typical application of plane stress problem, wherein the loading is in-plane.
- In case of thin-walled cylinders stresses may be assumed uniformly distributed over the wall thickness stresses in thin walled cylinders.

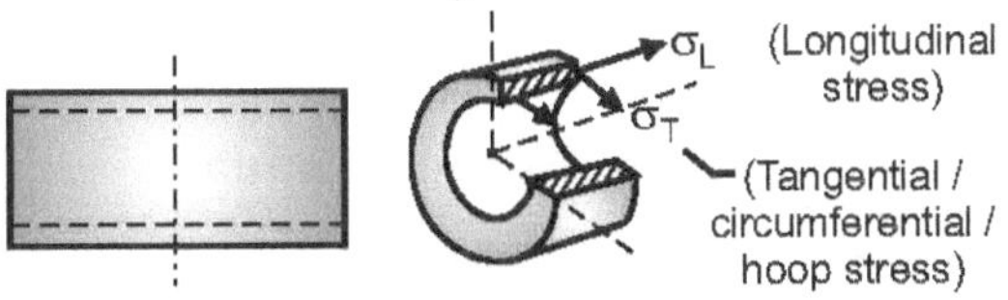

Fig. 10.2

(i) Circumferential/Tangential/Hoop Stress :

When a closed cylindrical shell is subjected to internal pressure (P_i), a tensile stress acts in the direction tangential to the circumference (or perpendicular to the axis) is called circumferential or hoop stress (σ_T).

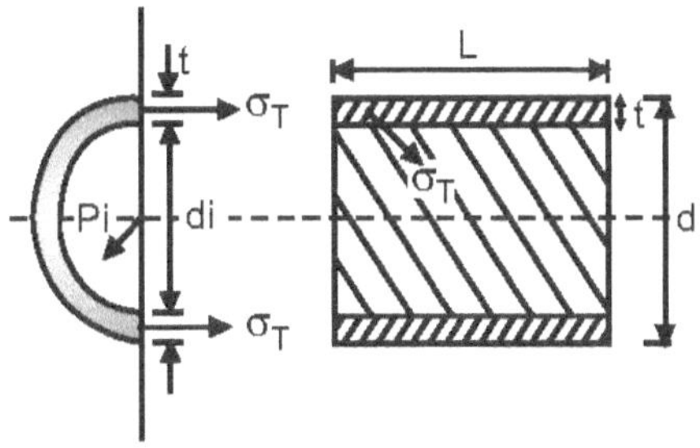

Fig. 10.3

P_i - Internal pressure

d_i - Internal diameter, I.D.

σ_T - Hoop stress

L - Length of cylinder

t - Thickness of shell

$$\text{Pressure force} = \text{Internal press} \times \text{Projected area} = P_i \times (L.d_i)$$

This force is resisted by longitudinal cross-sectional area.

This force resulted in circumferential stress to be induced.

$$\therefore \quad \text{Resisting force} = (\text{Hoop stress}) \times (\text{Resisting area}) = \sigma_T \times (L \cdot t + L \cdot t) = 2\,\sigma_T \cdot L \cdot t$$

For equilibrium,

$$\text{Pressure force} = \text{Resisting force}$$

$$P_i \times L \cdot d_i = 2\,\sigma_T \cdot L \cdot t$$

$$t = \frac{P_i\, d_i}{2\sigma_T}$$

(ii) Longitudinal Stress :

A tensile stress acting in the direction of the axis is called longitudinal stress.

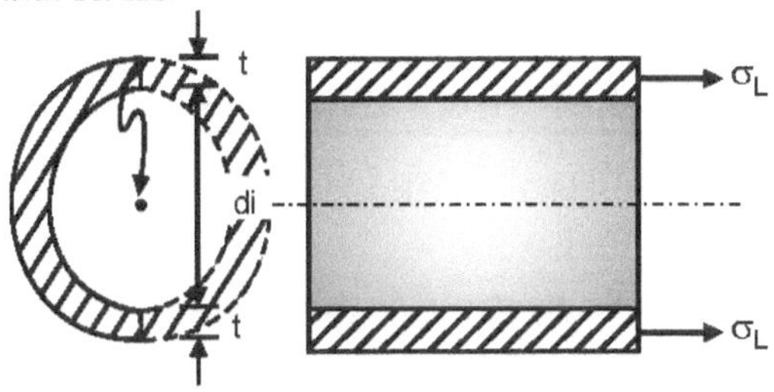

Fig. 10.4

$$\text{Pressure force} = P_i \times \left(\frac{\pi}{4} d_i^2\right)$$

$$\text{Resisting force} = (\sigma_L) \times (\pi\, d_i \cdot t)$$

For equilibrium,

$$\text{Pressure force} = \text{Resisting force}$$

$$P_i \times \frac{\pi}{4} d_i^2 = \sigma_L \times \pi\, d_i \cdot t$$

$$t = \frac{P_i \cdot d_i}{4\sigma_L}$$

Similar analysis is applicable to thin walled spherical shell.

(iii) Joint Efficiency :

$$t = \frac{P_i\, d_i}{2 \cdot J \cdot \sigma_T}$$

J - Joint efficiency, $t = \dfrac{P_i\, d_i}{4 \cdot J \cdot \sigma_L}$

- **For Engine Cylinders :**

$$t = \left[\frac{P_i\, d_i}{2\sigma_T}\right] + 6 \text{ to } 12 \text{ mm}$$

  To permit reboring after wear has taken place.

- When riveted joints/welded joints are used in construct-ing vessel.

$$t = \frac{P_i\, d_i}{2\sigma_T \cdot \eta_l}$$

  η_l = Efficiency of the longitudinal riveted joint

- **For Cylinder of Ductile Material :**

$$\sigma_T = 0.8\, \sigma_y \qquad \dots (10.1)$$

- **For Cylinders of Brittle Material :**

$$\sigma_T = 0.125\, \sigma_u \qquad \dots (10.2)$$

- **As per Boiler Code :**

Table 10.1

Boiler Diameter	Minimum Plate Thickness (t)
0.9 m or less	6 mm
$0.9 \le d < 1.35$ m	7.5 mm
$1.35 < d < 1.8$ m	9 mm
$1.8 < d$	12 mm

10.3 DESIGN OF THICK CYLINDERS (May 18)

- Thick cylinders are used generally for the high pressure applications.

  Herein, uniform stress distribution over the thickness is not valid.
- Radial stress is taken into account.
- $\left[\dfrac{t}{d} > \dfrac{1}{10}\right] \Rightarrow$ Analysis is done by considering vessel as thick.

  σ_R = Radial, σ_T = Hoop, σ_L = Longitudinal

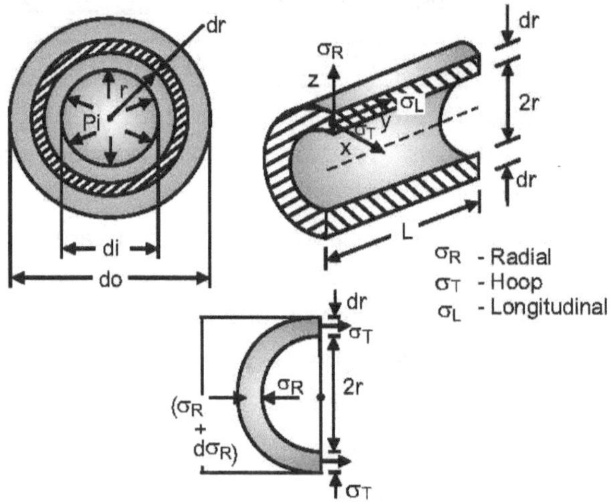

Fig. 10.5

Considering equilibrium, force acting on elemental ring.

$$+ (\sigma_R + d\sigma_R)(2r + 2dr)\, L - \sigma_R\,(2rL) + \sigma_T\,(2dr)\, L = 0$$

$$2r\, L\, \sigma_R + 2r\, L \cdot d\,\sigma_R + 2dr\, L \cdot \sigma_R + 2\, dr \cdot L\, d\,\sigma_R - \sigma_R \cdot 2r\, L + \sigma_T \cdot 2dr \cdot L = 0$$

Here in above equation '(2r + 2dr) L' is critical area.

Ignoring small products,

$$2rL \cdot d\sigma_R + 2dr \cdot L \cdot \sigma_R + \sigma_T \cdot 2dr \cdot L = 0$$

$$r \cdot d\sigma_R + L \cdot dr \cdot \sigma_R + \sigma_T \cdot dr = 0$$

$$\sigma_R + \sigma_T + r\frac{d\sigma_R}{dr} = 0 \qquad \dots (10.3)$$

Stress due to
σ_L = Tensile Positive
σ_T = Tensile – Produce negative strain in lateral plane
σ_R = Compressive – Producing positive strain in lateral plane

(i) Strain in Lateral Plane :

$$\epsilon_L = \frac{\sigma_L}{E} - \frac{\mu}{E}\sigma_T + \frac{\mu}{E}\sigma_R$$

(∵ Triaxial stress system)

$$E\,\epsilon_L = \sigma_L + \mu\,(\sigma_R - \sigma_T)$$

$$\mu\,(\sigma_R - \sigma_T) = E\left(\epsilon_L - \frac{\sigma_L}{E}\right)$$

$$\sigma_R - \sigma_T = \frac{E}{\mu}\left(\epsilon_L - \frac{\sigma_L}{E}\right)$$

Let $\sigma_R - \sigma_T = (-2A) = \text{Constant} \qquad \dots (10.4)$

Also (in transverse section),

$$\sigma_L \cdot \frac{\pi}{4}(d_o^2 - d_i^2) = P_i \frac{\pi}{4} d_i^2$$

$$\sigma_L = \frac{P_i d_i^2}{d_o^2 - d_i^2} = \frac{P_i}{k^2 - 1}$$

$$k = \frac{d_o}{d_i}$$

Adding equations (10.3) and (10.4).

$$\therefore \quad 2\sigma_R + r\frac{d\sigma_R}{dr} = -2A$$

Multiplying by 'r',

$$2\sigma_R \cdot r + r^2\frac{d\sigma_R}{dr} = -2A \cdot r$$

$$\frac{d}{dr}(r^2 \cdot \sigma_R) = -2A \cdot r$$

Integrating w.r.t. r.,

$$r^2 \sigma_R = -\frac{2Ar^2}{2} + B$$

$$\therefore \quad \sigma_R = -A + \frac{B}{r^2} \quad \ldots (10.5)$$

Using in equation (10.4),

$$\sigma_R - \sigma_T = -2A$$

$$\left(-A + \frac{B}{r^2}\right) - \sigma_T = -2A$$

$$\sigma_T = A + \frac{B}{r^2} \quad \ldots (10.6)$$

For cylinder subjected to internal pressure only. Boundary conditions are

At $r = \frac{d_i}{2} \Rightarrow \sigma_R = P_i$

$r = \frac{d_o}{2} \Rightarrow \sigma_R = 0$

Substituting in equation (10.5),

$$0 = -A + \frac{4B}{d_o^2} \Rightarrow A = \frac{4B}{d_o^2}$$

$$P_i = -A + \frac{4B}{d_i^2} = -\frac{4B}{d_o^2} + \frac{4B}{d_i^2}$$

$$= -4B\left[\frac{1}{d_o^2} - \frac{1}{d_i^2}\right] = -\frac{4B}{d_o^2}[1 - k^2]$$

$$= \frac{4B}{d_o^2}[k^2 - 1]$$

$$B = \frac{P_i d_o^2}{4(k^2 - 1)} \; ; \; A = \frac{4}{d_o^2}\left[\frac{P_i d_o^2}{4(k^2 - 1)}\right]$$

$$A = \frac{P_i}{k^2 - 1}$$

$$\therefore \quad \sigma_R = -\frac{P_i}{k^2 - 1} + \frac{P_i d_o^2}{4(k^2 - 1) r^2}$$

$$= -\frac{P_i}{k^2 - 1}\left[1 - \frac{d_o^2}{4r^2}\right]$$

As radial stress is always compressive.

$$\sigma_R = \frac{P_i}{k^2 - 1}\left[1 - \frac{d_o^2}{4r^2}\right]$$

$$\sigma_R = \frac{-P_i}{k^2 - 1}\left[\frac{d_o^2}{4r^2} - 1\right]$$

Similarly, $\sigma_T = A + \frac{B}{r^2}$

$$= \frac{P_i}{k^2 - 1} + \frac{P_i d_o^2}{4(k^2 - 1) r^2}$$

$$\sigma_T = \frac{P_i}{k^2 - 1}\left[1 + \frac{d_o^2}{4r^2}\right]$$

Now, $\sigma_T = \sigma_{T\,max}$ at $r = \frac{d_i}{2}$

$$\sigma_{T\,max} = \frac{P_i}{k^2 - 1}\left(1 + \frac{d_o^2}{d_i^2}\right)$$

$$\sigma_{T\,max} = P_i\left[\frac{k^2 + 1}{k^2 - 1}\right]$$

$$\sigma_R = \sigma_{R\,max} \text{ at } r = \frac{d_i}{2}$$

$$\sigma_{R\,max} = -P_i$$

At $r = \frac{d_o}{2}$, $\sigma_T = \frac{2P_i}{k^2 - 1}$ and $\sigma_R = 0$

(ii) Stress Distribution :

$1 \rightarrow \sigma_{R\,min}$ at $r = \frac{d_o}{2} = 0$

$2 \rightarrow \sigma_{R\,max}$ at $r = \frac{d_i}{2} = -P_i$

$3 \rightarrow \sigma_{T\,min}$ at $r = \frac{d_o}{2} = \frac{2P_i}{k^2 - 1} = P_i\left(\frac{2}{k^2 - 1}\right)$

$4 \rightarrow \sigma_{T\,max}$ at $r = \frac{d_i}{2} = P_i\left(\frac{k^2 + 1}{k^2 - 1}\right)$

$5 \rightarrow \sigma_L = \frac{P_i}{k^2 - 1}$

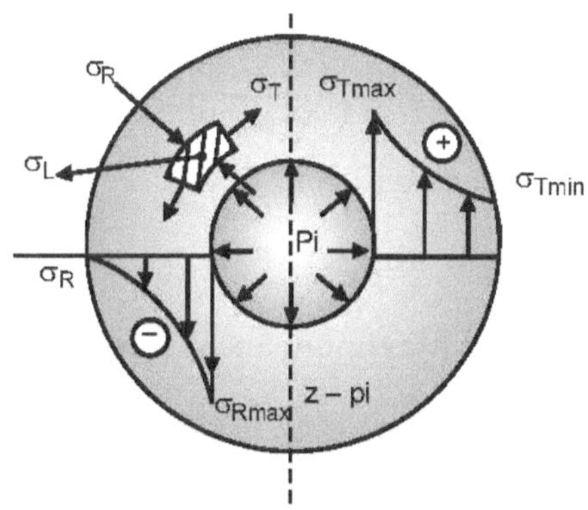

Fig. 10.6

Major Conclusions :

(i) $\sigma_{T\,max} > P_i$.

(ii) Maximum σ_T and σ_R occurs at $\frac{d_i}{2}$.

(iii) $\sigma_R = 0$ at $r = \frac{d_o}{2}$.

10.4 LAME'S THEORY (Nov. 16)

(i) Design Based on Maximum Principle Stress Theory :

Assumptions :

- Cylinder material is homogeneous, isotropic and obeys Hooke's law.
- Cylinder open at both ends.
- Failure occurs when,

  Maximum principle stress = Maximum permissible strength of material.
- It is applicable to brittle material CI, CS and CA.

$$\sigma_T = P_i\left(\frac{k^2+1}{k^2-1}\right)$$

$$\sigma_R = -P_i$$

$$\sigma_L = \frac{P_i}{k^2-1}$$

As $k > 1$, $\sigma_T > \sigma_R$ and $\sigma_T > \sigma_L$

∴ As per maximum principle stress theory,

$$\sigma = \sigma_T = P_i\left(\frac{k^2+1}{k^2-1}\right)$$

$$\frac{\sigma}{P_i} = \frac{k^2+1}{k^2-1}$$

$$\frac{\sigma+P_i}{\sigma-P_i} = \frac{k^2+1+k^2-1}{k^2+1-k^2+1}$$

$$k = \sqrt{\frac{\sigma+P_i}{\sigma-P_i}}$$

Also $$k = \frac{d_o}{d_i} = \frac{d_i+2t}{d_i} = 1+\frac{2t}{d_i}$$

$$\frac{2t}{d_i} = k-1$$

$$t = (k-1)\frac{d_i}{2}$$

$$t = \frac{d_i}{2}\left(\sqrt{\frac{\sigma+P_i}{\sigma-P_i}}-1\right)$$

This is called Lame's equation,

$$\sigma = \frac{S_{ut}}{F{\cdot}S}$$

(ii) Design Based on Maximum Shear Stress Theory :

Failure occurs when

Maximum shear stress = Maximum permissible shear strength

$$\tau = \frac{\sigma_T-\sigma_R}{2} \qquad (\text{As } \sigma_T > \sigma_R)$$

$$= \frac{P_i\left(\frac{k^2+1}{k^2-1}\right)-(-P_i)}{2}$$

$$= \frac{P_i}{2}\left[\frac{k^2+1}{k^2-1}+1\right] = \frac{P_i}{2}\left[\frac{2k^2}{k^2-1}\right]$$

$$= \frac{P_i\,k^2}{k^2-1}$$

$$\frac{\tau}{P_i} = \frac{k^2}{k^2-1}$$

$$\frac{\tau}{\tau-P_i} = \frac{k^2}{k^2-(k^2-1)} \Rightarrow k^2 = \frac{\tau}{\tau-P_i}$$

$$k = \sqrt{\frac{\tau}{\tau-P_i}}$$

$$k = \frac{d_o}{d_i}$$

$$t = (k-1)\frac{d_i}{2} = \frac{d_i+2t}{d_i}$$

$$t = \frac{d_i}{2}\left[\sqrt{\frac{\tau}{\tau-P_i}}-1\right] = 1+\frac{2t}{d_i}$$

where $\tau = \frac{S_{sy}}{F{\cdot}S}$ and $S_{sy} = \frac{S_{yp}}{2}\,(k-1)\frac{d_i}{2} = t$

(iii) Design Based on Maximum Distortion Energy Theory :

As per this theory failure occurs when

Strain energy of distortion per unit volume

= Strain energy of distortion per unit volume of a standard test specimen when yielding starts

For tri-axial system,

$$\sigma^2 = \frac{1}{2}\left[(\sigma_T-\sigma_R)^2+(\sigma_R-\sigma_L)^2+(\sigma_L-\sigma_T)^2\right]$$

$$t = \frac{d_i}{2}\left[\sqrt{\frac{\sigma}{\sigma-\sqrt{3}\,P_i}}-1\right]$$

where, $$\sigma = \frac{S_{yp}}{F{\cdot}S}$$

10.5 DESIGN BASED ON MAXIMUM STRAIN THEORY (ST. VENANT'S THEORY)

(i) Clavarino's Equation :

- For ductile material e.g. M.S. or alloy steel. Maximum strain energy is used as failure criterion.
- According to this theory, failure/yielding occurs when maximum strain equals yield point strain in a standard test specimen.

- Clavarino's equation is applicable to cylinders with closed ends and made of ductile materials.

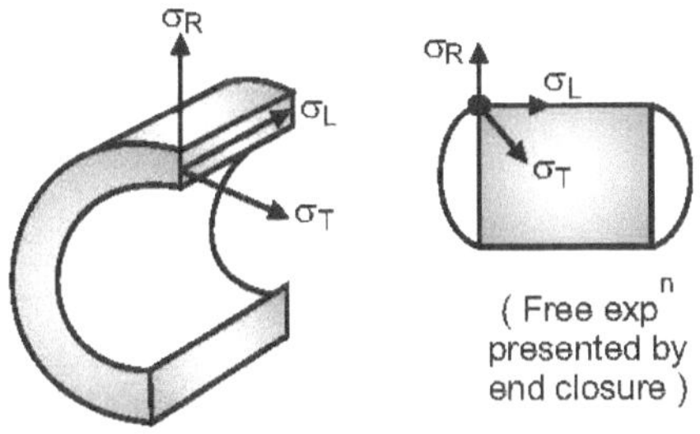

Fig. 10.7

$$\in = \frac{\sigma}{E} = \frac{\sigma_T}{E} - \mu\frac{\sigma_R}{E} - \mu\frac{\sigma_L}{E}$$

$$\sigma = \sigma_T - \mu(\sigma_R + \sigma_L)$$

$$\sigma = P_i\left(\frac{k^2+1}{k^2-1}\right) - \mu\left[-P_i + \frac{P_i}{k^2-1}\right]$$

$$= P_i\left(\frac{k^2+1}{k^2-1}\right) + \mu P\left[1 - \frac{1}{k^2-1}\right]$$

$$= P_i\left(\frac{k^2+1}{k^2-1}\right) + \mu P_i\left(\frac{k^2-1-1}{k^2-1}\right)$$

$$= \frac{P_i}{k^2-1}(k^2 + 1 + \mu k^2 - 2\mu)$$

$$\therefore \quad \sigma k^2 - \sigma = P_i[(1-2\mu) + k^2(1+\mu)]$$

$$\therefore \sigma k^2 - P_i \cdot k^2(1+\mu) = \sigma + P_i(1-2\mu)$$

$$k^2(\sigma - P_i(1+\mu)) = \sigma + P_i(1-2\mu)$$

$$\therefore \quad k = \sqrt{\frac{\sigma + P_i(1-2\mu)}{\sigma - P_i(1+\mu)}}$$

But $$k = \frac{d_o}{d_i} = \frac{d_i + 2t}{d_i} = 1 + \frac{2t}{d_i}$$

$$\therefore \quad \frac{2t}{d_i} = k - 1 \Rightarrow t = \frac{d_i}{2}(k-1)$$

$$\therefore \quad t = \frac{d_i}{2}\left[\sqrt{\frac{\sigma + P_i(1-2\mu)}{\sigma - P_i(1+\mu)}} - 1\right]$$

This is called Clavarino's equation,

$$\sigma = \frac{S_{yp}}{F \cdot S}$$

(ii) Bernie's Equation : (May 16, 17, Nov. 17)

This equation is applicable to the cylinder with open ends and made of ductile material.

As cylinder ends are open, $\sigma_L = 0$.

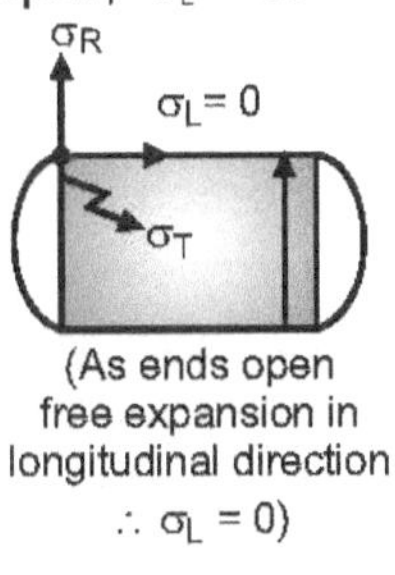

Fig. 10.8

$$\therefore \quad \in = \frac{\sigma}{E} = \left(\frac{\sigma_T}{E}\right) - \frac{\mu}{E}(\sigma_R + \sigma_L)$$

$$\sigma = \sigma_T - \mu\sigma_R$$

$$\sigma = P_i\left(\frac{k^2+1}{k^2-1}\right) - \mu(-P_i)$$

$$\sigma = P_i\left[\frac{k^2+1}{k^2-1} + \mu\right]$$

$$\frac{k^2+1}{k^2-1} = \frac{\sigma - \mu P_i}{P_i}$$

$$\therefore \quad \frac{(k^2+1)+(k^2-1)}{(k^2+1)-(k^2-1)} = \frac{\sigma - \mu P_i + P_i}{\sigma - \mu P_i - P_i}$$

$$\frac{2k^2}{2} = \frac{\sigma + P_i(1-\mu)}{\sigma - P_i(1+\mu)}$$

$$k = \sqrt{\frac{\sigma + P_i(1-\mu)}{\sigma - P_i(1+\mu)}}$$

But $$t = \frac{d_i}{2}(k-1)$$

$$\therefore \quad t = \frac{d_i}{2}\left[\sqrt{\frac{\sigma + P_i(1-\mu)}{\sigma - P_i(1+\mu)}} - 1\right]$$

This is Bernie's equation where $\sigma = \frac{S_{yp}}{F \cdot S}$.

(iii) Lame's Equation : Design of pressure vessel based on maximum principle stress theory.

(iv) Clavarino's and Bernie's Equation : Design based on maximum strain theory.

10.6 CYLINDER WITH EXTERNAL PRESSURE

- **Shells Subjected to External Pressure**

The thickness of thin-walled shells subjected to external pressure may be calculated either by the formulae given below or by the method given in Appendix F of IS : 2825 : 1969.

Notation : The following notation has been used in the formulae in this section :

t = minimum thickness of the shell material in mm

D_o = outer diameter of the shell in mm, and

L = effective length in mm.

In the case of cylindrical shells, it is the maximum of the following values and is measured parallel to the axis of the shell :

(a) The distance between head bend lines plus one-third the depth of each head, when no stiffening rings are present.

(b) the maximum centre distance between two adjacent stiffening rings and

(c) the distance from the centre of the nearest stiffening to the head bend line plus one third the depth of the head.

In the case of spherical shells, the effective length is equal to the inside radius of the spherical shell :

p = design pressure in kgf/cm^2 ;

σ = 0.2 percent proof stress in kgf/mm^2; and

K = the ratio of the elastic modulus E of the material at the design metal temperature to the room temperature elastic modulus.

- **Thickness of Cylindrical Shells Under External Pressure :** The thickness of cylindrical shells under external pressure is given by equations 10.7, 10.8, or 10.9 as is applicable :

(a) for $\frac{L}{D_o} < \frac{0.58\left(\frac{10p}{\sigma}\right)^{\frac{5}{2}}}{pK}$ or $< \frac{38\left(\frac{100t}{D_o}\right)^{\frac{3}{2}}}{K\sigma}$

$$t = \frac{D_o}{100}\left[\frac{1.15p}{\sigma} + 0.053\left(\frac{K\sigma L}{D_o}\right)^{\frac{2}{3}}\right] \quad \text{... (10.7)}$$

$$\text{or } p = \frac{\sigma}{1.15}\left[\frac{100t}{D_o} - 0.053\left(\frac{K\sigma L}{D_o}\right)^{\frac{2}{3}}\right] \quad \text{... (10.7a)}$$

(b) for $\frac{L}{D_o} > \frac{14.4}{(pK)^{\frac{1}{6}}}$ or $> \frac{14.6}{\sqrt{\frac{100t}{D_o}}}$; or

$$\frac{0.58\left(\frac{10p}{\sigma}\right)^{\frac{5}{2}}}{pK} > \frac{14.4}{(pK)^{\frac{1}{6}}};$$

$$\text{or } 38\frac{\left(\frac{100t}{D_o}\right)^{\frac{3}{2}}}{K\sigma} > \frac{14.6}{\sqrt{\frac{100t}{D_o}}}$$

$$t = 1.03 \times \frac{D_o}{100} \times (pK)^{\frac{1}{3}} \text{ but not less than } \frac{3.5\,p\,D_o}{200\sigma} \quad \text{... (10.8)}$$

$$\text{or } p = \frac{0.91}{K} \times \left(\frac{100t}{D_o}\right)^3 \text{ but not greater than } \frac{200t\sigma}{3.5\,D_o} \quad \text{... (10.8a)}$$

(c) In all other cases

$$t = \frac{D_o}{100}\left[0.075p \cdot \frac{L}{D_o} \cdot K\right]^{\frac{2}{5}} \quad \text{... (10.9)}$$

$$\text{or } P = \frac{13.3\left(\frac{100t}{D_o}\right)^{\frac{5}{2}}}{\frac{L}{D_o} \cdot K} \quad \text{... (10.9a)}$$

- **Thickness of Spherical Shells Under External Pressure :** The thickness of spherical shells under external pressure is given by equation 10.10 :

$$t = \frac{pD_o}{80\sigma} \quad \text{... (10.10)}$$

$$\text{or} \quad p = \frac{80\sigma t}{D_o} \quad \text{... (10.10a)}$$

10.7 AUTO FRETTAGE (PRESTRESSING OF THICK CYLINDERS)

- It is defined as a process of prestressing the cylinder before using it in service in order to increase its pressure capacity.
- When a cylinder is subjected to internal pressure P_i, hoop stress is maximum at inner radius and it decrease towards the outer radius and minimum at $d_o/2$. As this hoop stress at $d_i/2$ is tensile in nature and it limits the pressure capacity of cylinder.

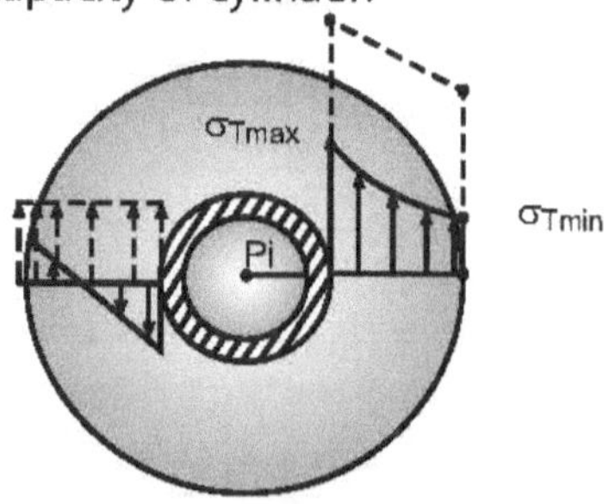

Fig. 10.9

- In prestressing process residual compressive stresses are developed at the inner surface, when the cylinder is loaded, the compressive stresses must be relieved before any tensile stress is developed, thus increasing pressure capacity of cylinder.
- There are three methods of prestressing the cylinder :
  1. **Shrink Fitting One Cylinder Over the Other :** This is achieved by heating outer cylinder and slipping it over the inner cylinder to achieve contact stresses when cooled. This induces **compressive stresses** in **inner cylinder** and **tensile stresses** in **outer cylinder (tangential).**
  2. **Overloading the Cylinder Before Using in Service :** The overloading pressure is adjusted in such a way that a portion of the cylinder near the inner diameter is subjected to stresses in plastic range, while the outer portion is still in the elastic range. When pressure is released, the outer portion of contracts exerting pressure on the inner portion which has undergone permanent deformation. This induced residual compressive stresses at the inner surface.

3. **Wire Wound Cylinder :**

In this method a wire under tension is closely wound around the cylinder, which results in residual compressive stresses.

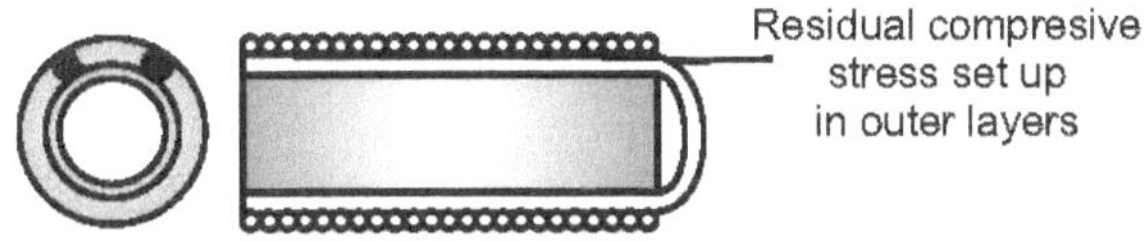

Fig. 10.10

Advantages :

- Thus autofrettage results in increased pressure capacity of cylindrical vessels by including prior residual compressive stresses at inner surface of cylinder, thus resulting vessel can withstand higher internal pressures.

10.8 COMPOUND CYLINDER ANALYSIS

- It consists of cylinder and outer jacket where inner diameter of jacket is slightly smaller than outer diameter of cylinder. This jacket is fitted onto cylinder by heating. After cooling, it tends to contract onto the inner cylinder, which results in residual compressive stresses to set over inner cylinder and tensile stresses to set on inner diameter of jacket.
- There is shrinkage pressure P between inner cylinder and the jacket which tends to contract the cylinder and expand the jacket.

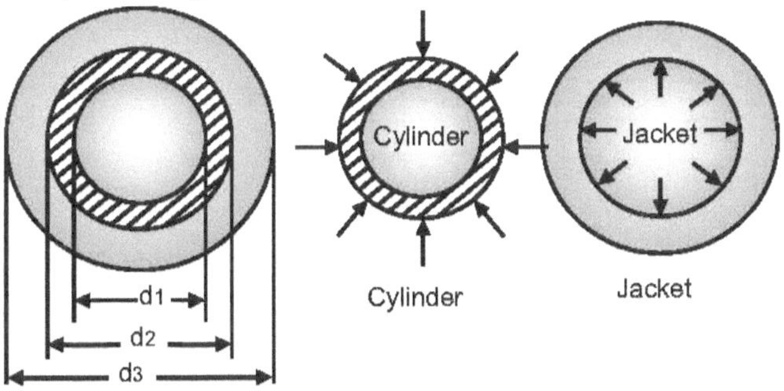

Fig. 10.11

Let δ_j be the increase in inner diameter of jacket

σ_c be the decrease in outer diameter of cylinder

Jacket Analysis :

Assuming it as thick cylinder with an internal pressure P.

Tangential strain,

$$\in_T = \frac{\text{Change in circumference}}{\text{Original circumference}}$$

$$\in_{T_j} = \frac{\pi(d_2+\delta_j) - \pi d_2}{\pi d_2} = \frac{\delta_j}{d_2}$$

Also, $$\in_{T_j} = \frac{\sigma_T}{E} - \mu\frac{\sigma_R}{E}$$

$$= \frac{P}{E}\left[\frac{k_1^2+1}{k_1^2-1}\right] + \frac{\mu}{E}\cdot P \quad \text{where } k_1 = \frac{d_3}{d_2}$$

$$= \frac{P}{E}\left[\frac{\frac{d_3^2}{d_2^2}+1}{\frac{d_3^2}{d_2^2}-1} + \mu\right]$$

$$\in_{T_j} = \frac{P}{E}\left[\frac{d_3^2+d_2^2}{d_3^2-d_2^2} + \mu\right] = \frac{\delta_j}{d_2}$$

$$\therefore \quad \delta_j = \frac{Pd_2}{E}\left[\frac{d_3^2+d_2^2}{d_3^2+d_2^2} + \mu\right]$$

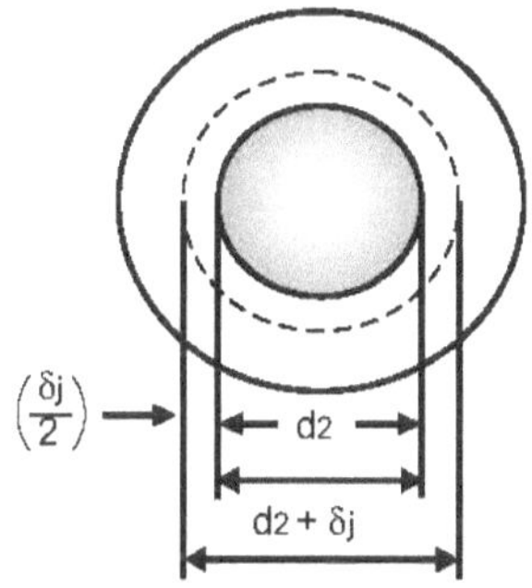

Fig. 10.12

Cylinder Analysis :

Analogous to the analysis of thick cylinder subjected to external pressure.

$$\in_{T_c} = \frac{\pi d_2 - \pi(d_2-\delta_c)}{\pi d_2}$$

$$= \frac{\delta_c}{d_2} = \frac{\sigma_T}{E} - \mu\frac{\sigma_R}{E}$$

$$= -\frac{P}{E}\left(\frac{k_2^2+1}{k_2^2-1}\right) - \mu\frac{(-P)}{E}$$

where, $$k_2 = \frac{d_2}{d_1}$$

$$\frac{\delta_c}{d_2} = -\frac{P}{E}\left(\frac{k_2^2+1}{k_2^2-1}\right) + \frac{\mu P}{E}$$

$$= -\frac{P}{E}\left(\frac{\frac{d_2^2}{d_1^2}+1}{\frac{d_2^2}{d_1^2}-1}\right) + \frac{\mu P}{E}$$

$$= -\frac{P}{E}\left(\frac{d_2^2+d_1^2}{d_2^2-d_1^2}\right) + \frac{\mu P}{E}$$

$$\therefore \quad \delta_c = -\frac{Pd_2}{E}\left[\frac{d_2^2+d_1^2}{d_2^2-d_1^2} - \mu\right]$$

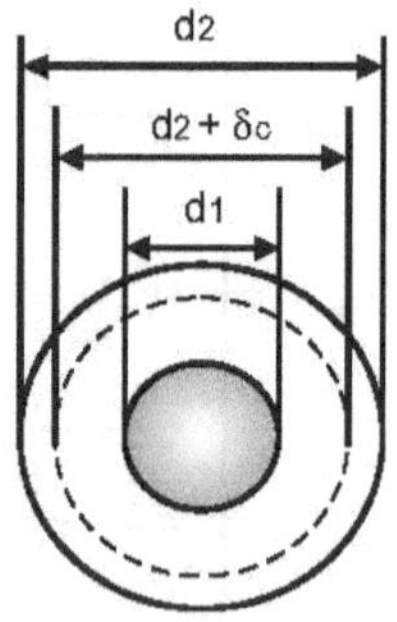

Fig. 10.13

Negative sign indicates **contraction**.

Total deformation,

$$\delta = |\delta_j| + |\delta_c|$$

$$= \frac{P \cdot d_2}{E}\left[\left(\frac{d_3^2 + d_2^2}{d_3^2 - d_2^2}\right) + \mu + \left(\frac{d_2^2 + d_1^2}{d_2^2 - d_1^2}\right) - \mu\right]$$

$$\therefore \quad \delta = \frac{P \cdot d_2}{E}\left[\left(\frac{d_3^2 + d_2^2}{d_3^2 - d_2^2}\right) + \left(\frac{d_2^2 + d_1^2}{d_2^2 - d_1^2}\right)\right]$$

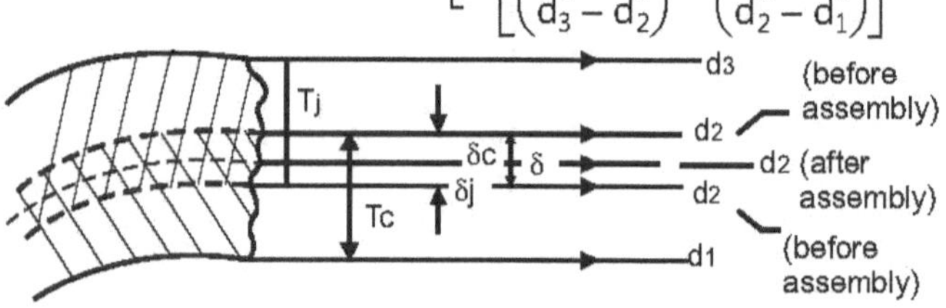

Fig. 10.14

10.9 GASKET FLANGED JOINT IN CYLINDER VESSEL

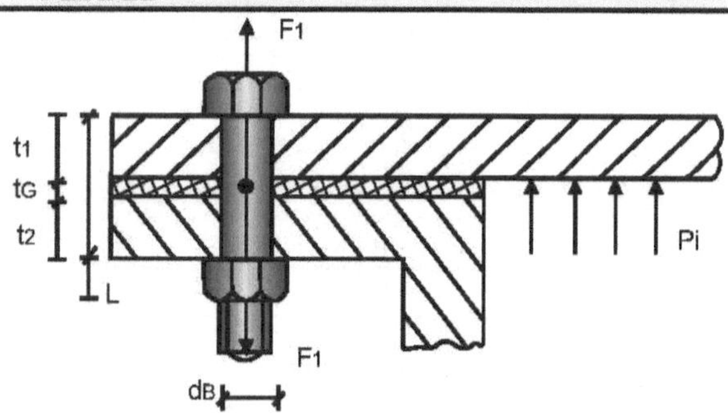

Fig. 10.15

F_1 : Preload in bolt after flange is initially tightened

t_1 : Thickness of head

t_2 : Thickness of flange

t_G : Thickness of gasket

d_B : Bolt diameter

When flange is tightened initially then there are following changes set.

(i) Preload F_1 (tension) acts in bolt.

(ii) Bolt elongates by δ_B.

(iii) Due to F_1, flanges and gasket compressed by δ_c.

$$\therefore \quad \delta = \frac{FL}{AE} \text{ and } k = \frac{F}{\delta} = \frac{AE}{L}$$

$$\text{Stiffness of bolt } k_B = \frac{AE_B}{L} = \frac{\pi d_3^2}{4} \cdot \frac{E_B}{L}$$

Combined stiffness (including stiffness of flanges and gasket in series)

$$\frac{1}{k_c} = \frac{1}{k_1} + \frac{1}{k_2} + \frac{1}{k_G}$$

Now annular area which is under grip of the bolt load.

$$A_F = \frac{\pi}{4}\left[\left(3d_B^2\right) - d_B^2\right] = \mathbf{2\pi\, d_B^2}$$

$$\therefore \quad k_1 = \frac{AE}{L} = \frac{2\pi\, d_B^2 \cdot E_1}{t_1}$$

$$k_2 = \frac{2\pi\, d_B^2\, E_2}{t_2} \text{ and } k_G = \frac{2\pi\, d_B^2\, E_G}{t_G}$$

$$\text{Due to preload } \delta_B = \frac{F_1}{k_B} \text{ and } \delta_c = \frac{F_1}{k_c}$$

Consider load-deflection diagram as shown in Fig. 10.16 :

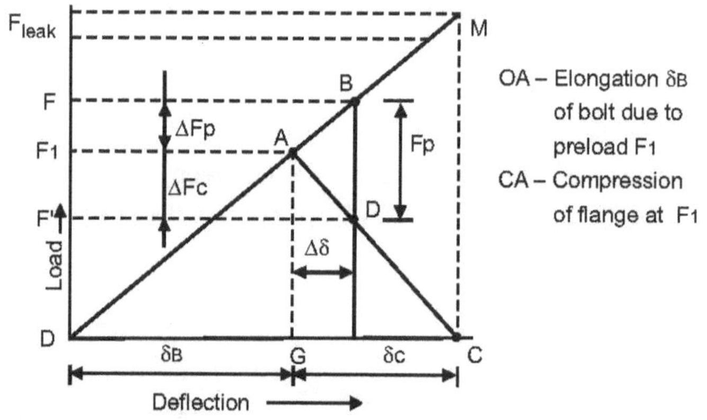

Fig. 10.16

OA–Elongation δ_B of bolt due to preload F_1

CA–Compression of flange at F_1

When P_i is applied – bolt further elongated by $\Delta\delta$ resulting in increase in load on the bolt by ΔF_p as shown by AB. But simultaneously this results in relieving the gasket force deflection by $\Delta\delta$ shown by AD.

$$k_B = \frac{\Delta F_p}{\Delta\delta} \text{ and } k_c = \frac{\Delta F_c}{\Delta\delta}$$

where ΔF_c is reduction in force in the compressed part.

$$\therefore \quad \frac{\Delta F_p}{\Delta F_c} = \frac{k_B}{k_C}$$

$$\therefore \quad \frac{\Delta F_p}{\Delta F_p + \delta F_c} = \frac{k_B}{k_B + k_C}$$

Total external load will be

$$F_p = \Delta F_p + \Delta F_c$$

$$\therefore \quad \frac{\Delta F_p}{F_p} = \frac{k_B}{k_B + k_C}$$

$$\therefore \quad \Delta F_p = \frac{F_p \cdot k_B}{k_B + k_C}$$

Also $$\Delta F_c = \frac{F_p \cdot k_C}{k_B + k_C}$$

∴ Resultant load on the bolt is,

$$F = F_1 + \Delta F_p \text{ (Tension)}$$

Force in the compressed flange,

$$F' = -(F_1 - \Delta F_c) \text{ (Compression)}$$

PART B : UNFIRED PRESSURE VESSELS

10.10 MODES OF FAILURES IN PRESSURE VESSELS

There are Following Modes of Failures in the Pressure Vessels Namely

1. Failure in circumferential direction (design of shell is discussed earlier in section 10.2 to 10.5)
2. Failure in longitudinal direction
3. Bursting of shell

10.10.1 Unfired Pressure Vessels

Unfired pressure vessels are the shells carrying, storing, receiving steam, gas or liquid above atmospheric pressure, and not fired inside.

Example of Unfired Pressure vessels:

1. Storage vessels
2. Reaction vessels
3. Heat exchangers
4. Evaporators
5. Crystallizers
6. Distillation columns

(Excluding -Nuclear vessels and steam boilers, which are fired vessels)

Codes: The design of such pressure vessels are governed by standard codes for safety purpose. Basically Codes specify requirements of design, fabrication, inspection and testing of pressure vessel.

Such vessels are classified based on IS 2825:1969, as well as ASME Sec VIII Div-1 in detail. Herein the classification based on IS code is presented in the following section.

10.11 CLASSIFICATION BASED ON IS 2825-1969 CODE

Case 1 :

- For storing lethal, toxic substances.
- Operating below – 20°C.
- For intended for any other operation not stipulated in code.

  Welded Joint : Double welded butt joint with full pene. / Single welded butt joint with backing strip.

  Inspection : Fully radiographed.

Case 2 :

- Maximum shell thickness limited to 38 mm which do not fall under class 1 and 3 are to be termed as class 2.

  Welded Joint : Same as 1.

  Inspection : Spot radiographed.

Case 3 :

- Used for relatively light duty.
- Plate thickness not exceed 16 mm and working pressure not exceeding 0.35 MPa vapour or 1.75 MPa hydrostatic design pressure at temperature not exceeding 250°C.
- Not recommended at temperature below 0°C.

Inspection : Visual inspection.

10.11.1 Categories of Welded Joint

Refers to location of the joint in P.V. not type of joint.

IS 2825 specifies four categories of joints.

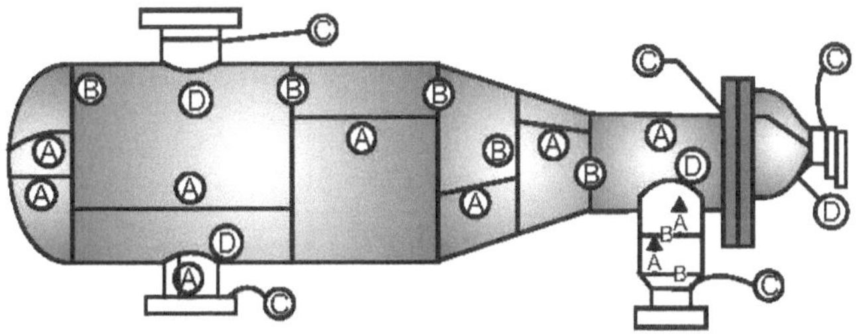

Fig. 10.17

Category A : Longitudinal welded joint within the main shell, communicating chambers, nozzles and any welded joint within a formed or flat head.

Category B : Circumferential welded joint within main shell, communicating chambers, nozzles and transitions in diameter.

- ➢ Connecting formed head to main shell.
- ➢ Connecting shell to shell.

Category C : Welded joints connecting flanges tube sheets and flat heads to main shells, to formed heads, to nozzles or to communicating chamber.

Category D : Welded joints connecting communication heads and to flat sided vessels.

10.11.2 Weld Joint Efficiencies for Different Types of Joints

Table 10.2

Types	Joint Efficiency		
	Full Radiation	Spot Radiation	No Radiation
B.W.B.J with full pene. (a) (b) (c) (d) 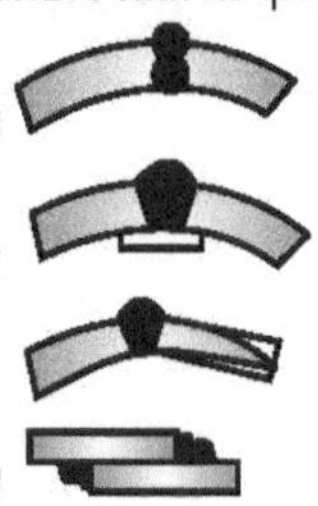	1.0	0.85	0.7
S.W.B.J. with back strip	0.9	0.80	0.65
S.W.B.J. without strip	–	–	0.6
S. Full fillet lap joint	–	–	0.55

10.11.3 Design of Shell for Internal Pressure

$$\sigma_T = \frac{P_i d}{2t}$$

$$\sigma_L = \frac{P_i d}{4t}$$

$$d = d_i + t$$

Design pressure = 5% to 10% more than (P_i)

Test pressure = 30% more than (P_i)

$$J\sigma = \frac{P_i (d_i + t)}{2t}$$

$$2t\, J \cdot \sigma = P_i d_i + P_i \cdot t$$

$$t\,(2J.\sigma - P_i) = P_i d_i$$

$$t = \frac{P_i d_i}{2T\sigma - P_i}$$

σ = Permissible des. stress at des. temperature

$$\sigma = \frac{S_y}{1.5} \text{(0.2\% of proof stress)}$$

$$= \frac{S_{ut}}{3}$$

10.12 PRESSURE VESSEL SUBJECTED TO COMBINED LOADINGS

(i) Stress in Longitudinal Axis Direction :

- Due to internal pressure

$$\sigma_{L_1} = \frac{P_i d_i}{4t} \text{ (Tensile)}$$

- Due to weight of vessel and its content

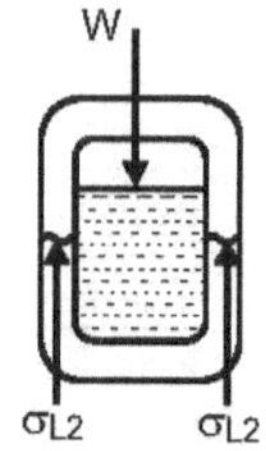

Fig. 10.18

$$\sigma_{L_2} = \frac{W}{\pi dt} = \frac{W}{\pi (d_i + t)\, t} \text{ (Compressive)}$$

- Due to bending by wind/(seismic load)

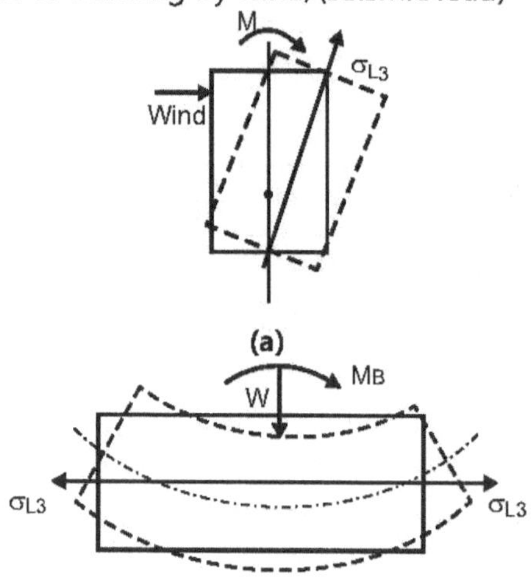

Fig. 10.19

$$\sigma_{L_3} = \pm \frac{M_B}{Z} \quad \text{Z-section modulus where} \quad Z = \frac{\pi d^2 \cdot t}{4}$$

$$\sigma_{L_3} = \frac{\pm 4 M_B}{\pi (d_i + t)^2\, t} \quad \text{(T or C)}$$

Total resultant in longitudinal direction.

$$\sigma_{LR} = \sigma_{L_1} + \sigma_{L_2} + \sigma_{L_3}$$

(ii) Stresses in Tangential Direction :

$$\sigma_T = \frac{P_i (d_i + t)}{2t} \text{ (T)}$$

(iii) Stresses Due to Torsion About Longitudinal Axis

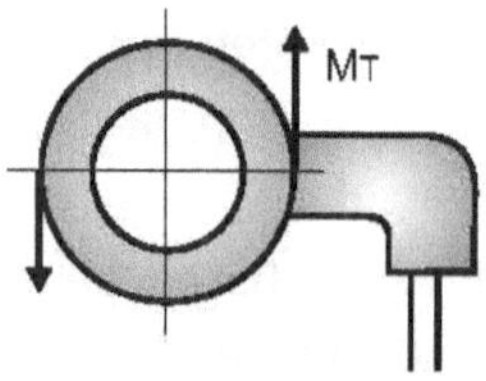

Fig. 10.20

$$\tau = \frac{2M_T}{\pi t d_i (d_i + t)}$$

Equivalent stress, $\sigma_R = \sqrt{\sigma_T^2 - \sigma_T \sigma_{LR} + \sigma_{LR}^2 + 3\tau^2}$

Using maximum strain energy criterion

$$\sigma_R \text{ (T)} \leq \text{Permissible T.S.}$$

$$\sigma_{LR}(T) \leq \text{Permissible T.S.}$$
$$\sigma_{LR}(C) \leq \text{Permissible C.S.}$$

Based on these calculation thickness of shell can be calculated.

10.13 SPHERICAL PRESSURE SHELLS

For analysis, it is assumed that the spherical shell is cut into two halves as shown in Fig. 10.21. Considering equilibrium of forces on each half.

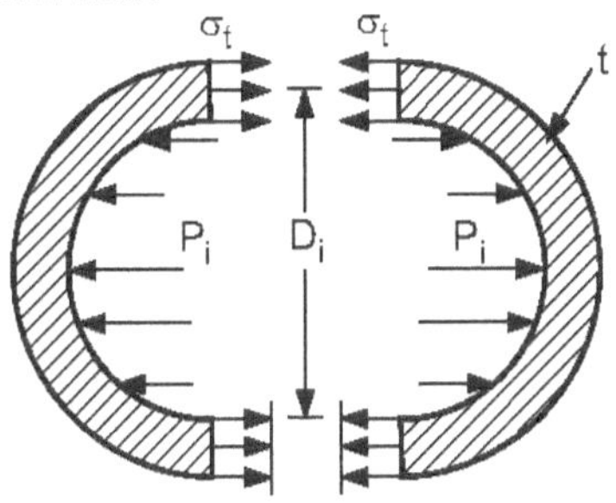

Fig. 10.21

$$p_i \cdot A_c = \sigma_t \cdot A$$
$$pi\left[\frac{\pi}{4} P_i^2\right] = \sigma_t\,[\pi D_i t]$$
$$\therefore \quad t = \frac{piD_i}{4\sigma_t}$$

...[Assuming that there are no any welded joints in the shell]

where p_i = internal pressure in N/mm²

D_i = inner diameter of spherical shell in mm

t = thickness of wall in mm

σ_t = permissible tensile stress in N/mm²

Volume of shell is

$$v = \frac{\pi}{6} D_i^3$$
$$D_i = \left(\frac{\sigma V}{\pi}\right)^{1/3}$$

10.14 END CLOSURE (FORMED HEADS)

1. Hemispherical :

- Strongest.
- Forming coast ↑ → expensive.

$$t = \frac{P_i d_i}{4\sigma J}$$
$$\text{Volume} = \left(\frac{\pi d_i^3}{6}\right) - \frac{1}{2} = 0.262\, d_i^3$$

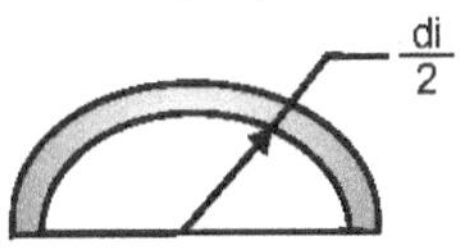

Fig. 10.22

2. Elliptical :

- Suitable for P.V. subjected to P ≥ 1.5 MPa.

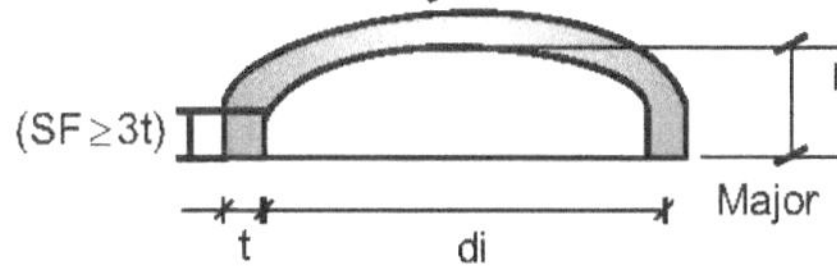

Fig. 10.23

- Shallow and deep reduces forming cost

$$t = \frac{P_i d_i w}{2\sigma J}$$
$$k = \frac{d_i}{d_m} \text{ generally } k = 2 \geq 2.6$$
$$w = \frac{2 + k^2}{6}$$
$$v = \frac{2\pi\, abc}{3} = 0.131\, d_i^3$$

3. Torispherical :

- Extensively used for P = 0.1 – 1.5 MPa.
- R_c – Crown/dish rad ≈ d or < d

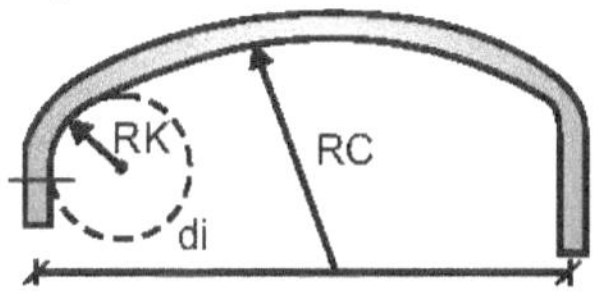

Fig. 10.24

R_k = Knuckle radius – 6% of R_c

– Requires less forming

– Have localized stresses at discontinuities

$$t = \frac{P_i R_c \cdot w}{2\sigma J}$$

w = Stress intensity factor

$$w = \frac{1}{4}\left[3 + \sqrt{\frac{R_c}{R_k}}\right] \qquad \frac{R_c}{R_k} = 16.667$$
$$\therefore \quad w = 1.7706$$
$$\therefore \quad t = \frac{0.885\, P_i R_c}{\sigma J}$$
$$h = 0.169\, d_i$$
$$v = 0.085\, d_i^3$$

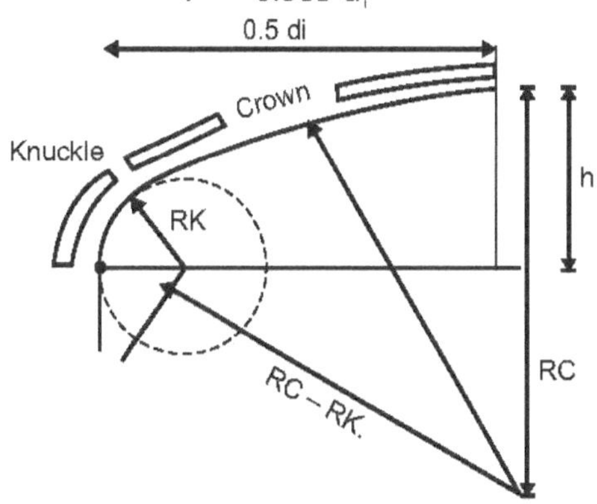

Fig. 10.25

4. Conical Head :

Widely used as bottom heads to facilitates removal of material.

$$\alpha \leq 30°$$

$$t = \frac{P_i\, d_i}{2\sigma J \cos \alpha}$$

Conical head

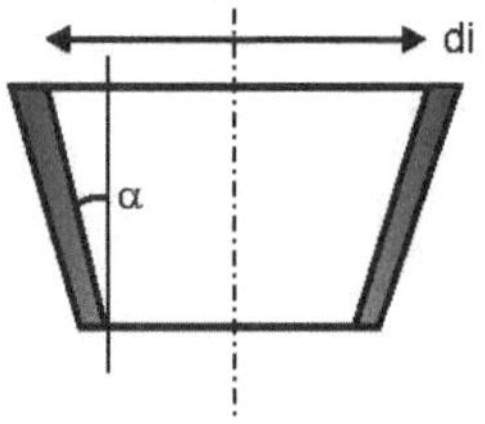

Fig. 10.26

10.15 NOZZLES AND OPENINGS (May 16)

- Inlet/outlet connections
- Manholes Located in shell or on the head
- Vents and drains
- Weakens the shells as they cause discontinuities.
- The weakening due to hole needs to be compensated by adding reinforcement/(pad) surrounding the opening.

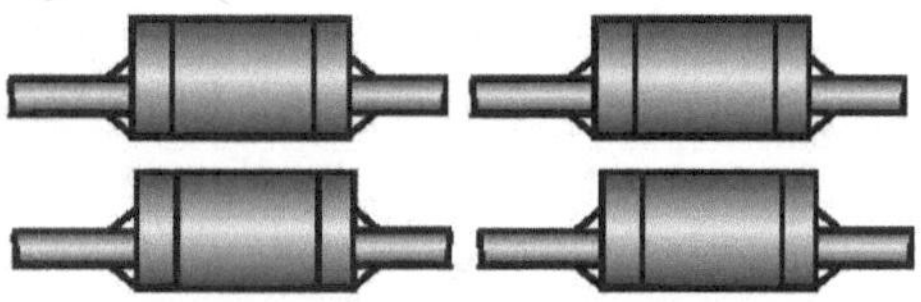

Fig. 10.27

10.16 AREA COMPENSATION METHOD

- ➢ Used to determine area of reinforcement.
- ➢ The actual area removed needs to be compensated by adding an equal amount by way of reinforcement.

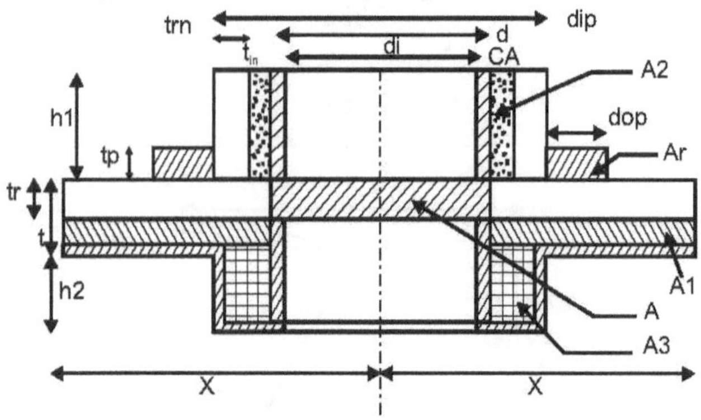

Fig. 10.28

d_i = Nozzle I.D.

$d = d_i + 2CA$ (Corrosion allowance)

t = Actual shell thickness in mm

t_r = Required thickness as per calculations

t_n = Actual thickness of nozzle

t_{rn} = Required thickness as per calculation

$$t_{rn} = \frac{P_i\, d_i}{2\sigma_J - P_i}$$

$h_{1\,act}$ = Height of nozzle above shell

$h_{2\,act}$ = Height of nozzle below shell

$$h = \underbrace{2.5\,(t - CA) \quad \text{OR} \quad 2.5\,(t_n - CA)}_{\text{whichever is smaller}}$$

h_1 = h OR $h_{1\,actual}$ → whichever is smaller

h_2 = h OR $h_{2\,actual}$ → whichever is smaller

Design till where effect of nozzle persists in mm on each side of centre line.

$$\text{Generally, } \left.\begin{array}{l} x = d \\ x = \left(\dfrac{d_i}{2} + t + t_n - 3CA\right) \end{array}\right\} \text{ whichever is maximum}$$

d_{op} = O.D. of reinforcing pad (mm)

d_{ip} = I.D. of reinforcing pad (mm)

t_p = Thickness of reinforcing pad

- **Area Calculations :**

$A = dt_r$ → Area pertaining to material removed

$A_1 = (2x - d)(t - t_r - CA)$ → Excess area in shell

$A_2 = 2h_1\,(t_n - t_{rn} - CA)$ → Excess area in nozzle

$A_3 = 2h_2\,(t_n - 2CA)$ → Excess area in nozzle inside shell

$A_r = (d_{op} - d_{ip})\, t_p$ → Area required

$A_r = A - (A_1 + A_2 + A_3)$ → When $A_r = 0$ or '–ve' no reinforcement is necessary

- **Flanged Joint :**
  - ➢ Specified by IS 4864-4867.
  - ➢ Design calculation not necessary for standard flanges.
  - ➢ For non-standard flanges design is required which involves proper selection of gasket, flange, facing, bolting and other geometrical proportion.
  - ➢ For leak proof joint gaskets are sandwiched between two flange facings which are then tightened by series of bolt.

- **Design Consideration :**
  - To prevent leakage by ensuring a positive contact pressure at gasket flange interface under all service conditions.
  - Create tightening force by bolts without overstressing it.
  - Ensure structural rigidity of flange faces and to minimize its deflection.

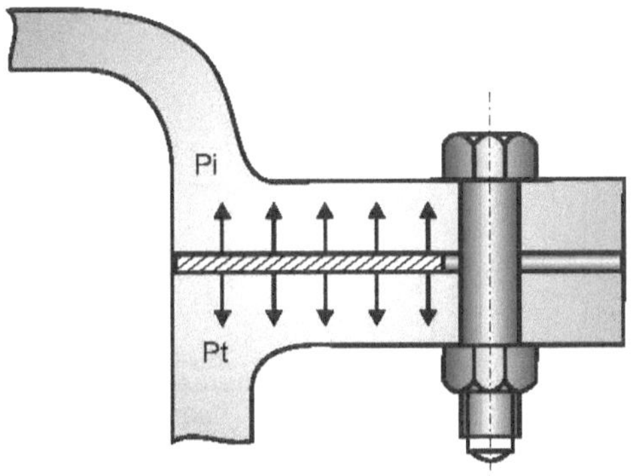

Fig. 10.29

- **Under Atmospheric Condition :**

$$W_A = A_G \cdot \sigma_A$$

where, W_A – Tightening bolt load

A_G – Gasket area under compression

σ_A – Gasket seating stress

- **After Applying Internal Press :**

Bolt elongates and gasket is released. This results in reduction of stress in gasket. Let this residual stress be σ_p.

New bolt load

$\mathbf{W_p = A_G \sigma_p + P_i A_F}$ } A_F – Flange area on which pressure exerted.

- It is difficult to determine σ_p under internal.
- Experimental verification reveals that for leak proof joint.

$$\sigma_p = m \cdot P_i$$

where, m = Gasket factor

(It is essentially a safety factor to increase the clamping load to such an extent that the hydrostatic end force does not unseat the gasket to the point of leakage).

- Gasket may cover the complete face of flange or may be narrow.
- To estimate gasket proportions.

Gasket seating force = Hydrostatic pressure force = Residual gasket force

Let G_o and G_i → O.D. and I.D. of gasket.

Then $\frac{\pi}{4}(G_o^2 - G_i^2)\,\sigma_A - \frac{\pi}{4} G_o^2 \cdot P_i = \frac{\pi}{4}(G_o^2 - G_i^2)\,\sigma_p$

where $\sigma_p = m \cdot P_i$

$$(G_o^2 - G_i^2)(\sigma_A - mP_i) = G_o^2 P_i$$

$$\frac{G_o^2}{G_o^2 - G_i^2} = \frac{\sigma_A - mP_i}{P_i}$$

$$\frac{G_o^2}{G_o^2 - G_i^2 - G_o^2} = \frac{\sigma_A - mP_i}{P_i - \sigma_A + m\,P_i}$$

$$\frac{G_o^2}{G_i^2} = \frac{\sigma_A - mP_i}{\sigma_A - P_i(m+1)}$$

$$\frac{G_o}{G_i} = \sqrt{\frac{\sigma_A - mP_i}{\sigma_A - P_i(m+1)}}$$

Based on this equation different gasket combination of seating stress and gasket factor may be tried to achieve suitable proportions.

$$W_A = A_G \cdot \sigma_A = (\pi\, b \cdot G)\,\sigma_A$$

b = Effective gasket width or joint contact surface width

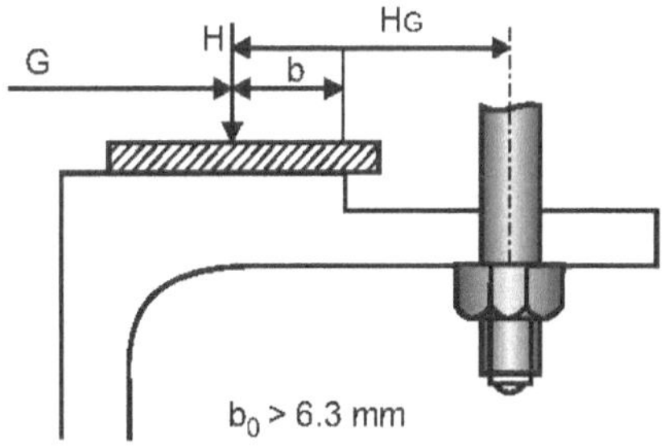

Fig. 10.30

$$b_o \le 6.3 \text{ mm}$$

- **Support for Pressure Vessel :**

Pending on type of vessels, supports can be classified as :

(a) Vertical Vessels : Thick vessel bracket or leg column/skirt.

Small → Leg support and leg support

Tall → Skirt.

(b) Horizontal Vessels : Saddle support.

(i) Backet/Leg Support :

- Fabricated from plates.
- Suitable for vessels with thick walls as they can absorb the bending stress due to eccentricity of loads.
- Main load on vessel = [Dead load of vessel] + [Content of vessel]

- Stresses induced due to wind/seismic load.

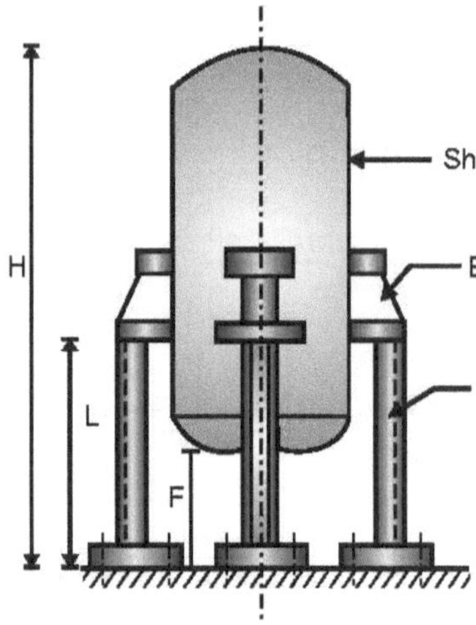

Fig. 10.31

- Total compressive load,

$$P = \frac{4P_w (H - F)}{n D_b} + \frac{\sum w}{n}$$

where, P_w = Wind force

$= KP_1 : HD_o$

K = Coefficient depending on shape factor = 0.7 (cylindrical vessels)

P_1 = Wind pressure

H = Height of vessel above foundation

D_o = O.D. of vessel

F = Vessel clearance

D_b = Bolt circle diameter

$\sum w$ = Dead weight

n = Number of brackets

(ii) Leg Support :

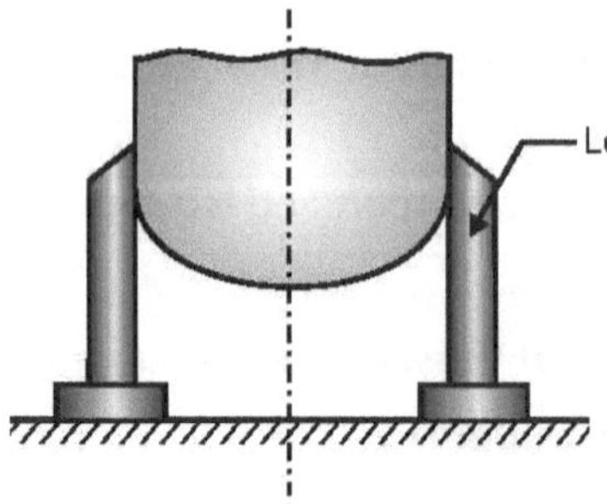

Fig. 10.32

- Legs can be made detachable to the vessel.
- Legs can be bolted to plates.
- Designation is similar to that of bracket S.
- If the legs are welded to the shell then shear stresses in the weld be,

$$\tau_w = \frac{\sum w}{0.707 \times t_w \times L_w \times n}$$

t_w = Weld height

L_w = Weld length

n = Number of legs

- Suitable for small vessels as there is a concentrated local stresses at the joint.

(iii) Skirt/Column S :

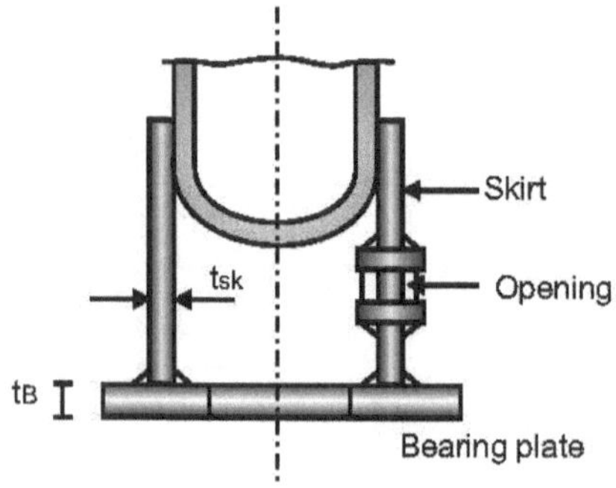

Fig. 10.33

- To support tall vertical vessels with cylindrical or conical shells called as skirts.
- Skirt has high section modulus.

  ∴ Better resistance to bending.
- Bearing plate is attached to the bottom to rest it on concrete foundation

Skirt designed for following loads :

1. Dead Weight :

Direct compression stress

$$\delta_D = \frac{\sum W}{\pi D_{osk} t_{sk}}$$

$\sum W$ = Total weight

D_{osk} = Skirt thickness

t_{sk} = O.D. of skirt

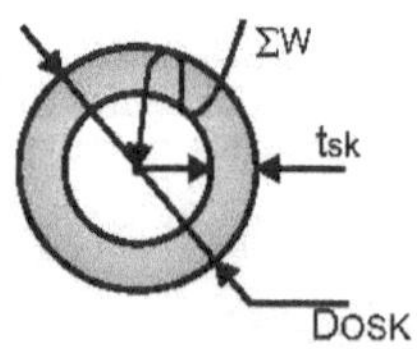

Fig. 10.34

2. Wind Load :

$$P_{w_1} = KP_1 \cdot H \cdot D_o$$

[Equation valid for H = 20 m]

For $H > 20$ m

$$P_{w_2} = KP_2 \cdot H \cdot D_o \qquad [H > 20 \text{ m}]$$

$$P_1 = 400 - 1000 \text{ N/m}^2$$

P_2 may be upto 2000 N/m^2.

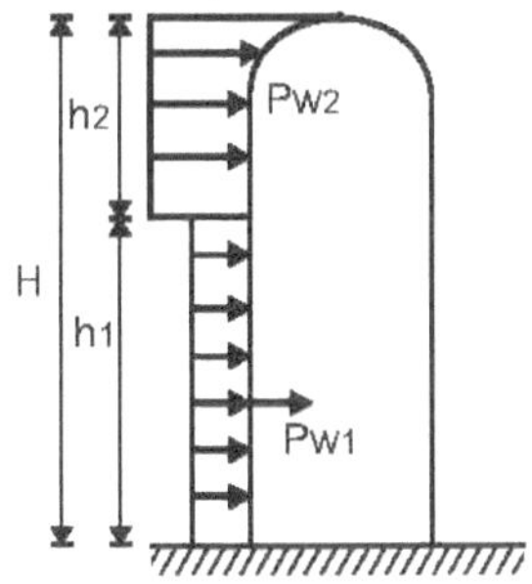

Fig. 10.35

3. **Bending Moment:**

$$M_w = \frac{P_{w1} \cdot h_1}{2} \qquad [H \leq 20 \text{ m}]$$

$$= \frac{P_{w_1} \cdot h_1}{2} + P_{w_2}\left(h_1 + \frac{h_2}{2}\right) \qquad [H > 20 \text{ m})$$

$$\therefore \quad \sigma_{bw} = \frac{M_w}{z} = \frac{(M_w)}{(\pi D_{osk}^2 t_{sk}/4)}$$

$$\sigma_{bw} = \frac{4 M_w}{\pi D_{osk}^2 t_{sk}}$$

Wind load will create tensile stress on wind side, compression stress on other side.

4. **Seismic Load :**

- Tall vessels are subjected to vibration due to earthquakes.
- Load distribution is linear with zero at bottom and maximum at top.
- Equivalent load.

F = CW

C = Seismic coefficient – Depending on geographical location and available data on seismic activity

W = Total weight of vessel

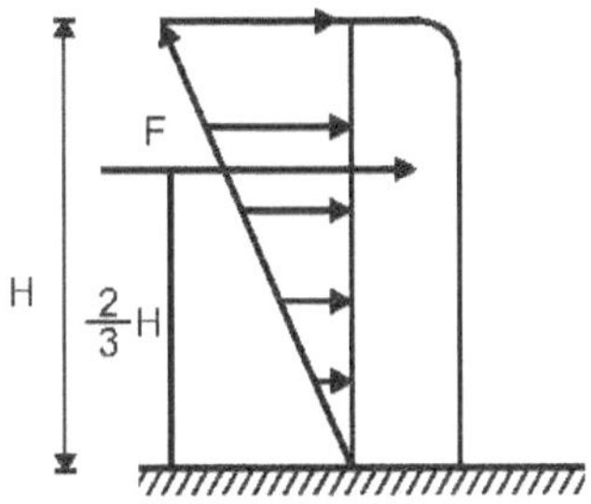

Fig. 10.36

Bending moment,

$$M_s = F\left(\frac{2}{3}H\right)$$

$$= \frac{2}{3}CWH$$

$\therefore$ Stress due to seismic load,

$$\sigma_{bs} = \frac{M_s}{Z} = \frac{4M_s}{\pi d_{osk}^2 \cdot t_{sk}}$$

For design most adverse loadings should be considered.

- Maximum tensile stress = (σ_{bw} or σ_{bs}) – σ_D.
- Maximum compression stress = (σ_{bw} or σ_{bs}) + σ_D.

5. **Saddle Support :**

(a) Ring type, (b) Plate type.

Horizontal vessels supported on saddle.

- Stiffness may be added.
- Designed as per Zick's criteria $\theta \not< 120$.

SOLVED EXAMPLES

Example 10.1 : *A cylinder made of steel has 400 mm inner diameter is used for carrying a fluid under 50 bar pressure. Ultimate strength of material is 400 MPa joint efficiencies*

η_l = 0.75 (longitudinal joint efficiency)

η_c = 0.7 (circumferential joint efficiency)

Factor of safety = 3

Find the cylinder thickness.

Solution :

d_i = 400 mm

σ_{nt} = 400 MPa

pi = 50 bar = 5 MPa

η_l = 0.75

η_c = 0.7

Thickness of cylinder

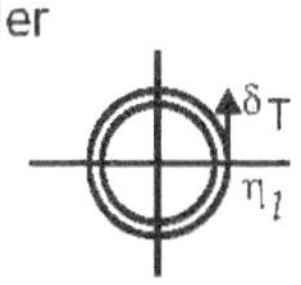

Fig. 10.37

$$t = \frac{p_i d_i}{2\sigma_{all} \cdot \eta_l} = \frac{5 \times 400}{2 \times \left(\frac{\sigma_{ut}}{Fos}\right) \cdot \eta_l}$$

$$= \frac{5 \times 400}{2 \times \frac{400}{3} \times 0.75}$$

$$t = 10 \text{ mm}$$

$$\therefore \quad \frac{t}{d_i} = \frac{10}{400} = 0.025 < 0.05$$

$\therefore$ (It is a thin cylinder) ... as

$$\frac{t}{d_i} < \frac{1}{20}$$

Example 10.2 : *A cylinder 200 mm in diameter, closed on both its ends is to be designed to withstand an internal pressure of 20 MPa. Material of construction is steel (σ_{nt} = 440 MPa, σ_{yt} = 240 MPa, μ = 0.3). Factor of safety based on yield strength is 1.5. Calculate cylinder thickness basedon.*

(i) Maximum principal stress theory

(ii) Maximum principal strain theory

(iii) Maximum shear stress theory and

(iv) Maximum distortion energy theory

Solution :

$$d_i = 200 \text{ mm}$$
$$p_i = 20 \text{ MPa}$$
$$\sigma_{ut} = 440 \text{ MPa}$$
$$\sigma_{yt} = 240 \text{ MPa}$$
$$\mu = 0.3$$
$$\text{Fos} = 1.5$$
$$\sigma_{sy} = 0.5, \sigma_{yt} = 0.5 \times 240 = 120 \text{ MPa}$$
$$\sigma_{all} = \frac{\sigma_{yt}}{\text{Fos}} = \frac{240}{1.5} = 160 \text{ MPa}$$
$$\tau_{all} = \frac{\sigma_{sy}}{\text{Fos}} = \frac{120}{1.5} = 80 \text{ MPa}$$

(i) Based on Maximum Principal Stress Theory

$$t = \frac{d_i}{2}\left(\sqrt{\frac{\sigma_{all} + P_i}{\sigma_{all} - P_i}} - 1\right) \quad \text{...(Lame's Equation)}$$
$$= \frac{200}{2}\left(\sqrt{\frac{160 + 20}{160 - 20}} - 1\right)$$
$$t = 100\ (0.13389)$$
$$t = 13.389 \text{ mm}$$

(ii) Based on Maximum Principal Strain Theory

$$t = \frac{d_i}{2}\left[\left(\sqrt{\frac{\sigma_{all} + p_i\ (1 - 2\mu)}{\sigma_{all} - p_i\ (1 + \mu)}}\right) - 1\right] \quad \text{...(Clavarino's equation)}$$
$$= \frac{200}{2}\left[\left(\sqrt{\frac{160 + 20\ (1 - 2 \times 0.3)}{160 - 20\ (1 + 0.3)}}\right) - 1\right]$$
$$= 100\ (0.1197)$$
$$t = 11.97 \text{ mm}$$

(iii) Based on Maximum Shear Stress Theory

$$t = \frac{d_i}{2}\left[\sqrt{\frac{\tau_{all}}{\tau_{all} - p_i}} - 1\right]$$
$$= \frac{200}{2}\left[\sqrt{\frac{80}{80 - 20}} - 1\right]$$
$$= 100\ (0.15469)$$
$$t = 15.469 \text{ mm}$$

(iv) Based on Distortion Energy Theory

$$t = \frac{d_i}{2}\left[\sqrt{\frac{\sigma_{all}}{\sigma_{all} - \sqrt{3}\ p_i}} - 1\right]$$
$$= \frac{200}{2}\left[\sqrt{\frac{160}{160 - \sqrt{3} \times 20}} - 1\right]$$
$$= 100\ (0.129748)$$
$$t = 12.9748 \text{ mm}$$

Example 10.3 : *A cylinder made of grey cast iron (FG 300) is subjected to internal pressure of 20 MPa. If cylinder inner diameter is 300 mm and factor of safety is 1.5. Determine the thickness of cylinder and its outer diameter.*

Solution : Material : Grey cast iron FG 300 it is brittle material

$$\sigma_{ut} = 300 \text{ MPa}$$
$$P_i = 20 \text{ MPa}$$
$$\text{Fos} = 1.5$$
$$d_i = 300 \text{ mm}$$

For brittle material, thickness of cylinder is based on Lame's equation

$$t = \frac{d_i}{2}\left(\sqrt{\frac{\sigma_{all} + p_i}{\sigma_{all} - p_i}} - 1\right) \qquad \sigma_{all} = \frac{\sigma_{ut}}{\text{Fos}}$$
$$= \frac{300}{2}\left(\sqrt{\frac{200 + 20}{200 - 20}} - 1\right) \qquad = \frac{300}{1.5} = 200$$
$$= 150\ (0.10554)$$
$$t = 10.554 \text{ mm} = 11 \text{ mm}$$

$\therefore$ Outer diameter of cylinder

$$d_o = d_i + 2t$$
$$= 300 + 2 \times 11$$
$$d_o = 322 \text{ mm}$$

Example 10.4 : *A storage vessel closed at its both ends and made of plain carbon steel 10C4 (σ_{ut} = 340 MPa, μ = 0.3) is to be designed for 120 bar internal pressure. The inner diameter and length of cylinder is 250 mm and 1.4 m respectively. Factor of safety is 3. Determine the thickness and outer diameter of vessel.*

Solution : Material : Plain carbon steel (10 C4)

$$\sigma_{ut} = 340 \text{ Mpa}$$

It is ductile material

$$d_i = 250 \text{ mm}$$
$$l = 1.4 \text{ m} = 1400 \text{ mm}$$
$$\text{Fos} = 3$$
$$p_i = 120 \text{ bar} = 120 \times 10^5 \text{ N/m}^2$$
$$= 12 \text{ MPa}$$
$$\mu = 0.3$$
$$\sigma_{all} = \frac{\sigma_{ut}}{3} = \frac{340}{3} = 113.33 \text{ MPa}$$

As material of construction is 10 C4 (ductile) with closed ends, clavarino's equation is applied for calculating vessel thickness.

$$\therefore \quad t = \frac{d_i}{2}\left[\sqrt{\frac{\sigma_{all} + p_i\ (1 - 2\mu)}{\sigma_{all} - p_i\ (1 + \mu)}} - 1\right]$$

$$= \frac{250}{2}\left[\sqrt{\frac{113.33 + 12\,(1 - 2 \times 0.3)}{113.33 - 12\,(1 + 0.3)}} - 1\right]$$

$$= 125\,[0.09943]$$

$$t = 12.4283 \text{ mm} = 13 \text{ mm}$$

Outer diameter $= d_i + 2t$

$$= 250 + 2 \times 13$$

$$d_o = 276 \text{ mm}$$

Example 10.5 : *A vessel open at both the ends is subjected to an internal pressure of 15 MPa. The material of vessel is plain carbon steel 15 C8 (σ_{ut} = 440 MPa, σ_{yt} = 220 MPa, μ = 0.3). Inner diameter of vessel shell is 200 mm. A factor of safety of 1.5 is used based on yield strength. Evaluate vessel thickness.*

Solution : Given : Material is plain carbon steel 15 C8 which is ductile material Vessel is open ended.

∴ Birnie's equation is used for calculating shell thickness.

$$\sigma_{all} = \frac{\sigma_{yt}}{Fos} = \frac{220}{1.5} = 146.667 \text{ MPa}$$

$$t = \frac{d_i}{2}\left[\sqrt{\frac{\sigma_{all} + p_i\,(1 - \mu)}{\sigma_{all} - p_i\,(1 + \mu)}} - 1\right]$$

$$= \frac{200}{2}\left[\sqrt{\frac{146.667 + 15\,(1 - 0.3)}{146.667 - 15\,(1 + 0.3)}} - 1\right]$$

$$= 100\,[0.117]$$

$$t = 11.17 \text{ mm} \approx 12 \text{ mm}$$

∴ Outer diameter of vessel shell

$$d_o = d_i + 2t$$

$$= 200 + 2 \times 12$$

$$d_o = 224 \text{ mm}$$

Example 10.6 : *A thick cylinder with inner diameter of 200 mm and outer diameter of 300 mm is subjected to an internal pressure of 100 MPa. Plot the distribution of radial stress, tangential stress across the wall thickness.*

Solution : Given : d_i = 200 mm

d_o = 300 mm

p_i = 100 MPa

Radial stress : $\sigma_R = -A + \dfrac{B}{r^2}$ ...(i)

Tangential stress : $\sigma_T = A + \dfrac{B}{r^2}$ ...(ii)

Here $A = \dfrac{p_i}{k^2 - 1}$; $B = \dfrac{p_i d_o^2}{4\,(k^2 - 1)}$

$$k = \frac{d_o}{d_i} = \frac{300}{200} = 1.5$$

$$A = \frac{100}{(1.5^2 - 1)} = 80$$

$$B = \frac{100 \times (300)^2}{4\,(1.5^2 - 1)} = 1800000$$

From equation (i) and (ii)

Sr. No.	r (mm) Radius	σ_R (MPa) Radial stress	σ_T (MPa) Tangential / Hoop stress
1.	100	100	260
2.	110	68.76	228.76
3.	120	45	205
4.	130	26.51	186.51
5.	140	11.84	171.84
6.	150	0	160

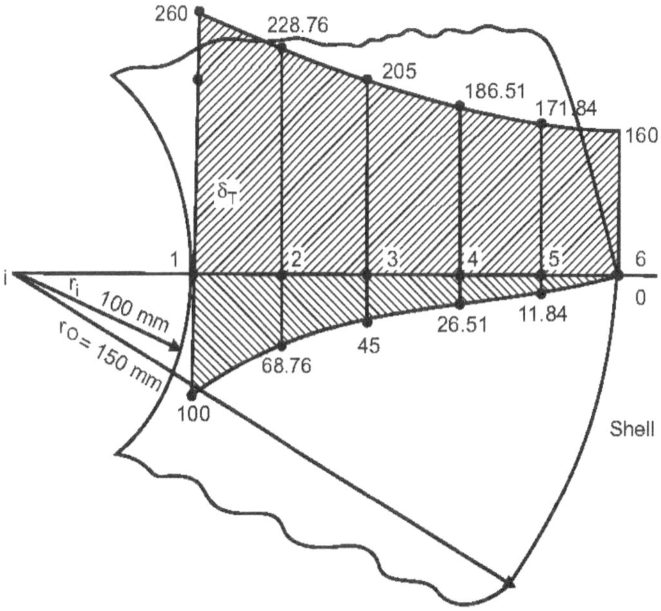

Fig. 10.38 : Distribution of radial stress (σ_R) and tangential stress (σ_T) across wall

EXERCISE

1. Discuss in brief classification of pressure vessels.
2. A closed vessel is to be designed to withstand an internal pressure of 58 MPa having inside diameter of 450 mm following properties may be assumed :

   Yield strength = 300 MPa, Ultimate tensile strength = 500 MPa, Poisson's ratio = 0.3.

   Estimate the wall thickness required by using a F.S 1.5 based on yield strength on the basis of :

   (a) Maximum principle stress theory.

   (b) Maximum shear stress theory.

   (c) Maximum principle strain theory.

Considering maximum thickness out of above three cases, plot the stress distribution in the wall thickness (plot minimum six values of stress as over the given thickness).

3. A tube with 50 mm and 75 mm as inner and outer diameters respectively is reinforced by shrink fitting another tube of outer diameter 100 mm.

   The compound tube is to withstand an internal pressure of 35 MPa. The shrinkage is such that the maximum tangential stress in both the tubes is the same. Calculate :

   (a) Contact pressure.

   (b) The dimensions of tube.

   Plot the important values of the resultant stresses.

4. Derive an equation for radial stress distribution across wall for thick cylinder and plot the same

$$\sigma_R = \frac{-P_i}{k^2-1}\left[\frac{d_o^2}{4r^2}-1\right]$$

where σ_R - Radial stress

p_i - Internal pressure

$k = \frac{d_o}{d_i}$

r = radius of shell

Based on this equation, find maximum radial stress $\sigma_{R\,max}$.

5. Derive an equation for tangential stress distribution across wall for thick cylinder and plot the same.

$$\sigma_T = \frac{p_i}{k^2-1}\left[1-\frac{d_o^2}{4r^2}\right]$$

where σ_T = Tangential or hoop stress

P_i = Internal pressure

$k = \frac{d_o}{d_i}$

r - is radius of shell

Based on this equation, find maximum tangential stress $\sigma_{T\,max}$.

6. Plot radial and tangential stress distribution across wall of thick cylinder and comment over this plot.
7. Explain design of thick cylinder based on maximum principle stress theory.
8. Derive Lame's equation.
9. Explain the design of thick cylinder based on maximum shear stress theory.
10. Explain the design of thick cylinder based on maximum distortion energy theory.
11. Based on St. venant's theory, derive clavarino's equation and state its applicability.
12. Explain the design of thick cylinder based on maximum strain theory.
13. Derive clavarino's and Birnies equation and state its applicability.
14. What is 'Autofrettage' ? Discuss its importance.
15. Explain the methods used for prestressing thick cylinders.
16. Discuss the analysis of compound cylinders.
17. Explain gasketed joint in the pressure vessle
18. Discuss the classification of unfired pressure vessel based on IS 2825 : 1969.
19. Explain various categories of welded joints as per the code.
20. Discuss design of unfired pressure vessel for combined loadings.
21. What are different types of heads used for unfired pressured vessels.
22. Discuss opening compensation method used for pressure vessel.
23. Explain the stresses induced as in case of thin cylinder.
24. Explain the stresses induced as in case of thin spherical pressure shells.

MODEL QUESTION PAPERS FOR
End-Semester Examination

Time : 3 Hours **Max. Marks : 100**

Instructions to the candidates :

(1) Each Question carries 20 Marks.

(2) Attempt any five questions to the following.

(3) Illustrate your answers with neat sketches, diagram etc., wherever necessary.

(4) If some part or parameter is noticed to be missing, you may appropriately assume it and should mention it clearly.

1. **(a)** Enlist applications of rolling contact bearing. Discuss Load-life relationship of rolling contact bearing.

A cylindrical roller bearing is subjected to a radial load of 5 kN. The desired life of the bearing with 90% reliability is 15000 hrs. The load application factor is 1.5. If the shaft rotates at 1440 rpm, calculate the required basic dynamic load rating of the bearing. **[10]**

(b) What is viscosity ? Dissuss method of measuring viscosity of an oil. Determine the viscosity of the lubricant in centipoise having viscosity 160 SUS and specific gravity 0.86. **[10]**

2. **(a)** Explain various assumptions made in the Lewis beam strength equation.

Design a spur gear pair with 20° involute teeth based on beam strength. The velocity factor is to be used to account for dynamic load. The pinion shaft is connected to a 12 kW, 1440 rpm motor. The reduction ratio is 4 : 1 and pinion has 18 teeth. Pinion as well as gear are made of 40C8 (σ_{ut} = 610 N/mm^2). The factor of safety and service factors both can be taken as 1.5. **[10]**

(b) Draw force analysis diagram for helical gear pair. Derive an equation of beam strength of a helical gear. **[10]**

3. **(a)** Discuss following terms in brief

(i) Virtual number of teeth for bevel gear.

(ii) Material for worm and worm gear.

(iii) Wear strength of bevel gear. **[10]**

(b) A pair of worm and worm gear is designated as 1/30/10/10. The input speed of worm is 1200 rpm. The worm wheel is centrifugal cast and made from phosphor bronze having ultimate strength 240 N/mm^2. Coefficient of friction between worm and worm teeth is 0.04. Application factor and factor of safety are 1.25 and 1.5 respectively. The tooth system is 20° full involute. Calculate power transmitting capacity based on beam strength. **[10]**

4. **(a)** Explain in brief with construction and use of

(i) Roller chain (ii) Silent chain

(iii) V-belt (iv) Flywheel

(v) Stresses in disc and armed flywheel. **[10]**

(b) A pulley of 1000 mm diameter is driven by an open type flat belt from 25 kW, 1440 rpm electric motor. The pulley on the motor shaft is 250 mm in diameter and the centre distance between the two shaft is 2.0m. The allowable tensile stress for the belt material is 2 N/mm^2 and the coefficient of friction between the belt and pulley is 0.28. The density of the belt material is 900 kg/m^3. If the width of the belt is 125 mm. Determine :

(i) Thickness of belt

(ii) Length of belt and

(iii) Initial tension required in the belt. **[10]**

5. **(a)** Write a brief note on **[10]**

(i) Friction materials

(ii) Cone clutch

(iii) Internal expanding shoe brake

(iv) Thermal considerations in the brakes

(v) Energy equation for brake

(b) Derive an equation for torque transmitting capacity of single plate clutch.

An automobile single plate clutch consists of a pair of contacting surfaces. The inner and outer radii of friction plate are 120 mm and 250 mm respectively. Coefficient of friction is 0.25, and the total axial force is 15 kN. Calculate the power transmitting capacity of the clutch at 500 rpm, by using

(i) Uniform wear theory and

(ii) Uniform pressure theory **[10]**

6. **(a)** Derive Lamis equation for pressure vessel. Based on this equation calculate cylinder thickness for following data.

(i) Cylinder diameter = 200 mm

(ii) Cylinder closed on both ends subjected internal pressure of 20 MPa

(iii) Cylinder material is steel with

$$\sigma_{ut} = 440 \text{ MPa}$$

$$\sigma_{yt} = 240 \text{ MPa}$$

and $\mu = 0.3$

(iv) Factor of safety based on yield strength = 0.3. **[10]**

(b) Discuss following in brief (Any three)

(i) Autofrettage

(ii) Compounding of cylinders

(iii) Classes of unfired pressure vessels

(iv) Area compensation method **[10]**

IMPORTANT NOTE

For the current semester the End Semester Examination will be conducted for 100 marks. The question paper template is being provided to all the paper setters. After the evaluation, marks obtained out of 100 in semester examination will be converted for 60 marks and then grade will be calculated.

◈◈◈